Applied Complex Variables

Applied Complex Variables

John W. Dettman

Professor of Mathematics, Oakland University

DOVER PUBLICATIONS, INC.
New York

This Dover edition, first published in 1984, is an unabridged and unaltered
republication of the fifth printing, 1970, of the work first published by The
Macmillan Company, N.Y., in 1965.

Manufactured in the United States of America
Dover Publications, Inc., 31 East 2nd Street, Mineola, N.Y. 11501

Library of Congress Cataloging in Publication Data

Dettman, John W. (John Warren)
    Applied complex variables.

    Reprint. Originally published: New York : Macmillan, 1965.
    Bibliography: p.
    Includes index.
    1. Functions of complex variables.   I. Title.
QA331.D55   1984   515.9   83-20604
ISBN 0-486-64670-X

Preface

There was a time when the mathematical needs of most engineers and physicists were fairly well met with a routine calculus course followed by an elementary course in ordinary differential equations. But nothing stands still, least of all science and mathematics. Now, we hear it suggested that every good engineer or physicist should know topics from advanced calculus, complex analysis, ordinary and partial differential equations, boundary value problems, Fourier and Laplace transforms, integral equations, asymptotic expansions, linear algebra, probability, statistics, numerical analysis, and so on. Thus a large body of mathematics which was formerly treated only in advanced mathematics books has become the province of engineers and physicists. This book is an attempt to make accessible to a larger audience a certain number of those topics which seem to be a natural outgrowth of the study of analytic functions of a complex variable.

Most students of applied mathematics want to get as quickly as possible to the applications. There is a danger in this, however, for if the foundation is weak the structure will be shaky and will not support further building. On the other hand, the student must be well motivated to make the effort to master the fundamentals and should not be thwarted in his goal to study the applications by too many frills. With this in mind, the book has been divided into two parts. The first part deals with the fundamentals of *analytic function theory* and should be studied sequentially before going on to the second part which deals with the *applications of analytic function theory*. The second part introduces five of the major applications, more or less independently of one another. In other words, once the first half of the book is mastered, one can pick any one of the five applications to study in detail. This makes it possible for the book to be used for a one-semester course, where one or more applications can be introduced, depending on the time available, or for a two-semester course, in which virtually all the applications are studied.

Although a fairly brisk pace is maintained throughout, I have assumed no advanced calculus background. A well-motivated student with a modern calculus course behind him should be able to study this material with profit. This should make it possible to use the book for a junior- or senior-level advanced engineering mathematics course or for an undergraduate introduction to complex variables for mathematics majors.

Complex analysis is a traditional subject going back to Cauchy and Riemann in the nineteenth century. Therefore, it would be impossible to cite original sources for the material in this book. For the benefit of the reader a list of references is included at the end.

The author is indebted to many for help in the completion of this work, not the least of whom is Professor Bernard Epstein of the University of New Mexico who read a large part of the manuscript and made many helpful suggestions. About one-third of the manuscript was written while the author was supported by a National Science Foundation fellowship at New York University.

JOHN W. DETTMAN
Oakland University
Rochester, Michigan

Contents

Applied Complex Variables

PART ONE

Analytic Function Theory

CHAPTER 1

The Complex Number Plane

1.1. INTRODUCTION

Probably the first time a school child comes across the need for any kind of number other than the real numbers is in an algebra course where he studies the quadratic equation $ax^2 + bx + c = 0$. He is shown by quite elementary methods that the roots of this equation must be of the form

$$-\frac{b}{2a} \pm \frac{\sqrt{b^2 - 4ac}}{2a},$$

the usual quadratic formula. He is then told that if the discriminant $b^2 - 4ac \geq 0$, then the roots of the quadratic equation are real. Otherwise, they are *complex*. It must seem very strange to the student, who has known only the real numbers and is told that a, b, c, and x in the equation are real, that he must be confronted with this strange new beast, the complex number. Actually, the difficulty could be avoided if the problem were stated in the following way: For what real values of x does the polynomial $P(x) = ax^2 + bx + c$, with a, b, and c real, and $a \neq 0$, take on the value zero? Then $P(x)$ can be written as follows:

$$P(x) = a\left(x^2 + \frac{b}{a}x + \frac{c}{a}\right) = a\left[\left(x + \frac{b}{2a}\right)^2 - \frac{b^2 - 4ac}{4a^2}\right].$$

Obviously, since $(x + b/2a)^2$ and $4a^2$ are nonnegative, if $b^2 - 4ac < 0$, then $P(x)$ is never zero, and the answer, in this case, is that there are no real values of x for which $P(x) = 0$. The notion of a complex number need not enter the discussion. Furthermore, if the student has any notion of "the graph of a function," then there is an obvious geometrical picture which makes the situation quite clear. Let us assume, without loss of generality, that $a > 0$. Then, clearly, $P(x)$ takes on a minimum value of $(4ac - b^2)/4a$ when $x = -(b/2a)$. If $b^2 - 4ac > 0$, there are two zeros of $P(x)$, if

3

$b^2 - 4ac = 0$, there is one zero, and if $b^2 - 4ac < 0$, there are no zeros. See Figure 1.1.1.

Actually, the discussion of complex roots of the quadratic equation need not come up at all unless we are trying to answer the more difficult question: For what values of the complex variable z does the polynomial $P(z) = az^2 + bz + c$ take on the value zero? Here, a, b, and c may be complex numbers. But this plunges us into the heart of our subject and raises many interesting questions: What is a complex number? What is a function of a complex variable? What are some of the more interesting properties of functions of a complex variable? Do polynomial functions have zeros? (Fundamental Theorem of Algebra), and so on. In trying to answer these questions and others which will most certainly come up, we shall uncover a very rich branch of mathematics, which has and will have far reaching implications and great applicability to other branches of science.

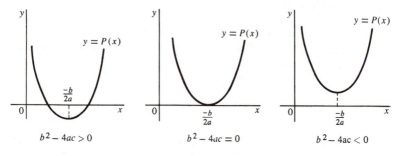

Figure 1.1.1.

The first notion of a complex number was discovered in connection with solving quadratic equations. Consider, for example, the equation $z^2 + 1 = 0$. Obviously, this has no real solutions, since for any real x, $x^2 \geq 0$ and $x^2 + 1 > 0$. We are tempted to write $z = \pm\sqrt{-1}$ but there is no real number which when squared yields -1. If the equation is to have a solution, it must be in some system of numbers larger than the set of reals. This then was the problem faced by seventeenth-century mathematicians: To extend the reals to a larger system of numbers in which the equation $z^2 + 1 = 0$ would have a solution. In the next section, we begin a systematic discussion of the complex numbers which is the result of this extension of the reals.

1.2. COMPLEX NUMBERS

Consider the collection (set) of all ordered pairs of real numbers (x, y). A particular member of the collection (x_1, y_1) we shall refer to as z_1. If

$z_1 = (x_1, y_1)$ and $z_2 = (x_2, y_2)$, then we say that $z_1 = z_2$ if and only if $x_1 = x_2$ *and* $y_1 = y_2$. We can prove immediately that

1. $z_1 = z_1$ for all z_1,
2. if $z_1 = z_2$, then $z_2 = z_1$,
3. if $z_1 = z_2$ and $z_2 = z_3$, then $z_1 = z_3$.

When a relation satisfies these three properties we say that it is an *equivalence relation*.

Next, we define *addition*. We say that $z_1 + z_2 = z_3$ if and only if $x_1 + x_2 = x_3$ and $y_1 + y_2 = y_3$. It is not difficult to prove the following properties:

1. $z_1 + z_2 = z_2 + z_1$ (commutativity),
2. $z_1 + (z_2 + z_3) = (z_1 + z_2) + z_3$ (associativity),
3. there exists a zero, 0, such that for every z, $z + 0 = z$,
4. for every z there exists a *negative* $-z$ such that $z + (-z) = 0$.

In proving Property 3, we must find some member of the collection $0 = (x_0, y_0)$ such that $z + 0 = z$, or $x + x_0 = x$ and $y + y_0 = y$. An obvious solution is $x_0 = 0$, $y_0 = 0$. The zero of the system is unique, for if $z + 0 = z$ and $z + 0' = z$, then

$$0' = 0' + 0 = 0 + 0' = 0.$$

The existence of a negative for each z is easily established. As a matter of fact, $-z = (-x, -y)$, since

$$z + (-z) = (x, y) + (-x, -y) = (x - x, y - y) = (0, 0) = 0.$$

We define *subtraction* in terms of addition of the negative, that is,

$$z_1 - z_2 = z_1 + (-z_2),$$
$$(x_1, y_1) - (x_2, y_2) = (x_1, y_1) + (-x_2, -y_2) = (x_1 - x_2, y_1 - y_2).$$

Next, we define *multiplication*. We say that $z_1 z_2 = z_3$ if and only if $x_3 = x_1 x_2 - y_1 y_2$, $y_3 = x_1 y_2 + x_2 y_1$. Again, it is easy to prove the following properties:

1. $z_1 z_2 = z_2 z_1$ (commutativity),
2. $z_1(z_2 z_3) = (z_1 z_2) z_3$ (associativity),
3. there exists an *identity* 1 such that for every z, $1z = z$,
4. for every z, other than zero, there exists an inverse z^{-1} such that $z^{-1}z = 1$.

If an identity $1 = (a, b)$ exists, then $1z = (ax - by, ay + bx) = (x, y)$. Hence, $x = ax - by$ and $y = ay + bx$. One solution of this is $a = 1$

and $b = 0$. This is the only solution since the identity is unique. Indeed, if $1'z = z$ and $1z = z$, then

$$1' = 11' = 1'1 = 1.$$

For z^{-1} we must have, if $z^{-1} = (\xi, \eta)$,

$$z^{-1}z = (\xi, \eta)(x, y) = (\xi x - \eta y, \xi y + \eta x) = (1, 0),$$
$$\xi x - \eta y = 1,$$
$$\xi y + \eta x = 0.$$

These equations have the solution $\xi = x/(x^2 + y^2)$, $\eta = -y/(x^2 + y^2)$. It is obvious that we must exclude zero as a solution, for $x^2 + y^2 = 0$ if and only if $x = 0$ and $y = 0$, which is the only case where the above equations do not have a unique solution. We define *division* in terms of multiplication by the inverse, that is,

$$\frac{z_1}{z_2} = z_1 z_2^{-1} = (x_1, y_1)\left(\frac{x_2}{x_2^2 + y_2^2}, -\frac{y_2}{x_2^2 + y_2^2}\right)$$
$$= \left(\frac{x_1 x_2 + y_1 y_2}{x_2^2 + y_2^2}, \frac{x_2 y_1 - x_1 y_2}{x_2^2 + y_2^2}\right).$$

Hence, division by any element *other than zero* is defined.

Definition 1.2.1. The complex number system C is the collection of all ordered pairs of real numbers along with the equality, addition operation, and multiplication operation defined above.

We see immediately that the algebraic structure of C is much like that of the real numbers. For example, Properties 1–4 for addition and multiplication hold also for the real number system. Both the reals and C are examples of an algebraic *field*.* This can be easily verified by establishing the *distributive law*,

$$z_1(z_2 + z_3) = z_1 z_2 + z_1 z_3,$$

in addition to the above listed properties. To show that we have not merely reproduced the reals, we should display at least one element of C which is not a real number. Consider the number $i = (0, 1)$; then $i^2 = (-1, 0) = -1$. Hence, the equation $z^2 = -1$ has at least one solution, namely, i, in C, and we have already verified that this equation has no solution in the system of reals. Actually, $z^2 = -1$ has two solutions $\pm i$.

---

* A *field* is a collection of elements together with two operations, "addition" and "multiplication," satisfying the above properties plus the distributive law. For example, the rational numbers form a field.

The real numbers behave as a *proper subset* of C. Consider the subset R in C consisting of all ordered pairs of the form $(x, 0)$, where x is real. There is a one-to-one correspondence between the reals and the elements of R, which preserves the operations of addition and multiplication,

$$(x_1, 0) + (x_2, 0) = (x_1 + x_2, 0),$$
$$(x_1, 0)(x_2, 0) = (x_1 x_2, 0).$$

We say that R is *isomorphic* to the system of reals. In other words, there are just as many elements in R as there are real numbers and the two systems behave algebraically alike. We shall henceforth simply refer to R as the system of reals.

There is another very important difference between C and R. R is an *ordered field*, whereas C is not. We know, for example, that for every pair of reals a and b either $a < b$, $a = b$, or $a > b$.* Furthermore, if $x_1 \neq 0$ then either $x_1 > 0$ or $-x_1 > 0$. Now consider i in C. If C were ordered like R either i or $-i$ would be positive. But $i^2 = (-i)^2 = -1$ and this would imply that -1 is positive. However, -1 and 1 cannot both be positive. Hence, it is meaningless to write $z_1 < z_2$ when z_1 and z_2 are complex numbers unless both z_1 and z_2 are in R.

As a matter of definition, we distinguish between the two numbers in our ordered pairs by calling the first the *real part* and the second the *imaginary part*; that is, if $z = (x, y)$ then $x = \text{Re}(z)$ and $y = \text{Im}(z)$. *It should be noted that* $\text{Im}(z)$ *is a real number.*

In C we also have the notion of *multiplication by a real scalar*. Let a be real; then

$$az = (a, 0)(x, y) = (ax, ay).$$

We see that multiplication by the real scalar has the effect of multiplying both the real and imaginary parts by the scalar. It is easy to verify that

1. $(a + b)z = az + bz$,
2. $a(z_1 + z_2) = az_1 + az_2$,
3. $a(bz) = (ab)z$,
4. $1z = z$.

These properties plus Properties 1–4 for addition imply that C is a *linear vector space over the field of reals.*

Let a and b be real. Then

$$(a, b) = a(1, 0) + b(0, 1) = a1 + bi = a + bi.$$

---

* We assume that the reader is familiar with the properties of the real numbers. If not he should refer to a calculus book like T. M. Apostol, *Calculus*, I. New York: Blaisdell Publishing Company, 1961, pp. 12–37.

Hence, in the future, we shall simply write $z = x + iy$, which is the more familiar notation for the complex number.

The *modulus* $|z|$ of a complex number z is defined as follows

$$|z| = \sqrt{x^2 + y^2} \, .$$

The modulus is obviously a nonnegative real number and $|z| = 0$ if and only if $z = 0$.

The *conjugate* $\bar{z}$ of a complex number $z = x + iy$ is obtained by changing the sign of the imaginary part,

$$\bar{z} = x - iy.$$

The obvious connection between the conjugate and the modulus is

$$z\bar{z} = x^2 + y^2 = |z|^2.$$

Exercises 1.2

1. Compute   (a) $(3 + 5i) + (2 - i)$;   (b) $(1 + i)(7 - 3i)$;
(c) $(3 + i)/(4 + 2i)$.
Ans. (a) $5 + 4i$; (b) $10 + 4i$; (c) $7/10 - (1/10)i$.

2. Prove the associative and commutative laws for addition of complex numbers.

3. Prove the associative and commutative laws for multiplication of complex numbers.

4. Prove that for a given complex number there is a unique negative.

5. (a)   Prove that if $z_2 \neq 0$, $\dfrac{z_1}{z_2} = \dfrac{z_1 \bar{z}_2}{|z_2|^2}$.

   (b)   Use part (a) to compute $\dfrac{2 + 3i}{1 - i}$.   Ans. $\dfrac{-1 + 5i}{2}$.

6. Prove the distributive law.

7. Prove the cancellation law for addition; that is, if $z_1 + z_2 = z_1 + z_3$, then $z_2 = z_3$.

8. Prove the cancellation law for multiplication; that is, if $z_1 z_2 = z_1 z_3$ and $z_1 \neq 0$, then $z_2 = z_3$.

9. Prove the following: (a) $|\bar{z}| = |z|$;   (b) $\overline{z_1 + z_2} = \bar{z}_1 + \bar{z}_2$;
(c) $\overline{z_1 z_2} = \bar{z}_1 \bar{z}_2$;   (d) $\overline{z_1/z_2} = \bar{z}_1/\bar{z}_2$, $z_2 \neq 0$.

10. Prove that $\bar{z} = z$ if and only if $\text{Im}(z) = 0$.

11. Prove that (a) $|z_1 z_2| = |z_1| \, |z_2|$, and (b) $|z_1/z_2| = |z_1|/|z_2|$, $z_2 \neq 0$.

12. Prove that $\text{Re}(z) = \dfrac{z + \bar{z}}{2}$, $\text{Im}(z) = \dfrac{z - \bar{z}}{2i}$.

1.3. THE COMPLEX PLANE

We mentioned in Section 1.2 that C is a linear vector space over the reals. This suggests the possibility of a geometrical interpretation of complex numbers. Indeed, we have characterized the complex numbers in terms of ordered pairs of reals just as one considers ordered pairs of reals in the two-dimensional analytic geometry. We see immediately that we can set up a one-to-one correspondence between the complex numbers and the points in a two-dimensional cartesian coordinate system (see Figure 1.3.1). We

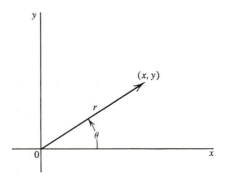

Figure 1.3.1.

call the x axis the *real axis* and the y axis the *imaginary axis* and we identify the complex number $z = x + iy$ with the *vector* drawn from the origin to the point with coordinates (x, y). The modulus $|z|$ is equal to $r = \sqrt{x^2 + y^2}$, where r and θ are the usual polar coordinates for the point (x, y), that is, $x = r \cos \theta$ and $y = r \sin \theta$. The angle θ measured in radians from the positive real axis is called the *argument of z*. Actually, because of the periodicity of the cosine and sine functions, $\theta \pm 2n\pi$, where n is any positive integer, would serve just as well as arguments for the same complex number. It is immediately apparent that if $z \neq 0 + 0i$, then

$$z = r(\cos \theta + i \sin \theta),$$
$$r = |z|, \ \theta = \arg z = \tan^{-1} \frac{y}{x}.$$

This is the *polar form* of the complex number. The argument of zero is undefined.

Let $z_1 = x_1 + iy_1$ and $z_2 = x_2 + iy_2$ be two complex numbers. Then $z_1 + z_2 = (x_1 + x_2) + i(y_1 + y_2)$. Consider the geometrical inter-

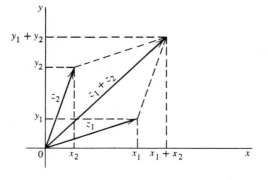

Figure 1.3.2.

pretation of this statement. See Figure 1.3.2. We see that $z_1 + z_2$ is the ordinary vector sum of the vectors corresponding to z_1 and z_2.

The negative, the difference, and the conjugate also have obvious geometrical interpretations (see Figure 1.3.3). The negative $-z$ corresponds to

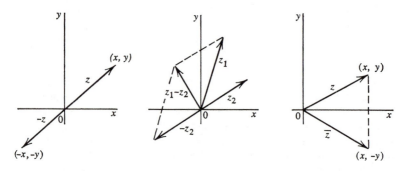

Figure 1.3.3.

a vector equal in magnitude but opposite in direction from that of z. The conjugate $\bar{z}$ corresponds to a vector equal in magnitude to that of z but reflected in the real axis. Obviously, then, we have

$$|z| = |-z| = |\bar{z}|, \quad \arg z = -\arg \bar{z}, \quad \arg (-z) = \pi + \arg z.$$

Let us define*

$$e^{i\theta} = \cos \theta + i \sin \theta.$$

* This definition is completely consistent with the definition of the exponential function e^z of Chapter 2.

Clearly,

$$e^{-i\theta} = \cos(-\theta) + i \sin(-\theta) = \cos\theta - i\sin\theta = \overline{e^{i\theta}}$$
$$|e^{i\theta}| = |e^{-i\theta}| = \sqrt{\cos^2\theta + \sin^2\theta} = 1$$
$$e^{i\theta_1}e^{i\theta_2} = (\cos\theta_1 + i\sin\theta_1)(\cos\theta_2 + i\sin\theta_2)$$
$$= (\cos\theta_1\cos\theta_2 - \sin\theta_1\sin\theta_2) + i(\cos\theta_1\sin\theta_2 + \sin\theta_1\cos\theta_2)$$
$$= \cos(\theta_1 + \theta_2) + i\sin(\theta_1 + \theta_2)$$
$$= e^{i(\theta_1+\theta_2)}$$
$$\frac{e^{i\theta_1}}{e^{i\theta_2}} = e^{i\theta_1}e^{-i\theta_2} = e^{i(\theta_1-\theta_2)}.$$

It follows that if $z_1 = r_1(\cos\theta_1 + i\sin\theta_1) = r_1 e^{i\theta_1}$ and $z_2 = r_2 e^{i\theta_2}$, then*

$$z_1 z_2 = (r_1 e^{i\theta_1})(r_2 e^{i\theta_2}) = r_1 r_2 e^{i(\theta_1+\theta_2)}$$
$$\frac{z_1}{z_2} = \frac{r_1 e^{i\theta_1}}{r_2 e^{i\theta_2}} = \frac{r_1}{r_2} e^{i(\theta_1-\theta_2)},$$

or

$$|z_1 z_2| = r_1 r_2 = |z_1||z_2|,$$
$$\left|\frac{z_1}{z_2}\right| = \frac{r_1}{r_2} = \frac{|z_1|}{|z_2|},$$
$$\arg(z_1 z_2) = \arg z_1 + \arg z_2,$$
$$\arg\left(\frac{z_1}{z_2}\right) = \arg z_1 - \arg z_2.$$

We conclude this section by deriving some important inequalities. Obviously,

$$\mathrm{Re}(z) \le |x| \le \sqrt{x^2 + y^2} = |z|,$$
$$\mathrm{Im}(z) \le |y| \le \sqrt{x^2 + y^2} = |z|.$$

Now, consider $|z_1 + z_2|$, and we have

$$|z_1 + z_2|^2 = (z_1 + z_2)(\bar{z}_1 + \bar{z}_2)$$
$$= z_1\bar{z}_1 + z_2\bar{z}_2 + z_1\bar{z}_2 + \bar{z}_1 z_2$$
$$= |z_1|^2 + |z_2|^2 + 2\,\mathrm{Re}(z_1\bar{z}_2)$$
$$\le |z_1|^2 + |z_2|^2 + 2|z_1|\,|z_2|$$
$$\le (|z_1| + |z_2|)^2.$$

Taking the positive square roots of both sides, we obtain the *triangle inequality*,

$$|z_1 + z_2| \le |z_1| + |z_2|.$$

This can be extended to any finite number of terms by mathematical induction, that is,

$$|z_1 + z_2 + \cdots + z_n| \le |z_1| + |z_2| + \cdots + |z_n|.$$

* Provided $r_2 \ne 0$ in z_1/z_2.

Finally, we have

$$|z_1| = |z_1 - z_2 + z_2| \le |z_1 - z_2| + |z_2|,$$
$$|z_2| = |z_2 - z_1 + z_1| \le |z_1 - z_2| + |z_1|,$$

or

$$|z_1 - z_2| \ge ||z_1| - |z_2||.$$

Exercises 1.3

1. If $z = (1 - i\sqrt{3})/2$, find $|z|$, arg z, arg $(-z)$, arg $\bar{z}$.
Ans. $|z| = 1$, arg $z = (5\pi/3) \pm 2n\pi$, arg $(-z) = (2\pi/3) \pm 2n\pi$, arg $\bar{z} = (\pi/3) \pm 2n\pi$.

2. Let $z_1 = (1 + i)/\sqrt{2}$, $z_2 = \sqrt{3} - i$. Compute $z_1 z_2$ and z_1/z_2. Make a sketch showing the numbers z_1, z_2, $z_1 z_2$, and z_1/z_2 showing the geometrical interpretation of the product and the quotient.
Ans. $z_1 z_2 = 2[\cos(\pi/12) + i\sin(\pi/12)]$, $(z_1/z_2) = (1/2)[\cos(5\pi/12) + i\sin(5\pi/12)]$.

3. Prove that for any positive integer n, $z^n = r^n e^{in\theta}$, where $r = |z|$ and $\theta = $ arg z. In other words, show that $|z^n| = |z|^n$ and arg $z^n = n$ arg z.

4. Show that the points 1, $e^{i\pi/2}$, -1, $e^{i3\pi/2}$ are points equally spaced on the unit circle and that they satisfy the equation $z^4 = 1$. Show in general that the solutions of $z^n = 1$ (the nth roots of unity), where n is a positive integer, are 1, $e^{i2\pi/n}$, $e^{i4\pi/n}, \ldots, e^{i(2n-2)\pi/n}$.

5. Let z_1 and z_2 be two points in the complex plane. Show that the distance between them is $d(z_1, z_2) = |z_1 - z_2|$. Also show that the distance function satisfies the following properties:

(a) $d(z_1, z_2) = d(z_2, z_1)$ (symmetry),

(b) $d(z_1, z_2) \le d(z_1, z_3) + d(z_2, z_3)$ (triangle inequality),

(c) $d(z_1, z_2) > 0$ if $z_1 \ne z_2$,

(d) $d(z_1, z_2) = 0$ if and only if $z_1 = z_2$.

Note: When these properties are satisfied, it is said that we are dealing with a *metric space*.

6. If z is a general point on the locus, show that $|z - z_0| = a$ is the equation of a circle with center at z_0 and radius a.

7. What is the locus of points z satisfying $|z - z_1| = |z - z_2|$?

8. What is the locus of points z satisfying $|z - z_1| = 2|z - z_2|$?

9. Prove the inequality

$$|z_1 w_1 + z_2 w_2| \le \sqrt{|z_1|^2 + |z_2|^2}\sqrt{|w_1|^2 + |w_2|^2}.$$

Generalize to the quantity $|z_1 w_1 + z_2 w_2 + \cdots + z_n w_n|$.

1.4. POINT SETS IN THE PLANE

Any collection of points in the complex plane is called a *point set*. The important thing is that in defining a given point set it can always be determined if a given point in the plane is in the set or not in the set. For example, $S = \{z \mid \operatorname{Re}(z) \geq 0\}$, which we read "$S$ is the set of points z such that $\operatorname{Re}(z) \geq 0$," denotes all the points on or to the right of the imaginary axis. Or $S = \{z \mid |z| < 1\}$ denotes the set of points inside the unit circle. We shall use the notation $z \in S$ to denote "z is an element of S" and $z \notin S$ to denote "z is not an element of S."

We begin by stating several definitions:

1. $C(S) = \{z \mid z \notin S\}$ is the *complement of S*.

2. $S_1 \subseteq S_2$ if $z \in S_1$ implies $z \in S_2$. We say that S_1 is a *subset* of S_2.

3. $S_1 \cup S_2 = \{z \mid z \in S_1 \text{ or } z \in S_2\}$. $S_1 \cup S_2$ is called the *union* of S_1 and S_2.

4. $S_1 \cap S_2 = \{z \mid z \in S_1 \text{ and } z \in S_2\}$. $S_1 \cap S_2$ is called the *intersection* of S_1 and S_2.

5. $N_\epsilon = \{z \mid |z - z_0| < \epsilon\}$ is called an *ϵ-neighborhood* of z_0.

6. A point z_0 is a *limit point* of S if every ϵ-neighborhood of z_0 contains at least one point of S different from z_0. Note that z_0 need not be in S.

7. A point $z_0 \in S$ is an *interior point* of S if some ϵ-neighborhood of z_0 contains only points in S.

8. A set S is *open* if it contains only interior points.

9. A set S is *closed* if it contains all of its limit points *or* if it has no limit points.

10. A point z_0 is a *boundary point* of S if every ϵ-neighborhood of z_0 contains both points in S and in $C(S)$. The *boundary* of S is the collection of boundary points of S.

11. Let S' be the set of limit points of a set S. Then the *closure* $\overline{S}$ of S is $S \cup S'$.

12. The *empty set* $\emptyset$ is the set containing no points.

13. S is *bounded* if there exists a number M such that for all $z \in S$ $|z| < M$.

Several comments about these definitions are now in order. We shall make these comments and at the same time give some examples illustrating the appropriate concepts.

1. If S_1 is a subset of S_2, but there is at least one point $z \in S_2$ such that $z \notin S_1$, then we say that S_1 is a *proper subset*. We also have the notion of *equality of sets*. $S_1 = S_2$ if every point in S_1 is also in S_2 and every point in S_2 is also in S_1, that is, $S_1 \subseteq S_2$ and $S_2 \subseteq S_1$ implies $S_1 = S_2$.

2. S_1 and S_2 are said to be *disjoint* if $S_1 \cap S_2 = \emptyset$.

3. If z_0 is a limit point of S, then every ϵ-neighborhood of z_0 must contain infinitely many points of S. For suppose $\epsilon_1 = \epsilon$. Then there exists a $z_1 \in S$

such that $z_1 \neq z_0$ and $|z_0 - z_1| < \epsilon_1$. Next take $\epsilon_2 = |z_0 - z_1|/2$. There exists a $z_2 \in S$ such that $z_2 \neq z_0$, $z_2 \neq z_1$, and $|z_0 - z_2| < \epsilon_2$. Continuing in this way we can construct a sequence of distinct points $z_1, z_2, z_3, \ldots$ all belonging to S contained in the original ϵ-neighborhood. This, of course, implies that finite sets can have no limit points and are therefore closed.

4. Some examples of open sets are, (a) the empty set, (b) the whole plane, (c) $S = \{z \mid |z| < 1\}$, that is, the interior of the unit circle, and (d) $S = \{z \mid \mathrm{Re}(z) > 0\}$.

5. *The complement of an open set is closed.* This can be seen as follows. Let S be open and consider the limit points of the complement $C(S)$. If $C(S)$ has no limit points, it is closed, and we are finished. If z_0 is a limit point of $C(S)$, then it cannot be in S, because if it were, it would not be an interior point of S and S contains only interior points. Therefore, $z_0 \in C(S)$, and $C(S)$ contains all of its limit points. Hence, $C(S)$ is closed.

6. Some examples of closed sets are, (a) the empty set, (b) the whole plane, (c) $S = \{z \mid |z| \leq 1\}$, and (d) $S = \{z \mid \mathrm{Re}(z) \geq 0\}$.

7. There are sets which are neither open nor closed, for example, $S = \{z \mid 1 < |z| \leq 2\}$. Here the points on the circle of radius 2 are not interior points but boundary points, whereas the points of the unit circle are limit points of S not contained in S. There are sets which are both open and closed, namely, the empty set and the whole plane.

8. A set S is *compact*, if every infinite subset of S has at least one limit point in S. *Compact sets must be closed and bounded.* We see this as follows. If the set is finite or empty, it has no infinite subsets, and hence no limit points. These sets are therefore trivially compact, closed, and bounded. If a set S is infinite and compact, consider z_0, a limit point of S. Consider $N_{\epsilon_1} = \{z \mid |z - z_0| < \epsilon_1\}$. There exists a $z_1 \neq z_0$ such that $|z_1 - z_0| < \epsilon_1$, with $z_1 \in S$. Let $\epsilon_2 = |z_1 - z_0|/2$. There exists a $z_2 \neq z_1$, $z_2 \neq z_0$ such that $z_2 \in S$ and $|z_2 - z_0| < \epsilon_2$. Continuing we construct a distinct infinite sequence $z_1, z_2, z_3, \ldots$ which is a subset of S, whose only limit point is z_0. But S is compact and therefore $z_0 \in S$. S must contain all of its limit points and is closed. If S is unbounded there exists a $z_1 \in S$ such that $|z_1| > 1$. Also there exists a $z_2 \in S$ such that $|z_2| > 2$. Continuing, we have an infinite sequence $z_1, z_2, z_3, \ldots$ such that $|z_n| > n$. But this infinite subset has no limit point in S contradicting the compactness of S. Therefore, S must be bounded. The converse of this theorem is also true, that is, bounded closed sets are compact. This can be stated as follows:

Theorem 1.4.1. *Bolzano-Weierstrass Theorem. Every bounded infinite set has at least one limit point.*

Proof: Consider Figure 1.4.1. Let S be a bounded infinite set. There exists an M such that for all $z \in S$, $|z| < M$. Consider the square* of side

---

* In this context "square" means the boundary plus the interior.

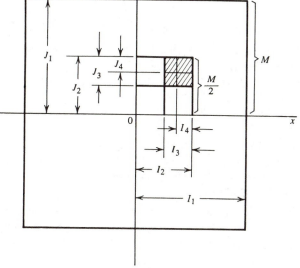

Figure 1.4.1.

$2M$ with center at the origin and sides parallel to the real and imaginary axes. There are an infinite number of points in S lying in this square. Next, consider the four closed squares in each of the four quadrants whose union is the original closed square. At least one of these squares has the property that it contains an infinite number of points in the set S within or on its boundary. Let us say it is the one in the first quadrant. We project each of the infinite number of points of S onto the real and imaginary axes. Then if $I_1 = \{x \mid 0 \leq x \leq M\}$ and $J_1 = \{y \mid 0 \leq y \leq M\}$, there are an infinite number of points of S such that $\text{Re}(z) \in I_1$ and $\text{Im}(z) \in J_1$. Next we divide up the square in the first quadrant into four closed squares of side $M/2$. At least one of these squares has the property that it contains an infinite number of points of S within or on its boundary. Let us say that it is the one touching both of the coordinate axes. If $I_2 = \{x \mid 0 \leq x \leq M/2\}$ and $J_2 = \{y \mid 0 \leq y \leq M/2\}$. Then there are an infinite number of points of S such that $\text{Re}(z) \in I_2$ and $\text{Im}(z) \in J_2$. Continuing in this way we obtain two nested sequences of closed intervals

$$I_1 \supseteq I_2 \supseteq I_3 \supseteq \cdots , J_1 \supseteq J_2 \supseteq J_3 \supseteq \cdots$$

where the lengths of I_n and J_n are $M/2^{n-1}$. By a well-known property of the real numbers, there is precisely one x_0 common to all the I's and one y_0 common to all the J's. We now assert that $z_0 = x_0 + iy_0$ is a limit point of S.

Consider an ϵ-neighborhood of z_0. Take n sufficiently large so that $M/2^{n-2} < \epsilon$. Then for every z with real part in I_n and imaginary part in J_n

$$|z - z_0| \leq |x - x_0| + |y - y_0| \leq M/2^{n-1} + M/2^{n-1} \leq M/2^{n-2} < \epsilon.$$

Therefore, in every ϵ-neighborhood of z_0 there are an infinite number of points in S. Hence, z_0 is a limit point of S.

9. In proving the Bolzano-Weierstrass Theorem we used the principle of nested sequences of closed intervals on the real line. There is a generalization of this to closed nested sets in the complex plane embodied in the following theorem.

Theorem 1.4.2. *Let* $S_1 \supseteq S_2 \supseteq S_3 \supseteq \cdots$ *be a sequence of bounded, nested, closed, nonempty sets. Let* d_n *be the least upper bound of the set of distances between pairs of points in* S_n* *and let* $\lim_{n\to\infty} d_n = 0$. *Then there is one and only one point in all the* S_n.

Proof: Let $z_k \in S_k$. The set $z_1, z_2, z_3, \ldots$ is a bounded infinite set. It therefore has at least one limit point z_0. All of the S_n are closed. Hence, $z_0 \in S_n$, for all n. No other point z_0' is common to all S_n, since, for some n,

$$d_n < |z_0 - z_0'|.$$

10. A *covering* of a set S is a collection of sets $\{C_\alpha\}$ such that for every $\alpha \in S$ there is a C_α for which $\alpha \in C_\alpha$. A very important theorem about coverings is:

Theorem 1.4.3. *Heine-Borel Theorem. If S is a bounded closed set with a covering $\{C_\alpha\}$ consisting of open sets, there is a finite subcovering $\{C_{\alpha_i}\}$ which also covers S.*

Proof: Since S is bounded there exists a number M such that $|z| < M$ for all $z \in S$. Consider the square $\sum$ with corners at $\pm M \pm Mi$. As in the proof of the Bolzano-Weierstrass theorem, we subdivide this square into four equal closed squares. We shall assume that S cannot be covered by a finite subcovering of $\{C_\alpha\}$. This implies that at least one of the four squares $\sum_1$ intersects S in a closed set S_1 which cannot be covered by a finite subset of the covering $\{C_\alpha\}$. Next, we subdivide $\sum_1$ into four equal closed squares. At least one of these squares $\sum_2$ intersects S in a closed set S_2 which cannot be covered by a finite subset of the covering $\{C_\alpha\}$. Continuing in this way we obtain a sequence of nested closed subsets of S none of which can be covered by a finite subset of $\{C_\alpha\}$. The diameter of $\sum_n$ is $\sqrt{2}M/2^{n-1}$. Since $S_n \subseteq \sum_n$, $d(S_n) \leq d(\sum_n)$. Therefore, $\lim_{n\to\infty} d(S_n) = 0$. By the previous theorem, there is one and only one point z_0 common to all S_n.

---

* d_n is called the diameter of S_n.

Now there is an open set C_{z_0} in the covering $\{C_\alpha\}$ containing an ϵ-neighborhood, $|z - z_0| < \epsilon$. If we take n sufficiently large that $\sqrt{2}M/2^{n-1} < \epsilon$, then S_n is contained in this ϵ-neighborhood and is covered by C_{z_0}. This contradicts the statement that S_n is not covered by a finite subset of $\{C_\alpha\}$. This proves the theorem.

Exercises 1.4

1. Prove that $S_1 \cap (S_2 \cup S_3) = (S_1 \cap S_2) \cup (S_1 \cap S_3)$ and $S_1 \cup (S_2 \cap S_3) = (S_1 \cup S_2) \cap (S_1 \cup S_3)$.

2. Prove that the complement of a closed set is open.

3. Prove that the boundary of a set S is $\overline{S} \cap \overline{C(S)}$.

4. Prove that the union of any number (finite or infinite) of open sets is open. Prove that the union of a finite number of closed sets is closed. Give an example to show that the union of an infinite number of closed sets may be open.

5. Prove that the intersection of a finite number of open sets is open. Prove that the intersection of any number of closed sets is closed. Give an example to show that the intersection of an infinite number of open sets may be closed.

6. Classify the following sets according to the properties open, closed, bounded, unbounded, compact.

(a) $S_1 = \{z \mid a \leq \text{Re}(z) \leq b\}$;

(b) $S_2 = \{z \mid |z| = 1\}$;

(c) $S_3 = \left(z \mid 0 < \arg z < \dfrac{\pi}{2}\right)$;

(d) $S_4 = \{z \mid z = e^{(2\pi k/5)i}, k = 0, 1, 2, \ldots\}$;

(e) $S_5 = \{z \mid |z^2 - 1| < 1\}$;

(f) $S_6 = \{z \mid |z| > 2|z - 1|\}$.

7. Given $S = \left\{z \mid z = \dfrac{1}{m} + \dfrac{i}{n}, \quad m = 1, 2, 3, \ldots; n = 1, 2, 3, \ldots\right\}$.

What is $\overline{S}$? Is S compact? Is $\overline{S}$ compact? What is the boundary of S?

1.5. STEREOGRAPHIC PROJECTION. THE EXTENDED COMPLEX PLANE

In the previous section, we discussed the topology of the complex plane in terms of the usual *euclidean metric* $d(z_1, z_2) = |z_1 - z_2|$. We defined open and closed sets in terms of the notion of an ϵ-neighborhood using this

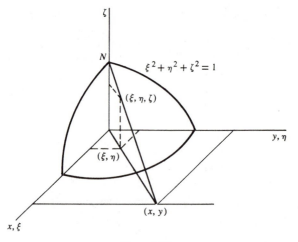

Figure 1.5.1.

metric and discussed compactness. It turned out that the compact sets were the bounded closed sets. However, the infinite set 1, 2, 3, ... has no limit point in the complex plane and is, therefore, not compact. Nevertheless, it will be useful to develop the notion of a limit point for such unbounded sets. Obviously, we will not be able to use the euclidean metric in this discussion. Therefore, we shall introduce a new metric and to this end we shall map the points in the complex plane into the surface of a sphere. This process is referred to as *stereographic projection*.

Through the origin of the complex plane construct a line perpendicular to the plane. Let this be the ζ-axis of a three-dimensional right-handed cartesian coordinate system. In this coordinate system we place a unit sphere known as the *Riemann sphere* with equation $\xi^2 + \eta^2 + \zeta^2 = 1$. See Figure 1.5.1. We call the point with coordinates $(0, 0, 1)$ the *north pole*. The point with coordinates $(0, 0, -1)$ we call the *south pole*. The sphere intersects the complex plane in the circle $x^2 + y^2 = 1$, which we call the *equator*. Next, we draw a line from the north pole to a point in the complex plane with coordinates (x, y). This line will intersect the sphere in some point with coordinates (ξ, η, ζ) which corresponds to $z = x + iy$ under stereographic projection. We say that the point z is *mapped* into the sphere according to this scheme and the point on the sphere is called the *image* of z. We immediately see that there is a one-to-one correspondence between points in the complex plane and points on the sphere with one exception, namely, the north pole itself. The interior of the unit circle maps into the southern hemisphere while points outside the unit circle map into the northern hemisphere. In fact, we see that the greater the distance from the point z to the origin the closer will be its image to the north pole on the sphere. If we wish to include the north pole as an image point, we will have to extend the complex plane by adding an idealized point (not a complex number) known as the *point at infinity*. We then say that the north pole is the image of the point at infinity under stereographic projection. If we wish to include the point at infinity in the discussion we shall refer to the *extended complex plane*. With the addition of this idealized element to the complex numbers we can state the following result. *There is a one-to-one correspondence between the points in the extended complex plane and the points on the Riemann sphere under stereographic projection.*

It is easy to obtain explicit relations describing the stereographic projection. Using similar triangles, we have

$$\frac{\xi}{x} = \frac{\eta}{y} = \frac{\rho}{r} = \frac{1 - \zeta}{1},$$

where $r = \sqrt{x^2 + y^2}$ and $\rho = \sqrt{\xi^2 + \eta^2}$. From this, we obtain

$$x = \frac{\xi}{1 - \zeta}, \; y = \frac{\eta}{1 - \zeta}, \; \xi = \frac{2x}{r^2 + 1}, \; \eta = \frac{2y}{r^2 + 1}, \; \zeta = \frac{r^2 - 1}{r^2 + 1}.$$

Using these relations we can show several interesting properties of the stereographic projection. For example, *straight lines and circles are mapped into circles.* The general equation of a circle in the complex plane is

$$\alpha x^2 + \alpha y^2 + \beta x + \gamma y + \delta = 0.$$

Under stereographic projection, we have

$$\alpha \frac{\xi^2 + \eta^2}{(1 - \zeta)^2} + \frac{\beta \xi}{1 - \zeta} + \frac{\gamma \eta}{1 - \zeta} + \delta = 0,$$

$$\alpha \frac{1 - \zeta^2}{(1 - \zeta)^2} + \frac{\beta \xi}{1 - \zeta} + \frac{\gamma \eta}{1 - \zeta} + \delta = 0,$$

$$\alpha(1 + \zeta) + \beta \xi + \gamma \eta + \delta(1 - \zeta) = 0,$$

$$\beta \xi + \gamma \eta + (\alpha - \delta)\zeta + \alpha + \delta = 0.$$

The last equation is the equation of a plane in the (ξ, η, ζ) coordinates, and it is well known that a plane and a sphere intersect in a circle. For example, if $\alpha = 0$, we have a straight line in the xy-plane. The image of this line is given by the intersection of

$$\xi^2 + \eta^2 + \zeta^2 = 1,$$
$$\beta \xi + \gamma \eta - \delta \zeta + \delta = 0,$$

which is a circle passing through the north pole. If $\delta = 0$ and $\alpha \neq 0$, $\alpha x^2 + \alpha y^2 + \beta x + \gamma y = 0$ is the equation of a circle passing through the origin, which has an image given by the intersection of

$$\xi^2 + \eta^2 + \zeta^2 = 1,$$
$$\beta \xi + \gamma \eta + \alpha \zeta + \alpha = 0.$$

This is a circle passing through the south pole.

Next, we introduce a new metric by considering the chordal distance between the images of two points on the Riemann sphere. See Figure 1.5.2.

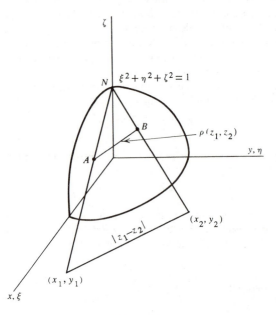

Figure 1.5.2.

Let A be the image of $z_1 = x_1 + iy_1$ and B be the image of $z_2 = x_2 + iy_2$. Let $\rho(z_1, z_2)$ be the distance between A and B. Then

$$\rho(z_1, z_2) = \sqrt{(\xi_1 - \xi_2)^2 + (\eta_1 - \eta_2)^2 + (\zeta_1 - \zeta_2)^2}$$

$$= \sqrt{\xi_1^2 + \eta_1^2 + \zeta_1^2 + \xi_2^2 + \eta_2^2 + \zeta_2^2 - 2\xi_1\xi_2 - 2\eta_1\eta_2 - 2\zeta_1\zeta_2}$$

$$= \sqrt{2(1 - \xi_1\xi_2 - \eta_1\eta_2 - \zeta_1\zeta_2)}$$

$$= \frac{\sqrt{2[(1 + r_1^2)(1 + r_2^2) - 4x_1x_2 - 4y_1y_2 - (r_1^2 - 1)(r_2^2 - 1)]}}{\sqrt{1 + r_1^2}\,\sqrt{1 + r_2^2}}$$

$$= \frac{2\sqrt{r_1^2 + r_2^2 - 2x_1x_2 - 2y_1y_2}}{\sqrt{1 + r_1^2}\,\sqrt{1 + r_2^2}}$$

$$= \frac{2|z_1 - z_2|}{\sqrt{1 + |z_1|^2}\,\sqrt{1 + |z_2|^2}}.$$

If z_2 is the point at infinity, then $\rho(z_1, \infty)$ is the distance between the image of z_1 and north pole; that is,

$$\rho(z_1, \infty) = \sqrt{\xi_1^2 + \eta_1^2 + (1 - \zeta_1)^2}$$
$$= \sqrt{2(1 - \zeta_1)}$$
$$= \frac{2}{\sqrt{1 + r_1^2}}$$
$$= \frac{2}{\sqrt{1 + |z_1|^2}}.$$

So, we see that the chordal metric is defined in the extended complex plane. We are permitted to call it a metric, because it satisfies the usual properties

1. $\rho(z_1, z_2) = \rho(z_2, z_1)$,
2. $\rho(z_1, z_2) \leq \rho(z_1, z_3) + \rho(z_3, z_2)$,
3. $\rho(z_1, z_2) \geq 0$,
4. $\rho(z_1, z_2) = 0$ if and only if $z_1 = z_2$.

The following inequalities hold between the chordal metric and the euclidean metric in any bounded set for which $|z| < M$

$$\rho(z_1, z_2) \leq 2|z_1 - z_2| \leq (1 + M^2)\rho(z_1, z_2).$$

This implies that for bounded sets discussions involving limit points, compactness, and so on, the two metrics can be used interchangeably. On the other hand, if the set is unbounded, it has the point at infinity as a limit point using the chordal metric. If the set is unbounded it contains an infinite subset $z_1, z_2, \ldots$, such that $\lim_{n \to \infty} z_n = \infty$. Hence,

$$\lim_{n \to \infty} \rho(z_n, \infty) = \lim_{n \to \infty} \frac{2}{\sqrt{1 + |z_n|^2}} = 0.$$

This implies that the *extended complex plane is compact in the chordal metric.* This is not surprising, since the extended complex plane is bounded in the chordal metric. In fact, $\rho(z_1, z_2) \leq 2$.

It is possible to give the point at infinity an interpretation in terms of the mapping of the complex z plane into the complex w plane by the following function

$$w = \frac{1}{z}.$$

This function is defined for every complex number z except $z = 0$. The unit circle $|z| = 1$ is mapped onto the unit circle $|w| = 1$. Points outside the unit circle in the z plane are mapped into points inside the unit circle in the w plane. Let us consider points such that $|z| > R$. The images of these points satisfy the inequality $|w| < 1/R$ and are thus in a neighborhood of the origin in the w plane. On the Riemann sphere the points for which $|z| > R$ have images for which $\zeta > (R^2 - 1)/(R^2 + 1)$. Hence, these points are in a neighborhood of the north pole in the chordal metric. Therefore, it is quite natural to define a neighborhood of the point at infinity as those points for which $|z| > R$. In fact, we could define the point at infinity as the preimage of $w = 0$ under the mapping $w = 1/z$. Since the inverse of this function is $z = 1/w$, the point at infinity in the w plane is the image of $z = 0$. By adding the points at infinity in both planes we have the following result: *there is a one-to-one correspondence between the extended z plane and the extended w plane under the mapping $w = 1/z$, with the origin mapping into the point at infinity and the point at infinity mapping into the origin.*

Exercises 1.5

1. Describe the relative positions of the images of z, $-z$, $\bar{z}$, $1/z$, and $1/\bar{z}$ on the Riemann sphere.

2. Prove that the function $w = 1/z$ maps circles and straight lines into circles and straight lines. What happens to circles through the origin and straight lines through the origin?

3. Prove that a set is open in the extended complex plane if and only if its image is open on the Riemann sphere. Hint: Define open sets in terms of the chordal metric.

1.6. CURVES AND REGIONS

Of great importance in the study of integration in the complex plane are sets of points called curves. We shall begin our discussion of curves with the definition of a *simple Jordan arc.*

Definition 1.6.1. A simple Jordan arc is a set of points defined by a parametric equation $z(t) = x(t) + iy(t)$, $0 \leq t \leq 1$, where $x(t)$ and $y(t)$ are continuous real-valued functions such that $t_1 \neq t_2$ implies that $z(t_1) \neq z(t_2)$.

Definition 1.6.2. A simple smooth arc is a Jordan arc defined by a parametric equation $z(t) = x(t) + iy(t)$, $0 \leq t \leq 1$, where $\dot{x}(t) = dx/dt$ and $\dot{y}(t) = dy/dt$ are continuous, and $\dot{x}^2 + \dot{y}^2 \neq 0$.

Actually, there is a bit of redundancy in the second definition since the existence of $\dot{x}$ and $\dot{y}$ implies the continuity of $x(t)$ and $y(t)$. A tangent vector to the arc has components $\dot{x}$ and $\dot{y}$. Hence, the simple smooth arc has a continuously turning tangent. Also since the simple smooth arc is also a

Jordan arc, implying that $z(t_1) \neq z(t_2)$ when $t_1 \neq t_2$, the arc does not go through the same point twice and does not close on itself. The condition $\dot{x}^2 + \dot{y}^2 \neq 0$ is included to insure that $\dot{x}$ and $\dot{y}$ are not both zero, in which case the tangent would not exist.

Definition 1.6.3.   A simple closed Jordan curve is a set of points defined by a parametric equation $z(t) = x(t) + iy(t)$, $0 \leq t \leq 1$, where $x(t)$ and $y(t)$ are continuous real-valued functions such that $z(t_1) = z(t_2)$ if and only if $t_1 = 0$, $t_2 = 1$, or $t_1 = 1$, $t_2 = 0$.

This definition differs from that of the simple Jordan arc only in that $z(0) = z(1)$ which implies that the first point and the last point are the same and hence that the curve closes on itself. There are no other multiple points.

An example of a simple closed Jordan curve is the unit circle given parametrically by

$$z(t) = e^{2\pi it} = \cos 2\pi t + i \sin 2\pi t.$$

The unit circle divides the plane into two disjoint open sets namely $|z| < 1$ and $|z| > 1$. The unit circle itself is the boundary of each of these sets. Actually, every simple closed Jordan curve has this same property as was indicated by Camille Jordan in 1887 in the now famous Jordan Curve Theorem. We state the theorem here without proof.

Theorem 1.6.1.   *Jordan Curve Theorem. Every simple closed Jordan Curve in the complex plane divides the plane into two disjoint open sets. The curve is the boundary of each of these sets. One set (the interior of the curve) is bounded and the other (the exterior of the curve) is unbounded.*

Definition 1.6.4.   A simple piecewise smooth curve is a simple Jordan arc whose parametric equation $z(t) = x(t) + iy(t)$, $0 \leq t \leq 1$, has piecewise continuous derivatives $\dot{x}$ and $\dot{y}$, where $\dot{x}^2 + \dot{y}^2 \neq 0$.

By piecewise continuity we imply that the interval $0 \leq t \leq 1$ can be divided up into a finite number of open intervals by the partition

$$0 < t_1 < t_2 < \cdots < t_n < 1$$

in each of which $\dot{x}$ and $\dot{y}$ are continuous, and furthermore that the limits of $\dot{x}$ and $\dot{y}$ exist as t approaches the ends of these intervals from their interiors. This does not imply that $\dot{x}$ and $\dot{y}$ are continuous, since it is possible that $\lim_{t \to t_k-} \dot{x} \neq \lim_{t \to t_k+} \dot{x}$ and/or $\lim_{t \to t_k-} \dot{y} \neq \lim_{t \to t_k+} \dot{y}$. However, the piecewise continuity does imply that $\dot{x}$ and $\dot{y}$ are bounded and that the tangent vector to the curve is piecewise continuous. An example of a simple piecewise smooth curve is the broken line segment given parametrically by

$$z = \begin{cases} z_0 + 2t(z_1 - z_0), & 0 \leq t \leq \frac{1}{2} \\ z_1 + 2(t - \frac{1}{2})(z_2 - z_1), & \frac{1}{2} \leq t \leq 1 \end{cases}.$$

Definition 1.6.5. A simple closed piecewise smooth curve is a simple closed Jordan curve whose parametric equation $z(t) = x(t) + iy(t)$, $0 \leq t \leq 1$, has piecewise continuous derivatives $\dot{x}$ and $\dot{y}$, where $\dot{x}^2 + \dot{y}^2 \neq 0$.

For brevity, we shall henceforth refer to a simple piecewise smooth curve as a *simple contour* and to a simple closed piecewise smooth curve as a *simple closed contour*.

In the case of simple smooth arcs, we are dealing with curves which have length. This goes back to the calculus, where it is shown that when $\dot{x}$ and $\dot{y}$ are continuous the length of the curve can be expressed as the ordinary Riemann integral

$$L = \int_0^1 \sqrt{\dot{x}^2 + \dot{y}^2} \, dt = \int_0^1 |\dot{z}| \, dt.$$

Furthermore, since simple contours are just the union of a finite number of simple smooth arcs the length of a simple contour is

$$L = \sum_{k=0}^n \int_{t_k}^{t_{k+1}} \sqrt{\dot{x}^2 + \dot{y}^2} \, dt,$$

where $0 = t_0 < t_1 < t_2 < \cdots < t_n < t_{n+1} = 1$ is the partition needed to break up the interval into subintervals in which $\dot{x}$ and $\dot{y}$ are continuous.

Definition 1.6.6. A set S in the complex plane is *connected* if every pair of points in S can be joined by a simple Jordan arc lying entirely in S.

For example, the set

$$S_1 = \{z \mid |z - 1| \leq 1 \text{ or } |z + 1| \leq 1\}$$

is connected, for it consists of the interior plus the boundary of the two circles of radius one centered at plus one and minus one. Any pair of points inside the right-hand circle can be joined by a straight-line segment. The same is true for pairs inside the left-hand circle. If one point is inside the left-hand circle and one is inside the right-hand circle they may be joined by the curve consisting of straight line segments joining the points to the centers of their respective circles plus the segment $-1 \leq x \leq 1$. On the other hand, the set

$$S_2 = \{z \mid |z - 1| < 1 \text{ or } |z + 1| < 1\}$$

is not connected, since a curve joining a pair of points, one inside the left-hand circle and the other inside the right, would have to pass through a point which is not in the set. The reader can show that the set

$$S_3 = \{z \mid |z - 1| \leq 1 \text{ or } |z + 1| < 1\}$$

is connected.

If a set is open and connected then every pair of points can be joined by a *polygonal line*, that is, a simple contour consisting of a finite union of line segments, lying in the set. This can be proved as follows. Let the set S be open and connected. If z_1 and z_2 are two distinct points in S, then there is a simple Jordan arc lying in S given parametrically by $z(t) = x(t) + iy(t)$, $0 \leq t \leq 1$, such that $z_1 = z(0)$ and $z_2 = z(1)$. Now, $x(t)$ and $y(t)$ are continuous and hence bounded. Therefore, the arc is a bounded closed set. Furthermore, each point on the arc is an interior point of S since S is open. If $z_\alpha = z(t_\alpha)$ is a point on the arc there is an open set

$$C_\alpha = \{z \mid |z - z_\alpha| < \delta_\alpha\}$$

lying in S. The collection $\{C_\alpha\}$ is an open covering of the arc. By the Heine-Borel Theorem there exists a finite subcovering $\{C_{\alpha_i}\}$, $i = 1, 2, 3, \ldots,$ n of the arc, where we have ordered the index according to the ordering in t,

$$0 = t_{\alpha_1} < t_{\alpha_2} < t_{\alpha_3} < \cdots < t_{\alpha_{n-1}} < t_{\alpha_n} = 1.$$

Now, the set C_{α_1} must intersect at least one other C_{α_i}, $i \neq 1$. Otherwise, the arc would not be connected. We join z_1 to z_{α_k} with a straight line segment, where k is the largest integer for which $C_{\alpha_1} \cap C_{\alpha_i} \neq \emptyset$. This line segment lies in $C_{\alpha_1} \cup C_{\alpha_k}$. We continue in this way until after a finite number of steps we reach z_2. The union of the line segments thus drawn lies in S, and if it is a simple contour, the proof is complete. If not, there must be some multiple points, that is, intersections of line segments. But in this case we can construct a simple contour as follows. We first number the end points of segments in ascending order according to the order in which the straight line segments were drawn. See Figure 1.6.1. Starting with the first end point we proceed to the second along the segment joining them unless we are stopped by another end point with a higher number or an intersection with a line segment with end points with higher numbers. In the latter case, we

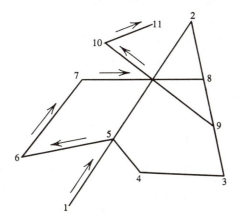

Figure 1.6.1.

always move along the new segment intersected in the direction which takes us toward the end point with the higher number. Proceeding in this way we eventually arrive at z_2 and the path we have traversed is a simple contour.

Definition 1.6.7. A nonempty open connected set of points is a *domain*.

Definition 1.6.8. A *region* is a domain together with all, some, or none of its boundary points.

Definition 1.6.9. A domain D is *simply connected* if every simple closed Jordan curve lying in D has its interior lying in D.

For example, the set $S_1 = \{z \mid |z| < 1\}$ is a simply connected domain but the set $S_2 = \{z \mid 0 < |z| < 1\}$ is not. In the latter case some simple closed Jordan curves have interiors in S_2, for example, $|z - \frac{1}{2}| = \frac{1}{4}$, but others do not, for example, $|z| = \frac{1}{2}$.

If a domain is not simply connected, it is said to be multiply connected.

Exercises 1.6

1. Classify the following sets according to the terms simple Jordan arc, simple closed Jordan curve, simple contour, simple closed contour:

(a) $z(t) = t + i \sin 1/t, 0 \leq t \leq 1$;

(b) $z(t) = t + it \sin 1/t, 0 \leq t \leq 1$;

(c) $S = \{z \mid |z^2 - 1| = 1\}$;

(d) $|z| = 1 - \cos(\arg z), 0 \leq \arg z \leq 2\pi$.

Ans. (a) none; (b) simple Jordan arc; (c) none; (d) simple closed contour.

2. Find the lengths of the following curves:

(a) $z(t) = a \cos 2\pi t + ib \sin 2\pi t$;

(b) $|z| = 1 - \cos(\arg z), 0 \leq \arg z \leq 2\pi$;

(c) $z = te^{it}, 0 \leq t \leq 2\pi$;

(d) $z = (t - \sin t) + i(1 - \cos t), 0 \leq t \leq 2\pi$.

3. Classify the following sets according to the terms domain, region, connected, simply connected, multiply connected.

(a) $S = \{z \mid 1 < |z| < 2\}$;   (b) $S = \{z \mid 1 < \mathrm{Re}(z) \leq 2\}$;

(c) $S = \{z \mid |z^2 - 1| < 1\}$;   (d) $S = \left\{z \mid 0 \leq \arg z \leq \frac{\pi}{2}\right\}$.

Ans. (a) multiply connected domain; (b) region; (c) not connected; (d) region.

4. Let $z(t) = x(t) + iy(t), a \leq t \leq b$, be the parametric equation for a curve. Is there a parameter τ for the curve such that $0 \leq \tau \leq 1$? What if $a = -\infty$ and $b = \infty$?

CHAPTER 2

Functions of a
Complex Variable

2.1. FUNCTIONS AND LIMITS

We are now ready to begin the development of the calculus of functions of a complex variable. We begin the discussion with a definition of function.

Let S be a nonempty set of points in the complex z plane. If for each $z \in S$ there corresponds a uniquely determined number in the complex w plane, then we say that a *function of the complex variable z* is defined in S into the w plane. In other words, a function of the complex variable z is a collection of ordered pairs (z, w), the first, designating the value of the *independent variable z*, and the second, designating the corresponding value of the *dependent variable w*. We shall use the conventional functional notation; that is, if $z_0 \in S$ and (z_0, w_0) is a pair in the function, then we write $w_0 = f(z_0)$. We say that z_0 is *mapped into* w_0, w_0 is the *image* of z_0 under the *mapping* (function), and the function *maps* S into the w plane. The set S is called the *domain** of the function and the set of image points of S is called the *range* of the function. We say that the function maps S *onto* the range R since every $w \in R$ is an image point of at least one point in S. If each $w \in R$ is the image of precisely one point in S, that is, $z_1 \neq z_2$ implies $f(z_1) \neq f(z_2)$, then the mapping is *one-to-one*. In this case, for each $w \in R$ there corresponds a uniquely determined $z \in S$. Hence, a function is defined in R into the z-plane with R as the domain and S as the range. This function is called the *inverse* of the original function. If $w = f(z)$, $z \in S$, is a one-to-one function in S onto R, then there exists an inverse function $z = g(w)$ in R onto S such that $z = g[f(z)]$. The following examples will serve to illustrate these ideas.

EXAMPLE 2.1.1. The function $w = z^2$, $|z| \leq 1$, maps the unit circle plus its interior onto the unit circle plus its interior in the w-plane. The domain of the function is $S = \{z \mid |z| \leq 1\}$. The range is $R = \{w \mid |w| \leq 1\}$.

---

* The use of the term domain here is different from the usage of Section 1.6, where domain refers to a nonempty open connected set of points.

The function is not one-to-one. This can be proved as follows: Let $z = re^{i\theta}$. Then $w = \rho e^{i\phi} = z^2 = r^2 e^{2i\theta}$. Hence, $\rho = r^2$ and $\phi = 2\theta$. This shows that the function maps S onto R twice. This function, therefore, has no inverse.

EXAMPLE 2.1.2.   The function $w = z^2$, $|z| \leq 1$, $0 \leq \arg z < \pi$.

This is not the same function as in Example 2.1.1, since the domain is now $S = \{z \mid |z| \leq 1, 0 \leq \arg z < \pi\}$. The range is the same however, $R = \{w \mid |w| \leq 1\}$. This time the function is one-to-one and the inverse function is defined by $|z| = r = \sqrt{\rho}$, $0 \leq \rho \leq 1$, and $\arg z = \theta = \frac{1}{2}\phi$, $0 \leq \phi < 2\pi$.

EXAMPLE 2.1.3. $w = \bar{z}$ maps the entire z-plane onto the entire w-plane. The mapping consists of a reflection in the real axis. The function is one-to-one and the inverse is $z = \bar{w}$.

EXAMPLE 2.1.4. $w = |z|$ maps the entire z plane onto the nonnegative real axis of the w plane. The function is not one-to-one and hence no inverse exists.

EXAMPLE 2.1.5. $w = 1/z$, $z \neq 0$ maps the entire z plane except for the origin onto the entire w plane except for the origin. We could extend the definition of the function to the extended z plane by defining the function at the point at infinity to have the value zero. Then the extended z plane, without the origin, is mapped onto the w plane. The inverse $z = 1/w$ is defined in the w plane onto the extended z plane without the origin. If we associate the origin of the z plane with the point at infinity of the w plane then we have a one-to-one function in the extended z plane onto the extended w plane.

EXAMPLE 2.1.6. $w = (az + b)/(cz + d)$, where a, b, c, and d are complex constants such that $ad - bc \neq 0$, and $z \neq -d/c$, maps the z plane without $-d/c$ onto the w plane without the point a/c. Let

$$\frac{az_1 + b}{cz_1 + d} = \frac{az_2 + b}{cz_2 + d}.$$

Then

$$(az_1 + b)(cz_2 + d) = (az_2 + b)(cz_1 + d)$$
$$(ad - bc)(z_1 - z_2) = 0$$
$$z_1 = z_2$$

Therefore, the function is one-to-one and the inverse is $z = (dw - b)/(-cw + a)$, $w \neq a/c$. If we associate $z = -d/c$ with $w = \infty$ and $z = \infty$

with $w = a/c$ we have a one-to-one function defined in the extended z plane onto the extended w plane.

Let $w = f(z)$ be a complex-valued function of z defined for $z \in S$. Let $u = \text{Re}(w)$ and $v = \text{Im}(w)$. Then we can write $w = u(x, y) + iv(x, y)$, where $u(x, y)$ and $v(x, y)$ are two real-valued functions of the two real variables x and y both defined in S. For example, if $w = z^2$, then $w = x^2 - y^2 + 2ixy$. Hence, $u(x, y) = x^2 - y^2$ and $v = 2xy$. Suppose, $u = 1$ and $v = 2$. Then $x^2 - y^2 = 1$ and $xy = 1$. These are the equations of hyperbolas. See Figure 2.1.1. Similarly, $u = 4$ and $v = 8$ are also hyperbolas. Hence, the shaded region in the z plane is mapped onto the shaded region in the w plane.

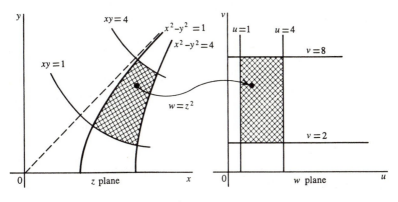

Figure 2.1.1.

Just as in the real calculus, we shall define differentiability of a function of a complex variable in terms of a limit. Hence, we shall have to begin with a careful definition of limit of a function. Let $w = f(z)$ be defined in a set S. Let z_0 be a limit point of S. If there exists a number L such that for every $\epsilon > 0$ there exists a δ such that $|f(z) - L| < \epsilon$ for all $z \in S$ satisfying $0 < |z - z_0| < \delta$, then L is the limit as z approaches z_0 in S, written

$$L = \lim_{z \to z_0} f(z).$$

Several comments on this definition are in order. First, we see that the function need not be defined at z_0 in order for it to have a limit at z_0. Second, the function need not be defined everywhere in an annulus $0 < |z - z_0| < \alpha$, since it is only required that $f(z)$ approach L on those points near z_0 *in* S. Third, we can easily give meaning to the statement

$$\lim_{z \to \infty} f(z) = L,$$

provided ∞ is a limit point of S on the Riemann sphere, by merely replacing $0 < |z - z_0| < \delta$ by $0 < \rho(z, z_0) < \delta$, where ρ is the chordal metric. Finally, we can give meaning to the statement

$$\lim_{z \to z_0} f(z) = \infty,$$

if we replace $|f(z) - L| < \epsilon$ by $\rho[f(z), L) < \epsilon$. Let us consider some examples.

EXAMPLE 2.1.7. Let $f(z) = z^2$ in $S = \{z \mid |z| \leq 1\}$. Then $\lim_{z \to 1} f(z) = 1$. This can be proved as follows. Consider $|z^2 - 1| < \epsilon$ or $|z - 1||z + 1| < \epsilon$. When $0 < |z - 1| < \delta$ and $z \in S$, $|z + 1| < 2$ and $|z - 1||z + 1| < 2\delta$. Hence, if we take $\delta = \epsilon/2$, $|z^2 - 1| < \epsilon$. For $\epsilon > 0$ we have found an appropriate δ. Therefore, the limit is 1.

EXAMPLE 2.1.8. Let $f(z) = \dfrac{z^2 - 1}{z - 1}$, in $S = \{z \mid z \neq 1\}$. Then

$$\lim_{z \to 1} f(z) = 2.$$

This time the function is not defined at $z = 1$ but the limit still exists. When $z \neq 1$, $f(z) = z + 1$, and $|f(z) - 2| = |z - 1|$. Hence, in this case we can take $\delta = \epsilon$ since $0 < |z - 1| < \delta$ will imply that

$$|f(z) - 2| < \epsilon.$$

EXAMPLE 2.1.9. Let $f(z) = 1 - z$ in $S = \{z \mid z = (1/n), n = 1, 2, 3, \ldots\}$. Then $\lim_{z \to 0} f(z) = 1$. This follows because $f(1/n) = 1 - (1/n)$ and this clearly approaches 1, as n approaches infinity.

The reader can easily verify the following statements:

EXAMPLE 2.1.10. Let $f(z) = 1 - (1/z)$, $S = \{z \mid z = n, n = 1, 2, 3, \ldots\}$. Then $\lim_{z \to \infty} f(z) = 1$.

EXAMPLE 2.1.11. Let $f(z) = (az + b)/(cz + d)$, in $S = \{z \mid z \neq -d/c, ad - bc \neq 0\}$. Then $\lim_{z \to -d/c} f(z) = \infty$. Hint: $\rho[f(z), \infty] = 2/(1 + |f(z)|^2)^{1/2}$.

Definition 2.1.1. The function $w = f(z)$ with domain S is continuous at $z_0 \in S$ if $f(z_0) \neq \infty$ and for every $\epsilon > 0$ there exists a δ such that

$$|f(z) - f(z_0)| < \epsilon$$

for all $z \in S$ satisfying $|z - z_0| < \delta$. The function is continuous in S if it is continuous at every point in S.

Theorem 2.1.1. *The function $w = f(z)$ with domain S is continuous at z_0, a limit point of S, if and only if $\lim_{z \to z_0} f(z) = f(z_0)$.*

Proof: The proof will be left for the exercises.

EXAMPLE 2.1.12. The function $w = z^2$ is continuous everywhere in the finite complex plane since $f(z_0) = z_0^2$ and $\lim_{z \to z_0} z^2 = z_0^2$.

EXAMPLE 2.1.13. The function $w = 1/z$ is continuous everywhere in the extended plane except at $z = 0$, if we define $f(\infty) = 0$. We use the chordal metric in treating the point $z_0 = \infty$, that is, $\rho(z, \infty) = 2/(1 + |z|^2)^{1/2} < 2/(1 + \epsilon^{-2})^{1/2}$ implies that $|z| > 1/\epsilon$ or $|1/z| < \epsilon$. Therefore, we can take $\delta = 2/(1 + \epsilon^{-2})^{1/2}$.

EXAMPLE 2.1.14. The function $w = (z^2 - 1)/(z - 1)$ is not continuous at $z = 1$ since it is not defined there. However, if $f(1)$ is defined to be equal to $\lim_{z \to 1} (z^2 - 1)/(z - 1) = 2$ then the function becomes continuous at $z = 1$.

EXAMPLE 2.1.15. The function $f(z) = 1 - z$, $f(0) = 1$,

$$S = \left\{ z \mid z = 0, z = \frac{1}{n}, n = 1, 2, 3, \ldots \right\}$$

is continuous everywhere in S. The only limit point in S is $z = 0$, where $\lim_{z \to 0} f(z) = f(0) = 1$. All other points in S are isolated points and by definition a function defined at an isolated point is continuous there. See Exercises 2.1.

The statement $f(z_0) \neq \infty$ is an important part of the definition. This requirement seems a little arbitrary, since in the definition of limit we allowed the limit to be the point at infinity. However, we wish to relate continuity to boundedness in the following theorem, which would not be possible if we allowed $f(z_0) = \infty$.

Theorem 2.1.2. *If $w = f(z)$ is continuous in a closed set S, it is bounded in S; that is, there exists an $M > 0$ such that $|f(z)| < M$ for all $z \in S$.*

Proof: If $f(z)$ is unbounded, then there exists a $z_n \in S$ such that

$$|f(z_n)| > n \text{ for } n = 1, 2, 3, \ldots.$$

This defines an infinite set of points $z_1, z_2, z_3, \ldots$, in S. But S is compact with respect to the chordal metric. Therefore, the set $z_1, z_2, z_3, \ldots$, has a limit point $z_0 \in S$. The function is continuous at z_0 and, therefore, $f(z_0) \neq \infty$ and $\lim_{z_n \to z_0} f(z) = f(z_0)$. But this is contradicted by $|f(z_n)| > n$. Hence, $f(z)$ must be bounded.

In proving that a function is continuous at a point z_0 it is necessary to find an appropriate δ for each ϵ in the definition of limit. In general, the δ will depend on z_0 as well as on ϵ. If a δ, depending on ϵ but not on z_0, can be found which works for all $z_0 \in S$, the domain of the function, then we say that the function is *uniformly continuous* in S.

Definition 2.1.2. A function $w = f(z)$, defined in S, is uniformly continuous in S if it is continuous in S and for every $\epsilon > 0$ there exists a δ independent of z_0 such that $|f(z) - f(z_0)| < \epsilon$ when $|z - z_0| < \delta$, $z_0 \in S$, $z \in S$.

Consider, for example, the function $w = 1/z$ in $S = \{z \mid 0 < |z| \leq 1\}$. Let $z_0 \in S$. Then $w_0 = 1/z_0$ and

$$\left| \frac{1}{z} - \frac{1}{z_0} \right| = \frac{|z - z_0|}{|z| \, |z_0|} .$$

If $\delta < |z_0|$ and $|z - z_0| < \delta$, then $|z| = |z - z_0 + z_0| \geq |z_0| - \delta$, and

$$\left| \frac{1}{z} - \frac{1}{z_0} \right| = \frac{|z - z_0|}{|z| \, |z_0|} \leq \frac{\delta}{|z_0|^2 - \delta |z_0|} .$$

Hence, we can take $\delta/(|z_0|^2 - \delta|z_0|) = \epsilon$ or $\delta = \epsilon|z_0|^2/(1 + \epsilon|z_0|)$. We see that δ depends on z_0 as well as ϵ. Clearly, in S we do not have uniform continuity. The difficulty here is that S is not closed. If, on the other hand, we consider the closed set $S_1 = \{z \mid \alpha \leq |z| \leq 1\}$, $\alpha > 0$, then for a given ϵ we can take $\delta = \epsilon\alpha^2/(1 + \epsilon)$ which is independent of z_0. Therefore, in S_1 we have uniform continuity. This example illustrates the following theorem.

Theorem 2.1.3. *If $w = f(z)$ is continuous in the closed set S, it is uniformly continuous in S.*

Proof: We shall prove the theorem for a bounded set S where we can use the euclidean metric. The same argument will prove the theorem for unbounded sets closed on the Riemann sphere if the euclidean metric is replaced by the chordal metric.

Let z_α be a point in S. Since $f(z)$ is continuous at z_α, given an ϵ there exists a δ_α such that $|f(z) - f(z_\alpha)| < \epsilon/2$ when $z \in S$ and $|z - z_\alpha| < \delta_\alpha$. Now consider the open sets $C_\alpha = \{z \mid |z - z_\alpha| < \frac{1}{2}\delta_\alpha\}$. The union of all the C_α is an open covering of S. By the Heine-Borel theorem there is a finite subcovering by sets $C_1, C_2, \ldots, C_n$. Let $2\delta = \min [\delta_1, \delta_2, \ldots, \delta_n]$ and let $|z - z_0| < \delta$. There exists a z_j such that $|z_0 - z_j| < \frac{1}{2}\delta_j$. Then

$$|z - z_j| \leq |z - z_0| + |z_0 - z_j| < \tfrac{1}{2}\delta_j + \tfrac{1}{2}\delta_j = \delta_j,$$

and

$$|f(z) - f(z_0)| \leq |f(z) - f(z_j)| + |f(z_j) - f(z_0)| < \frac{\epsilon}{2} + \frac{\epsilon}{2} = \epsilon.$$

This completes the proof.

Exercises 2.1

1. Find the image of $|z - 1| \leq 1$ under the mapping $w = (1/z)$.

2. Find the image of $1 \leq \text{Re}(z) \leq 2$, $1 \leq \text{Im}(z) \leq 2$ under the mapping $w = z^2$.

3. Prove that the mapping $w = z^3 + 3z + 1$ is one-to-one for $|z| < 1$.

4. Prove that $w = z^3$ in $S = \{z \mid |z| \leq 1\}$ is continuous at $z = 1$.

5. Find the $\lim\limits_{z \to 2} \dfrac{z^2 + z - 6}{z - 2}$.

6. Find the $\lim\limits_{z \to \infty} \dfrac{z^2 + z + 1}{2z^2 + 3z - 2}$.

7. Find the $\lim\limits_{z \to 0} \dfrac{az + (b/z)}{cz + (d/z)}$.

8. Find the image of the unit circle under the mapping $w = z + (1/z)$.

9. Assuming that $\lim\limits_{z \to z_0} f(z)$ and $\lim\limits_{z \to z_0} g(z)$ exist, prove

(a) $\lim\limits_{z \to z_0} [f(z) \pm g(z)] = \lim\limits_{z \to z_0} f(z) \pm \lim\limits_{z \to z_0} g(z)$;

(b) $\lim\limits_{z \to z_0} f(z)g(z) = \lim\limits_{z \to z_0} f(z) \lim\limits_{z \to z_0} g(z)$;

(c) $\lim\limits_{z \to z_0} f(z)/g(z) = \lim\limits_{z \to z_0} f(z) / \lim\limits_{z \to z_0} g(z)$, if $\lim\limits_{z \to z_0} g(z) \neq 0$.

10. Prove that $\lim\limits_{z \to z_0} f(z) = \lim\limits_{\substack{x \to x_0 \\ y \to y_0}} u(x, y) + i \lim\limits_{\substack{x \to x_0 \\ y \to y_0}} v(x, y)$ if and only if the right-hand side exists.

11. Prove that if a function is defined at an isolated point z_0 of its domain where $f(z_0) \neq \infty$, then it is continuous at z_0.

12. Prove Theorem 2.1.1.

13. Prove that $w = (1/z^2)$ is continuous, but not uniformly continuous in

$$S = \{z \mid 0 < |z| \leq 1\}.$$

2.2. DIFFERENTIABILITY AND ANALYTICITY

In this section we introduce the concept of derivative of a function of a complex variable. We shall find that the differentiability of a function will have very far reaching consequences. For example, if $f(z)$ is differentiable in some ϵ-neighborhood of a point z_0, it will have *all* derivatives at z_0. This is of course not true of functions of a real variable. But we are getting ahead of the story. Let us start with the definition of derivative.

Definition 2.2.1. Let $w = f(z)$ be defined in some ϵ-neighborhood of z_0 in the finite complex plane with $f(z_0) \neq \infty$. Then the *derivative* of $f(z)$ at z_0 is

$$f'(z_0) = \lim_{h \to 0} \frac{f(z_0 + h) - f(z_0)}{h},$$

provided this limit exists and is not equal to infinity. If $f'(z_0)$ exists then $f(z)$ is said to be *differentiable* at z_0.

Let us consider several examples to illustrate this definition.

EXAMPLE 2.2.1. Let $w = z^2$. Then

$$f'(z_0) = \lim_{h \to 0} \frac{(z_0 + h)^2 - z_0^2}{h} = \lim_{h \to 0} (2z_0 + h) = 2z_0.$$

This function is defined everywhere in the unextended z plane and therefore is differentiable without exception.

EXAMPLE 2.2.2. Let $w = \bar{z}$. Consider the limit

$$\lim_{h \to 0} \frac{\overline{z_0 + h} - \bar{z}_0}{h} = \lim_{h \to 0} \frac{\bar{h}}{h}.$$

This limit does not exist since if it did, its value must be independent of the manner in which $h \to 0$. Yet if h is real, $\bar{h}/h = 1$ and if h is imaginary $\bar{h}/h = -1$. Therefore, the function is nowhere differentiable.

EXAMPLE 2.2.3. Let $w = |z|^2$. Consider the limit

$$\lim_{h \to 0} \frac{|z_0 + h|^2 - |z_0|^2}{h} = \lim_{h \to 0} \frac{(z_0 + h)(\bar{z}_0 + \bar{h}) - z_0 \bar{z}_0}{h}$$

$$= \lim_{h \to 0} \left(\bar{z}_0 + z_0 \frac{\bar{h}}{h} \right).$$

If $z_0 = 0$, then the limit is zero and $f'(0) = 0$. Otherwise, the limit does not exist since $\bar{h}/h$ will take on different values depending on the manner in which h approaches zero. See Example 2.2.2.

The usual differentiation formulas of the elementary calculus hold for derivatives of functions of a complex variable. For example,

$$\frac{d}{dz} z^n = nz^{n-1}, n \text{ a positive integer.}$$

This can be proved by making use of the binomial expansion,

$$\frac{d}{dz} z^n = \lim_{h \to 0} \frac{(z + h)^n - z^n}{h}$$

$$= \lim_{h \to 0} \left[nz^{n-1} + \frac{hn(n-1)}{2} z^{n-2} + \cdots + h^{n-1} \right]$$

$$= nz^{n-1}.$$

We also have, assuming that $f(z)$ and $g(z)$ are both differentiable at z,

1. $\dfrac{d}{dz} cf(z) = cf'(z)$, c a constant;

2. $\dfrac{d}{dz} [f(z) \pm g(z)] = f'(z) \pm g'(z)$;

3. $\dfrac{d}{dz} [f(z)g(z)] = f(z)g'(z) + g(z)f'(z)$;

4. $\dfrac{d}{dz} \left[\dfrac{f(z)}{g(z)} \right] = \dfrac{g(z)f'(z) - g'(z)f(z)}{[g(z)]^2}$, at points where $g(z) \neq 0$.

By way of illustration, we shall prove Formula 4. We shall have to make use of the following theorem.

Theorem 2.2.1. *If $w = f(z)$ is differentiable at z_0, then it is continuous at z_0.*

Proof: If $f(z)$ is differentiable at z_0 then it must be defined in some ϵ-neighborhood of z_0, and

$$f(z) = f(z_0) + \frac{f(z_0 + h) - f(z_0)}{h} h, 0 < |h| < \epsilon;$$

$$\lim_{z \to z_0} f(z) = f(z_0) + \lim_{h \to 0} \frac{f(z_0 + h) - f(z_0)}{h} h = f(z_0) \neq \infty.$$

Using this theorem, we see that if $g(z_0) = g_0 \neq 0$, then

$$|g(z_0 + h)| = |g(z_0) - g(z_0 + h) - g(z_0)|$$
$$\geq |g_0 - \epsilon| > 0,$$

provided $|h| < \delta(\epsilon)$ and $\epsilon < |g_0|$. Now

$$\frac{f(z_0 + h)}{g(z_0 + h)} - \frac{f(z_0)}{g(z_0)}$$

$$= \frac{g(z_0)[f(z_0 + h) - f(z_0)] - f(z_0)[g(z_0 + h) - g(z_0)]}{g(z_0)g(z_0 + h)};$$

$$\lim_{h \to 0} \frac{1}{h}\left[\frac{f(z_0 + h)}{g(z_0 + h)} - \frac{f(z_0)}{g(z_0)}\right]$$

$$= \lim_{h \to 0} \frac{g(z_0)\dfrac{f(z_0 + h) - f(z_0)}{h} - f(z_0)\dfrac{g(z_0 + h) - g(z_0)}{h}}{g(z_0)g(z_0 + h)}$$

$$= \frac{g(z_0)f'(z_0) - f(z_0)g'(z_0)}{[g(z_0)]^2}.$$

This proves Formula 4.

EXAMPLE 2.2.4.   Let $w = P(z) = a_0 + a_1z + a_2z^2 + a_3z^3 + \cdots + a_nz^n$.

Using the above formulas, we see that the *polynomial $P(z)$* is differentiable everywhere in the unextended plane and

$$P'(z) = a_1 + 2a_2z + 3a_3z^2 + \cdots + na_nz^{n-1}.$$

EXAMPLE 2.2.5.   Let

$$w = \frac{P(z)}{Q(z)} = \frac{a_0 + a_1z + a_2z^2 + \cdots + a_nz^n}{b_0 + b_1z + b_2z^2 + \cdots + b_mz^m}.$$

This is a *rational function* of z. Using the above formulas, we see that it is differentiable everywhere in the unextended plane except where $Q(z) = 0$. We shall see later that there are at most m distinct points where $Q(z) = 0$.

Let $w = f(z)$ be a function differentiable at z_0. Then $f(z)$ is defined in some ϵ-neighborhood of z_0, $|z - z_0| < \epsilon$. Let

$$\eta(h, z_0) = \begin{cases} \dfrac{f(z_0 + h) - f(z_0)}{h} - f'(z_0), & 0 < |h| < \epsilon \\ 0, & h = 0. \end{cases}$$

Then η is a continuous function of h at $h = 0$, since

$$\lim_{h \to 0} \eta = \lim_{h \to 0} \left[\frac{f(z_0 + h) - f(z_0)}{h} - f'(z_0) \right] = 0.$$

Now, we can write

$$\Delta w = f(z_0 + h) - f(z_0) = f'(z_0)h + \eta h.$$

We call $f'(z_0)h$ the *principal part* of Δw or *differential*

$$dw = f'(z_0)h.$$

Actually, the differential is defined as a linear function of h for all h, but the statement that the differential is *close to* Δw only holds for h *sufficiently small*.

Suppose $z = g(\zeta)$ is differentiable in some ϵ-neighborhood of ζ_0. Then by the continuity of $g(\zeta)$ at ζ_0, $g(\zeta)$ is in some ϵ-neighborhood of $z_0 = g(\zeta_0)$ when ζ is near ζ_0. Let $w = f(z)$ be differentiable in some ϵ-neighborhood of z_0. Then $w = f[g(\zeta)]$ is a *composite* function of ζ defined in some ϵ-neighborhood of ζ_0 with $w_0 = f(z_0) = f[g(\zeta_0)]$. Now

$$\Delta z = g(\zeta_0 + h) - g(\zeta_0);$$
$$\Delta w = f(z_0 + \Delta z) - f(z_0);$$
$$\frac{\Delta w}{h} = f'(z_0)\frac{\Delta z}{h} + \eta(z_0, \Delta z)\frac{\Delta z}{h};$$
$$\left(\frac{dw}{d\zeta}\right)_{\zeta = \zeta_0} = \lim_{h \to 0} \left[f'(z_0)\frac{g(\zeta_0 + h) - g(\zeta_0)}{h} + \eta\frac{g(\zeta_0 + h) - g(\zeta_0)}{h} \right]$$
$$= f'(z_0)g'(\zeta_0).$$

This is called the *chain rule* for differentiating a function of a function. We have also proved that a *differentiable function of a differentiable function is differentiable*.

EXAMPLE 2.2.6. Let $w = |z|^{1/2} e^{i(\arg z)/2}$, $0 \leq \arg z < 2\pi$. This function is not continuous on the positive real axis, since $\mathrm{Re}(w) = |z|^{1/2} \cos \dfrac{\arg z}{2}$ is discontinuous, there being a jump in $\cos \dfrac{\arg z}{2}$ from -1 to $+1$ as $\arg z$ changes from 2π to 0. If the function were differentiable on the positive real axis then it would be continuous there but this is not the case. Suppose $z_0 \neq 0$, $z_0 \neq x_0 > 0$. Then

$$w^2 = |z|e^{i \arg z} = z;$$

$$\lim_{z \to z_0} \frac{w - w_0}{z - z_0} = \lim_{w \to w_0} \frac{w - w_0}{w^2 - w_0^2} = \lim_{w \to w_0} \frac{1}{w + w_0} = \frac{1}{2w_0}.$$

Another way to proceed to get the derivative, once we know that the function is differentiable, is by *implicit* differentiation. Since $w^2 = z$, by the chain rule, we have

$$2w_0 w'(z_0) = 1, \ w'(z_0) = \frac{1}{2w_0} = \frac{e^{-i(\arg z_0)/2}}{2|z_0|^{1/2}}.$$

This function is not differentiable at $z = 0$, since

$$\lim_{z \to 0} \frac{w}{z} = \lim_{z \to 0} \frac{|z|^{1/2} e^{i(\arg z)/2}}{|z|e^{i \arg z}} = \lim_{z \to 0} \frac{e^{-i(\arg z)/2}}{|z|^{1/2}},$$

and this limit obviously does not exist.

EXAMPLE 2.2.7. Let $w = |z|^{p/q} e^{i(p \arg z)/q}$, p and q integers different from zero and p not divisible by q, $0 \leq \arg z < 2\pi$. This function is discontinuous on the positive real axis and hence not differentiable there. Let $z_0 \neq 0$, $z_0 \neq x_0 > 0$. Then if $\zeta = |z|^{1/q} e^{i(\arg z)/q}$, $w = \zeta^p$, $\zeta^q = z$.

$$\lim_{z \to z_0} \frac{\zeta - \zeta_0}{z - z_0} = \lim_{\zeta \to \zeta_0} \frac{\zeta - \zeta_0}{\zeta^q - \zeta_0^q}$$

$$= \lim_{\zeta \to \zeta_0} \frac{1}{\zeta^{q-1} + \zeta^{q-2}\zeta_0 + \cdots + \zeta\zeta_0^{q-2} + \zeta_0^{q-1}}$$

$$= \frac{1}{q\zeta_0^{q-1}}$$

$$= \frac{1}{q|z_0|^{1-1/q} e^{i(1-1/q)\arg z_0}}.$$

By the chain rule

$$\frac{dw}{dz} = \frac{dw}{d\zeta} \frac{d\zeta}{dz} = \frac{p\zeta^{p-1}}{q\zeta^{q-1}} = \frac{p}{q} |z_0|^{(p/q)-1} e^{i[(p/q)-1] \arg z}.$$

If $z_0 = 0$ and $p/q > 1$, then $f(0) = 0$ and

$$\lim_{z \to 0} \frac{w}{z} = \lim_{z \to 0} |z|^{(p/q)-1} e^{i[(p/q)-1] \arg z} = 0.$$

Otherwise, the function is not differentiable at the origin.

Definition 2.2.2. If $w = f(z)$ is defined in some ϵ-neighborhood of z_0, then $f(z)$ is *analytic* at z_0 if and only if it is differentiable in some ϵ-neighborhood of z_0.

Other terms used for this concept are *regular, holomorphic,* and *monogenic.* The function in Example 2.2.1 is analytic everywhere in the unextended plane. Example 2.2.2 gives a function which is nowhere differentiable and hence, nowhere analytic. In Example 2.2.3, we have a function which is nowhere analytic even though it is differentiable at the origin. The reason is that there is no ϵ-neighborhood of the origin in which the function is differentiable. Polynomials are analytic everywhere in the unextended plane. Rational functions are analytic everywhere in the unextended plane except at the points where the denominator is zero. The function of Example 2.2.7 is analytic everywhere in the unextended plane except at the origin and on the positive real axis. This is the case even though for $p/q > 1$ the function is differentiable at the origin.

EXAMPLE 2.2.8.   Let $w = f(z) = \dfrac{az + b}{cz + d}$, $c \neq 0$ and $ad - bc \neq 0$.

Then

$$w' = \frac{ad - bc}{(cz + d)^2},$$

provided $z \neq -d/c$. Let $z = 1/\zeta$. Then

$$f(1/\zeta) = \frac{a(1/\zeta) + b}{c(1/\zeta) + d}, \; \zeta \neq 0.$$

Let us define

$$f(\infty) = \lim_{\zeta \to 0} \frac{a(1/\zeta) + b}{c(1/\zeta) + d} = \frac{a}{c}.$$

In other words, $g(\zeta) = f(1/\zeta) = \dfrac{b\zeta + a}{d\zeta + c}$. Now,

$$g'(\zeta) = \frac{bc - ad}{(d\zeta + c)^2}.$$

Hence, $g(\zeta)$ is analytic at $\zeta = 0$. We say that $f(z)$ is *analytic at infinity* if $f(1/\zeta)$ is analytic at $\zeta = 0$. Hence, $f(z) = (az + b)/(cz + d)$ is analytic at infinity. Note that, even though we define $f(\infty)$, the difference quotient is not defined at infinity, and hence we do not define differentiability at infinity.

Exercises 2.2

1. Prove that if $f(z) = c$, a constant, then $f'(z) = 0$.

2. Prove formulas 1, 2, and 3 of this section.

3. Prove that $f(z) = \arg z$ is nowhere differentiable.

4. Prove that $f(z) = |z|$ is nowhere differentiable.

5. Prove that $f(z) = \text{Re}(z)$ is nowhere differentiable.

6. Assuming that $w = |z|^{p/q}e^{i(p/q)\arg z}$, $0 \le \arg z < 2\pi$, is differentiable at z_0, find the derivative by writing $w^q = z^p$ and using implicit differentiation.

7. Assuming that $w = \sqrt{z^2 - 1} = |z^2 - 1|^{1/2}e^{i[\arg(z^2-1)]/2}$ is differentiable at z_0, find the derivative by implicit differentiation.

8. Let $w = f(z)$ be defined in some ϵ-neighborhood of z_0 which is mapped into some ϵ-neighborhood of $w_0 = f(z_0)$. Suppose $f(z)$ has an inverse $g(w)$ in some ϵ-neighborhood of w and that $f'(z_0) \ne 0$. Show that

$$g'(w_0) = \frac{1}{f'(z_0)}.$$

9. Prove that a rational function is analytic at infinity if the degree of the numerator is less than or equal to the degree of the denominator.

2.3. THE CAUCHY-RIEMANN CONDITIONS

We have seen before that we can always write $w = f(z) = u(x, y) + iv(x, y)$ in terms of two real-valued functions of the real variables x and y. We can relate the differentiability of $f(z)$ to conditions on $u(x, y)$ and $v(x, y)$ and their first partial derivatives.

Theorem 2.3.1. *The Necessary Cauchy-Riemann Conditions for Differentiability. Let $w = f(z) = u(x, y) + iv(x, y)$ be differentiable at $z_0 = x_0 + iy_0$. Then the partial derivatives $u_x(x_0, y_0)$, $u_y(x_0, y_0)$, $v_x(x_0, y_0)$, and $v_y(x_0, y_0)$ exist and $u_x(x_0, y_0) = v_y(x_0, y_0)$ and $u_y(x_0, y_0) = -v_x(x_0, y_0)$. Also $f'(z_0) = u_x(x_0, y_0) + iv_x(x_0, y_0) = v_y(x_0, y_0) - iu_y(x_0, y_0)$.*

Proof: Since $\lim\limits_{h \to 0} \dfrac{f(z_0 + h) - f(z_0)}{h}$ exists we can compute it by letting h approach zero in any manner whatsoever. In particular, if $h = \Delta x$, then

$$f'(z_0) = \lim_{\Delta x \to 0} \frac{u(x_0 + \Delta x, y_0) - u(x_0, y_0)}{\Delta x} + i\frac{v(x_0 + \Delta x, y_0) - v(x_0, y_0)}{\Delta x}$$

$$= u_x(x_0, y_0) + iv_x(x_0, y_0).$$

If $h = i\Delta y$, then

$$f'(z_0) = \lim_{\Delta y \to 0} \frac{v(x_0, y_0 + \Delta y) - v(x_0, y_0)}{\Delta y} - i\frac{u(x_0, y_0 + \Delta y) - u(x_0, y_0)}{\Delta y}$$

$$= v_y(x_0, y_0) - iu_y(x_0, y_0).$$

Equating the two expressions for $f'(z_0)$ we obtain the desired result, $u_x(x_0, y_0) = v_y(x_0, y_0)$, $v_x(x_0, y_0) = -u_y(x_0, y_0)$. These are called the *Cauchy-Riemann equations*.

EXAMPLE 2.3.1. We saw that $w = z^2$ is differentiable everywhere in the unextended plane. Now, $z^2 = x^2 - y^2 + i2xy$. Hence, $u = x^2 - y^2$ and $v = 2xy$, and $u_x = v_y = 2x, v_x = -u_y = 2y$ and $dw/dz = 2x + i2y = 2(x + iy) = 2z$.

EXAMPLE 2.3.2. Let $w = |z|^2 = x^2 + y^2$. Then $u(x, y) = x^2 + y^2$ and $v = 0$. $u_x = 2x, u_y = 2y, v_x = 0, v_y = 0$. We see that the Cauchy Riemann equations are only satisfied at $z = 0$. As we saw in the last section, this is the only point where $w = |z|^2$ is differentiable.

EXAMPLE 2.3.3. To prove that $w = \bar{z}$ is nowhere differentiable we invoke the Cauchy Riemann equations. If the function were differentiable at a point, the Cauchy Riemann equations would be satisfied there. In this case, $u = x$ and $v = -y$. Therefore, $u_x = 1$, $u_y = 0$, $v_x = 0$, $v_y = -1$, and we see that the Cauchy Riemann equations are never satisfied.

EXAMPLE 2.3.4. Let $w = z^5/|z|^4$ when $z \neq 0$, and $w = 0$ when $z = 0$. Then $\lim_{h\to0} \frac{w}{h} = \lim_{h\to0} \frac{h^4}{|h|^4}$ and this limit clearly does not exist. On the other hand,

$$u_x(0, 0) = \lim_{\Delta x\to0} \left(\frac{\Delta x}{|\Delta x|}\right)^4 = 1, v_y = \lim_{\Delta y\to0} \left(\frac{i\Delta y}{|\Delta y|}\right)^4 = 1,$$

$$u_y(0, 0) = \lim_{\Delta y\to0} \mathrm{Re}\left[\frac{(i\Delta y)^5}{\Delta y |\Delta y|^4}\right] = 0, v_x = \lim_{\Delta x\to0} \mathrm{Im}\left[\left(\frac{\Delta x}{|\Delta x|}\right)^4\right] = 0.$$

Hence, the Cauchy Riemann equations are satisfied at the origin even though the function is not differentiable there.

The last example shows that the necessary Cauchy-Riemann conditions are *not* sufficient. However, if we assume that $u(x, y)$ and $v(x, y)$ have continuous first partial derivatives in some ϵ-neighborhood of (x_0, y_0) and satisfy the Cauchy-Riemann equations there, then the function

$$w = u(x, y) + iv(x, y)$$

is differentiable at the point (x_0, y_0).

Theorem 2.3.2. *The Sufficient Cauchy-Riemann Conditions. Let $w = f(z) = u(x, y) + iv(x, y)$ be defined in some ϵ-neighborhood of $z_0 = x_0 + iy_0$. Let u_x, u_y, v_x, v_y be continuous in this same ϵ-neighborhood and*

$$u_x(x_0, y_0) = v_y(x_0, y_0), u_y(x_0, y_0) = -v_x(x_0, y_0).$$

Then $f(z)$ is differentiable at z_0 and $f'(z_0) = u_x(x_0, y_0) + iv_x(x_0, y_0) = v_y(x_0, y_0) - iu_y(x_0, y_0)$.

Proof: By the continuity of the first partial derivatives, $u(x, y)$ and $v(x, y)$ are continuous in some ϵ-neighborhood of z_0 and hence the mean value theorem applies, that is,

$$u(x_0 + \Delta x, y_0 + \Delta y) - u(x_0, y_0) = u_x(x_0, y_0)\Delta x + u_y(x_0, y_0)\Delta y \\ + \xi_1\Delta x + \eta_1\Delta y,$$
$$v(x_0 + \Delta x, y_0 + \Delta y) - v(x_0, y_0) = v_x(x_0, y_0)\Delta x + v_y(x_0, y_0)\Delta y \\ + \xi_2\Delta x + \eta_2\Delta y,$$

where $\xi_1, \xi_2, \eta_1, \eta_2$ all go to zero as Δx and Δy go to zero. Using the Cauchy-Riemann equations, we have

$$\lim_{\Delta z \to 0} \frac{f(z_0 + \Delta z) - f(z_0)}{\Delta z}$$
$$= \lim_{\Delta z \to 0} \left[u_x(x_0, y_0) + iv_x(x_0, y_0) + \xi_1 \frac{\Delta x}{\Delta z} + \eta_1 \frac{\Delta y}{\Delta z} + i\xi_2 \frac{\Delta x}{\Delta z} + i\eta_2 \frac{\Delta y}{\Delta z} \right]$$
$$= u_x(x_0, y_0) + iv_x(x_0, y_0),$$

where we have used the facts that $|\Delta x| \leq |\Delta z|$, $|\Delta y| \leq |\Delta z|$ and ξ_1, ξ_2, η_1, and η_2 go to zero as Δx and Δy go to zero.

EXAMPLE 2.3.5. Let $f(z) = e^x \cos y + ie^x \sin y$. Then $u = e^x \cos y$ and $v = e^x \sin y$ have continuous first partial derivatives everywhere in the unextended plane and

$$u_x = e^x \cos y = v_y,$$
$$u_y = -e^x \sin y = -v_x.$$

Hence, the Cauchy-Riemann equations are satisfied, the function is differentiable everywhere and

$$\jmath'(z) = e^x \cos y + ie^x \sin y = f(z).$$

EXAMPLE 2.3.6. Let $f(z) = \ln\sqrt{x^2 + y^2} + i\tan^{-1}(y/x)$,* where $-\pi/2 < \tan^{-1}(y/x) < \pi/2$. We have $u = \ln\sqrt{x^2 + y^2}$ and $v = \tan^{-1}(y/x)$. Hence,

$$u_x = \frac{x}{x^2 + y^2} = v_y,$$
$$u_y = \frac{y}{x^2 + y^2} = -v_x,$$

provided $\text{Re}(z) > 0$. The given function is differentiable in the right half-plane where

$$f'(z) = \frac{x - iy}{x^2 + y^2} = \frac{\bar{z}}{z\bar{z}} = \frac{1}{z}.$$

---

* ln stands for the natural logarithm or logarithm to the base e.

If a function $f(z) = u(x, y) + iv(x, y)$ is differentiable in some open set, then the Cauchy-Riemann equations are satisfied; that is, $u_x = v_y$ and $u_y = -v_x$. If we differentiate the first with respect to x and the second with respect to y and add, we have

$$u_{xx} + u_{yy} = v_{yx} - v_{xy}.$$

If the second partial derivatives of u and v exist and are continuous then $v_{xy} = v_{yx}$ and we have

$$u_{xx} + u_{yy} = \nabla^2 u = 0.$$

In other words, the real part of an analytic function satisfies *Laplace's equation* where it is differentiable. We shall see later that analyticity at a point implies that the function has all derivatives at the point and that all the partial derivatives of u and v exist and are continuous. Hence, we need not assume the existence and the continuity of the second partial derivatives. Similarly, we can show that v also satisfies Laplace's equations. We say that v is the *conjugate harmonic function* of u.

EXAMPLE 2.3.7.   Let $u(x, y) = \sin x \cosh y$. By straightforward differentiation we can show that u is harmonic; that is, it satisfies Laplace's equation.

$$u_x = \cos x \cosh y \qquad\qquad u_y = \sin x \sinh y$$
$$u_{xx} = -\sin x \cosh y \qquad\qquad u_{yy} = \sin x \cosh y$$
$$u_{xx} + u_{yy} = 0.$$

We can determine a conjugate harmonic function from the Cauchy-Riemann equations, as follows

$$u_x = \cos x \cosh y = v_y,$$
$$v(x, y) = \cos x \int \cosh y \, dy + g(x)$$
$$= \cos x \sinh y + g(x),$$

where $g(x)$ is an arbitrary function of x. But then

$$u_y = \sin x \sinh y = \sin x \sinh y + g'(x).$$

Therefore, $g'(x) \equiv 0$ and $g(x)$ is any constant. In particular, if we take $g(x) = 0$, we have

$$f(z) = \sin x \cosh y + i \cos x \sinh y,$$

an analytic function of z.

Exercises 2.3

1. Use the necessary Cauchy-Riemann conditions to show that the following are not differentiable:

(a) $f(z) = |z|$;   (b) $f(z) = \arg z$;

(c) $f(z) = \mathrm{Re}(z)$;   (d) $f(z) = \mathrm{Im}(z)$.

2. Use the sufficient Cauchy-Riemann conditions to find where the following are differentiable:

(a) $f(z) = \cos |z|^2$

(b) $f(z) = e^{-y} \cos x + i e^{-y} \sin x$

(c) $f(z) = \cosh x \cos y + i \sinh x \sin y$

(d) $f(z) = e^{|z|}$

3. Let $f(z) = u(x, y) + iv(x, y)$ be analytic at $z_0 = x_0 + iy_0$, with $f'(z_0) \neq 0$. Prove that the curves $u(x, y) = u(x_0, y_0), v(x, y) = v(x_0, y_0)$ intersect at right angles.

4.* Let $u(x, y)$ be harmonic everywhere in the unextended z plane. Prove that $v(x, y) = \int_{(x_0, y_0)}^{(x, y)} (-u_y \, dx + u_x \, dy)$ defined by line integration along a simple contour, is independent of the path and is a conjugate harmonic function for u.

5. Show that $u = x^3 - 3xy^2$ is harmonic in the unextended z plane and find a conjugate harmonic function.

6. Develop the polar coordinate form for the Cauchy-Riemann equations $ru_r = v_\theta, rv_r = -u_\theta$.

7. Show that if $f(re^{i\theta}) = u(r, \theta) + iv(r, \theta)$ is analytic at a point then both u and v satisfy the equation $r^2 \phi_{rr} + r\phi_r + \phi_{\theta\theta} = 0$.

8. Where do the following satisfy the polar coordinate form of the Cauchy-Riemann equations:

(a) $f(re^{i\theta}) = \ln r + i\theta, r > 0, 0 \leq \theta < 2\pi$.

(b) $f(re^{i\theta}) = \sqrt{r} \left(\cos \dfrac{\theta}{2} + i \sin \dfrac{\theta}{2} \right), r \geq 0, 0 \leq \theta < 2\pi$.

(c) $f(re^{i\theta}) = r^2$

(d) $f(re^{i\theta}) = e^{-\theta} \cos (\ln r) + i \sin (\ln r), r > 0$.

2.4. LINEAR FRACTIONAL TRANSFORMATIONS

In this section, we shall consider a simple class of functions, the linear fractional transformations, which are especially useful because of their special mapping properties. But first let us look at some of the general mapping properties shared by all analytic functions.

---

* For those unfamiliar with line integration this exercise should be postponed until after Section 3.1 is studied.

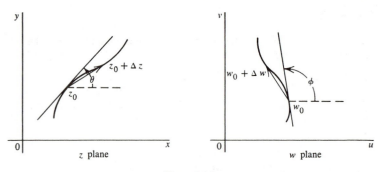

Figure 2.4.1.

Consider $w = f(z)$ analytic in a domain D. Let $z(t) = x(t) + iy(t)$ be the parametric equation of a simple smooth curve passing through $z_0 = x(t_0) + iy(t_0)$ in D. The curve has a tangent vector at z_0 with components $\dot{x}(t_0)$, $\dot{y}(t_0)$. See Figure 2.4.1. Now we know from our discussion of the derivative that

$$\Delta w = f'(z_0)\Delta z + \eta(z_0, \Delta z)\Delta z,$$

where

$$\lim_{\Delta z \to 0} \eta(z_0, \Delta z) = 0.$$

Therefore,

$$\arg \Delta w = \arg \Delta z + \arg [f'(z_0) + \eta].$$

If $f'(z_0) \neq 0$, then

$$\lim_{\Delta z \to 0} \arg [f'(z_0) + \eta] = \arg f'(z_0),$$

and

$$\phi = \lim_{\Delta z \to 0} \arg \Delta w = \theta + \arg f'(z_0).$$

In other words, the tangent vector at z_0 has been rotated by the angle $\arg f'(z_0)$ provided $f'(z_0) \neq 0$. Let two smooth curves pass through z_0 making angles of θ_1 and θ_2 with the positive x axis. Then the angle between the images at w_0 is

$$\phi_2 - \phi_1 = \theta_2 + \arg f'(z_0) - \theta_1 - \arg f'(z_0) = \theta_2 - \theta_1.$$

Hence, the angle between two curves is preserved by the mapping in magnitude as well as sense. A mapping which preserves angles in magnitude and sense is said to be *conformal*. Therefore, the mapping $w = f(z)$ is conformal at z_0 if it is analytic at z_0 and $f'(z_0) \neq 0$.

Again consider

$$\lim_{\Delta z \to 0} \frac{|\Delta w|}{|\Delta z|} = \lim_{\Delta z \to 0} |f'(z_0) + \eta| = |f'(z_0)|.$$

This shows that small distances are changed by the scale factor $|f'(z_0)|$ by the mapping. We shall show later that if $f'(z_0) \neq 0$ at a point where $f(z)$ is analytic, there exists an ϵ-neighborhood of z_0 throughout which $f'(z_0) \neq 0$. Hence, since near z_0 angles are preserved and distances are scaled by nearly the same factor, we conclude that *small figures near z_0 are mapped into similar small figures near w_0.*

We saw in the last section that the real and imaginary parts of an analytic function satisfy Laplace's equation. This fact in itself will prove to be useful in solving boundary value problems in physics and engineering, but what is even more interesting is that Laplace's equation is preserved under a mapping by an analytic function. Let $\phi(x, y)$ be a real-valued function satisfying Laplace's equation in a domain D in the z plane. Let $w = f(z) = u(x, y) + iv(x, y)$ be a function analytic in D defining a mapping of D into the w plane. Let $\Phi(u, v) = \phi(x, y)$ be the image of ϕ defined by the mapping on the range of $f(z)$. Then

$$\phi_x = \Phi_u u_x + \Phi_v v_x, \; \phi_y = \Phi_u u_y + \Phi_v v_y,$$
$$\phi_{xx} = \Phi_{uu}(u_x)^2 + \Phi_{uv} u_x v_x + \Phi_u u_{xx} + \Phi_{vv}(v_x)^2 + \Phi_{vu} v_x u_x + \Phi_v v_{xx},$$
$$\phi_{yy} = \Phi_{uu}(u_y)^2 + \Phi_{uv} u_y v_y + \Phi_u u_{yy} + \Phi_{vv}(v_y)^2 + \Phi_{vu} v_y u_y + \Phi_v v_{yy},$$
$$\phi_{xx} + \phi_{yy} = [(u_x)^2 + (v_x)^2][\Phi_{uu} + \Phi_{vv}],$$

since $u_{xx} + u_{yy} = 0, v_{xx} + v_{yy} = 0, u_x v_x = -u_y v_y$. We therefore have

$$\nabla^2 \Phi = \nabla^2 \phi / |f'(z)|^2.$$

Hence, if $f'(z) \neq 0$ in D, then $\nabla^2 \Phi = 0$ in the range of D. This gives us a means of transforming the domain over which we must solve Laplace's equation. We shall pursue this point further in Chapter 6 for the purpose of solving boundary value problems in potential theory.

Now let us consider the *linear fractional transformation*

$$w = \frac{az + b}{cz + d}, \; ad - bc \neq 0.$$

We have seen (Example 2.1.6) that this is a one-to-one mapping of the extended z plane onto the extended w plane. Also this is an analytic function in the extended z plane except at $z = -d/c$ (Example 2.2.8).

EXAMPLE 2.4.1. Let $w = az = |a| \, |z| e^{i(\arg z + \arg a)}$. Then this represents an expansion or contraction by the factor $|a|$ of the z plane plus a rotation through the angle $\arg a$.

EXAMPLE 2.4.2. Let $w = z + b = [x + \mathrm{Re}(b)] + i[y + \mathrm{Im}(b)]$. This represents a translation of the z plane by the amount $|b|$ at the angle $\arg b$.

EXAMPLE 2.4.3. Let $w = 1/z = (1/|z|)e^{-i \arg z}$. This maps the unit circle onto the unit circle, the inside of the unit circle onto the outside, and the outside of the unit circle onto the inside. Consider the family of curves given by the equation

$$\alpha(x^2 + y^2) + \beta x + \gamma y + \delta = 0.$$

This is a circle or a straight line depending on the values of α, β, γ, and δ. Writing the equation in terms of the complex variable z, we have

$$\alpha|z|^2 + \beta\left(\frac{z + \bar{z}}{2}\right) + \gamma\left(\frac{z - \bar{z}}{2i}\right) + \delta = 0.$$

Under the transformation $w = 1/z$ the equation becomes

$$\frac{\alpha}{|w|^2} + \frac{\beta}{2}\left(\frac{w + \bar{w}}{|w|^2}\right) - \frac{\gamma}{2i}\left(\frac{w - \bar{w}}{|w|^2}\right) + \delta = 0,$$

$$\delta|w|^2 + \frac{\beta}{2}(w - \bar{w}) - \frac{\gamma}{2i}(w - \bar{w}) + \alpha = 0,$$

$$\delta(u^2 + v^2) + \beta u - \gamma v + \alpha = 0.$$

This is the equation of a circle or a straight line in the w plane. Hence, *circles and straight lines map into circles or straight lines.*

Following these three examples, it is easy to show that in general the linear fractional transformation maps circles and straight lines into circles or straight lines. Indeed, we can write

$$w = \frac{az + b}{cz + d} = \frac{a}{c} + \frac{bc - ad}{c}\frac{1}{cz + d}.$$

This shows that the whole transformation can be built up of a sequence of transformations of the types considered in Examples 2.4.1–3; that is, $\zeta_1 = cz$, $\zeta_2 = \zeta_1 + d$, $\zeta_3 = 1/\zeta_2$, $\zeta_4 = (bc - ad)\zeta_3/c$, and $w = \zeta_4 + a/c$. Clearly, each separate transformation maps circles and straight lines into circles or straight lines and hence so does the whole.

EXAMPLE 2.4.4. Find a linear fractional transformation which maps the upper half plane onto the interior of the unit circle. Since the upper half plane is bounded by a straight line we wish to determine a transformation which will map the line $y = 0$ onto the unit circle. It takes three points to uniquely determine a circle. Therefore we will set up the correspondence

$$z_1 = -1 \rightarrow w_1 = -i,$$
$$z_2 = 0 \quad \rightarrow w_2 = 1,$$
$$z_3 = 1 \quad \rightarrow w_3 = i.$$

This implies

$$-i = \frac{-a + b}{-c + d}, \quad 1 = \frac{b}{d}, \quad i = \frac{a + b}{c + d}, \text{ or } i = \frac{(a/d) - 1}{-(c/d) + 1} = \frac{(a/d) + 1}{(c/d) + 1}.$$

Solving these equations, we have $a/d = i$, $b/d = 1$, and $c/d = -i$ and

$$w = \frac{\dfrac{a}{d}z + \dfrac{b}{d}}{\dfrac{c}{d}z + 1} = \frac{iz + 1}{-iz + 1}.$$

To show that this is *onto*, we obtain the inverse

$$z = \frac{w - 1}{iw + i},$$

and let $w = \rho e^{i\phi}$, where $\rho < 1$. Then

$$\text{Im}(z) = \text{Im}\left[\frac{\rho e^{i\phi} - 1}{i\rho e^{i\phi} + i}\right] = \frac{1}{2i}\left[\frac{\rho e^{i\phi} - 1}{i(\rho e^{i\phi} + 1)} + \frac{\rho e^{-i\phi} - 1}{i(\rho e^{-i\phi} + 1)}\right]$$

$$= -\frac{1}{2}\frac{(\rho e^{i\phi} - 1)(\rho e^{-i\phi} + 1) + (\rho e^{i\phi} + 1)(\rho e^{-i\phi} - 1)}{1 + \rho^2 + 2\rho \cos\phi}$$

$$= \frac{1 - \rho^2}{1 + \rho^2 + 2\rho \cos\phi}.$$

Obviously, $\text{Im}(z) > 0$, which shows that every point in the interior of the unit circle in the w plane is the image of some point in the upper half-plane.

The most general mapping of the upper half-plane onto the interior of the unit circle by a linear fractional transformation can be found as follows. Since the real axis maps onto the unit circle, $|ax + b|/|cx + d| = 1$. In particular, if $x = 0$, $|b|/|d| = 1$, if $x = 1$, $|a + b|/|c + d| = 1$, and if $x \to \infty$, $|a|/|c| = 1$, then

$$w = \frac{a}{c}\left(\frac{z + \dfrac{b}{a}}{z + \dfrac{d}{c}}\right) = e^{i\alpha}\frac{z + \beta}{z + \gamma}, \quad \alpha \text{ real.}$$

Now,

$$\frac{|a + b|}{|c + d|} = \frac{|1 + \beta|}{|1 + \gamma|} = 1,$$

so that

$$|1 + \beta| = |1 + \gamma|,$$

and β and γ are both on the same circle centered at -1. Furthermore,

$$\left.\frac{|b|}{|d|}\middle/\frac{|a|}{|c|}\right. = \left.\frac{|b|}{|a|}\middle/\frac{|d|}{|c|}\right. = 1.$$

Hence, $|\beta| = |\gamma|$, which implies that β and γ are both on the same circle centered at the origin. Together we must conclude that $\gamma = \bar{\beta}$. See Figure 2.4.2. Finally, we must require that $\text{Im}(\beta) < 0$, since $z = -\beta$ is the preimage of $w = 0$ which is a point inside the unit circle. The result is that a linear fractional transformation which maps the upper half-plane onto the unit circle must be of the form

$$w = e^{i\alpha} \frac{z + \beta}{z + \bar{\beta}},$$

where α is real and $\text{Im}(\beta) < 0$.

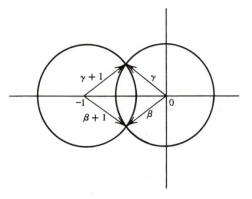

Figure 2.4.2.

We saw in Example 2.4.4 that we could specify the images of three given points. This is, in general, the case since the general transformation has basically three independent parameters. We have already seen that we can write the transformation in the form

$$w = \frac{a}{c} + \frac{bc - ad}{c^2} \frac{1}{z + \dfrac{d}{c}} = \alpha + \beta \frac{1}{z + \gamma}$$

If we specify three images say $z_1 \to w_1$, $z_2 \to w_2$, and $z_3 \to w_3$, then we have

$$w_1 = \alpha + \beta \frac{1}{z_1 + \gamma}, \quad w_2 = \alpha + \beta \frac{1}{z_2 + \gamma}, \quad w_3 = \alpha + \beta \frac{1}{z_3 + \gamma},$$

$$w - w_1 = \beta \left[\frac{1}{z + \gamma} - \frac{1}{z_1 + \gamma} \right] = \beta \frac{z_1 - z}{(z + \gamma)(z_1 + \gamma)},$$

$$\frac{(w - w_1)(w_2 - w_3)}{(w - w_3)(w_2 - w_1)} = \frac{(z - z_1)(z_2 - z_3)}{(z - z_3)(z_2 - z_1)}.$$

The last line shows that the *cross ratio* $(z - z_1)(z_2 - z_3)/(z - z_3)(z_2 - z_1)$ is preserved by the linear fractional transformation. This also gives us a convenient form for displaying the transformation which maps three given points z_1, z_2, z_3 into three given points w_1, w_2, w_3.

EXAMPLE 2.4.5. Find a linear fractional transformation which maps the interior of the unit circle onto the interior of the circle $|w - 1| = 1$ so that $z_1 = -1$, $z_2 = -i$, $z_3 = i$ map into $w_1 = 0$, $w_2 = 2$, $w_3 = 1 + i$, respectively. We use the cross ratio form of the transformation giving us

$$\frac{w}{w - 1 - i} \frac{1 - i}{2} = \frac{z + 1}{z - 1} \frac{-i - i}{-i + 1} = \frac{-2i}{1 - i} \frac{z + 1}{z - i}.$$

Cross multiplying and solving for w, we have

$$w = (2 + 2i)\frac{z + 1}{z + 2 + i}.$$

Finally, we should check that the interiors of the two circles correspond. Let $z = 0$ inside the unit circle. Then $w = (2 + 2i)/(2 + i) = (6/5) + (2/5)i$ which is in the interior of the image circle.

EXAMPLE 2.4.6. Find a linear fractional transformation which will map the right half-plane $\text{Re}(z) > 0$ onto the interior of the unit circle so that $z_1 = \infty$, $z_2 = i$, $z_3 = 0$ map into $w_1 = -1$, $w_2 = -i$, $w_3 = 1$. To handle the point at infinity, we write the cross ratio form as

$$\frac{(w - w_1)(w_2 - w_3)}{(w - w_3)(w_2 - w_1)} = \frac{\left(\frac{z}{z_1} - 1\right)(z_2 - z_3)}{(z - z_3)\left(\frac{z_2}{z_1} - 1\right)},$$

and let z_1 approach infinity, giving

$$\frac{(w - w_1)(w_2 - w_3)}{(w - w_3)(w_2 - w_1)} = \frac{z_2 - z_3}{z - z_3}.$$

Substituting the given points, we have

$$\frac{(w + 1)(-i - 1)}{(w - 1)(-i + 1)} = \frac{i}{z}.$$

Solving for w, we obtain $w = \dfrac{1 - z}{1 + z}$.

A *fixed point* of a transformation is a point which is mapped into itself. If z_0 is a fixed point of a linear fractional transformation, then

$$z_0 = \frac{az_0 + b}{cz_0 + d},$$

or

$$cz_0^2 + (d - a)z_0 - b = 0.$$

If $c = 0$ this equation has but one solution $z_0 = b/(d - a)$ provided $d \neq a$. However, in this case infinity is mapped into infinity, so we may call infinity a fixed point. If $d = a$ and $b = 0$, we have the identity transformation $w = z$ which has infinitely many fixed points. If $d = a$ and $b \neq 0$, we have a pure translation which has only infinity as a fixed point.

If $c \neq 0$ there may be one or two fixed points depending on whether $(d - a)^2 + 4bc = 0$ or not. If there are two fixed points z_1 and z_2, the transformation can be written

$$\frac{w - z_1}{w - z_2} = \alpha \frac{z - z_1}{z - z_2},$$

where $\alpha = (d + cz_2)/(d + cz_1)$. If there is one fixed point z_0, the transformation can be written

$$\frac{1}{w - z_0} = \frac{1}{z - z_0} + \alpha,$$

where $a = 1 + \alpha z_0$, $b = -\alpha z_0^2$, $c = \alpha$, $d = 1 - \alpha z_0$. The proofs of these statements will be left as exercises.

EXAMPLE 2.4.7. Show that $\phi = (1/\pi) \tan^{-1}(y/x)$, where the values of the inverse tangent are taken between 0 and π, is harmonic in the upper half-plane, and takes on the values 0 and 1, respectively, on the positive and negative real axes. Find a function harmonic inside the unit circle which takes on the values 0 and 1, respectively, on the lower and upper unit semicircles.

The first part can be handled by direct verification.

$$\phi_x = -\frac{1}{\pi} \frac{y}{x^2 + y^2}, \qquad \phi_y = \frac{1}{\pi} \frac{x}{x^2 + y^2},$$

$$\phi_{xx} = \frac{2}{\pi} \frac{xy}{(x^2 + y^2)^2}, \qquad \phi_{yy} = -\frac{2}{\pi} \frac{xy}{(x^2 + y^2)^2},$$

$$\phi_{xx} + \phi_{yy} = 0.$$

For $x > 0$ and y approaching zero through positive values,

$$\lim_{\substack{y \to 0 \\ x > 0}} \frac{1}{\pi} \tan^{-1} \frac{y}{x} = 0.$$

For $x < 0$ and y approaching zero through positive values,

$$\lim_{\substack{y \to 0 \\ x < 0}} \frac{1}{\pi} \tan^{-1} \frac{y}{x} = 1.$$

To do the second part, we find a conformal linear fractional transformation which maps the upper half-plane onto the interior of the unit circle so that infinity maps into $w = 1$ and the origin maps into $w = -1$. Such a transformation is

$$w = \frac{z - i}{z + i},$$

$$z = -i\frac{w + 1}{w - 1},$$

$$x + iy = -i\frac{u + 1 + iv}{u - 1 + iv} = \frac{-2v + i(1 - u^2 - v^2)}{(u - 1)^2 + v^2}.$$

Hence,

$$x = \frac{-2v}{(u - 1)^2 + v^2}, \quad y = \frac{1 - u^2 - v^2}{(u - 1)^2 + v^2},$$

and

$$\Phi(u, v) = \frac{1}{\pi}\tan^{-1}\left(\frac{u^2 + v^2 - 1}{2v}\right)$$

is harmonic inside the unit circle and takes on the required boundary values, as can be verified directly.

This example illustrates the conformal mapping technique for solving boundary value problems in potential theory. This subject will be developed much further in Chapter 6, but first we must build up a larger repertoire of mapping functions and develop further the general theory of analytic functions, so that we may make good use of their interesting properties.

Exercises 2.4

1. Show that $w = z^2$ is conformal except at $z = 0$. Show that at $z = 0$ angles are not preserved but are doubled in size.

2. Show that the most general linear fractional transformation which maps the interior of the unit circle onto the interior of the unit circle is

$$w = e^{i\alpha}\frac{z - \beta}{\bar{\beta}z - 1},$$

where α is real and $|\beta| < 1$. Hint: Map the interior of the unit circle onto the upper half-plane and then map the upper half-plane onto the interior of the unit circle.

3. Find a linear fractional transformation which maps the left half-plane onto the interior of the unit circle, so that $z_1 = -i$, $z_2 = 0$, $z_3 = i$ map respectively into $w_1 = i$, $w_2 = -i$, $w_3 = 1$.

4. Find a linear fractional transformation which maps the interior of the unit circle onto the lower half-plane so that $z_1 = 1$, $z_2 = i$, $z_3 = -1$ map respectively into $w_1 = 1$, $w_2 = -1$, $w_3 = \infty$.

5. Find a transformation which maps the first quadrant $0 < \arg z < \pi/2$, $|z| > 0$, onto the interior of the unit circle, so that $z_1 = i$, $z_2 = 0$, $z_3 = 1$ map respectively into $w_1 = -1$, $w_2 = -i$, $w_3 = 1$. Hint: First map the first quadrant onto the upper half-plane.

6. Prove the statements made in the text about the forms of the transformations in the case of two fixed points and one fixed point.

7. Find a function harmonic in the first quadrant which takes on the value zero on the positive real axis and the value one on the positive imaginary axis.

2.5. TRANSCENDENTAL FUNCTIONS

Most of the functions considered so far, for example,

$$z^2, z + \frac{1}{z}, \frac{az + b}{cz + d}, z^{1/2},$$

and so on, are algebraic. A function $w = f(z)$ is *algebraic*, if w and z satisfy an equation $P(z, w) = 0$, where P is a polynomial in z and w. If a function is not algebraic it is *transcendental*. We begin our discussion of transcendental functions by constructing the *exponential function*.

Let us find a function $w = f(z)$ which satisfies the following three properties:

1. $f(z)$ is analytic;
2. $f'(z) = f(z)$;
3. $f(x) = e^x$.

From the first and second properties we have

$$u_x + iv_x = v_y - iu_y = u + iv.$$

Therefore, $u_x = u$, $v_x = v$ which means that $u = e^x g(y)$, $v = e^x h(y)$. Also since $v_y = u$, $-u_y = v$, we have $h'(y) = g(y)$ and $-g'(y) = h$. Differentiating, we infer that

$$-g'' = h' = g, \quad g'' + g = 0,$$
$$g = A \cos y + B \sin y,$$
$$h = -g' = A \sin y - B \cos y.$$

Now, Property 3 implies that

$$u(x, 0) = e^x = e^x g(0) = Ae^x,$$
$$v(x, 0) = 0 = e^x h(0) = -Be^x.$$

Therefore, $A = 1$, $B = 0$, and

$$f(z) = e^x \cos y + ie^x \sin y.$$

We define this to be the *exponential* function and write

$$\exp z = e^z = e^x \cos y + i e^x \sin y.$$

By the sufficient Cauchy-Riemann conditions we see that this is analytic in the entire z plane. Such a function is called an *entire function*. We shall show later (Chapter 4) that this is the only entire function which satisfies Property 3.

Let us examine some of the common properties of the exponential function.

1. $e^z = e^x(\cos y + i \sin y) = e^x e^{iy}$.

2. $e^{z_1} e^{z_2} = e^{x_1} e^{iy_1} e^{x_2} e^{iy_2} = e^{x_1+x_2} e^{i(y_1+y_2)} = e^{z_1+z_2}$.

3. $e^{z_1}/e^{z_2} = e^{x_1} e^{iy_1}/e^{x_2} e^{iy_2} = e^{x_1-x_2} e^{i(y_1-y_2)} = e^{z_1-z_2}$.

4. $|e^z| = e^x |e^{iy}| = e^x$ (exp z has no zeros).

5. $e^{z+2\pi i} = e^z e^{2\pi i} = e^z$ (exp z is periodic with period $2\pi i$).

6. $\overline{e^z} = e^x(\cos y - i \sin y) = e^x e^{-iy} = e^{x-iy} = e^{\bar{z}}$.

If x is real, we have $e^{ix} = \cos x + i \sin x$ and $e^{-ix} = \cos x - i \sin x$. Therefore,

$$\cos x = \frac{e^{ix} + e^{-ix}}{2},$$

$$\sin x = \frac{e^{ix} - e^{-ix}}{2i}.$$

We generalize these relations to define the *trigonometric functions*

$$\cos z = \frac{e^{iz} + e^{-iz}}{2},$$

$$\sin z = \frac{e^{iz} - e^{-iz}}{2i}.$$

It follows from the fact that the exponential function is entire that $\cos z$ and $\sin z$ are entire fractions and

$$\frac{d}{dz} \cos z = \frac{i e^{iz} - i e^{-iz}}{2} = -\frac{e^{iz} - e^{-iz}}{2i} = -\sin z,$$

$$\frac{d}{dz} \sin z = \frac{i e^{iz} + i e^{-iz}}{2i} = \frac{e^{iz} + e^{-iz}}{2} = \cos z.$$

Some of the common properties of the $\cos z$ and $\sin z$ functions are:

1. $\cos z = \frac{1}{2} e^{-y}(\cos x + i \sin x) + \frac{1}{2} e^{y}(\cos x - i \sin x)$

$$= \cos x \frac{e^y + e^{-y}}{2} - i \sin x \frac{e^y - e^{-y}}{2}$$

$$= \cos x \cosh y - i \sin x \sinh y.$$

2. $\sin z = \dfrac{-i}{2} e^{-y}(\cos x + i \sin x) + \dfrac{i}{2} e^{y}(\cos x - i \sin x)$

$= \sin x \dfrac{e^{y} + e^{-y}}{2} + i \cos x \dfrac{e^{y} - e^{-y}}{2}$

$= \sin x \cosh y + i \cos x \sinh y.$

3. $\cos(-z) = \dfrac{e^{-iz} + e^{iz}}{2} = \cos z.$

4. $\sin(-z) = \dfrac{e^{-iz} - e^{iz}}{2i} = -\sin z.$

5. $\cos(z_1 + z_2) = \dfrac{e^{i(z_1+z_2)} + e^{-i(z_1+z_2)}}{2}$

$= \dfrac{e^{iz_1}e^{iz_2} + e^{-iz_1}e^{-iz_2}}{2}$

$= \dfrac{e^{iz_1} + e^{-iz_1}}{2} \dfrac{e^{iz_2} + e^{-iz_2}}{2}$

$\qquad\qquad\qquad - \dfrac{e^{iz_1} - e^{-iz_1}}{2i} \dfrac{e^{iz_2} - e^{-iz_2}}{2i}$

$= \cos z_1 \cos z_2 - \sin z_1 \sin z_2.$

6. $\sin(z_1 + z_2) = \dfrac{e^{i(z_1+z_2)} - e^{-i(z_1+z_2)}}{2i}$

$= \dfrac{e^{iz_1}e^{iz_2} - e^{-iz_1}e^{-iz_2}}{2i}$

$= \sin z_1 \cos z_2 + \cos z_1 \sin z_2.$

7. $|\cos z| = (\cos^2 x \cosh^2 y + \sin^2 x \sinh^2 y)^{1/2}$

$= [\cos^2 x (1 + \sinh^2 y) + \sin^2 x \sinh^2 y]^{1/2}$

$= (\cos^2 x + \sinh^2 y)^{1/2}.$

8. $|\sin z| = (\sin^2 x \cosh^2 y + \cos^2 x \sinh^2 y)^{1/2}$

$= [\sin^2 x (1 + \sinh^2 y) + \cos^2 x \sinh^2 y]^{1/2}$

$= (\sin^2 x + \sinh^2 y)^{1/2}.$

Properties 7 and 8 show that, although $\sin x$ and $\cos x$ are bounded functions, $\sin z$ and $\cos z$ are not bounded since $\sinh^2 y$ approaches infinity as y approaches infinity.

9. $\overline{\cos z} = \cos x \cosh y + i \sin x \sinh y$

$= \cos x \cosh(-y) - i \sin x \sinh(-y)$

$= \cos(x - iy) = \cos \bar{z}.$

10. $\overline{\sin z} = \sin x \cosh y - i \cos x \sinh y$

$\qquad = \sin x \cosh (-y) + i \cos x \sinh (-y)$

$\qquad = \sin (x - iy) = \sin \bar{z}.$

11. $\cos^2 z + \sin^2 z = \left(\dfrac{e^{iz} + e^{-iz}}{2}\right)^2 + \left(\dfrac{e^{iz} - e^{-iz}}{2i}\right)^2 = 1.$

The other trigonometric functions can be defined by analogy with their definitions for the real variable case, for example,

$$\tan z = \frac{\sin z}{\cos z}, \qquad \cot z = \frac{\cos z}{\sin z},$$

$$\sec z = \frac{1}{\cos z}, \qquad \csc z = \frac{1}{\sin z}.$$

These functions are analytic except where the denominators vanish. In the case of the tangent and secant functions they fail to be analytic where

$$\cos z = \cos x \cosh y - i \sin x \sinh y = 0,$$
$$\cos x \cosh y = 0, \qquad \sin x \sinh y = 0.$$

Since $\cosh y$ is never zero, we must place $\cos x = 0$. This implies

$$x = \frac{2n + 1}{2} \pi, n = 0, \pm 1, \pm 2, \ldots.$$

But $\sin (n + 1/2)\pi \neq 0$. Therefore, we must have $\sinh y = 0$. This requires that $y = 0$. In other words, $\tan z$ and $\sec z$ are analytic except at

$$z = \frac{2n + 1}{2} \pi, n = 0, \pm 1, \pm 2, \ldots.$$

Where they are analytic, we have

$$\frac{d}{dz} \tan z = \frac{\cos^2 z + \sin^2 z}{\cos^2 z} = \frac{1}{\cos^2 z} = \sec^2 z,$$

$$\frac{d}{dz} \sec z = \frac{\sin z}{\cos^2 z} = \tan z \sec z.$$

Similarly, we can show that $\cot z$ and $\csc z$ are analytic except at $z = n\pi$, $n = 0, \pm 1, \pm 2, \ldots.$ Where they are analytic, we have

$$\frac{d}{dz} \cot z = \frac{-\sin^2 z - \cos^2 z}{\sin^2 z} = -\frac{1}{\sin^2 z} = -\csc^2 z,$$

$$\frac{d}{dz} \csc z = -\frac{\cos z}{\sin^2 z} = -\cot z \csc z.$$

The *hyperbolic sine* and *hyperbolic cosine* are defined from the exponential function as follows:

$$\sinh z = \frac{e^z - e^{-z}}{2},$$

$$\cosh z = \frac{e^z + e^{-z}}{2}.$$

They are obviously entire functions and their derivatives can be computed from the exponential function

$$\frac{d}{dz} \sinh z = \frac{e^z + e^{-z}}{2} = \cosh z,$$

$$\frac{d}{dz} \cosh z = \frac{e^z - e^{-z}}{2} = \sinh z.$$

The obvious connection between the hyperbolic sine and cosine and the trigonometric functions is

$$\sinh iz = \frac{e^{iz} - e^{-iz}}{2} = i \sin z,$$

$$\cosh iz = \frac{e^{iz} + e^{-iz}}{2} = \cos z.$$

We also have the following properties:

1. $\sinh z = \frac{1}{2}e^x(\cos y + i \sin y) - \frac{1}{2}e^{-x}(\cos y - i \sin y)$

$$= \cos y \frac{e^x - e^{-x}}{2} + i \sin y \frac{e^x + e^{-x}}{2}$$

$$= \sinh x \cos y + i \cosh x \sin y.$$

2. $\cosh z = \frac{1}{2}e^x(\cos y + i \sin y) + \frac{1}{2}e^{-x}(\cos y - i \sin y)$

$$= \cos y \frac{e^x + e^{-x}}{2} + i \sin y \frac{e^x - e^{-x}}{2}$$

$$= \cosh x \cos y + i \sinh x \sin y.$$

3. $\sinh(-z) = \dfrac{e^{-z} - e^z}{2} = -\sinh z.$

4. $\cosh(-z) = \dfrac{e^{-z} + e^z}{2} = \cosh z.$

5. $\sinh(z_1 + z_2) = \sinh i(-iz_1 - iz_2)$

$$= i \sin(-iz_1 - iz_2)$$

$$= i \sin(-iz_1) \cos(-iz_2) + i \cos(-iz_1) \sin(-iz_2)$$

$$= \sinh z_1 \cosh z_2 + \cosh z_1 \sinh z_2.$$

6. $\cosh(z_1 + z_2) = \cosh i(-iz_1 - iz_2)$
$$= \cos(-iz_1 - iz_2)$$
$$= \cos(-iz_1)\cos(-iz_2) - \sin(-iz_1)\sin(-iz_2)$$
$$= \cosh z_1 \cosh z_2 + \sinh z_1 \sinh z_2.$$

7. $|\sinh z| = [\sinh^2 x \cos^2 y + \cosh^2 x \sin^2 y]^{1/2}$
$$= [\sinh^2 x \cos^2 y + (1 + \sinh^2 x)\sin^2 y]^{1/2}$$
$$= [\sinh^2 x + \sin^2 y]^{1/2}.$$

8. $|\cosh z| = [\cosh^2 x \cos^2 y + \sinh^2 x \sin^2 y]^{1/2}$
$$= [(1 + \sinh^2 x)\cos^2 y + \sinh^2 x \sin^2 y]^{1/2}$$
$$= [\sinh^2 x + \cos^2 y]^{1/2}.$$

9. $\overline{\sinh z} = \sinh x \cos y - i \cosh x \sin y$
$$= \sinh x \cos(-y) + i \cosh x \sin(-y)$$
$$= \sinh(x - iy) = \sinh \bar{z}.$$

10. $\overline{\cosh z} = \cosh x \cos y - i \sinh x \sin y$
$$= \cosh x \cos(-y) + i \sinh x \sin(-y)$$
$$= \cosh(x - iy) = \cosh \bar{z}.$$

11. $\cosh^2 z - \sinh^2 z = \left(\dfrac{e^z + e^{-z}}{2}\right)^2 - \left(\dfrac{e^z - e^{-z}}{2}\right)^2 = 1.$

The other hyperbolic functions can be defined by analogy with their definitions for the real variable case, for example,

$$\tanh z = \frac{\sinh z}{\cosh z}, \qquad \coth z = \frac{\cosh z}{\sinh z},$$

$$\operatorname{sech} z = \frac{1}{\cosh z}, \qquad \operatorname{csch} z = \frac{1}{\sinh z}.$$

These functions are analytic except where the denominators vanish. Unlike $\cosh x$ which never is zero $\cosh z$ has infinitely many zeros. Let

$$\cosh z = \cosh x \cos y + i \sinh x \sin y = 0,$$
$$\cosh x \cos y = 0, \qquad \sinh x \sin y = 0.$$

Now, $\cosh x \neq 0$. Therefore, $\cos y = 0$, which implies that $y = (n + 1/2)\pi$, $n = 0, \pm 1, \pm 2, \ldots$. However, $\sin(n+1/2)\pi \neq 0$. This implies that $\sinh x = 0$ or that $x = 0$. Hence, $\cosh z$ vanishes at $z = (n + 1/2)\pi i$, $n = 0, \pm 1, \pm 2, \ldots$, and $\tanh z$ and $\operatorname{sech} z$ are analytic except at these points. Their derivatives are

$$\frac{d}{dz}\tanh z = \frac{\cosh^2 z - \sinh^2 z}{\cosh^2 z} = \operatorname{sech}^2 z,$$

$$\frac{d}{dz}\operatorname{sech} z = \frac{-\sinh z}{\cosh^2 z} = -\tanh^2 z \operatorname{sech} z.$$

Similarly, we can show that sinh z vanishes at $z = n\pi i, n = 0, \pm1, \pm2, \ldots$ and, therefore, that coth z and csch z are analytic except at these points. Their derivatives are

$$\frac{d}{dz}\coth z = \frac{\sinh^2 z - \cosh^2 z}{\sinh^2 z} = -\csch^2 z,$$

$$\frac{d}{dz}\csch z = \frac{-\cosh z}{\sinh^2 z} = -\coth z \csch z.$$

The *logarithm function* is an inverse of the exponential function just as in the case of functions of the real variable. There is difficulty, however, in trying to define an inverse function for the exponential which stems from the periodicity of exp z; that is,

$$e^{z+2n\pi i} = e^z,$$

where $n = 0, \pm1, \pm2, \ldots$. This means that the strips $0 \le \text{Im}(z) < 2\pi$, $2\pi \le \text{Im}(z) < 4\pi$, $-2\pi \le \text{Im}(z) < 0$, and so on, all map into the same points in the w plane. The function is therefore many-to-one and an inverse does not exist. However, if we were to restrict the domain of the function to the strip $-\pi < \text{Im}(z) \le \pi$, the function is one-to-one and an inverse does exist.

Reversing the roles of the z plane and w plane, suppose we take

$$z = e^w = e^u(\cos v + i \sin v),$$

with $-\pi < v \le \pi$. Then we have $|z| = e^u$, arg $z = v$, or

$$w = u + iv = \ln |z| + i \arg z,$$

$-\pi < \arg z \le \pi$. We call this function the *principal value* of log z and write

$$\text{Log } z = \ln \sqrt{x^2 + y^2} + i \tan^{-1}\frac{y}{x}$$

$-\pi < \tan^{-1}(y/x) \le \pi$. This implies that there are many logarithm functions depending on what restriction is placed on the arg z to make the function single-valued. In our original definition of function we did not allow for the possibility of having multiple-valued functions. However, we can conceive of a function which has many *branches*, each of which is single-valued and fits the original definition of function. Suppose, in the present case, we let $\theta_0 = \arg z$ which satisfies the above restriction for the principal value, then

$$\log z = \ln |z| + i(\theta_0 \pm 2n\pi),$$

$n = 0, 1, 2, \ldots$, is a multiple-valued function with infinitely many branches, each for a different n, and each single-valued.

Another way out of the dilemma is to define the logarithm function on a more complicated surface (Riemann surface) consisting of infinitely many planes joined together so that the function varies continuously as one passes from one plane to the next. This notion will be pursued in the next section.

Returning to the principal value,* we see that it is not defined at $z = 0$ and is discontinuous on the negative real axis. This means that the function cannot be analytic at these points. Elsewhere, the sufficient Cauchy-Riemann conditions show us that the function is differentiable and

$$\frac{d}{dz} \text{Log } z = \frac{\partial}{\partial x} \ln \sqrt{x^2 + y^2} + i \frac{\partial}{\partial x} \tan^{-1} \frac{y}{x}$$

$$= \frac{x - iy}{x^2 + y^2} = \frac{\bar{z}}{z\bar{z}} = \frac{1}{z}.$$

As a matter of fact, the other branches have the same derivative where they are differentiable. This is not surprising since the different branches only differ by a constant.

The points where Log z is not analytic, $z = 0$ where the function is un-defined and the negative real axis where it is discontinuous, play distinctly different roles in the study of log z. The negative real axis was introduced in a rather arbitrary way as a restriction on the arg z to make the function single-valued. Another branch of log z could be defined with $0 \le \text{arg } z < 2\pi$ and then the discontinuity would be on the positive real axis and these points would become points where the function fails to be analytic. This line cuts the plane so that when it and the origin are deleted we are left with a domain in which the function is single-valued and analytic. The line is called a *branch cut*. We have already seen that the branch cut can be moved at will by defining different branches of the function. As a matter of fact, the cut need not be a straight line. For example, we could delete the points on the parabolic arc $x = t$, $y = t^2$, $0 \le t < \infty$ by properly restricting the arg z for various values of $|z|$ and thus define a single-valued analytic branch of log z in the plane with these points deleted. In the next section, we shall define log z on a Riemann surface and the branch cut will com-pletely lose its identity. On the other hand, in defining various branches of log z every branch cut starts at $z = 0$. No matter where the branch cut is moved the point $z = 0$ is still a point where the function fails to be analytic. This point we call a *branch point* and we see that it is a more basic type of singularity than the points on a particular branch cut. We will illustrate this concept more thoroughly in the next section with other types of functions which have branch points.

The usual properties of log z hold, provided one exercises care in selecting branches. For example,

$$\log (z_1 z_2) = \ln |z_1| \, |z_2| + i \arg (z_1, z_2)$$
$$= \ln |z_1| + i \arg z_1 + \ln |z_2| + i \arg z_2$$
$$= \log z_1 + \log z_2.$$

---

* One advantage to working with the principal value is that Log $z = \ln x$ when $z = x > 0$, and hence reduces to the natural logarithm when z is real and positive.

EXAMPLE 2.5.1. Let $z_1 = -1$ and $z_2 = -1$. Then $\text{Log } z_1 = \text{Log } z_2 = \pi i$ and $\text{Log } z_1 + \text{Log } z_2 = 2\pi i = \log z_1 z_2 = \log 1$, provided we pick a branch such that $\log 1 = 2\pi i$.

We are now able to define the function z^a where a is any complex constant. We define it as follows

$$z^a = e^{a \log z}.$$

If $a = m$, an integer, $z^m = e^{m \log z} = \exp[m \ln |z| + mi \arg z \pm 2\pi mni]$, where $n = 0, 1, 2, \ldots$. The function, in this case, is single-valued, since mn is an integer and the exponential function is periodic with period $2\pi i$.

If $a = p/q$, p and q integers, then

$$z^{p/q} = \exp[(p/q) \ln |z| + (p/q)i \arg z] \exp[\pm 2\pi npi/q].$$

In this case, there are only q distinct values since if $n = \pm q, \pm(q + 1)$, $\pm(q + 2), \ldots$, then $e^{+2\pi npi/q}$ just repeats the values for $n = 0, \pm 1$, $\pm 2, \ldots, \pm(q - 1)$. Also, if $n = -k$, $e^{2\pi npi/q} = e^{-2\pi kpi/q} = e^{2\pi(q-k)pi/q}$. Hence, there is no need to take negative n.

If a is irrational or complex, there are infinitely many values of z^a. There is a distinct branch for each distinct branch of $\log z$ and branch cuts are introduced in the same way they were introduced for $\log z$. Except at the branch point $z = 0$ and on a branch cut, z^a is analytic and

$$\frac{d}{dz} z^a = \frac{d}{dz} e^{a \log z} = \frac{a}{z} e^{a \log z} = ae^{a \log z - \log z} = ae^{(a-1)\log z} = az^{a-1}$$

EXAMPLE 2.5.2. Find all possible distinct values of i^i. By definition

$$i^i = e^{i \log i} = e^{i \ln 1} e^{-\arg i} = e^{-(4n+1)\pi/2}, n = 0, \pm 1, \pm 2, \ldots.$$

The *inverse trigonometric functions* are multiple-valued just as the logarithmic function is. This is obvious when we realize that $\cos z$ and $\sin z$ are many-to-one functions. For example, we already know that $\cos(-z) = \cos z$ and, therefore, the first and third quadrants, and the second and fourth quadrants map into the same points. But what is more, if $m = 1, 2, 3, \ldots$

$$\cos(z \pm 2m\pi) = \cos z \cos(\pm 2m\pi) - \sin z \sin(\pm 2m\pi)$$
$$= \cos z.$$

Also,

$$\cos(2\pi - z) = \cos(z - 2\pi) = \cos z.$$

If we restrict z so that $0 < \text{Re}(z) < \pi$, then $w = \cos z$ is one-to-one into the entire w plane with the half-lines $v = 0$, $u \geq 1$ and $v = 0$, $u \leq -1$ deleted. With this restriction we can define the inverse cosine. Reversing the roles of the z plane and the w plane we write

$$z = \cos w = \frac{e^{iw} + e^{-iw}}{2},$$
$$e^{iw} + e^{-iw} - 2z = 0,$$
$$e^{2iw} - 2ze^{iw} + 1 = 0.$$

Solving for e^{iw}, we have

$$e^{iw} = z + (z^2 - 1)^{1/2}.$$

Then,

$$w = \cos^{-1} z = -i \log [z + (z^2 - 1)^{1/2}].$$

We must define $(z^2 - 1)^{1/2}$ and the logarithm so that the function is single-valued. For example, if $z = 0$

$$w = -i \log (\pm i).$$

If we take $+i$ rather than $-i$ and the principal value of $\log i$ we obtain $w = -i (i\pi/2) = \pi/2$ which is in the proper range. We take

$$(z^2 - 1)^{1/2} = (z - 1)^{1/2}(z + 1)^{1/2} = (r_1 r_2)^{1/2} e^{i(\theta_1 + \theta_2)/2},$$
$$r_1 = |z - 1|, r_2 = |z + 1|, \theta_1 = \arg (z - 1), \theta_2 = \arg (z + 1),$$

with $-\pi \le \theta_1 < \pi$ and $0 \le \theta_2 < 2\pi$. Then $(z^2 - 1)^{1/2}$ has the value i at $z = 0$, has branch points at $z = \pm 1$, and branch cuts $y = 0$, $|x| > 1$. See Figure 2.5.1. We take the principal value of the logarithm. This does

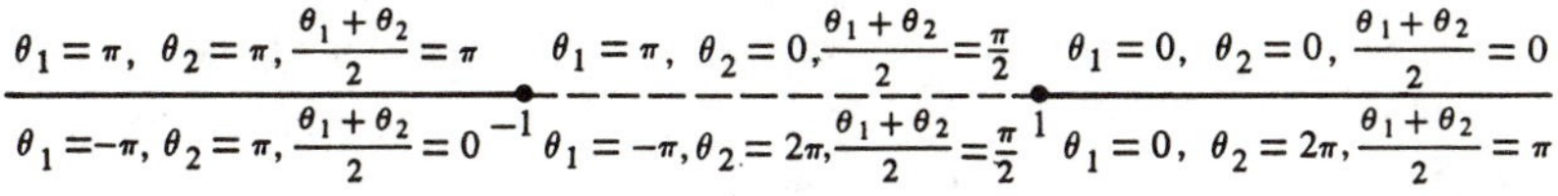

$$\theta_1 = \pi, \ \theta_2 = \pi, \frac{\theta_1 + \theta_2}{2} = \pi \quad \theta_1 = \pi, \ \theta_2 = 0, \frac{\theta_1 + \theta_2}{2} = \frac{\pi}{2} \quad \theta_1 = 0, \ \theta_2 = 0, \frac{\theta_1 + \theta_2}{2} = 0$$

$$\theta_1 = -\pi, \theta_2 = \pi, \frac{\theta_1 + \theta_2}{2} = 0 \quad {}^{-1}\theta_1 = -\pi, \theta_2 = 2\pi, \frac{\theta_1 + \theta_2}{2} = \frac{\pi}{2} \quad {}^{1}\theta_1 = 0, \ \theta_2 = 2\pi, \frac{\theta_1 + \theta_2}{2} = \pi$$

Figure 2.5.1.

not introduce any new branch points or cuts, since $z + (z^2 - 1)^{1/2} \ne 0$ and $\arg [z + (z^2 - 1)^{1/2}] = \pi$, when $z = x < -1$.

As long as we stay away from the branch points and cuts the function is differentiable. We can obtain the derivative by direct differentiation, that is,

$$\frac{dw}{dz} = \frac{-i}{z + (z^2 - 1)^{1/2}} \left[1 + \frac{z}{(z^2 - 1)^{1/2}} \right] = \frac{-i}{(z^2 - 1)^{1/2}},$$

where the square root is defined precisely as in the definition of the function.

The inverse sine function can be developed in a similar manner. We write $z = \sin w$, with $-\pi/2 < \mathrm{Re}(w) < \pi/2$. The function is one-to-one in this strip and therefore has an inverse.

$$z = \frac{e^{iw} - e^{-iw}}{2i},$$
$$e^{-2iw} + 2ize^{-iw} - 1 = 0,$$
$$e^{-iw} = -i[z + (z^2 - 1)^{1/2}],$$
$$w = i \log [z + (z^2 - 1)^{1/2}] + \frac{\pi}{2}$$
$$= \frac{\pi}{2} - \cos^{-1} z,$$

where the $\cos^{-1} z$ is the branch defined above. The branch points are again $z = \pm 1$ and the branch cuts are $z = x$, $|x| > 1$. Except at the branch points and on the branch cuts the function is differentiable and

$$\frac{d}{dz} \sin^{-1} z = \frac{i}{(z^2 - 1)^{1/2}}.$$

The other inverse trigonometric functions can be defined in a similar way if due care is taken with multiple values, branch points, branch cuts, and so on. However, it would become tedious to catalogue them all. Furthermore, there are more convenient and interesting definitions in terms of definite integrals which we will introduce in the next chapter.

Exercises 2.5

1. Find all possible solutions of the equation $e^z = i$.

2. Show that $\lim\limits_{n \to \infty} \left| \left(1 + \dfrac{z}{n} \right)^n \right| = e^x$ and $\lim\limits_{n \to \infty} \arg \left(1 + \dfrac{z}{n} \right)^n = y$.

3. What is the range of $f(z) = e^z$, $\mathrm{Re}(z) \le 0$, $-\pi < \mathrm{Im}(z) \le \pi$? Show that the function is one-to-one and onto.

4. Show that $f(z) = e^{1/z}$ is analytic at infinity but not at the origin. What is its derivative?

5. Find all possible solutions of $\sin z = i$.

6. Is $\cos \bar{z}$ anywhere an analytic function of z?

7. Is $\sin |z|^2$ anywhere differentiable? Is it anywhere analytic?

8. Prove the following identities:

(a) $\sin 2z = 2 \sin z \cos z$;

(b) $2 \sin^2 \dfrac{z}{2} = 1 - \cos z$;

(c) $1 + \tan^2 z = \sec^2 z$;

(d) $2 \sin z_1 \sin z_2 = \cos (z_1 - z_2) - \cos (z_1 + z_2)$.

9. Find the range of $f(z) = \cos z$, $0 \le \mathrm{Re}(z) < \pi$. Show that the function is one-to-one and onto.

10. Find all possible solutions of $\sinh z = 0$.

11. Prove that $\log (z_1/z_2) = \log z_1 - \log z_2$ provided that the right branches are chosen.

2.6. RIEMANN SURFACES

As we have mentioned earlier, this section will deal with "multiple-valued" functions from the point of view of Riemann surfaces. It should be obvious by now that one cannot define inverses of many-to-one functions on a single complex plane without resorting to the definition of various branches and introducing artificial branch cuts, in order to define single-valued analytic functions on some open set. Another approach is to define functions on many complex planes (sheets) attached to one another to form an open connected set as the domain of definition. These many sheeted surfaces are known as Riemann surfaces. By giving up the simplicity of defining a function on a set of points in a single complex plane we gain in the simplicity of not having branch cuts, we remove artificial restrictions, and we allow for a more general definition of function, which is cast in a framework in which it is possible to bring to bear the methods of modern analysis and topology.

As our first example, let us consider the logarithm function. The Riemann surface consists of infinitely many sheets. Let S_n be an extended complex z plane in which $(2n - 1)\pi < \arg z \le (2n + 1)\pi, n = 0, \pm 1, \pm 2, \ldots$. Imagine each of these planes cut along the negative real axis from infinity to the origin. Finally, imagine these planes attached to one another along the negative real axes in such a way that if one were on sheet S_k walking around the origin toward the negative real axis with increasing $\arg z$ he could pass onto S_{k+1} with the $\arg z$ changing continuously through the value $(2k + 1)\pi$. On the other hand, if one were walking around the origin toward the negative real axis with decreasing $\arg z$, he could pass onto sheet S_{k-1} with continuously changing argument. Notice that the origin and the point at infinity are common to all the sheets.

Having defined the Riemann surface for the logarithm function we can now define the function as follows:

$$\log z = \ln |z| + i(\theta_0 + 2n\pi)$$

on S_n, where $-\pi < \theta_0 = \arg z \le \pi$. Notice that this defines the principal value on S_0. We have consequently defined the logarithm function uniquely at every point on the Riemann surface except at the origin and the point at infinity.

Pictorially, we can view the situation by considering a continuous curve in the Riemann surface wrapping around the origin. See Figure 2.6.1.

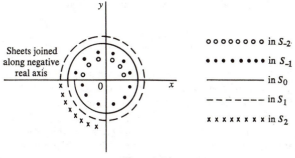

Figure 2.6.1.

Clearly, the function log z has infinitely many values in a neighborhood of the origin depending on what sheet you happen to be on. In this case, we call the origin a *branch point of infinite order*. A branch point of order n is one for which the given function has precisely $n + 1$ distinct values in some neighborhood of the point. We shall see branch points of finite order.

In the case of log z the point at infinity also is a branch point of infinite order. One can easily visualize a continuous curve winding around the point at infinity in the Riemann surface, if we first imagine each sheet mapped stereographically onto a Riemann sphere and then imagine the spheres joined together along a great circle joining the south pole (origin) and the north pole (infinity). Then a neighborhood of the point at infinity is a neighborhood of the north pole on the Riemann sphere.

In order to define continuity and differentiability of a function on a Riemann surface, we have to define the concept of neighborhood on a Riemann surface. We can do this for the logarithm function in a very natural way. If a point z_0 is an interior point of one of the sheets, then there exists an $\epsilon > 0$ such that $|z - z_0| < \epsilon$ defines an open set of points in the same sheet. We define this as a neighborhood of z_0 on the Riemann surface. If a point z_0 is on sheet S_k with arg $z_0 = (2k + 1)\pi$, then we define a neighborhood of z_0 as follows: all points z such that $|z - z_0| < \epsilon$ in S_k for Im(z) ≥ 0 and in S_{k+1} for Im(z) < 0. A neighborhood of the origin is defined by all points on the Riemann surface for which $|z| < \epsilon$, while a neighborhood of infinity is defined by all points such that $|z| > R$.

With these definitions of neighborhood, we find the log z is continuous and differentiable everywhere on the Riemann surface except at the origin and infinity. The derivative is

$$\frac{d}{dz} \log z = \frac{1}{z},$$

which in this case is a single-valued function.

Now, let us consider a Riemann surface for a different function $w = z^{1/2}$. We know already that there are two distinct solutions of $w^2 = z$ for each $z \neq 0$. For example, if $\theta_0 = $ arg z and $0 \leq \theta_0 < 2\pi$, then

$$w_1 = |z|^{1/2} e^{i\theta_0/2},$$
$$w_2 = |z|^{1/2} e^{i(\theta_0/2)+\pi} = -w_1,$$

are the two distinct solutions. Let S_1 be an extended complex plane cut along the positive real axis from infinity to the origin where in S_1 we define $0 \leq$ arg $z < 2\pi$. Let S_2 be an exact copy of S_1 except that in S_2 we define $2\pi \leq$ arg $z < 4\pi$. Now imagine S_1 and S_2 joined together so that the upper side of the positive real axis in S_1 is joined to the lower side of the positive real axis in S_2 and the lower side of the positive real axis in S_1 is joined to the upper side of the positive real axis in S_2. We can visualize the

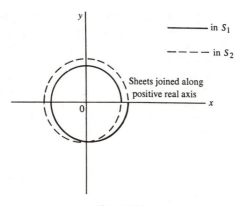

Figure 2.6.2.

surface by considering a continuous curve lying in it winding around the origin twice. See Figure 2.6.2. This is a Riemann surface for the function $w = z^{1/2}$ where we define the function to have the value w_1 on S_1 and w_2 on S_2. The origin and the point at infinity are considered in both sheets and both are branch points of order one. The function is defined as zero at the origin but is undefined at infinity.

In the present case, the concept of neighborhood is clear except possibly for points on the positive real axis. We have to make a distinction between points with argument zero in S_1 and points with argument 2π in S_2. If a point z_0 in S_1 has argument zero then we define a neighborhood of it as follows: those points satisfying $|z - z_0| < \epsilon$ in S_1 for $\mathrm{Im}(z) \geq 0$ and in S_2 for $\mathrm{Im}(z) < 0$. If a point z_0 in S_2 has argument 2π we define a neighborhood of it as follows: Those points satisfying $|z - z_0| < \epsilon$ in S_2 for $\mathrm{Im}(z) \geq 0$ and in S_1 for $\mathrm{Im}(z) < 0$.

With these definitions of neighborhood we find that $w = z^{1/2}$ is continuous and differentiable everywhere on the Riemann surface except at the origin and infinity. The derivative is

$$\frac{d}{dz}(z)^{1/2} = \begin{cases} \dfrac{1}{2}\dfrac{1}{w_1} & \text{on } S_1, \\[2ex] \dfrac{1}{2}\dfrac{1}{w_2} & \text{on } S_2. \end{cases}$$

Finally, let us consider a function with two finite branch points, $w = (z^2 - 1)^{1/2}$. Let $\zeta_1 = |z - 1|^{1/2}e^{i\theta_0/2}$, where $\theta_0 = \arg(z - 1)$ and $0 \leq \theta_0 < 2\pi$. Let $\zeta_2 = |z - 1|^{1/2}e^{i(\theta_0 + 2\pi)/2} = -\zeta_1$. Let $\eta_1 = |z + 1|^{1/2}e^{i\phi_0/2}$, where $\phi_0 = \arg(z + 1)$ and $0 \leq \phi_0 < 2\pi$. Let $\eta_2 = |z + 1|^{1/2}e^{i(\phi_0 + 2\pi)/2} = -\eta_1$. Then w has two distinct values $\zeta_1\eta_1 = \zeta_2\eta_2$, $\zeta_1\eta_2 = \zeta_2\eta_1 = -\zeta_1\eta_1$. Consider a Riemann surface consisting of two sheets. Let S_1 be an extended complex plane cut along the real axis from -1 to $+1$. Let S_2 be an identical extended plane. In S_1 we take $0 \le \arg(z - 1) < 2\pi$, $0 \le \arg(z + 1) < 2\pi$. In S_2 we take $2\pi \le \arg(z - 1) < 4\pi$, $2\pi \le \arg(z + 1) < 4\pi$. Imagine S_1 and S_2 joined along the segment of the real axis between -1 and $+1$ in such a way that the top side of the segment in S_1 is joined to the bottom side of the segment in S_2 and the bottom side of the segment in S_1 joined to the top side of the segment in S_2. The situation can be visualized by considering two smooth curves winding about the points -1 and $+1$ as follows: See Figure 2.6.3.

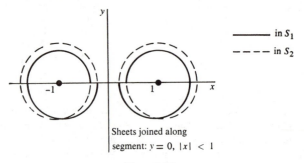

Sheets joined along
segment: $y = 0$, $|x| < 1$

Figure 2.6.3.

We define the function on the Riemann surface as

$$w = \begin{cases} \zeta_1 \eta_1 \text{ on } S_1, \\ -\zeta_1 \eta_1 \text{ on } S_2. \end{cases}$$

The function is continuous everywhere on the Riemann surface except at infinity. It is differentiable everywhere except at infinity and at the branch points -1 and $+1$.

$$\frac{dw}{dz} = \begin{cases} \dfrac{z}{\zeta_1 \eta_1} \text{ on } S_1, \\ \dfrac{-z}{\zeta_1 \eta_1} \text{ on } S_2. \end{cases}$$

The branch points are of order one in each case.

A Riemann surface for the function $w = (z^2 - 1)^{1/2}$ could have been constructed for which the sheets were cut along the real axis from $-\infty$ to -1 and from $+\infty$ to $+1$. In S_1 we take $0 \le \arg(z - 1) < 2\pi$ and $-\pi \le \arg(z + 1) < \pi$. In S_2 we take $2\pi \le \arg(z - 1) < 4\pi$ and $\pi \le \arg(z + 1) < 3\pi$ and the sheets are joined appropriately across the cuts. The situation seems quite different until viewed in terms of stereographic projection. In the original case the Riemann spheres are cut through the south pole and in the latter case are cut through the north pole. Otherwise, there is no essential difference.

Exercises 2.6

1. Construct a Riemann surface for the function $w = z^{1/3}$.

2. Construct a Riemann surface for the function $w = (z^3 - 1)^{1/2}$. What are the branch points and their orders?

3. Construct a Riemann surface for the function $w = \cos^{-1} z$. What are the branch points and their orders?

CHAPTER 3

Integration in the Complex Plane

3.1. LINE INTEGRALS

Our ultimate goal in this chapter will be to discuss the concept of integration of a function of a complex variable. We shall see that this can be related to the concept of line integral of a real-valued function of two real variables. Since line integration may be unfamiliar to some readers, we shall devote the present section to this topic. Anyone who is familiar with line integrals can start at Section 3.2.

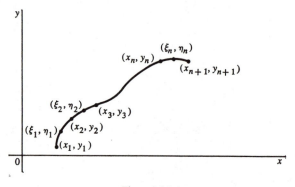

Figure 3.1.1.

Let $P(x, y)$ and $Q(x, y)$ be real-valued functions defined on a simple smooth arc C. We subdivide C into n subarcs by picking $n + 1$ points on C as shown in Figure 3.1.1. The points (x_1, y_1) and (x_{n+1}, y_{n+1}) are chosen as the end points of C, and the $n - 1$ other points are chosen on C between the end points, ordered by subscript along the arc. Let C_k be the subarc of C lying between (x_k, y_k) and (x_{k+1}, y_{k+1}). Let (ξ_k, η_k) be any point on C_k. We form the sums

$$\sum_{k=1}^{n} P(\xi_k, \eta_k)\, \Delta x_k, \qquad \sum_{k=1}^{n} Q(\xi_k, \eta_k)\, \Delta y_k,$$

where $\Delta x_k = x_{k+1} - x_k$ and $\Delta y_k = y_{k+1} - y_k$. Next, we make a similar subdivision of C with a larger n and recompute these sums. We continue this process letting n increase without bound, being careful to subdivide C so that as n approaches infinity, $|\Delta x_k|$ and $|\Delta y_k|$ approach zero for all k. If these sums approach definite limits as n approaches infinity irrespective of the particular set of subdivisions used, we define these limits as the line integrals

$$\int_C P(x, y)\, dx = \lim_{\substack{n \to \infty \\ \max|\Delta x_k| \to 0}} \sum_{k=1}^{n} P(\xi_k, \eta_k)\, \Delta x_k,$$

$$\int_C Q(x, y)\, dy = \lim_{\substack{n \to \infty \\ \max|\Delta y_k| \to 0}} \sum_{k=1}^{n} Q(\xi_k, \eta_k)\, \Delta y_k.$$

Theorem 3.1.1. *If $P(x, y)$ and $Q(x, y)$ are continuous on C, a simple smooth arc, then the line integrals $\int_C P(x, y)\, dx$ and $\int_C Q(x, y)\, dy$ exist and*

$$\int_C P(x, y)\, dx = \int_0^1 P[x(t), y(t)]\dot{x}\, dt,$$

$$\int_C Q(x, y)\, dy = \int_0^1 Q[x(t), y(t)]\dot{y}\, dt,$$

where $z(t) = x(t) + iy(t)$, $0 \leq t \leq 1$, is a parametric representation of C, $\dot{x} = dx/dt$ and $\dot{y} = dy/dt$.

Proof: Since C is a simple smooth arc, it has a parametric representation $z(t) = x(t) + iy(t)$, $0 \leq t \leq 1$, where $\dot{x}(t)$ and $\dot{y}(t)$ are continuous. Then, by the mean value theorem

$$\Delta x_k = x_{k+1} - x_k = x(t_{k+1}) - x(t_k) = \dot{x}(\tau_k)(t_{k+1} - t_k) = \dot{x}(\tau_k)\, \Delta t_k,$$

where $t_k < \tau_k < t_{k+1}$. Let τ_k' be chosen so that $t_k \leq \tau_k' \leq t_{k+1}$. Then $\xi_k = x(\tau_k')$, $\eta_k = y(\tau_k')$ are the coordinates of a point on C_k, the kth subarc of C. We form the sum

$$\sum_{k=1}^{n} P(\xi_k, \eta_k)\, \Delta x_k = \sum_{k=1}^{n} P[x(\tau_k'), y(\tau_k')]\dot{x}(\tau_k)\, \Delta t_k$$

$$= \sum_{k=1}^{n} P[x(\tau_k), y(\tau_k)]\dot{x}(\tau_k)\, \Delta t_k + \sum_{k=1}^{n} \Delta P_k \dot{x}(\tau_k)\, \Delta t_k,$$

where $\Delta P_k = P[x(\tau_k'), y(\tau_k')] - P[x(\tau_k), y(\tau_k)]$. Now, $P[x(t), y(t)]$ is a continuous function of t on the interval $0 \leq t \leq 1$, and hence is uniformly continuous. Therefore, given $\epsilon > 0$, there exists a $\delta(\epsilon)$ independent of t, such that $|\Delta P_k| < \epsilon$ when $|\tau_k - \tau_k'| \leq \Delta t_k < \delta$. Furthermore, $\dot{x}(t)$ is continuous and hence bounded, that is, $|\dot{x}(t)| \leq M$. We can now conclude that

$$\left| \sum_{k=1}^{n} \Delta P_k \dot{x}(\tau_k) \Delta t_k \right| \leq \sum_{k=1}^{n} |\Delta P_k||\dot{x}(\tau_k)| \Delta t_k < \epsilon M$$

provided max $|\Delta t_k| < \delta$. Therefore,

$$\lim_{\substack{n \to \infty \\ \max|\Delta t_k| \to 0}} \sum_{k=1}^{n} \Delta P_k \dot{x}(\tau_k) \Delta t_k = 0.$$

Finally, by the continuity of P and $\dot{x}$ we know that

$$\int_0^1 P[x(t), y(t)]\dot{x} \, dt = \lim_{\substack{n \to \infty \\ \max|\Delta t_k| \to 0}} \sum_{k=1}^{n} P[x(\tau_k), y(\tau_k)]\dot{x}(\tau_k) \Delta t_k,$$

Hence

$$\int_C P(x, y) \, dx = \lim_{\substack{n \to \infty \\ \max|\Delta x_k| \to 0}} \sum_{k=1}^{n} P(x_k, y_k) \Delta x_k = \int_0^1 P[x(t), y(t)]\dot{x} \, dt.$$

The result for $\int_C Q(x, y) \, dy$ can be proved in a similar manner.

Corollary 3.1.1. If $P(x, y)$ and $Q(x, y)$ are continuous on a contour C, then the line integrals $\int_C P(x, y) \, dx$ and $\int_C Q(x, y) \, dy$ exist and can be computed by summing the contributions of each of a finite number of simple smooth arcs.

EXAMPLE 3.1.1. Compute $\int_{C_i} (x^2 + y^2) \, dx - 2xy \, dy$ between the points $(1, 0)$ and $(0, 1)$ along each of the contours in the first quadrant.

C_1: $x + y = 1$ (a straight line segment).

C_2: $x^2 + y^2 = 1$ (an arc of the unit circle).

C_3: $z(t) = 1 + it, 0 \leq t \leq 1, z(t) = (2 - t) + i, 1 \leq t \leq 2$
    (a broken line)

(a) On C_1 $x = 1 - t, y = t, \dot{x} = -1$, and $\dot{y} = 1$. Hence,

$$\int_{C_1} (x^2 + y^2) \, dx - 2xy \, dy = -\int_0^1 (1 - 2t + 2t^2) \, dt - \int_0^1 (2t - 2t^2) \, dt$$

$$= -\int_0^1 dt = -1.$$

(b) On C_2 $x = \cos \theta, y = \sin \theta, 0 \le \theta \le \pi/2$. Hence,

$$\int_{C_2} (x^2 + y^2)\, dx - 2xy\, dy = -\int_0^{\pi/2} \sin \theta\, d\theta - 2\int_0^{\pi/2} \cos^2 \theta \sin \theta\, d\theta$$

$$= \left[\cos \theta\right]_0^{\pi/2} + \left[\frac{2}{3} \cos^3 \theta\right]_0^{\pi/2} = -\frac{5}{3}.$$

(c) On C_3, we integrate over two line segments and subtract.

$$\int_{C_3} (x^2 + y^2)\, dx = -\int_1^2 [(2 - t)^2 + 1]\, dt = -\int_1^2 (5 - 4t + t^2)\, dt$$

$$= \left[-5t + 2t^2 - \frac{t^3}{3}\right]_1^2 = -\frac{4}{3},$$

$$\int_{C_3} 2xy\, dy = \int_0^1 2t\, dt = [t^2]_0^1 = 1,$$

$$\int_{C_3} (x^2 + y^2)\, dx - 2xy\, dy = -\tfrac{7}{3}.$$

EXAMPLE 3.1.2. Find the work done on a particle by a force

$$\mathbf{F} = F_1(x, y)\mathbf{i} + F_2(x, y)\mathbf{j}$$

while the particle moves in the xy plane along the contour C.

Referring to Figure 3.1.2, let $z(t) = x(t) + iy(t)$, $0 \le t \le 1$, be a parametric representation of the contour C. Let $\mathbf{R}(t) = x(t)\mathbf{i} + y(t)\mathbf{j}$ be the displacement vector from the origin to the position of the particle on C at time t. Assuming that the force $\mathbf{F}$ is continuous, in a small interval of time

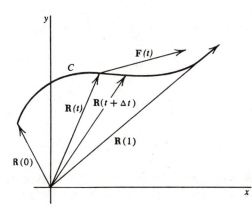

Figure 3.1.2.

F is very nearly constant. If the particle moved through the displacement ΔR in the time Δt, the work done by F would be approximately

$$dW = \mathbf{F} \cdot \Delta \mathbf{R} = F_1 \Delta x + F_2 \Delta y.$$

We *define* the total work done by F along C as

$$\lim_{\substack{n \to \infty \\ \max|\Delta x_k| \to 0 \\ \max|\Delta y_k| \to 0}} \sum_{k=1}^{n} [F_1(\xi_k, \eta_k) \Delta x_k + F_2(\xi_k, \eta_k) \Delta y_k],$$

provided this limit exists. Here, ξ_k and η_k are the coordinates of a point on the path which the particle occupies sometime during the time interval $\Delta t_k = t_{k+1} - t_k$. Clearly, if the work done exists, it is given by the line integral

$$W = \int_C F_1(x, y)\, dx + F_2(x, y)\, dy = \int_C \mathbf{F} \cdot d\mathbf{R}.$$

An examination of the definition gives us the following self-evident properties of the line integral:

1. $\displaystyle\int_C k_1 P(x, y)\, dx + k_2 Q(x, y)\, dy$

$$= k_1 \int_C P(x, y)\, dx + k_2 \int_C Q(x, y)\, dy$$

where k_1 and k_2 are any real constants.

2. $\displaystyle\int_C [P_1(x, y) + P_2(x, y)]\, dx + [Q_1(x, y) + Q_2(x, y)]\, dy$

$$= \int_C P_1(x, y)\, dx + Q_1(x, y)\, dy + \int_C P_2(x, y)\, dx + Q_2(x, y)\, dy.$$

3. $\displaystyle\int_{C_1 + C_2} P(x, y)\, dx + Q(x, y)\, dy$

$$= \int_{C_1} P(x, y)\, dx + Q(x, y)\, dy + \int_{C_2} P(x, y)\, dx + Q(x, y)\, dy.$$

4. $\displaystyle\int_{AC}^{B} P(x, y)\, dx + Q(x, y)\, dy = -\int_{BC}^{A} P(x, y)\, dx + Q(x, y)\, dy,$

where on the left we mean the line integral along C is taken from the point A to the point B, while on the right, the line integral along C is taken from B to A.

EXAMPLE 3.1.3. Compute $\int_{C_i} (x^2 - y^2)\, dx - 2xy\, dy$ between $(0, 0)$ and $(1, 1)$ along each of the following contours:

$C_1: x = y$; $C_2: y = x^2$; $C_3: z(t) = t, 0 \le t \le 1$, and $z(t) = 1 + i(t - 1)$, $1 \le t \le 2$.

(a) On C_1, $x = y$ and $dy = dx$. Therefore,

$$\int_{C_1} (x^2 - y^2)\, dx - 2xy\, dy = -2 \int_0^1 x^2\, dx = -\left[\frac{2x^3}{3}\right]_0^1 = -\frac{2}{3}.$$

(b) On C_2, $y = x^2$ and $dy = 2x\, dx$. Therefore,

$$\int_{C_2} (x^2 - y^2)\, dx - 2xy\, dy = \int_0^1 (x^2 - 5x^4)\, dx = \left[\frac{x^3}{3} - x^5\right]_0^1 = -\frac{2}{3}.$$

(c) On C_3, we can integrate over each of two line segments and add.

$$\int_{C_3} (x^2 - y^2)\, dx = \int_0^1 t^2\, dt = \left[\frac{t^3}{3}\right]_0^1 = \frac{1}{3},$$

$$-\int_{C_3} 2xy\, dy = -\int_1^2 (2t - 2)\, dt = -[t^2 - 2t]_1^2 = -1,$$

$$\int_{C_3} (x^2 - y^2)\, dx - 2xy\, dy = \tfrac{1}{3} - 1 = -\tfrac{2}{3}.$$

Unlike Example 3.1.1, in Example 3.1.3 the line integral seems to be independent of the contour. In fact, for this case we would get the same result for *any* contour joining $(0, 0)$ and $(1, 1)$. However, to show this by exhausting all possible contours would be an endless task. What is needed is a general criterion for determining when a line integral is independent of the contour joining the two fixed end points. The answer to this question is found partly in the next theorem.

Theorem 3.1.2. *Green's Lemma. Let C be a simple closed contour with the property that every straight line parallel to either coordinate axis intersects C in at most two points. Let P(x, y) and Q(x, y) be real-valued functions which are continuous and have continuous first partial derivatives in the closed set R consisting of C plus its interior. Then*

$$\int_C P(x, y)\, dx + Q(x, y)\, dy = \iint_R \left(\frac{\partial Q}{\partial x} - \frac{\partial P}{\partial y}\right) dx\, dy,$$

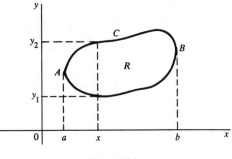

Figure 3.1.3.

where the line integral is evaluated in the counterclockwise direction, that is, in the direction which keeps the interior of C on the left of an observer moving along the curve in the direction of integration.

Proof: Consider all possible vertical lines in the plane. See Figure 3.1.3. Those which intersect the interior of C intersect C in two points. There is a "last line" to the left which intersects C in one point A, and a "last line" to the right which intersects C in one point B. The curve C has a parametric representation $z(t) = x(t) + iy(t)$, $0 \leq t \leq 1$, where the coordinates of A are $x(0) = x(1)$ and $y(0) = y(1)$, with a point on the curve moving first along the lower arc of C from A to B and then along the upper arc of C from B to A as t increases from 0 to 1.

Consider a vertical line which intersects the x axis in x and the curve C in two points with ordinates y_1 and y_2, where $y_1 < y_2$. Then (x, y_1) must be on the lower arc and (x, y_2) must be on the upper arc. Under the given hypotheses on continuity of the integrand, the integral

$$- \iint\limits_{R} \frac{\partial P}{\partial y} \, dy \, dx$$

can be computed by iteration, that is,

$$- \iint\limits_{R} \frac{\partial P}{\partial y} \, dy \, dx = - \int_{a}^{b} \left[\int_{y_1}^{y_2} \frac{\partial P}{\partial y} \, dy \right] dx$$

$$= - \int_{a}^{b} [P(x, y_2) - P(x, y_1)] \, dx$$

$$= \int_{a}^{b} P(x, y_1) \, dx + \int_{b}^{a} P(x, y_2) \, dx.$$

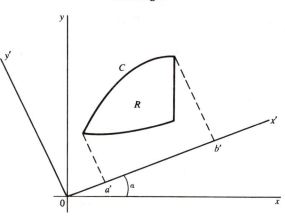

Figure 3.1.4.

The first term in this result is obviously the line integral along the lower arc of C from A to B, while the second term is the line integral along the upper arc of C from B to A. Combining, we have

$$- \iint_R \frac{\partial P}{\partial y}\, dx\, dy = \int_C P(x, y)\, dx.$$

A similar argument using horizontal lines and iteration of the integral first with respect to x and then with respect to y will establish the result

$$\iint_R \frac{\partial Q}{\partial x}\, dx\, dy = \int_C Q(x, y)\, dy.$$

Combining, we have

$$\iint_R \left(\frac{\partial Q}{\partial x} - \frac{\partial P}{\partial y} \right) dx\, dy = \int_C P(x, y)\, dx + Q(x, y)\, dy.$$

This form of Green's Lemma is entirely too restrictive to be useful. However, it can be generalized to apply to more general contours. Suppose, for example, that by a simple rotation of coordinate axes, we obtain coordinates x' and y' with respect to which Green's Lemma applies. See Figure 3.1.4. We can prove that the lemma applies in the original coordinates.

The coordinate transformation is

$$x' = x \cos \alpha + y \sin \alpha, \qquad x = x' \cos \alpha - y' \sin \alpha,$$
$$y' = -x \sin \alpha + y \cos \alpha, \qquad y = x' \sin \alpha + y' \cos \alpha.$$

Green's Lemma applies in x'-y' coordinates. Hence, letting

$$P'(x', y') = P(x' \cos \alpha - y' \sin \alpha, x' \sin \alpha + y' \cos \alpha),$$
$$Q'(x', y') = Q(x' \cos \alpha - y' \sin \alpha, x' \sin \alpha + y' \cos \alpha),$$

$$\int_C P(x, y) \, dx + Q(x, y) \, dy$$

$$= \int_C P'(x', y')(\cos \alpha \, dx' - \sin \alpha \, dy') + Q'(x', y')(\sin \alpha \, dx' + \cos \alpha \, dy')$$

$$= \cos \alpha \int_C P' \, dx' + Q' \, dy' + \sin \alpha \int_C Q' \, dx' - P' \, dy'$$

$$= \cos \alpha \iint_R \left(\frac{\partial Q'}{\partial x'} - \frac{\partial P'}{\partial y'} \right) dx' \, dy' - \sin \alpha \iint_R \left(\frac{\partial P'}{\partial x'} + \frac{\partial Q'}{\partial y'} \right) dx' \, dy'$$

$$= \cos \alpha \iint_R \left(\frac{\partial Q}{\partial x} \cos \alpha + \frac{\partial Q}{\partial y} \sin \alpha + \frac{\partial P}{\partial x} \sin \alpha - \frac{\partial P}{\partial y} \cos \alpha \right) dx \, dy$$

$$- \sin \alpha \iint_R \left(\frac{\partial P}{\partial x} \cos \alpha + \frac{\partial P}{\partial y} \sin \alpha - \frac{\partial Q}{\partial x} \sin \alpha + \frac{\partial Q}{\partial y} \cos \alpha \right) dx \, dy$$

$$= \iint_R \left(\frac{\partial Q}{\partial x} - \frac{\partial P}{\partial y} \right) dx \, dy.$$

Green's Lemma can also be extended to contours whose interiors can be divided into a finite number of regions satisfying one of the simpler forms by simple arcs lying in the interior of C. See Figure 3.1.5. Since the lemma

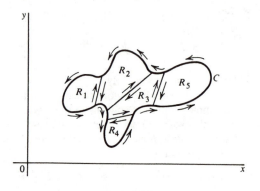

Figure 3.1.5.

applies to each of the regions $R_1, R_2, R_3, R_4,$ and $R_5,$ we have

$$\iint_R \left(\frac{\partial Q}{\partial x} - \frac{\partial P}{\partial y}\right) dx\, dy = \iint_{R_1+R_2+R_3+R_4+R_5} \left(\frac{\partial Q}{\partial x} - \frac{\partial P}{\partial y}\right) dx\, dy$$

$$= \iint_{C_1+C_2+C_3+C_4+C_5} P\, dx + Q\, dy,$$

where C_i is the complete boundary of R_i taken in the counterclockwise sense. Clearly, those arcs which lie in the interior of C are covered twice in this integration, once in each direction. Hence, they drop out of the calculation and we are left with just the line integral around C in the counterclockwise direction.

This discussion still leaves unanswered the issue of just how general C can be and still have Green's Lemma apply. Further, extensions are possible to contours which can be approximated by contours with simple properties, as in the above cases, but it is not our purpose to discuss here line integration in its most general form.

EXAMPLE 3.1.4. Show that the line integral of Example 3.1.3 is independent of the contour joining $(0, 0)$ and $(1, 1)$. Consider any two contours joining the end points which together form a simple closed contour for which one of the forms of Green's Lemma applies. See Figure 3.1.6. In this case, $P(x, y) = x^2 - y^2$ and $Q(x, y) = -2xy$. Hence,

$$\frac{\partial Q}{\partial x} = -2y, \frac{\partial P}{\partial y} = -2y, \text{ and } \frac{\partial Q}{\partial x} - \frac{\partial P}{\partial y} = 0.$$

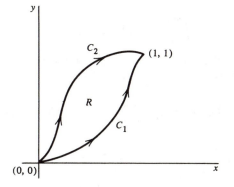

Figure 3.1.6.

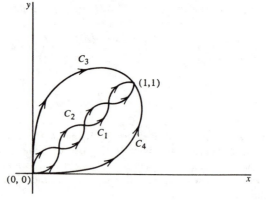

Figure 3.1.7.

Therefore,

$$\int_{C_1 - C_2} P(x, y)\, dx + Q(x, y)\, dy = \iint_R \left(\frac{\partial Q}{\partial x} - \frac{\partial P}{\partial y} \right) dx\, dy = 0,$$

$$\int_{C_1} P(x, y)\, dx + Q(x, y)\, dy = \int_{C_2} P(x, y)\, dx + Q(x, y)\, dy.$$

If the contours C_1 and C_2 do not form a simple closed contour (see Figure 3.1.7), we can introduce other contours C_3 and C_4 and show that

$$\int_{C_1} = \int_{C_3} = \int_{C_4} = \int_{C_2}$$

If within a certain domain D, the line integral

$$\int_{(x_0, y_0)}^{(x, y)} P(x, y)\, dx + Q(x, y)\, dy$$

is independent of the contour joining the end points, then the line integral defines a function

$$F(x, y) = \int_{(x_0, y_0)}^{(x, y)} P(x, y)\, dx + Q(x, y)\, dy.$$

Then,

$$\frac{\partial F}{\partial x} = \lim_{\Delta x \to 0} \frac{F(x + \Delta x, y) - F(x, y)}{\Delta x}$$

$$= \lim_{\Delta x \to 0} \frac{1}{\Delta x} \int_{(x, y)}^{(x + \Delta x, y)} P(x, y)\, dx + Q(x, y)\, dy.$$

Without loss of generality we can take a straight line path of integration between (x, y) and $(x + \Delta x, y)$. On this path, $dy = 0$, and hence,

$$\frac{\partial F}{\partial x} = \lim_{\Delta x \to 0} \frac{1}{\Delta x} \int_{(x,y)}^{(x+\Delta x,y)} P(x, y)\, dx$$

$$= \lim_{\Delta x \to 0} \frac{1}{\Delta x} P(x + \theta \Delta x, y)\, \Delta x,$$

using the mean value theorem for integrals. Here, $0 \leq \theta \leq 1$. Taking the limit, we have

$$\frac{\partial F}{\partial x} = P(x, y).$$

Similarly,

$$\frac{\partial F}{\partial y} = Q(x, y).$$

Therefore,

$$dF = \frac{\partial F}{\partial x}\, dx + \frac{\partial F}{\partial y}\, dy$$

$$= P(x, y)\, dx + Q(x, y)\, dy,$$

and

$$F(x, y) = \int_{(x_0,y_0)}^{(x,y)} dF$$

Let $f(z) = u(x, y) + iv(x, y)$ be analytic in some simply connected domain D, where we shall assume u and v to have continuous first partial derivatives. Then in D, $u_x = v_y$ and $u_y = -v_x$. Hence,

$$dv = v_x\, dx + v_y\, dy$$

$$= -u_y\, dx + u_x\, dy$$

and

$$v(x, y) = \int_{(x_0,y_0)}^{(x,y)} (-u_y\, dx + u_x\, dy).$$

Therefore, if the real part of $f(z)$ is given, the conjugate function can be found by line integration. Of course, $v(x, y)$ is only determined to within an additive constant. We shall show later that a necessary condition for $u(x, y)$ to be the real part of an analytic function in D is that it be harmonic in D.

Exercises 3.1

1. Show that for a particle in a force field $\mathbf{F} = F_1(x, y)\mathbf{i} + F_2(x, y)\mathbf{j}$, if the work done is independent of the path, the force is derivable from a potential function $\phi(x, y)$ such that $\mathbf{F} = \nabla\phi$.

2. Let C be a simple closed smooth curve. Let $\mathbf{V} = V_1(x, y)\mathbf{i} + V_2(x, y)\mathbf{j}$ be a vector with continuous first partial derivatives within and on C. Show that

$$\iint_R \nabla \cdot \mathbf{V} \, dx \, dy = \int_C \mathbf{V} \cdot \mathbf{N} \, ds,$$

where R is the interior of C, $\mathbf{N}$ is the unit outward normal on C and ds is the element of arc length on C. Hint: In Green's Lemma let $V_1 = Q$, $V_2 = -P$.

3. Let C be a simple closed smooth curve with interior R. Show that

$$A = \frac{1}{2} \int_C x \, dy - y \, dx$$

is the area enclosed by C.

4. Using the results of Exercise 2, show that

$$\iint_R (u\nabla^2 v + \nabla u \cdot \nabla v) \, dx \, dy = \int_C u\nabla v \cdot \mathbf{N} \, ds,$$

$$\iint_R (u\nabla^2 v - v\nabla^2 u) \, dx \, dy = \int_C (u\nabla v \cdot \mathbf{N} - v\nabla u \cdot \mathbf{N}) \, ds,$$

where u and v are functions with continuous first and second partial derivatives within and on C, a simple closed smooth curve with interior R.

5. Compute $\int_C (x + y) \, dx + (x - y) \, dy$ along the parabolic arc $y = x^2$ joining $(0, 0)$ and $(1, 1)$.

6. Compute $\displaystyle\int_{(0,-1)}^{(0,1)} \frac{x}{x^2 + y^2} \, dx + \frac{y}{x^2 + y^2} \, dy$ along (a)   the arc of the unit circle lying to the right of the origin, and (b)   the arc of the unit circle lying to the left of the origin.

7. Compute $\displaystyle\int_{(0,-1)}^{(0,1)} \frac{-y \, dx}{x^2 + y^2} + \frac{x \, dy}{x^2 + y^2}$ along the same two arcs as in Problem 6.

3.2. THE DEFINITE INTEGRAL

We are now ready to define the definite integral of a function of a complex variable. Let $f(z)$ be a function of the complex variable z defined on the contour C with end points at a and b. Let $a = z_1$ and $b = z_{n+1}$. Select

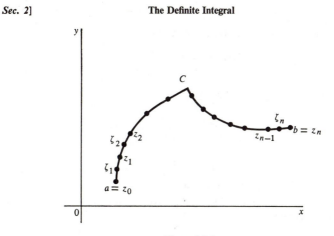

Figure 3.2.1.

$n - 1$ distinct points on C between a and b. See Figure 3.2.1. Let $\Delta z_k = z_{k+1} - z_k$ and let ζ_k be any point on C between z_k and z_{k+1}. We form the sum

$$\sum_{k=1}^{n} f(\zeta_k)\, \Delta z_k.$$

Next, we further subdivide C letting n increase without bound and consider

$$\lim_{n \to \infty} \sum_{k=1}^{n} f(\zeta_k)\, \Delta z_k,$$

wherein as n approaches infinity, $|\Delta z_k|$ approaches zero for all k. If this limit exists and is independent of the particular set of subdivisions used, we define its value to be the definite integral of $f(z)$ along C, that is,

$$\int_{a}^{b} f(z)\, dz = \lim_{\substack{n \to \infty \\ \max|\Delta z_k| \to 0}} \sum_{k=1}^{n} f(\zeta_k)\, \Delta z_k.$$

Theorem 3.2.1. *Let $f(z) = u(x, y) + iv(x, y)$ be continuous on a simple smooth arc C. Then $\int_C f(z)\, dz$ exists and is given by*

$$\int_C u\, dx - v\, dy + i \int_C u\, dy + v\, dx = \int_0^1 (u\dot{x} - v\dot{y})\, dt$$

$$+ i \int_0^1 (u\dot{y} + v\dot{x})\, dt,$$

where $z = x(t) + iy(t)$, $0 \le t \le 1$ is a parametric representation of C.

Proof: Let $z_k = x_k + iy_k$ and $\zeta_k = \xi_k + i\eta_k$. Then

$$f(\zeta_k)\,\Delta z_k = [u(\xi_k, \eta_k) + iv(\xi_k, \eta_k)](\Delta x_k + i\Delta y_k)$$
$$= u(\xi_k, \eta_k)\,\Delta x_k - v(\xi_k, \eta_k)\,\Delta y_k + i[u(\xi_k, \eta_k)\,\Delta y_k + v(\xi_k, \eta_k)\,\Delta x_k],$$

and since $|\Delta x_k| \leq |\Delta z_k|$ and $|\Delta y_k| \leq |\Delta z_k|$,

$$\lim_{\substack{n\to\infty \\ |\Delta z_k|\to 0}} \sum_{k=1}^{n} f(\zeta_k)\,\Delta z_k = \lim_{\substack{n\to\infty \\ |\Delta x_k|\to 0 |\Delta y_k|\to 0}} \sum_{k=1}^{n} [u(\xi_k, \eta_k)\,\Delta x_k - v(\xi_k, \eta_k)\,\Delta y_k]$$

$$+ i \lim_{\substack{n\to\infty \\ |\Delta x_k|\to 0 |\Delta y_k|\to 0}} \sum_{k=1}^{n} [u(\xi_k, \eta_k)\,\Delta y_k + v(\xi_k, \eta_k)\,\Delta x_k].$$

By Theorem 3.1.1, the limits on the right exist and are equal to line integrals which can be evaluated as follows:

$$\int_C f(z)\,dz = \int_C u\,dx - v\,dy + i\int_C u\,dy + v\,dx = \int_0^1 (u\dot{x} - v\dot{y})\,dt$$
$$+ i\int_0^1 (u\dot{y} + v\dot{x})\,dt.$$

EXAMPLE 3.2.1. Compute $\int_C \bar{z}\,dz$, where C is   (a) the straight line segment joining $(0, 0)$ and $(1, 1)$ and   (b) the arc of the circle $x^2 + (y - 1)^2 = 1$ joining $(0, 0)$ and $(1, 1)$.

(a) In this case, $x = y$, $\bar{z} = x - ix$, and $dz = dx + i\,dx$. Hence,

$$\int_C \bar{z}\,dz = \int_0^1 (x - ix)(dx + i\,dx) = \int_0^1 2x\,dx = 1.$$

(b) In this case, let us first translate the origin to $(0, 1)$, that is, let $x' = x$ and $y' = y - 1$. Then $x^2 + (y - 1)^2 = (x')^2 + (y')^2 = 1$, $\bar{z} = x' - i(y' + 1)$, and $dz = dz' = dx' + i\,dy'$. Therefore,

$$\int_C \bar{z}\,dz = \int_C [x' - i(y' + 1)]\,(dx' + i\,dy').$$

Now, introduce polar coordinates $x' = r\cos\theta$ and $y' = r\sin\theta$. On C $(x')^2 + (y')^2 = r^2 = 1$, $x' = \cos\theta$, $y' = \sin\theta$, and we have

$$\int_C \bar{z}\,dz = \int_{-\pi/2}^0 [\cos\theta - i(\sin\theta + 1)]\,(-\sin\theta + i\cos\theta)\,d\theta$$
$$= \int_{-\pi/2}^0 \cos\theta\,d\theta + i\int_{-\pi/2}^0 (1 + \sin\theta)\,d\theta$$
$$= 1 + i\left(\frac{\pi}{2} - 1\right).$$

Note that $\int \bar{z}\,dz$ is *not* independent of the path of integration.

EXAMPLE 3.2.2. Integrate $\int_a^b dz$. By definition

$$\int_a^b dz = \lim_{n \to \infty} [(z_2 - a) + (z_3 - z_2) + \cdots + (b - z_n)]$$

$$= b - a.$$

The integral is obviously independent of the path. Note that $\int_C dz = 0$ if C is *any* simple closed contour.

EXAMPLE 3.2.3. Integrate $\int_a^b z\, dz$. The integral along any simple contour exists by Theorem 3.2.1. Let Γ be any simple smooth arc joining α and β. Then if $z(t) = x(t) + iy(t)$, $0 \le t \le 1$, is a parameterization of Γ

$$\int_\alpha^\beta z(t)\, dz(t) = \int_0^1 [x(t)\dot{x} - y(t)\dot{y}]\, dt + i \int_0^1 [x(t)\dot{y} + y(t)\dot{x}]\, dt$$

$$= [x(1)]^2 - [x(0)]^2 - [y(1)]^2 + [y(0)]^2 - \int_0^1 [\dot{x}x - \dot{y}y]\, dt$$

$$+ 2i[x(1)y(1) - x(0)y(0)] - i \int_0^1 [\dot{x}y + \dot{y}x]\, dt$$

$$= \beta^2 - \alpha^2 - \int_\alpha^\beta z(t)\, dz(t),$$

where we have integrated by parts. Hence,

$$\int_\alpha^\beta z(t)\, dz(t) = \frac{\beta^2 - \alpha^2}{2}.$$

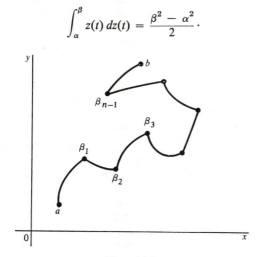

Figure 3.2.2.

Now, if the contour (see Figure 3.2.2) is made up of a finite number of simple smooth arcs with end points $a = \alpha_1, \beta_1; \alpha_2, \beta_2$; and so on, then

$$\int_a^b z \, dz = \tfrac{1}{2}(\beta_1^2 - a^2) + \tfrac{1}{2}(\beta_2^2 - \beta_1^2) + \cdots + \tfrac{1}{2}(b^2 - \beta_{n-1}^2)$$
$$= \tfrac{1}{2}(b^2 - a^2).$$

The result is independent of the simple contour joining a and b. Note that as a formal procedure, we could have written

$$\int_a^b z \, dz = \left[\frac{z^2}{2}\right]_a^b = \tfrac{1}{2}(b^2 - a^2).$$

This will be justified in Section 3.5.

EXAMPLE 3.2.4. Integrate

$$\int_C \frac{dz}{z},$$

where C is the unit circle taken in the counterclockwise direction. We may write

$$1/z = \bar{z}/|z|^2 = (x - iy)/(x^2 + y^2).$$

Then,

$$\int_C \frac{dz}{z} = \int_C \frac{x \, dx + y \, dy}{x^2 + y^2} + i \int_C \frac{x \, dy - y \, dx}{x^2 + y^2} = 2\pi i,$$

recalling the results of Exercises 3.1.6 and 3.1.7. Alternatively, we could have introduced polar coordinates $z = r(\cos\theta + i\sin\theta)$. On C, $r = 1$, and $z = \cos\theta + i\sin\theta = e^{i\theta}$. Then,

$$dz = (-\sin\theta + i\cos\theta)\, d\theta = i(\cos\theta + i\sin\theta)\, d\theta = ie^{i\theta}\, d\theta.$$

Therefore,

$$\int_C \frac{dz}{z} = \int_0^{2\pi} \frac{ie^{i\theta}\, d\theta}{e^{i\theta}} = i\int_0^{2\pi} d\theta = 2\pi i.$$

EXAMPLE 3.2.5. Let $f(z) = u(x, y) + iv(x, y)$ be analytic in a simply connected domain D, where u and v have continuous first partial derivatives. Prove that $\int_C f(z)\, dz = 0$ if C is any simple closed contour in D for which Green's Lemma applies. Let R be the interior of C. Then since $R \subseteq D$,

$$\int_C f(z)\, dz = \int_C u\, dx - v\, dy + i\int_C u\, dy + v\, dx$$
$$= \iint_R \left(\frac{\partial u}{\partial y} + \frac{\partial v}{\partial x}\right) dx\, dy + i\iint_R \left(\frac{\partial u}{\partial x} - \frac{\partial v}{\partial y}\right) dx\, dy = 0,$$

because u and v satisfy the Cauchy-Riemann equations in R.

Example 3.2.5 gives a weak form of Cauchy's theorem, which is one of the main results of the theory of analytic functions of a complex variable. At first sight the weakness of the present formulation might appear to be the restrictions on the contour. Closer examination would show that this is not the case. As a matter of fact, the theorem could be proved in its present form for any simple closed contour lying in D, by showing that the integral can be approximated arbitrarily well by a sequence of integrals over contours for which Green's Lemma applies. The real weakness in the present formulation is in the assumption that the first partial derivatives of u and v are continuous in D. The analyticity of $f(z)$ implies only the existence of the first partial derivatives. It is crucial to the development of the following that the hypothesis of continuity of the first partial derivatives be given up. We shall prove a stronger form of Cauchy's theorem in the next section.

We conclude this section by listing some obvious properties of the definite integral which follow immediately from the definition assuming that all functions involved are continuous on the simple contours of integration.

1. $\displaystyle\int_C k f(z)\,dz = k \int_C f(z)\,dz$, where k is a complex constant.

2. $\displaystyle\int_C [f(z) \pm g(z)]\,dz = \int_C f(z)\,dz \pm \int_C g(z)\,dz$.

3. $\displaystyle\int_{C_1+C_2} f(z)\,dz = \int_{C_1} f(z)\,dz + \int_{C_2} f(z)\,dz$.

4. $\displaystyle\int_{aC}^b f(z)\,dz = -\int_{bC}^a f(z)\,dz$.

5. $\displaystyle\left|\int_C f(z)\,dz\right| \le \int_C |f(z)|\,|dz| \le ML$, where $|dz| = \sqrt{\dot{x}^2 + \dot{y}^2}\,dt$,

$|f(z)| \le M$ on C, and L is the length of C.

The proof of Property 5 proceeds as follows. For a given n and a given subdivision of C

$$\left|\sum_{k=1}^n f(\zeta_k)\,\Delta z_k\right| \le \sum_{k=1}^n |f(\zeta_k)|\sqrt{(\Delta x_k)^2 + (\Delta y_k)^2}$$

$$\le \sum_{k=1}^n |f(\zeta_k)|\sqrt{\left(\frac{\Delta x_k}{\Delta t_k}\right)^2 + \left(\frac{\Delta y_k}{\Delta t_k}\right)^2}\,\Delta t_k.$$

Now, the function $|f(z)|(\dot{x}^2 + \dot{y}^2)^{1/2}$ is piecewise continuous on the contour C. Therefore, the sum on the right converges to $\int_C |f(z)|\sqrt{\dot{x}^2 + \dot{y}^2}\,dt$ as n approaches infinity and $\max \Delta t_k$ approaches zero. Furthermore, $\Delta t_k \to 0$ implies that $|\Delta z_k| \to 0$. Hence, as n approaches infinity

$$\sum_{k=1}^n f(\zeta_k)\,\Delta z_k \to \int_C f(z)\,dz.$$

Therefore,

$$\left| \int_C f(z)\, dz \right| \le \int_C |f(z)|\, |dz|.$$

Finally, since $|f(z)|$ is continuous on C, a bounded closed subset of the unextended plane, there is a bound M such that $|f(z)| \le M$ on C. Thus,

$$\int_C |f(z)|\, |dz| \le M \int_0^1 \sqrt{\dot{x}^2 + \dot{y}^2}\, dt = ML.$$

Exercises 3.2

1. Integrate each of the following along the straight line segment joining $(0, 0)$ and $(1, 1)$:

(a) $\displaystyle\int x\, dz;$ (b) $\displaystyle\int y\, dz;$ (c) $\displaystyle\int |z|\, dz;$ (d) $\displaystyle\int |z|^2\, dz.$

Ans. (a) $\frac{1}{2} + i\frac{1}{2};$ (b) $\frac{1}{2} + i\frac{1}{2};$ (c) $\sqrt{2}/2 + \sqrt{2}i/2$ (d) $\frac{2}{3} + 2i/3.$

2. Integrate $\int \bar{z}^2\, dz$ along each of the following contours joining $(0, 0)$ and $(1, 1)$:

(a) $y = x;$ (b) $y = x^2;$ (c) $x^2 + (y - 1)^2 = 1;$ (d) the broken line through $(0, 0)$, $(1, 0)$, and $(1, 1)$.

Ans. (a) $2/3 - 2i/3;$ (b) $14/15 - i/3;$ (c) $\pi - 2;$ (d) $4/3 + 2i/3.$

3. Show that $\int_C z^n\, dz = 0$, where C is the unit circle and n is any integer except -1.

4. Show that $\int_a^b \cos z\, dz = \sin b - \sin a$, independent of the simple contour joining a and b.

5. Prove that $f(z) = u(x, y) + iv(x, y)$, where u and v have continuous first partial derivatives in a simply connected domain D, is analytic if $\int_C f(z)\, dz = 0$ for any simple closed contour in D. Hint: Using Green's Lemma, prove that u and v satisfy the Cauchy-Riemann equations.

6. Evaluate $\displaystyle\int_C \frac{dz}{z - z_0}$, where z_0 is a constant and C is a circle $|z - z_0| = \rho$. Integrate in the counterclockwise direction.

3.3 CAUCHY'S THEOREM

In this section we shall prove Cauchy's theorem, which is perhaps the most important single result in the whole theory of analytic functions of a complex variable. The theorem can be stated as follows:

Theorem 3.3.1. *Cauchy's Theorem. Let $f(z)$ be analytic in a simply connected domain D. Then $\int_C f(z)\, dz = 0$, where C is any simple closed contour in D.*

We shall complete the proof in three steps. In Lemma 3.3.1, we shall establish that Cauchy's theorem holds in the special case where C is a triangle. We shall prove in Lemma 3.3.2 that the theorem holds for any closed polygon, a path consisting of a finite number of straight line segments, not necessarily simple. Finally, for a general contour, we shall approximate the path of integration by a sequence of polygonal paths. Passing to the limit as the number of sides of the polygon approaches infinity, we then establish the result for a general contour.

Lemma 3.3.1. Cauchy's theorem holds for any contour C consisting of a triangle lying in D.

Proof: Let C denote the triangular path of integration and let L be its length. We assume that $\int_C f(z)\,dz = I \neq 0$. We join the midpoints of the sides of C forming four smaller triangles as shown in Figure 3.3.1. Clearly,

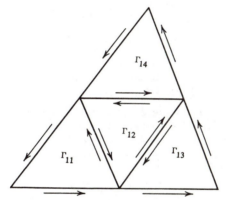

Figure 3.3.1.

$$\int_C f(z)\,dz = \int_{\Gamma_{11}} f(z)\,dz + \int_{\Gamma_{12}} f(z)\,dz + \int_{\Gamma_{13}} f(z)\,dz + \int_{\Gamma_{14}} f(z)\,dz,$$

where $\Gamma_{11}, \Gamma_{12}, \Gamma_{13}, \Gamma_{14}$ are the four small triangles. Since $\int_C f(z)\,dz \neq 0$, at least one of these integrals is not zero. Suppose J_1 is the maximum modulus of these four integrals. Then $\left|\int_C f(z)\,dz\right| \leq 4J_1$. Let Γ_1 be the triangle with modulus J_1. Then

$$\left|\int_C f(z)\,dz\right| \leq 4\left|\int_{\Gamma_1} f(z)\,dz\right|.$$

Next, we subject the triangle Γ_1 to the same treatment as C. After n steps, we arrive at the inequality

$$\left|\int_C f(z)\,dz\right| \leq 4^n\left|\int_{\Gamma_n} f(z)\,dz\right|.$$

Let $C^*, \Gamma_1^*, \Gamma_2^*, \ldots$ be the interiors of $C, \Gamma_1, \Gamma_2, \ldots$, respectively. Let $R_0 = C \cup C^*$, $R_1 = \Gamma_1 \cup \Gamma_1^*$, and so on. Then $R_0 \supseteq R_1 \supseteq R_2 \supseteq \ldots$, and the R's form a collection of closed nested sets. If $d(R_0)$ is the diameter of R_0, then $d(R_0) \leq L/2$. Likewise, $d(R_1) \leq L/4$, $d(R_2) \leq L/8, \ldots$. Clearly, $d(R_n) \leq L/2^{n+1} \to 0$, as $n \to \infty$. By Theorem 1.4.2, there is precisely one point z_0 common to all the R_n. Since $z_0 \in D$

$$f(z) = f(z_0) + f'(z_0)(z - z_0) + \eta(z, z_0)(z - z_0),$$

where given $\epsilon > 0$ there exists a $\delta(\epsilon)$ such that $|\eta| < \epsilon$ when $|z - z_0| < \delta$. Using the results of Examples 3.2.2 and 3.2.3, we have

$$\left| \int_{\Gamma_n} f(z)\, dz \right| = \left| \int_{\Gamma_n} \eta(z, z_0)(z - z_0)\, dz \right| \leq \epsilon \cdot \frac{L}{2^{n+1}} \cdot \frac{L}{2^n} = \frac{\epsilon L^2}{2 \cdot 4^n},$$

provided n is large enough that $d(R_n) < \delta$. Hence,

$$|I| \leq 4^n \left| \int_{\Gamma_n} f(z)\, dz \right| \leq \frac{\epsilon L^2}{2}.$$

But this contradicts the assumption that $I \neq 0$, because ϵ can be taken arbitrarily small.

Lemma 3.3.2. Cauchy's theorem holds for any contour C consisting of a closed polygon (not necessarily simple) lying in D.

Proof: It is sufficient to prove the lemma for a simple closed polygon lying in D, since the integral around any closed polygon in D reduces to a finite sum of integrals around simple closed polygons plus a finite number of integrals along straight line segments taken twice but in opposite directions. See Figure 3.3.2.

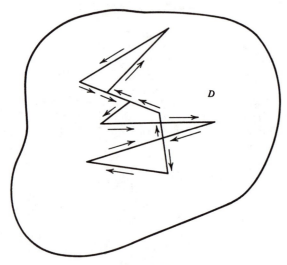

Figure 3.3.2.

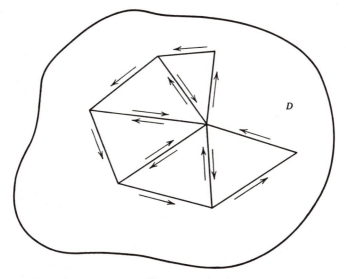

Figure 3.3.3.

For an arbitrary simple closed polygon the lemma follows by reducing the integration to a finite sum of integrals around triangles lying in D, thus reducing this case to Lemma 3.3.1. Suppose the polygon of n sides has been "triangulated." See Figure 3.3.3. By a triangulation we shall mean the joining of vertices of the polygon with line segments lying in the interior of the polygon until the integral around the polygon can be written as the finite sum of integrals around triangles. We see that in the process, the internal line segments are covered twice but in opposite directions, thus dropping out of the calculation, leaving just the integral around the original polygon.

It remains to show that every simple closed polygon can be triangulated. We shall do this by an induction on the number of sides of the polygon. First, we show that the triangulation is possible for $n = 4$, a quadrilateral. Let V_1 be any vertex of the quadrilateral. See Figure 3.3.4. If the quadri-

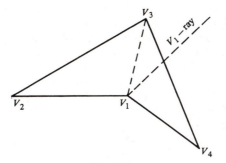

Figure 3.3.4.

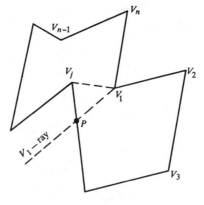

Figure 3.3.5.

lateral is nonconvex, the interior angle at V_1 should be $> \pi$. Let V_2, V_3, V_4 be the other vertices numbered consecutively around the quadrilateral. The sides $[V_1, V_2]$ and $[V_4, V_1]$ are the terminal sides of the interior angle at V_1. Consider all the V_1-rays, half-lines lying inside the interior angle at V_1. The union of all V_1-rays intersects the quadrilateral with $[V_1, V_2]$ and $[V_4, V_1]$ deleted. Therefore, there is a V_1-ray passing through V_3. The line segment $[V_1, V_3]$ along this ray triangulates the quadrilateral.

Next, we assume that the triangulation is possible for polygons of up to $n - 1$ sides. Consider a polygon of n sides. See Figure 3.3.5. If there exists a vertex V_1 from which another vertex V_k, $2 < k < n$ is visible, that is, there exists a V_1-ray along which a line segment $[V_1, V_k]$ can be drawn which intersects the polygon only at V_1 and V_k, then the polygon plus the line segment $[V_1, V_k]$ is the union of two polygons each of fewer than n sides. Since each of the smaller polygons can be triangulated, all we have to show is that there exists a V_k visible from V_1.

If the polygon is convex, all the interior angles are less than π, and numbering the vertices consecutively, we find that V_3 is visible from V_1. If the polygon is nonconvex, there is at least one interior angle greater than π. Assume that this angle is at V_1. Consider any V_1-ray contained in the interior angle at V_1. This ray must intersect the polygon in at least one point other than V_1. Let P be the nearest point of the polygon to V_1 on this ray. If P is a vertex, then there is a vertex visible from V_1. If P is an interior point of a side of the polygon, rotate the ray clockwise until either P becomes a vertex or until the segment $[V_1, P]$ intersects the polygon in some point other than V_1 and P. If this intersection is $[V_1, V_2]$ or $[V_n, V_1]$, then reverse the direction of rotation of the ray. The first intersection in the opposite direction cannot be $[V_n, V_1]$ or $[V_1, V_2]$, respectively, because the interior angle at V_1 is greater than π. Therefore, upon rotating the ray through

$[V_1, P]$ in one direction or the other, the first intersection with the polygon will be a vertex, a whole side, or a combination of vertices and/or whole sides. In any case, there is a vertex, not V_2 or V_n, on the ray which is visible from V_1. The line segment joining V_1 with this vertex divides the polygon into two polygons of fewer than n sides. This completes the proof of the lemma.

Proof of Cauchy's Theorem: Given $\epsilon > 0$ and a point ζ on C, there exists a δ such that $|f(z) - f(\zeta)| < \epsilon$ when $|z - \zeta| < \delta$. Also since ζ is an interior point of D, we can pick δ small enough that $\{z \mid |z - \zeta| < \delta\} \in D$. We construct an open covering of C as follows. Let $S(t) = \{z \mid |z - \zeta(t)| < \delta(t)/2\}$, where $\zeta(t)$, $0 \le t \le 1$, is a parametric representation of C. Of course, δ depends on ϵ. Now, $S = \bigcup_{0 \le t \le 1} S(t)$ is an open covering of C. By the Heine-Borel theorem there is a finite subcovering

$$S^* = \bigcup_{j=1}^{n} S(t_j), \quad 0 = t_1 < t_2 < \cdots < t_n < 1.$$

Let $\zeta_1 = \zeta(t_1)$, $\zeta_2 = \zeta(t_2), \ldots, \zeta_n = \zeta(t_n)$, $\zeta_{n+1} = \zeta(1) = \zeta(0)$. Consider ζ_k and ζ_{k+1}. Since $|\zeta_{k+1} - \zeta_k| < \delta_k/2 + \delta_{k+1}/2 < \max[\delta_k, \delta_{k+1}]$, for every point ζ on the line segment joining ζ_k and ζ_{k+1}, either $|f(\zeta) - f(\zeta_k)| < \epsilon$ or $|f(\zeta) - f(\zeta_{k+1})| < \epsilon$ depending on which of δ_{k+1} or δ_k is larger. Let

$$\zeta_k^* = \begin{cases} \zeta_k \text{ if } \delta_k \ge \delta_{k+1}, \\ \zeta_{k+1} \text{ if } \delta_{k+1} > \delta_k. \end{cases}$$

Now, $\sum_{k=1}^{n} f(\zeta_k^*)(\zeta_{k+1} - \zeta_k)$ is an approximating sum for $\int_C f(z)\,dz$. Consider the polygon Γ with vertices at $\zeta_1, \zeta_2, \ldots, \zeta_n$. By Lemma 3.3.2, $\int_\Gamma f(z)\,dz = 0$. Also,

$$\sum_{k=1}^{n} f(\zeta_k^*)(\zeta_{k+1} - \zeta_k) - \int_\Gamma f(z)\,dz = \sum_{k=1}^{n} \int_{\zeta_k}^{\zeta_{k+1}} [f(\zeta_k^*) - f(\zeta)]\,d\zeta.$$

Hence,

$$\left| \sum_{k=1}^{n} f(\zeta_k^*)(\zeta_{k+1} - \zeta_k) \right| \le \sum_{k=1}^{n} \int_{\zeta_k}^{\zeta_{k+1}} |f(\zeta_k^*) - f(\zeta)| \, |d\zeta| \le \epsilon L,$$

where L is the length of C. Since ϵ is arbitrary, this shows that

$$\int_C f(z)\,dz = \lim_{\substack{n \to \infty \\ |\zeta_{k+1} - \zeta_k| \to 0}} \sum_{k=1}^{n} f(\zeta_k^*)(\zeta_{k+1} - \zeta_k) = 0.$$

3.4. IMPLICATIONS OF CAUCHY'S THEOREM

As we indicated in the last section, Cauchy's theorem is one of the most important results in the theory of analytic functions of a complex variable. It has very far reaching implications which will take many pages to bring out. In a sense almost everything to follow will depend in one way or another on it. There are, however, many important immediate implications which we shall look at in this section.

Theorem 3.4.1. *If $f(z)$ is analytic within and on a simple closed contour C, then $\int_C f(z)\, dz = 0$.*

Proof: This result will follow if we can show that there is an open simply connected set D, containing C plus its interior, in which $f(z)$ is analytic. Let $z_0 \in C$. Then there exists a positive $\rho(z_0)$ such that $f(z)$ is analytic for $|z - z_0| < \rho(z_0)$. The collection of open sets of this form for all $z_0 \in C$ forms an open covering of the bounded closed set C. By the Heine-Borel theorem, there exists a finite subcovering of C,

$$S_i = \{z \mid |z - z_i| < \rho(z_i), z_i \in C\}, i = 1, 2, 3, \ldots, n,$$

such that

$$C \subseteq \bigcup_{i=1}^{n} S_i.$$

Let $D = C^* \cup \left(\bigcup_{i=1}^{n} S_i\right)$, where C^* is the interior of C. D is open, simply connected if the ρ's are not too large, and is a domain of analyticity of $f(z)$ which contains C. Therefore, by Cauchy's theorem $\int_C f(z)\, dz = 0$.

Actually, Theorem 3.4.1 is a special case of a more general theorem. If $f(z)$ is analytic within and on C, it is continuous in $R = C \cup C^*$. The following theorem can be proved.

Theorem 3.4.2. *If $f(z)$ is analytic within a simple closed contour C and is continuous in the closed region consisting of the union of C and its interior, then $\int_C f(z)\, dz = 0$.*

We shall not give the proof of this theorem, since it is rather involved and requires some delicate arguments involving approximation of the integral by sequences of contour integrals for contours lying inside C. For a proof for "starshaped" regions see E. Hille, *Analytic Function Theory*, I. Boston: Ginn and Company, 1959. We shall, nevertheless, use this result whenever it is needed.†

Theorem 3.4.3. *If $f(z)$ is analytic in a simply connected domain D, then $\int_a^b f(z)\, dz$ is independent of the simple contour joining a and b in D.*

† Also see Exercise 3.4.3.

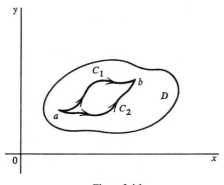

Figure 3.4.1.

Proof: Let C_1 and C_2 be any two simple contours joining a and b in D such that $C_1 - C_2$ forms a simple closed contour in D. See Figure 3.4.1. Then by Cauchy's theorem

$$\int_{C_1-C_2} f(z)\,dz = \int_{C_1} f(z)\,dz - \int_{C_2} f(z)\,dz = 0,$$

$$\int_{C_1} f(z)\,dz = \int_{C_2} f(z)\,dz.$$

If $C_1 - C_2$ does not form a simple closed contour, then we approximate C_1 and C_2 by polygonal paths P_1 and P_2 joining a and b. Then

$$\int_{P_1-P_2} f(z)\,dz = \int_{P_1} f(z)\,dz - \int_{P_2} f(z)\,dz = 0,$$

$$\int_{P_1} f(z)\,dz = \int_{P_2} f(z)\,dz,$$

by Lemma 3.3.2. Then by an argument similar to that used in the proof of Cauchy's theorem, we prove that

$$\int_{C_1} f(z)\,dz = \int_{C_2} f(z)\,dz,$$

by showing that $\int_{P_1} f(z)\,dz$ can be made arbitrarily close to $\int_{C_1} f(z)\,dz$ and similarly, for P_2 and C_2. Since C_1 and C_2 are *any* two contours joining a and b in D, the theorem is proved.

EXAMPLE 3.4.1. Compute $\int_{-i}^{i} \frac{1}{z}\,dz$ along the sides of the square passing through $(0, -1)$, $(1, -1)$, $(1, 1)$, $(0, 1)$. See Figure 3.4.2. It would be tedious

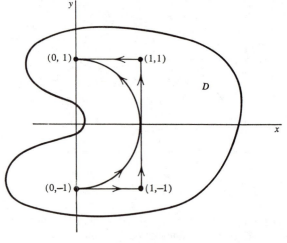

Figure 3.4.2.

to compute the integral along the actual contour proposed. However, there exists a simply connected domain D of analyticity of the integrand, containing $-i$ and i in which the integral is independent of the path. Therefore, we may compute the integral along another contour. For simplicity we take the arc of the unit circle joining $-i$ and i lying to the right of the origin. Letting $z = e^{i\theta}$, we have $z^{-1} = e^{-i\theta}$ and $dz = ie^{i\theta}\, d\theta$. Hence,

$$\int_{-i}^{i} \frac{1}{z}\, dz = i \int_{-\pi/2}^{\pi/2} d\theta = \pi i.$$

Note that D does not contain the origin. This prevents us from replacing the given contour by the arc of the unit circle lying to the left of the origin. As a matter of fact, on such an arc

$$\int_{-i}^{i} \frac{1}{z}\, dz = i \int_{3\pi/2}^{\pi/2} d\theta = -\pi i,$$

so that some care must be taken to make sure that the given contour and the actual contour of integration both lie in a domain of analyticity of the integrand which is *simply connected*.

EXAMPLE 3.4.2. Compute $\int_{a}^{b} e^{z}\, dz$. Since e^{z} is analytic everywhere in the unextended complex plane, we need not specify the path of integration. Instead, we may pick any contour which is convenient. For example, if

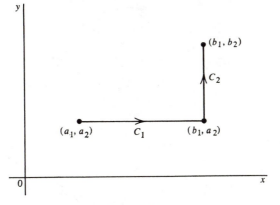

Figure 3.4.3.

$a = a_1 + ia_2$ and $b = b_1 + ib_2$, let us take the contour shown in Figure 3.4.3. On C_1, $z = x + ia_2$, $dz = dx$, and

$$\int_{C_1} e^z \, dz = e^{ia_2} \int_{a_1}^{b_1} e^x \, dx = e^{ia_2}(e^{b_1} - e^{a_1}).$$

On C_2, $z = b_1 + iy$, $dz = i \, dy$, and

$$\int_{C_2} e^z \, dz = ie^{b_1} \int_{a_2}^{b_2} (\cos y + i \sin y) \, dy$$

$$= ie^{b_1}(\sin b_2 - \sin a_2 - i \cos b_2 + i \cos a_2)$$

$$= e^{b_1}(e^{ib_2} - e^{ia_2}).$$

Therefore,

$$\int_a^b e^z \, dz = \int_{C_1} e^z \, dz + \int_{C_2} e^z \, dz = e^{b_1 + ib_2} - e^{a_1 + ia_2} = e^b - e^a.$$

Theorem 3.4.4. *Let C be a simple closed contour and let z_0 be any point in the interior of C. Then $\int_C \dfrac{1}{z - z_0} \, dz = \pm 2\pi i$ depending on the direction of integration around C.*

Proof: Refer to Figure 3.4.4. We shall define the direction which produces $2\pi i$ as the positive orientation of C and denote this by C_+. The

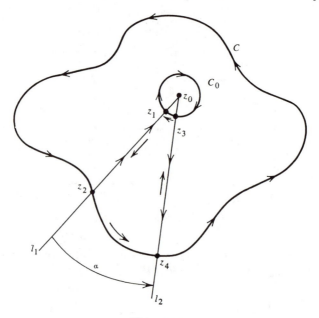

Figure 3.4.4.

negative orientation C_- will refer to the direction of integration which produces $-2\pi i$, that is,

$$\int_{C_+} \frac{1}{z - z_0} dz = 2\pi i, \qquad \int_{C_-} \frac{1}{z - z_0} dz = -2\pi i.$$

Since z_0 is an interior point of C^*, there exists a circle $C_0 = \{z \mid |z - z_0| = \rho_0\}$ lying entirely in C^*. We take any half-line l_1 originating in z_0. This half-line intersects C_0 in a point z_1. It also intersects C in at least one point. The point of C on l_1 nearest to z_0 we call z_2. Next, we rotate l_1 through a small angle α into the half-line l_2. Let z_3 be the intersection of l_2 with C_0 and z_4 be the nearest point of C to z_0 on l_2. We integrate $(z - z_0)^{-1}$ around two simple closed contours within and on which the integrand is analytic. The result for each will be zero by Theorem 3.4.1. Adding the two integrals will give us our result.

Starting at z_2 we integrate along l_1 to z_1. We then integrate along C_0 from z_1 to z_3 in the clockwise direction. Next, we integrate from z_3 to z_4 along l_2. Finally, we close the contour by integrating from z_4 to z_2 along C in such a way that z_0 is outside. This induces a definite orientation on C which we shall call the positive orientation for reasons which will appear later.

The result of the integration

$$\int_{[z_2,z_1]} \frac{dz}{z-z_0} + \oint_{z_1}^{z_3} \frac{dz}{z-z_0} + \int_{[z_3,z_4]} \frac{dz}{z-z_0} + \int_{z_4C_+}^{z_2} \frac{dz}{z-z_0} = 0.$$

Next, starting at z_2 we integrate to z_4 along C in the same direction induced by the above integration. Then we integrate from z_4 to z_3 along l_2. We integrate from z_3 to z_1 along C_0 in the clockwise direction, and complete the circuit by integrating from z_1 to z_2 along l_1. Again z_0 is outside this simple closed contour and the result is

$$\int_{[z_1,z_2]} \frac{dz}{z-z_0} + \oint_{z_3C_0}^{z_1} \frac{dz}{z-z_0} + \int_{[z_4,z_3]} \frac{dz}{z-z_0} + \int_{z_2C_+}^{z_4} \frac{dz}{z-z_0} = 0.$$

Adding the two results, the integrals along the straight line segments cancel and we have

$$\oint_{C_0} \frac{dz}{z-z_0} + \int_{C_+} \frac{dz}{z-z_0} = 0.$$

We know from Exercise 3.2.6 that the first of these integrals has the value $-2\pi i$. Hence,

$$\int_{C_+} \frac{dz}{z-z_0} = 2\pi i.$$

If we had integrated around C_0 in the counterclockwise direction, we would have induced the opposite direction on C and our result would be

$$\int_{C_-} \frac{dz}{z-z_0} = -2\pi i.$$

In most cases it will be completely obvious which is the positive direction on a simple closed contour. It is, in general, the direction which tends to increase the argument of $z - z_0$ when z_0 is a fixed point inside the curve and z moves along C. As a matter of fact, we shall later prove that $\int_C \frac{dz}{z-z_0}$ is equal to i times the total change in the argument of $z - z_0$ as z moves along C.

Theorem 3.4.5. *Let C_2 be a simple closed contour. Let C_1 be a simple closed contour lying in C_2^*. If $f(z)$ is analytic on C_1, C_2, and between C_1 and C_2, that is, the intersection of the interior of C_2 with the exterior of C_1, then*

$$\int_{C_{1+}} f(z)\, dz = \int_{C_{2+}} f(z)\, dz.$$

Proof: Let z_0 be a point in C_1^*. There exists a circle

$$C_0 = \{z \mid |z - z_0| = \rho\}$$

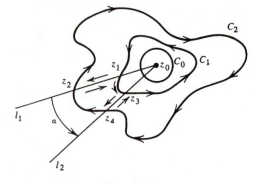

Figure 3.4.5.

lying entirely in C_1^*. Consider any half-line l_1 originating at z_0. See Figure 3.4.5. Let z_1 be the point of C_1 on l_1 farthest from z_0. Let z_2 be the point of C_2 on l_1 nearest to z_0. We now rotate l_1 through a small angle α in the counterclockwise direction into the half-line l_2. Let z_3 be the point of C_1 on l_2 farthest from z_0 and z_4 be the point of C_2 on l_2 nearest to z_0. As in the proof of Theorem 3.4.4, the counterclockwise direction of integration on C_0 induces a positive direction on C_1 and C_2. We perform two contour integrations each of which gives the result zero since in each case $f(z)$ is analytic within and on the contour. Adding the integrals we get the desired result.

Starting at z_1 we integrate to z_3 along C_1 in the positive sense. Then we integrate from z_3 to z_4 along l_2. Next, we integrate from z_4 to z_2 along C_2 in the negative sense. Finally, we integrate from z_2 to z_1 along l_1. The result is

$$\int_{z_1 C_{1+}}^{z_3} f(z)\, dz + \int_{[z_3, z_4]} f(z)\, dz + \int_{z_4 C_{2-}}^{z_2} f(z)\, dz + \int_{[z_2, z_1]} f(z)\, dz = 0.$$

The second integration proceeds as follows: from z_3 to z_1 along C_1 in the positive sense, from z_1 to z_2 along l_1, from z_2 to z_4 along C_2 in the negative sense, and from z_4 to z_3 along l_2. The result is

$$\int_{z_3 C_{1+}}^{z_1} f(z)\, dz + \int_{[z_1, z_2]} f(z)\, dz + \int_{z_2 C_{2-}}^{z_4} f(z)\, dz + \int_{[z_4, z_3]} f(z)\, dz = 0.$$

Adding, the integrals along the straight line segments drop out, and we have

$$\int_{C_{1+}} f(z)\, dz + \int_{C_{2-}} f(z)\, dz = 0,$$

$$\int_{C_{1+}} f(z)\, dz = - \int_{C_{2-}} f(z)\, dz = \int_{C_{2+}} f(z)\, dz.$$

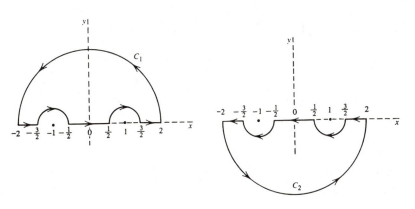

Figure 3.4.6.

EXAMPLE 3.4.3. Integrate $\int_{C_+} \dfrac{dz}{z^2 + 2z + 2}$, where C is a square with corners at $(0, 0)$, $(-2, 0)$, $(-2, -2)$, $(0, -2)$. We notice that the integrand is analytic except at $z = -1 \pm i$. The singularity at $-1 - i$ is inside the contour. By Theorem 3.4.5, we can replace the contour by the circle $|z + 1 + i| = 1/2$ taken in the positive sense. We can write the integrand as follows:

$$\frac{1}{z^2 + 2z + 2} = \frac{1}{(z + 1 - i)(z + 1 + i)}$$

$$= \frac{1}{2}\left[\frac{-i}{z + 1 - i} + \frac{i}{z + 1 + i}\right].$$

If Γ is the circle, $\dfrac{1}{2} \int_{\Gamma} \dfrac{i}{z + 1 - i} dz = 0$ by Cauchy's theorem. On Γ, $z = -1 - i + (1/2)e^{i\theta}$, $0 \le \theta \le 2\pi$, $dz = (1/2)ie^{i\theta}\, d\theta$ and

$$\int_{C_+} \frac{dz}{z^2 + 2z + 2} = \frac{1}{2} i \int_{\Gamma_+} \frac{1}{z + 1 + i} dz = \frac{1}{2} i \int_0^{2\pi} i\, d\theta = -\pi.$$

EXAMPLE 3.4.4. Integrate $\int_{C_+} \dfrac{dz}{z^2 - 1}$, where C is the circle $|z| = 2$. The singularities of the integrand are at $z = \pm 1$, both of which are inside the contour. Consider the two contours shown in Figure 3.4.6. From Cauchy's theorem, we have

$$\int_{C_{1+}} \frac{1}{z^2 - 1}\, dz = \int_{C_{2+}} \frac{1}{z^2 - 1}\, dz = 0.$$

Adding, we have

$$\int_{C_1+C_2} \frac{1}{z^2-1}\,dz = \int_{C_+} \frac{1}{z^2-1}\,dz + \int_{\Gamma_{1-}} \frac{1}{z^2-1}\,dz$$
$$+ \int_{\Gamma_{2-}} \frac{1}{z^2-1}\,dz = 0,$$

$$\int_{C_+} \frac{1}{z^2-1}\,dz = \int_{\Gamma_{1+}} \frac{1}{z^2-1}\,dz + \int_{\Gamma_{2+}} \frac{1}{z^2-1}\,dz,$$

where $\Gamma_1 = \{z \mid |z+1| = 1/2\}$ and $\Gamma_2 = \{z \mid |z-1| = 1/2\}$. We can write the integrand as follows:

$$\frac{1}{z^2-1} = \frac{1}{(z-1)(z+1)} = \frac{1}{2}\left[\frac{1}{z-1} - \frac{1}{z+1}\right].$$

By Cauchy's theorem

$$\int_{\Gamma_1} \frac{1}{z-1}\,dz = \int_{\Gamma_2} \frac{1}{z+1}\,dz = 0.$$

Hence,

$$\int_{C_+} \frac{1}{z^2-1}\,dz = -\frac{1}{2}\int_{\Gamma_{1+}} \frac{1}{z+1}\,dz + \frac{1}{2}\int_{\Gamma_{2+}} \frac{1}{z-1}\,dz$$
$$= -\pi i + \pi i = 0.$$

Exercises 3.4

1. Without calculating explicitly, verify that each of the following integrals is zero where the path of integration in each case is the unit circle in the positive sense:

(a) $\displaystyle\int_C \cos z\,dz$;

(b) $\displaystyle\int_C \frac{e^z}{z+2}\,dz$;

(c) $\displaystyle\int_C \tanh z\,dz$;

(d) $\displaystyle\int_C \frac{1}{z^2+5z+6}\,dz$;

(e) $\displaystyle\int_C \operatorname{Log}(z+2i)\,dz$;

(f) $\displaystyle\int_C \sec z\,dz$.

2. Evaluate (a) $\int_a^b (z^2+2z+2)\,dz$; (b) $\int_a^b \cos z\,dz$.

3. Prove that if $f(z)$ is analytic for $|z| < 1$ and continuous for $|z| \le 1$, then $\int_C f(z)\,dz = 0$, where C is the unit circle. Hint: Show that

$$\int_0^{2\pi} f(re^{i\theta})rie^{i\theta}\,d\theta = 0 \qquad \text{for} \quad 0 \le r < 1$$

and pass to the limit as r approaches one.

4. Show by counterexample that the following is *not* a correct statement: If $f(z)$ is analytic on and outside of a simple closed contour C, including infinity, then $\int_C f(z)\, dz = 0$. Find a sufficient condition to be added to this statement to make it a correct theorem. Hint: Replace the contour by a circle $z = Re^{i\theta}$ with R large and then make the change of variables $\zeta = \dfrac{1}{z}$.

5. Evaluate $\displaystyle\int_C \frac{dz}{z(z - z_0)}$, where C is the unit circle taken in the positive sense and (a) $|z_0| < 1$; (b) $|z_0| > 1$.

6. Evaluate $\displaystyle\int_C \frac{dz}{z^3 - 1}$, where C is the circle $|z - 1| = 1$ taken in the positive sense.

3.5. FUNCTIONS DEFINED BY INTEGRATION

In considering functions defined by integration, we shall consider two kinds, those with the variable as a limit of integration and those with the variable as a parameter in the integrand. For the first kind we shall want the value of the integral to be independent of the path joining the end points. The simplest set of sufficient conditions for this will be to require that the integrand be analytic in a simply connected domain containing the path of integration. We have the following theorem:

Theorem 3.5.1. *Let $f(z)$ be analytic in a simply connected domain D. Then $\int_{z_0}^{z} f(\zeta)\, d\zeta$ defines a function $F(z)$ in D provided z_0 is a fixed point and the path of integration lies in D. Furthermore, $F(z)$ is analytic in D, where $F'(z) = f(z)$. If a and b are any points in D, then $\int_a^b f(z)\, dz = F(b) - F(a)$, provided the path of integration lies in D.*

Proof: Under the hypothesis that $f(z)$ is analytic in the simply connected domain D, $\int_{z_0}^{z} f(\zeta)\, d\zeta$ is independent of the path and therefore defines a single-valued function of the upper limit. Let $F(z) = \int_{z_0}^{z} f(\zeta)\, d\zeta$. Then

$$\frac{F(z + \Delta z) - F(z)}{\Delta z} = \frac{1}{\Delta z}\left[\int_{z_0}^{z+\Delta z} f(\zeta)\, d\zeta - \int_{z_0}^{z} f(\zeta)\, d\zeta\right]$$

$$= \frac{1}{\Delta z}\int_{z}^{z+\Delta z} f(\zeta)\, d\zeta,$$

provided $|\Delta z|$ is sufficiently small that $z + \Delta z$ is in D. Without loss of generality, we may take a straight line path between z and $z + \Delta z$ if $|\Delta z|$ is small enough. Now, by an obvious calculation

$$f(z) = \frac{1}{\Delta z}\int_{z}^{z+\Delta z} f(z)\, d\zeta.$$

Therefore,

$$\left| \frac{F(z + \Delta z) - F(z)}{\Delta z} - f(z) \right| \le \frac{1}{|\Delta z|} \int_z^{z+\Delta z} |f(\zeta) - f(z)| \, |d\zeta|.$$

Given $\epsilon > 0$ there exists a δ such that $|f(\zeta) - f(z)| < \epsilon$ for all ζ satisfying $|\zeta - z| \le |\Delta z| < \delta$. Therefore, when $|\Delta z| < \delta$,

$$\left| \frac{F(z + \Delta z) - F(z)}{\Delta z} - f(z) \right| \le \frac{1}{|\Delta z|} \epsilon |\Delta z| = \epsilon.$$

But ϵ is arbitrary, which proves that

$$F'(z) = \lim_{\Delta z \to 0} \frac{F(z + \Delta z) - F(z)}{\Delta z} = f(z).$$

Finally, since D is simply connected, we can write

$$\int_a^b f(z) \, dz = \int_a^{z_0} f(z) \, dz + \int_{z_0}^b f(z) \, dz$$

$$= \int_{z_0}^b f(z) \, dz - \int_{z_0}^a f(z) \, dz = F(b) - F(a).$$

The discerning reader will no doubt compare this result with the "Fundamental Theorem of the Integral Calculus." Indeed, the function $F(z) = \int_{z_0}^z f(\zeta) \, d\zeta$ is the indefinite integral of $f(z)$, and the result states that to evaluate the definite integral of a function analytic in a simply connected domain one may find another function, the indefinite integral, analytic in the same domain, whose derivative is the original function, and then compute the difference between the values of the indefinite integral at the upper and lower limits.

EXAMPLE 3.5.1. Compute $\int_a^b z \cos z^2 \, dz$. The function $z \cos z^2$ is analytic in the unextended plane where $(1/2) \sin z^2$ is also analytic and

$$\frac{d}{dz} (\tfrac{1}{2} \sin z^2) = z \cos z^2.$$

Hence, $\int_a^b z \cos z^2 \, dz = \tfrac{1}{2}[\sin b^2 - \sin a^2]$.

EXAMPLE 3.5.2. Compute $\displaystyle\int_{-i}^i \frac{dz}{z}$ along the arc of the unit circle lying to the right of the origin. This can be done using the result of Theorem 3.5.1. The function $\log z$ has the derivative $1/z$. However, $\log z$ is infinitely many valued. We must pick a branch of $\log z$ which is analytic on the path of integration. We therefore take the principal value $\text{Log } z = \ln |z| + i \arg z$, $-\pi < \arg z \le \pi$. This branch has its branch cut along the negative real

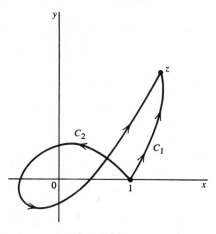

Figure 3.5.1.

axis and is analytic on the path of integration. Therefore, we have

$$\int_{-i}^{i} \frac{dz}{z} = \operatorname{Log} z \Big|_{-i}^{i} = \ln 1 + i\frac{\pi}{2} - \left(\ln 1 - i\frac{\pi}{2}\right) = i\pi.$$

Note that the branch $\log z = \ln |z| + i \arg z$, $0 \le \arg z < 2\pi$, with its branch cut crossing the path of integration, would have given the incorrect result

$$\int_{-i}^{i} \frac{dz}{z} = \log z \Big|_{-i}^{i} = \ln 1 + i\frac{\pi}{2} - \left(\ln 1 + 3i\frac{\pi}{2}\right) = -i\pi.$$

However, this would have been a suitable branch if the path of integration had been to the left of the origin. In other words, care must be taken to select an indefinite integral which is analytic in a simply connected domain containing the path of integration.

EXAMPLE 3.5.3. Identify the function $\displaystyle\int_{1}^{z} \frac{d\zeta}{\zeta}$, where the path of integration is any piecewise smooth curve which does not pass through the origin. Consider the values of the integral for two different paths, one which loops the origin in the positive sense and one which does not. See Figure 3.5.1. As in Example 3.5.2, we can evaluate the integral along C_1 using the principal value of the logarithm function, that is,

$$\int_{1 \ C_1}^{z} \frac{d\zeta}{\zeta} = \operatorname{Log} \zeta \Big|_{1}^{z} = \operatorname{Log} z - \operatorname{Log} 1 = \operatorname{Log} z.$$

Clearly,

$$\int_{1 \ C_1}^{z} \frac{d\zeta}{\zeta} = \int_{1 \ C_2}^{z} \frac{d\zeta}{\zeta} - 2\pi i.$$

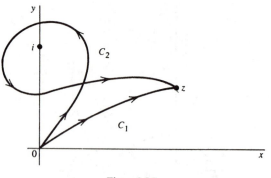

Figure 3.5.2.

In fact, a little reflection will convince the reader that the value for any other path will differ from Log z by a multiple of $2\pi i$ depending on how many times and in what sense the path loops the origin. Hence, the given function is an integral representation of the logarithm function. The branch point at the origin is the only singularity of the integrand.

EXAMPLE 3.5.4. Identify the function $\int_0^z \dfrac{d\zeta}{1 + \zeta^2}$ where the path of integration is any piecewise smooth curve not passing through $\pm i$. The only singularities of the integrand are $\pm i$. If D is a simply connected domain not containing $\pm i$, then $F(z) = \int_1^z \dfrac{d\zeta}{1 + \zeta^2}$ is analytic in D with derivative $F'(z) = 1/(1 + z^2)$. Therefore, in D, $F(z)$ and $\tan^{-1} z$ differ by at most a constant. If $z = x > 0$ and the path of integration does not loop either singularity, then $F(x) = \int_0^x \dfrac{1}{1 + \xi^2}\, d\xi = \tan^{-1} x$, where we have taken the principal value of the real function $\tan^{-1} x$. Now, consider two paths which connect the origin to z, one which loops i once in the positive direction and one which does not loop either singularity. See Figure 3.5.2. The integrals along C_1 and C_2 differ by

$$\int_{|z-i|=1/2} \frac{d\zeta}{1 + \zeta^2} = \frac{1}{2}\int_{|z-i|=1/2} \frac{i\,d\zeta}{\zeta + i} - \frac{1}{2}\int_{|z-i|=1/2} \frac{i\,d\zeta}{\zeta - i}$$
$$= 0 + \pi = \pi.$$

A loop around $-i$ in the positive direction would contribute

$$\int_{|z+i|=1/2} \frac{d\zeta}{1 + \zeta^2} = \frac{1}{2}\int_{|z+i|=1/2} \frac{i\,d\zeta}{\zeta + i} - \frac{1}{2}\int_{|z+i|=1/2} \frac{i\,d\zeta}{\zeta - i}$$
$$= -\pi + 0 = -\pi.$$

Therefore, the values for the integral will differ from the principal value of the $\tan^{-1} z$ by a multiple of π, depending on how many times and in what sense the path of integration loops i, $-i$, or both. Hence, the integral represents the many-valued function $\tan^{-1} z$. The branch points at $\pm i$ are the singularities of $1/(1 + z^2)$.

We now turn to functions defined by integrals of functions containing a parameter. Let $f(z, \zeta)$ be a complex-valued function of two complex variables when z is in some domain D and ζ is on some simple contour C. Then, clearly,

$$F(z) = \int_C f(z, \zeta)\, d\zeta$$

is a function of z in D provided that for each $z \in D$, $f(z, \zeta)$ is continuous as a function of ζ on C. We shall say that $f(z, \zeta)$ is continuous in D and on C if C has a continuous parametric representation $\zeta(t) = \xi(t) + i\eta(t)$, $0 \leq t \leq 1$, and $f(z, \zeta) = u[x, y, \xi(t), \eta(t)] + iv[x, y, \xi(t), \eta(t)]$, where u and v are continuous functions of x, y, and t when $(x, y) \in D$ and $0 \leq t \leq 1$.

Theorem 3.5.2. *Let $f(z, \zeta)$ be continuous for z in a domain D and ζ on a simple contour C. Then $F(z) = \int_C f(z, \zeta)\, d\zeta$ is continuous in D.*

Proof: Let $z_0 \in D$. Then there exists a ρ such that $\{z \mid |z - z_0| \leq \rho\}$ is in D. Let R be a region in euclidean three space with coordinates (x, y, t) bounded by the cylinder $(x - x_0)^2 + (y - y_0)^2 = \rho^2$ and the planes $t = 0$ and $t = 1$. This region is a compact set in three space and hence $u[x, y, \xi(t), \eta(t)]$ and $v[x, y, \xi(t), \eta(t)]$ are uniformly continuous in R. Hence, given $\epsilon > 0$, there exists a $\delta(\epsilon)$ independent of t such that

$$|u[x, y, \xi(t), \eta(t)] - u[x_0, y_0, \xi(t), \eta(t)]| < \epsilon$$
$$|v[x, y, \xi(t), \eta(t)] - v[x_0, y_0, \xi(t), \eta(t)]| < \epsilon,$$

for all (x, y) such that $(x - x_0)^2 + (y - y_0)^2 < \delta^2$. Then, letting $u_0 = u[x_0, y_0, \xi, \eta]$, $v_0 = v[x_0, y_0, \xi, \eta]$, and $M = \max [|\xi|, |\dot\eta|]$,

$$|F(z) - F(z_0)| = \left| \int_C [f(z, \zeta) - f(z_0, \zeta)]\, d\zeta \right|$$

$$= \left| \int_0^1 [(u - u_0)\dot\xi - (v - v_0)\dot\eta]\, dt + i \int_0^1 [(v - v_0)\dot\xi \right.$$
$$\left. + (u - u_0)\dot\eta]\, dt \right|$$

$$\leq 2M \left| \int_0^1 |u - u_0|\, dt + \int_0^1 |v - v_0|\, dt \right|$$

$$\leq 4M\epsilon.$$

But since ϵ is arbitrary, this shows that

$$\lim_{z \to z_0} F(z) = \lim_{z \to z_0} \int_C f(z, \zeta) \, d\zeta = \int_C f(z_0, \zeta) \, d\zeta = F(z_0).$$

Since z_0 was *any* point of the open set D, we have shown that $F(z)$ is continuous in D.

A partial derivative of a function of two complex variables is defined in the obvious way. Let $f(z, \zeta)$ be defined in some ϵ-neighborhood of z_0 for a fixed value of ζ. Then

$$\left(\frac{\partial f}{\partial z} \right)_{z=z_0} = f_z(z_0, \zeta) = \lim_{z \to z_0} \frac{f(z, \zeta) - f(z_0, \zeta)}{z - z_0},$$

provided this limit exists and is finite.

Theorem 3.5.3. *Let $f(z, \zeta)$ and $f_z(z, \zeta)$ be continuous for z in a domain D and ζ on a simple contour C. Then $F(z) = \int_C f(z, \zeta) \, d\zeta$ is analytic in D, and*

$$F'(z) = \int_C \frac{\partial f}{\partial z} \, d\zeta.$$

Proof: Let $F(z) = U(x, y) + iV(x, y)$ and $f(z, \zeta) = u[x, y, \xi(t), \eta(t)] + iv[x, y, \xi(t), \eta(t)]$. Then

$$U(x, y) = \int_0^1 (u\dot{\xi} - v\dot{\eta}) \, dt,$$

$$V(x, y) = \int_0^1 (u\dot{\eta} + v\dot{\xi}) \, dt.$$

If $z_0 \in D$

$$
\begin{aligned}
U_x(x_0, y_0) &= \lim_{\Delta x \to 0} \frac{U(x_0 + \Delta x, y_0) - U(x_0, y_0)}{\Delta x} \\
&= \lim_{\Delta x \to 0} \frac{1}{\Delta x} \int_0^1 [u(x_0 + \Delta x, y_0, \xi, \eta)\dot{\xi} - v(x_0 + \Delta x, y_0, \xi, \eta)\dot{\eta} \\
&\qquad\qquad\qquad - u(x_0, y_0, \xi, \eta)\dot{\xi} + v(x_0, y_0, \xi, \eta)\dot{\eta}] \, dt \\
&= \lim_{\Delta x \to 0} \int_0^1 [u_x(x_0 + \theta_1 \Delta x, y_0, \xi, \eta)\dot{\xi} \\
&\qquad\qquad\qquad - v_x(x_0 + \theta_2 \Delta x, y_0, \xi, \eta)\dot{\eta}] \, dt \\
&= \int_0^1 [u_x(x_0, y_0, \xi, \eta)\dot{\xi} - v_x(x_0, y_0, \xi, \eta)\dot{\eta}] \, dt.
\end{aligned}
$$

The justification for taking the limit under the integral sign is supplied by the continuity of the first partial derivatives and Theorem 3.5.2. Similarly,

$$U_y(x_0, y_0) = \int_0^1 [u_y(x_0, y_0, \xi, \eta)\dot{\xi} - v_y(x_0, y_0, \xi, \eta)\dot{\eta}] \, dt,$$

$$V_x(x_0, y_0) = \int_0^1 [u_x(x_0, y_0, \xi, \eta)\dot{\eta} + v_x(x_0, y_0, \xi, \eta)\dot{\xi}] \, dt,$$

$$V_y(x_0, y_0) = \int_0^1 [u_y(x_0, y_0, \xi, \eta)\dot{\eta} + v_y(x_0, y_0, \xi, \eta)\dot{\xi}] \, dt.$$

Since $f(z, \zeta)$ is analytic at z_0, $u_x = v_y$, $u_y = -v_x$, and

$$U_x(x_0, y_0) = \int_0^1 (u_x\dot{\xi} - v_x\dot{\eta}) \, dt = \int_0^1 (v_y\dot{\xi} + u_y\dot{\eta}) \, dt = V_y(x_0, y_0),$$

$$U_y(x_0, y_0) = \int_0^1 (u_y\dot{\xi} - v_y\dot{\eta}) \, dt = - \int_0^1 (v_x\dot{\xi} + u_x\dot{\eta}) \, dt = -V_x(x_0, y_0).$$

Hence, $F(z)$ is analytic at z_0 and

$$\begin{aligned}
F'(z_0) &= U_x(x_0, y_0) + iV_x(x_0, y_0) \\
&= \int_0^1 (u_x\dot{\xi} - v_x\dot{\eta}) \, dt + i \int_0^1 (u_x\dot{\eta} + v_x\dot{\xi}) \, dt \\
&= \int_0^1 (u_x + iv_x)(\dot{\xi} + i\dot{\eta}) \, dt \\
&= \int_C f_z(z_0, \zeta) \, d\zeta.
\end{aligned}$$

Since z_0 is *any* point in D, the result holds throughout D.

EXAMPLE 3.5.5. Evaluate $\displaystyle\int_C \frac{\zeta \, d\zeta}{\zeta^2 - z^2}$, whenever it exists, if C is the unit circle in the positive sense. The integrand can be written as follows:

$$\frac{\zeta}{\zeta^2 - z^2} = \frac{1}{2}\frac{1}{\zeta - z} + \frac{1}{2}\frac{1}{\zeta + z}.$$

If z is inside the unit circle, then so is $-z$. If $z \neq 0$ then $z \neq -z$ and the integration can be carried out over two small circles with centers at z and $-z$. The value for each circle is πi and the value of the function is $2\pi i$. If $z = 0$ then the integrand is $1/\zeta$ and the value of the integral is again $2\pi i$.

If z is outside the unit circle, the function $\zeta/(\zeta^2 - z^2)$ is analytic within and on C, and by Cauchy's theorem the integral is zero. Summarizing, we have

$$F(z) = \int_C \frac{\zeta \, d\zeta}{\zeta^2 - z^2} = \begin{cases} 2\pi i, \text{ if } |z| < 1, \\ \text{undefined, if } |z| = 1, \\ 0, \text{ if } |z| > 1. \end{cases}$$

Note that $F(z)$ is analytic both inside and outside the unit circle, but not on the unit circle.

EXAMPLE 3.5.6. Let $f(\zeta)$ be continuous on the simple closed contour C. Show that $\int_C \dfrac{f(\zeta)}{\zeta - z} \, d\zeta$ is analytic both inside and outside C. Let $f(\zeta) = u(t) + iv(t)$, where $\zeta(t) = \xi(t) + i\eta(t)$, $0 \leq t \leq 1$, is a continuous parametric representation of C. Then

$$\begin{aligned} \frac{f(\zeta)}{\zeta - z} &= \frac{u(t) + iv(t)}{[\xi(t) - x] + i[\eta(t) - y]} \\ &= \frac{[u(t) + iv(t)]\{[\xi(t) - x] - i[\eta(t) - y]\}}{[\xi(t) - x]^2 + [\eta(t) - y]^2} \\ &= \frac{u(t)[\xi(t) - x] + v(t)[\eta(t) - y]}{[\xi(t) - x]^2 + [\eta(t) - y]^2} \\ &\quad + i \frac{v(t)[\xi(t) - x] - u(t)[\eta(t) - y]}{[\xi(t) - x]^2 + [\eta(t) - y]^2}. \end{aligned}$$

Clearly, $f(\zeta)/(\zeta - z)$ is continuous in x, y, and t if $u(t)$, $v(t)$, $\xi(t)$, and $\eta(t)$ are continuous and (x, y) is not on C. Furthermore,

$$\frac{\partial}{\partial z} \frac{f(\zeta)}{\zeta - z} = \frac{f(\zeta)}{(\zeta - z)^2}$$

and is continuous when (x, y) is not on C. Therefore, by Theorem 3.5.3, $F(z) = \int_C \dfrac{f(\zeta)}{\zeta - z} \, d\zeta$ is analytic both inside and outside C, where

$$F'(z) = \int_C \frac{f(\zeta)}{(\zeta - z)^2} \, d\zeta.$$

An interesting question can be raised in connection with the function of Example 3.5.6, that is, what happens if z is on C? Clearly, if $f(z) \neq 0$, then $\dfrac{f(\zeta)}{\zeta - z}$ behaves like $\dfrac{1}{\zeta - z}$ near $\zeta = z$. Hence, $\int_C \dfrac{f(\zeta)}{\zeta - z} \, d\zeta$ cannot exist in the ordinary sense since $\int \dfrac{1}{\zeta - z} \, d\zeta = \log(\zeta - z)$. However, if we define the integral in a more restrictive sense, we can give meaning to the function when z is on C. Let C be smooth at z. Then the curve has a unique tangent

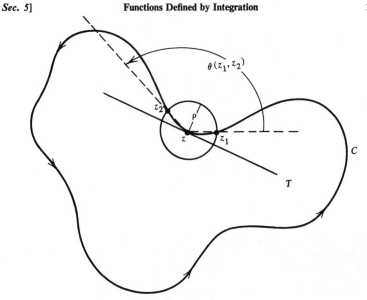

Figure 3.5.3.

at z. We draw this tangent T and also a small circle of radius ρ with center at z. See Figure 3.5.3. Let z_1 be the point on C, where the small circle cuts C nearest T on one side of z and let z_2 be the point on C where the small circle cuts C nearest T on the opposite side of z from z_1. Let $\theta(z_1, z_2)$ be $\arg(z_2 - z) - \arg(z_1 - z)$. As ρ approaches zero, z_1 and z_2 approach the tangent and $\theta(z_1, z_2)$ approaches π.

Now, we define the *Cauchy principal value* of $\int_C \dfrac{1}{\zeta - z}\, d\zeta,\ z \in C$, as $\lim\limits_{\rho \to 0} \int_{C'} \dfrac{1}{\zeta - z}\, d\zeta$, where C' is that part of C outside the small circle taken in the positive sense. Let γ_ρ be the arc of the small circle between z_1 and z_2 which along with C' gives us a simple closed contour with z on the inside. Then

$$\int_{C'+\gamma_\rho} \frac{1}{\zeta - z}\, d\zeta = 2\pi i,$$

$$\int_{\gamma_\rho} \frac{1}{\zeta - z}\, d\zeta = i\theta(z_1, z_2) \to \pi i \text{ as } \rho \to 0.$$

Therefore, the Cauchy principal value of $\int_C \dfrac{1}{\zeta - z}\, d\zeta = \pi i$ and

$$\int_C \frac{f(z)}{\zeta - z}\, d\zeta = \pi i f(z).$$

Let us now define the Cauchy principal value of $\displaystyle\int_C \frac{f(\zeta)}{\zeta - z}\, d\zeta$ as

$$\lim_{\rho \to 0} \int_{C'} \frac{f(\zeta)}{\zeta - z}\, d\zeta.$$

Then

$$\int_C \frac{f(\zeta)}{\zeta - z}\, d\zeta = \lim_{\rho \to 0} \int_{C'} \frac{f(\zeta)}{\zeta - z}\, d\zeta$$

$$= \lim_{\rho \to 0} \int_{C'} \frac{f(\zeta) - f(z)}{\zeta - z}\, d\zeta + \lim_{\rho \to 0} \int_{C'} \frac{f(z)}{\zeta - z}\, d\zeta$$

$$= \lim_{\rho \to 0} \int_{C'} \frac{f(\zeta) - f(z)}{\zeta - z}\, d\zeta + \pi i f(z).$$

Definition 3.5.1. The function $f(\zeta)$ satisfies a Lipschitz condition in the neighborhood of z on C, if there exist positive numbers A, α, δ such that $|f(\zeta) - f(z)| \le A|\zeta - z|^\alpha$ for all ζ on C satisfying $|\zeta - z| < \delta$.

Note that if $f(\zeta)$ satisfies a Lipschitz condition at z, it is continuous but not necessarily differentiable. However, if $f(\zeta)$ is differentiable at z, it satisfies a Lipschitz condition with $\alpha = 1$.

Theorem 3.5.4. *If $f(z)$ is continuous on a simple closed contour which is smooth at $z_0 \in C$, where $f(\zeta)$ satisfies a Lipschitz condition, then the Cauchy principal value of $\displaystyle\int_C \frac{f(\zeta)}{\zeta - z_0}\, d\zeta$ exists and*

$$\int_C \frac{f(\zeta)}{\zeta - z_0}\, d\zeta = F(z_0^-) - \pi i f(z_0) = F(z_0^+) + \pi i f(z_0),$$

where $F(z) = \displaystyle\int_C \frac{f(\zeta)}{\zeta - z}\, d\zeta$, $z \notin C$, and $F(z_0^-)$ and $F(z_0^+)$ are the limits of $F(z)$ as z approaches z_0 from the inside and outside of C, respectively.

Proof: Using the above notation, we have to show that

$$\lim_{\rho \to 0} \int_{C'} \frac{f(\zeta) - f(z_0)}{\zeta - z_0}\, d\zeta$$

exists. Since the integrand is continuous for $\zeta \ne z_0$, $\displaystyle\int_{C'} \frac{f(\zeta) - f(z_0)}{\zeta - z_0}\, d\zeta$ exists in the Riemann sense. To show the existence of the limit we have to show that $\displaystyle\int \frac{f(\zeta) - f(z_0)}{\zeta - z_0}\, d\zeta$ is absolutely integrable in the neighborhood of z_0. Clearly, the result should be independent of translation, rotation, and parameterization of C. Therefore, we assume $z_0 = 0$ and $\zeta = \xi(s) + i\eta(s)$

is a parameterization of C with s the arc length and $s = 0$ at $\zeta = 0$. We can also assume that $|\xi|$ and $|\dot{\eta}|$ are bounded. Then

$$\int_{s=0}^{s=\sigma} \frac{|f(\zeta) - f(0)|}{|\zeta|} \, |d\zeta| \leq \int_{s=0}^{s=\sigma} AK \, |\zeta|^{\alpha-1} \, ds,$$

where $K = 2 \max [|\xi|, |\dot{\eta}|]$ for $0 \leq s \leq \sigma$. Now, $\zeta(s) = \zeta(0) + \dot{\zeta}s$, where $\dot{\zeta}$ is evaluated between zero and s. Hence, $|\zeta(s)| \leq Ks$ and we have

$$\int_{s=0}^{s=\sigma} \frac{|f(\zeta) - f(0)|}{|\zeta|} \, |d\zeta| \leq AK^{\alpha} \int_0^{\sigma} s^{\alpha-1} \, ds = AK^{\alpha} \frac{\sigma^{\alpha}}{\alpha}.$$

If z is inside C then $F(z) = \int_C \frac{f(\zeta)}{\zeta - z} \, d\zeta$ and this integral differs by an arbitrarily small amount from $\int_{C'+\gamma_\rho} \frac{f(\zeta)}{\zeta - z} \, d\zeta$ if ρ is sufficiently small.*

We can now let z approach z_0 from inside C and the $\int_{C'+\gamma_\rho} \frac{f(\zeta)}{\zeta - z_0} \, d\zeta$ exists for all sufficiently small positive ρ. We also know that

$$\lim_{\rho \to 0} \int_{C'+\gamma_\rho} \frac{f(\zeta)}{\zeta - z_0} \, d\zeta$$

exists, since

$$\lim_{\rho \to 0} \int_{C'+\gamma_\rho} \frac{f(\zeta)}{\zeta - z_0} \, d\zeta = \lim_{\rho \to 0} \int_{C'} \frac{f(\zeta)}{\zeta - z_0} \, d\zeta + \lim_{\rho \to 0} \int_{\gamma_\rho} \frac{f(\zeta) - f(z_0)}{\zeta - z_0} \, d\zeta$$

$$+ \lim_{\rho \to 0} \int_{\gamma_\rho} \frac{f(z_0)}{\zeta - z_0} \, d\zeta.$$

$$F(z_0^-) = \int_C \frac{f(\zeta)}{\zeta - z_0} \, d\zeta + \pi i f(z_0).$$

The second limit on the right is zero by the continuity of $f(\zeta)$ at z_0. Similarly, by indenting the contour C and letting z approach z_0 from the outside, we have

$$F(z_0^+) = \int_C \frac{f(\zeta)}{\zeta - z_0} \, d\zeta - \pi i f(z_0).$$

Exercises 3.5

1. Evaluate each of the following by using the indefinite integral.

(a) $\displaystyle\int_a^b \sin z \, dz$;   (b) $\displaystyle\int_a^b z e^{z^2} \, dz$;   (c) $\displaystyle\int_a^b \sin^2 z \, dz$.

---

* Here we assume that $f(\zeta)$ is defined as a continuous function in some ϵ-neighborhood of z_0.

2. Let $f(z)$ and $g(z)$ be analytic in a simply connected domain D. Prove that

$$\int_a^b f(z)g(z)\,dz \;=\; f(z)h(z)\bigg|_a^b \;-\; \int_a^b f'(z)h(z)\,dz,$$

if the path of integration is in D and $h(z)$ is an indefinite integral of $g(z)$ in D.

3. Identify the function $\displaystyle\int^z \frac{d\zeta}{\zeta^2 - 1}$, where the path of integration is any piecewise smooth curve connecting 0 and z not passing through ± 1.

4. Evaluate $\displaystyle\int_{C_+} \frac{\zeta^2 + z^2}{\zeta^2 - z^2}\,d\zeta$, where C is the unit circle and $|z| \neq 1$.

5. Prove that $\displaystyle w(z) = \int_C \frac{e^{\zeta z}}{P(\zeta)}\,d\zeta$, where $P(\zeta)$ is a polynomial, is analytic if C is any simple contour not passing through a zero of $P(\zeta)$. Prove that this function satisfies the differential equation $P(d/dz)\,w(z) = 0$.

6. Referring to Theorem 3.5.4, suppose $f(\zeta)$ is analytic within and on C, a simple closed contour. If $\displaystyle F(z) = \int_{C_+} \frac{f(\zeta)}{\zeta - z}\,d\zeta$, $z \notin C$, find $F(z_0^-)$ and $F(z_0^+)$, $z_0 \in C$. Verify

$$F(z_0^-) = \int_{C_+} \frac{f(\zeta)}{\zeta - z_0}\,d\zeta + \pi i f(z_0),$$

$$F(z_0^+) = \int_{C_+} \frac{f(\zeta)}{\zeta - z_0}\,d\zeta - \pi i f(z_0),$$

if C is smooth at z_0. What if C is not smooth at z_0, that is, the tangent to C undergoes an abrupt change of direction of θ?

3.6. CAUCHY FORMULAS

In Section 3.5, we encountered functions defined by integrals of the type $\displaystyle\int_{C_+} \frac{f(\zeta)\,d\zeta}{\zeta - z}$, where C is a simple closed contour and $f(\zeta)$ is continuous on C. We saw that such a function is analytic inside C and outside C but not defined on C except as a Cauchy principal value. What if $f(\zeta)$ is defined inside C and is analytic within and on C? Then what value does the integral give? First, we note that if z is inside C, then the integration on C can be replaced by integration around an arbitrarily small circle with center at z. Obviously,

$$f(z)\int_{C_+} \frac{d\zeta}{\zeta - z} = \int_{C_+} \frac{f(z)\,d\zeta}{\zeta - z} = 2\pi i f(z),$$

and since $f(\zeta) \to f(z)$ as $\zeta \to z$, the following result is reasonable:

$$f(z) = \frac{1}{2\pi i} \int_{C_+} \frac{f(\zeta)\, d\zeta}{\zeta - z}.$$

This is known as Cauchy's integral formula. We now establish this result rigorously.

Theorem 3.6.1. *Let $f(z)$ be analytic within and on a simple closed contour C. Then if z is inside C*

$$f(z) = \frac{1}{2\pi i} \int_{C_+} \frac{f(\zeta)\, d\zeta}{\zeta - z}.$$

Proof: Since z is inside C, there exists a ρ such that $C_0 = \{\zeta \mid |\zeta - z| = \rho\}$ is inside C. Then by Theorem 3.4.5,

$$\frac{1}{2\pi i} \int_{C_+} \frac{f(\zeta)\, d\zeta}{\zeta - z} = \frac{1}{2\pi i} \int_{C_{0+}} \frac{f(\zeta)\, d\zeta}{\zeta - z}.$$

Also,

$$f(z) = \frac{f(z)}{2\pi i} \int_{C_{0+}} \frac{d\zeta}{\zeta - z} = \frac{1}{2\pi i} \int_{C_{0+}} \frac{f(z)\, d\zeta}{\zeta - z}.$$

We establish Cauchy's formula by showing that

$$\frac{1}{2\pi i} \int_{C_{0+}} \frac{f(\zeta)\, d\zeta}{\zeta - z} - \frac{1}{2\pi i} \int_{C_{0+}} \frac{f(z)\, d\zeta}{\zeta - z} = 0.$$

If this was not so, the modulus of this difference would be positive. However, given $\epsilon > 0$, there exists a $\delta(\epsilon)$ such that $|f(\zeta) - f(z)| < \epsilon$ for all ζ such that $|\zeta - z| < \delta$. If $\rho < \delta$, then

$$\left| \frac{1}{2\pi i} \int_{C_{0+}} \frac{f(\zeta)\, d\zeta}{\zeta - z} - \frac{1}{2\pi i} \int_{C_{0+}} \frac{f(z)\, d\zeta}{\zeta - z} \right|$$

$$= \left| \frac{1}{2\pi i} \int_{C_{0+}} \frac{f(\zeta) - f(z)}{\zeta - z}\, d\zeta \right| \le \frac{\epsilon}{2\pi \rho}\, 2\pi \rho = \epsilon.$$

But ϵ is arbitrary, showing that

$$f(z) = \frac{1}{2\pi i} \int_{C_{0+}} \frac{f(z)}{\zeta - z}\, d\zeta = \frac{1}{2\pi i} \int_{C_{0+}} \frac{f(\zeta)\, d\zeta}{\zeta - z} = \frac{1}{2\pi i} \int_{C_+} \frac{f(\zeta)\, d\zeta}{\zeta - z}.$$

At first sight, this integral representation seems of little significance. However, it is quite a breakthrough, because by Theorem 3.5.3 and Example 3.5.6 we can now differentiate under the integral sign as long as z is inside C. This gives a representation for the derivative

$$f'(z) = \frac{1}{2\pi i} \int_{C_+} \frac{f(\zeta)\, d\zeta}{(\zeta - z)^2}.$$

Moreover, we can differentiate again, giving us the second derivative

$$f''(z) = \frac{2}{2\pi i} \int_{C_+} \frac{f(\zeta)}{(\zeta - z)^3} \, d\zeta.$$

As a matter of fact, we can differentiate as many times as we please. We thus obtain, by an obvious induction, the following theorem:

Theorem 3.6.2. *Let $f(z)$ be analytic within and on a simple closed contour C. Then at z inside $C, f(z)$ has all derivatives given by the Cauchy formulas*

$$f^{(n)}(z) = \frac{n!}{2\pi i} \int_{C_+} \frac{f(\zeta)}{(\zeta - z)^{n+1}} \, d\zeta.$$

Corollary 3.6.1. Let $f(z)$ be analytic at z_0. Then $f(z)$ has all derivatives at z_0.

Proof: Since $f(z)$ is analytic at z_0, there exists an ϵ-neighborhood throughout which $f(z)$ is analytic. Then if $C_\epsilon = \{z \mid |z - z_0| = \epsilon/2\}$

$$f(z_0) = \frac{1}{2\pi i} \int_{C_{\epsilon+}} \frac{f(z)}{z - z_0} \, dz$$

and

$$f^{(n)}(z_0) = \frac{n!}{2\pi i} \int_{C_{\epsilon+}} \frac{f(z)}{(z - z_0)^{n+1}} \, dz.$$

We have now reached the surprising result that the existence of the first derivative of a function in some ϵ-neighborhood of z_0 implies the existence of all the derivatives at z_0. This is sharply contrasted with the behavior of functions of the real variable x, where for example, $x^{4/3}$ has a first derivative $(4/3)x^{1/3}$ for all x, but has no second derivative at $x = 0$. The existence of all derivatives in an ϵ-neighborhood of z_0 implies the existence and continuity of all partial derivatives of u and v. Since we know, for example, that

$$f'(z) = u_x + iv_x = v_y - iu_y,$$

then

$$f''(z) = u_{xx} + iv_{xx} = v_{yx} - iu_{yx} = v_{xy} - iu_{xy} = -u_{yy} - iv_{yy}.$$

This shows that all the second partial derivatives exist and

$$u_{xx} = -u_{yy} = v_{xy} = v_{yx},$$
$$v_{xx} = -v_{yy} = -u_{xy} = -u_{yx}.$$

Therefore, u and v satisfy Laplace's equation and also the mixed second partial derivatives are equal. The existence of $f''(z)$ in an ϵ-neighborhood of a point implies the continuity of the first partial derivatives of u and v at the point. Likewise, the existence of $f'''(z)$ implies the continuity of the second partial derivatives. Continuing in this way, we can establish the existence and continuity of the partial derivatives of u and v of all orders at a point of analyticity.

We next prove a set of important inequalities which also bear Cauchy's name.

Theorem 3.6.3. *Cauchy's Inequalities. Let $f(z)$ be analytic for $|z - z_0| < R$ and continuous for $|z - z_0| \leq R$. Then*

$$|f^{(n)}(z_0)| \leq n!\, M/R^n, \qquad n = 0, 1, 2, \ldots,$$

where $M = $ maximum of $|f(z)|$ on $C = \{z \mid |z - z_0| = R\}$.

Proof: Let $|z - z_0| \leq r < R$ and $C_r = \{z \mid |z - z_0| = r\}$. Using the Cauchy integral formulas, we have

$$f^{(n)}(z_0) = \frac{n!}{2\pi i} \int_{C_{r+}} \frac{f(z)}{(z - z_0)^{n+1}}\, dz,$$

$$|f^{(n)}(z_0)| \leq \frac{n!\, M(r)}{2\pi r^{n+1}} 2\pi r = \frac{n!\, M(r)}{r^n},$$

where $M(r) = \max |f(z_0 + re^{i\theta})|$. By continuity, $\lim\limits_{r \to R} M(r) = M$, and since the inequality holds for all $r < R$, we have

$$|f^{(n)}(z_0)| \leq \lim_{r \to R} \frac{n!\, M(r)}{r^n} = \frac{n!\, M}{R^n}.$$

Definition 3.6.1. $f(z)$ is an *entire* function if and only if it is analytic everywhere in the unextended plane.

All polynomials are entire functions. In addition such functions as e^z, $\sin z$, $\cos z$, $\sinh z$, $\cosh z$, and so on, are entire functions. The following theorem about entire functions follows directly from Cauchy's inequalities.

Theorem 3.6.4. *Liouville's Theorem. The only bounded entire functions are constants.*

Proof: Let z_0 be any complex number. Then if $f(z)$ is bounded and entire

$$|f'(z_0)| \leq \frac{M}{R},$$

where z_0 is the center of a circle of radius R and M is a bound for $|f(z)|$ everywhere in the unextended plane. But since $f(z)$ is entire, R can be made arbitrarily large. Therefore, $f'(z_0) = 0$. However, z_0 is *any* point, which implies that $f'(z) \equiv 0$. Now, let z_1 and z_2 be any two points in the unextended plane. Then

$$f(z_1) - f(z_2) = u(x_1, y_1) - u(x_2, y_2) + i[v(x_1, y_1) - v(x_2, y_2)]$$
$$= u_x(x_1 + \theta_1 \Delta x, y_1)\Delta x + u_y(x_1 + \Delta x, y_1 + \theta_2 \Delta y)\Delta y$$
$$+ i[v_x(x_1 + \theta_3 \Delta x, y_1)\Delta x + v_y(x_1 + \Delta x, y_1 + \theta_4 \Delta y)\Delta y],$$

where $0 < \theta_i < 1$, $i = 1, 2, 3, 4$, $\Delta x = x_1 - x_2$, $\Delta y = y_1 - y_2$. In using the mean value theorem, we have used the continuity of u_x, u_y, v_x, and v_y. Finally, $u_x \equiv u_y \equiv v_x \equiv v_y \equiv 0$, which proves that $f(z_1) = f(z_2)$.

A direct application of Liouville's theorem will give us the following algebraic result.

Theorem 3.6.5. *Fundamental Theorem of Algebra. Let $P(z) = a_0 + a_1 z + a_2 z^2 + \cdots + a_n z^n$ be a polynomial of degree $n \geq 1$ with real or complex coefficients. Then $P(z)$ has at least one root (zero); that is, there exists a complex number r such that $P(r) = 0$.*

Proof: Suppose $P(z)$ is never zero in the unextended plane. In this case $1/P(z)$ is an entire function. It is also bounded. If not, there would exist a sequence $\{z_n\}$ such that $P(z_n) \to 0$ as $n \to \infty$. If $\{z_n\}$ is bounded, it has at least one limit z_0 in the unextended plane and by continuity

$$P(z_0) = \lim_{n_k \to \infty} P(z_{n_k}) = 0.$$

If $\{z_n\}$ is unbounded, then there exists a subsequence $\{z_{n_j}\}$ such that

$$\lim_{n_j \to \infty} z_{n_j} = \infty \text{ and } \lim_{n_j \to \infty} P(z_{n_j}) = 0.$$

However,

$$|P(z)| \geq |a_n| R^n \left\{ 1 - \left[\frac{|a_{n-1}|}{|a_n| R} - \frac{|a_{n-2}|}{|a_n| R^2} - \cdots - \frac{|a_1|}{|a_n| R^{n-1}} - \frac{|a_0|}{|a_n| R^n} \right] \right\}$$
$$\geq \tfrac{1}{2} |a_n| R^n,$$

for all z outside of a circle of radius R sufficiently large that

$$\frac{|a_{n-1}|}{|a_n| R} + \frac{|a_{n-2}|}{|a_n| R^2} + \cdots + \frac{|a_1|}{|a_n| R^{n-1}} + \frac{|a_0|}{|a_n| R^n} < \frac{1}{2}.$$

Hence, the latter case is ruled out, and $1/P(z)$ is a bounded entire function if $P(z)$ has no root. By Liouville's theorem $1/P(z)$ is a constant and $P(z)$ is a constant. But this is impossible if $n \geq 1$. This completes the proof.

Let $P(z)$ be a polynomial of degree $n \geq 1$. Then there is at least one complex number r_1 such that $P(r_1) = 0$. If we divide $P(z)$ by $z - r_1$, by long division, we obtain

$$P(z) = (z - r_1) Q(z) + R,$$

where the quotient $Q(z)$ is a polynomial of degree $n - 1$ and R, the remainder, is a constant. Hence, $P(r_1) = R = 0$. This shows that $P(z)$ has a linear factor $z - r_1$. Now, $Q(z)$ is a polynomial of degree $n - 1$ and if $n \geq 2$, there is a complex number r_2 such that $Q(r_2) = 0$, and $Q(z)$ has a linear factor $z - r_2$. Continuing in this way, we can express $P(z)$ as follows:

$$P(z) = a_n (z - r_1)(z - r_2) \cdots (z - r_n),$$

where $r_1, r_2, \ldots,$ are roots of $P(z)$, not necessarily distinct. If the factor $z - r_k$ appears m times in the factorization, then we say that r_k is a zero of order m.

The factorization of $P(z)$ into linear factors is unique except possibly for the order of the factors. For suppose we arrive at a different factorization of $P(z)$ as follows:

$$P(z) = a'_n(z - r'_1)(z - r'_2) \cdots (z - r'_n).$$

First there must be precisely n linear factors. Otherwise, the polynomial would not be of degree n. In multiplying out the linear factors we find the term $a'_n z^n$ plus terms of lower degree. Differentiating n times, we have

$$P^{(n)}(0) = n!a'_n = n!a_n.$$

Therefore, $a'_n = a_n$. Now, if r_1 does not appear in the second factorization, then

$$P(r_1) = a_n(r_1 - r'_1)(r_1 - r'_2) \cdots (r_1 - r'_n) \neq 0,$$

contradicting the fact that r_1 is a root. Therefore, $r_1 = r'_k$ for some k. Similarly, $r_2 = r'_m$ for some m, and so on. The last question remaining is whether a repeated root appears the same number of times in each factorization. Suppose r_1 is an mth order zero in the first factorization. Then

$$P(z) = a_n(z - r_1)^m Q(z),$$

where $Q(z)$ is a polynomial of degree $n - m$ and $Q(r_1) \neq 0$. Differentiating $P(z)$ m times, we have

$$P^{(1)}(z) = ma_n(z - r_1)^{m-1}Q(z) + a_n(z - r_1)^m Q^{(1)}(z),$$
$$P^{(2)}(z) = m(m - 1)a_n(z - r_1)^{m-2}Q(z) + 2ma_n(z - r_1)^{m-1}Q^{(1)}(z)$$
$$+ a_n(z - r_1)^m Q^{(2)}(z),$$
$$\vdots$$
$$P^{(m)}(z) = m!a_n Q(z) + m(m - 1) \cdots 2a_n(z - r_1)Q^{(1)}(z) + \cdots$$
$$+ a_n(z - r_1)^m Q^{(m)}(z).$$

Hence,

$$P^{(1)}(r_1) = P^{(2)}(r_1) = \cdots = P^{(m-1)}(r_1) = 0$$

and

$$P^{(m)}(r_1) = m!a_n Q(r_1) \neq 0.$$

If there is another factorization $P(z) = a_n(z - r_1)^{m'}Q'$ with $m' \neq m$, there are two cases (1) $m < m'$ or (2) $m > m'$. However, this implies (1) $P^{(m)}(r_1) = 0$ or (2) $P^{(m-1)}(r_1) \neq 0$. We have proved the following theorem:

Theorem 3.6.6. *If* $P(z) = a_n z^n + a_{n-1}z^{n-1} + \cdots + a_1 z + a_0$ *is a polynomial of degree* $n \geq 1$, *it can be written as*

$$P(z) = a_n(z - r_1)^{m_1}(z - r_2)^{m_2} \cdots (z - r_k)^{m_k},$$

where $r_1, r_2, \ldots, r_k$ are the distinct roots with multiplicities $m_1, m_2, \ldots, m_k$, respectively. The factorization is unique except possibly for the order of the factors. The sum of the multiplicities is n. If r is a root of multiplicity m, then

$$P(z) = c_n(z - r)^m Q(z),$$

where $Q(z)$ is a polynomial of degree $n - m$ such that $Q(r) \neq 0$.

We conclude this section with a theorem which plays the role of a converse of Cauchy's theorem.

Theorem 3.6.7. *Morera's Theorem. Let $f(z)$ be continuous in domain D, where $\int_C f(z)\,dz = 0$ for every simple closed contour. Then $f(z)$ is analytic in D.*

Proof: We first show that $\int_a^b f(z)\,dz$ is independent of the simple contour joining a and b in D using the argument of Theorem 3.4.3. Then we define

$$F(z) = \int_{z_0}^{z} f(\zeta)\,d\zeta,$$

where z_0, z, and the path of integration lie in D. We prove that $F'(z) = f(z)$ by the argument of Theorem 3.5.1, which does not use the analyticity of $f(z)$. Then by Corollary 3.6.1, $F''(z) = f'(z)$ exists in D. This proves the analyticity of $f(z)$ in D.

Exercises 3.6

1. Prove the following theorem: Let $f(z)$ be analytic within and on a simple closed contour C except for z_0 inside C. Let $F(z)$ be a function analytic within and on C defined as follows:

$$F(z) = (z - z_0)^n f(z), \qquad z \neq z_0$$
$$= \lim_{z \to z_0} (z - z_0)^n f(z), z = z_0,$$

then

$$\int_{C_+} f(z)\,dz = \frac{2\pi i F^{(n-1)}(z_0)}{(n-1)!}.$$

2. Using the result of Exercise 3.6.1, evaluate each of the following integrals:

(a) $\displaystyle \int_{C_+} \frac{\sin z}{z^2}\,dz$, where C is the unit circle.

(b) $\displaystyle \int_{C_+} \frac{1}{z(z^3 + 1)^2}\,dz$, where C is the rectangle with corners at $1 \pm \frac{1}{2}i$, $-2 \pm \frac{1}{2}i$.

3. Prove that an analytic function satisfies the mean-value property; that is if $f(z)$ is analytic for $|z - z_0| \le r$, then

$$f(z_0) = \frac{1}{2\pi} \int_0^{2\pi} f(z_0 + re^{i\theta})\, d\theta.$$

4. Let $f(z)$ be entire and $|f(re^{i\theta})| \le Mr$, where M is a constant. Prove that $f(z)$ is a polynomial of degree at most one. Can this result be generalized to polynomials of higher degree?

3.7. MAXIMUM MODULUS PRINCIPLE

Suppose $f(z)$ is analytic for $|z - z_0| < \rho_0$ and continuous for $|z - z_0| \le \rho_0$. Then using the result of Exercises 3.6.1, we have

$$f(z_0) = \frac{1}{2\pi i} \int_{|z-z_0|=\rho_0} \frac{f(z)}{z - z_0}\, dz.$$

Making the change of variable $z = z_0 + \rho_0 e^{i\theta}$, $0 \le \theta \le 2\pi$, we obtain

$$f(z_0) = \frac{1}{2\pi} \int_0^{2\pi} f(z_0 + \rho_0 e^{i\theta})\, d\theta,$$

$$|f(z_0)| \le \frac{1}{2\pi} M 2\pi = M,$$

where $M = \max |f(z_0 + \rho_0 e^{i\theta})|$. In other words, the modulus of the function at the center of the circle is not greater than the maximum modulus of the function on the circle. Actually, a much deeper result than this can be proved. The modulus of a function analytic inside a simple closed contour C and continuous within and on C is continuous in a bounded closed set and, therefore, must take on a maximum value. We can prove that the maximum modulus occurs on the boundary C and never inside, unless the function is a constant.

Theorem 3.7.1. *Maximum Modulus Principle. Let $f(z)$ be analytic in a bounded domain D and continuous in the closure $\overline{D}$. If $M = \max |f(z)|$ in $\overline{D}$, then either $f(z) = Me^{i\gamma}$, a constant, or $|f(z)| < M$ in D.*

Proof: Assume that at some point $z_0 \in D$ $|f(z_0)| = M$. There exists a disk $|z - z_0| \le \rho_0$ in D. Hence,

$$f(z_0) = \frac{1}{2\pi} \int_0^{2\pi} f(z_0 + \rho_0 e^{i\theta})\, d\theta,$$

$$M = |f(z_0)| \le \frac{1}{2\pi} \int_0^{2\pi} |f(z_0 + \rho_0 e^{i\theta})|\, d\theta,$$

where $|f(z_0 + \rho_0 e^{i\theta})| \leq M$. Let $g(\theta) = |f(z_0 + \rho_0 e^{i\theta})|$. Then $g(\theta)$ is continuous, $g(\theta) \leq M$, and its average value

$$\frac{1}{2\pi} \int_0^{2\pi} g(\theta) \, d\theta \geq M.$$

This implies that $g(\theta) \equiv M$, for suppose $g(\alpha) < M$. Then by continuity there exists a $\delta > 0$ such that $g(\theta) \leq [M + g(\alpha)]/2$ for $\alpha - \delta \leq \theta \leq \alpha + \delta$, and

$$\frac{1}{2\pi} \int_0^{2\pi} g(\theta) \, d\theta \leq \frac{2\delta}{2\pi} \left[\frac{M + g(\alpha)}{2} \right] + \frac{2\pi - 2\delta}{2\pi} M,$$

$$\frac{1}{2\pi} \int_0^{2\pi} g(\theta) \, d\theta \leq M - \frac{\delta}{2\pi} [M - g(\alpha)] < M,$$

which is impossible. Hence, $g(\theta) = |f(z_0 + \rho_0 e^{i\theta})| \equiv M$. Therefore,

$$f(z_0 + \rho_0 e^{i\theta}) = M e^{i\phi(\theta)},$$

where $\phi(\theta)$ is a continuous real-valued function of θ. Now

$$f(z_0) = M e^{i\gamma} = \frac{1}{2\pi} \int_0^{2\pi} M e^{i\phi(\theta)} \, d\theta$$

$$1 = \frac{1}{2\pi} \int_0^{2\pi} e^{i(\phi - \gamma)} \, d\theta.$$

Taking the real part of this equation, we have

$$1 = \frac{1}{2\pi} \int_0^{2\pi} \cos(\phi - \gamma) \, d\theta.$$

But $\cos(\phi - \gamma) \leq 1$. Therefore, by an argument similar to the one for $g(\theta)$, we can show that $\cos(\phi - \gamma) \equiv 1$. This means that $f(z_0 + \rho_0 e^{i\theta}) = M e^{i\gamma}$. But if $f(z)$ is constant on the boundary of the disk $|z - z_0| \leq \rho_0$, by Cauchy's formula it is constant inside.

Let ζ be any point of D other than z_0. Since D is connected, there exists a simple curve C in D joining z_0 and ζ. Such a curve is a bounded closed set. Since D is open, for every point on C there is a circle with center at the point and with interior in D. The collection of the interiors of all such circles is an open covering of C. By the Heine-Borel theorem there is a finite subcovering of C. Using this finite covering we can show that $f(\zeta) = M e^{i\gamma}$, or, in other words, that $f(z) = M e^{i\gamma}$ throughout D if $|f(z_0)| = M$.

Let $S_0 = \{z \mid |z - z_0| < \rho_0\}$ be an element of the finite covering. We already know that $f(z) = M e^{i\gamma}$ in S_0. Let

$$S_1 = \{z \mid |z - z_1| < \rho_1, z_1 \in C\}$$

be another element of the finite covering which intersects S_0. Let

$$z^* = z_0 + \frac{\rho_0(z_1 - z_0)}{|z_1 - z_0|}.$$

Since $z^* \in \bar{S}_0, f(z^*) = Me^{i\gamma}$. Also $z^* \in S_1$ and there is a positive δ such that $S^* = \{z \mid |z - z^*| \leq \delta\} \subseteq D$. By the same argument as above, we can show that $f(z) = Me^{i\gamma}$ in S^*. If $z_1 \in S^*$, then we have shown that $f(z_1) = Me^{i\gamma}$. If not, we can continue the process toward z_1, eventually, after a finite number of steps, to show that $f(z_1) = Me^{i\gamma}$. Finally, having reached z_1, we see that the whole process can be continued to ζ, using other elements of the finite covering, showing that $f(\zeta) = Me^{i\gamma}$. But ζ is any point in D. Hence, either $f(z) = Me^{i\gamma}$ throughout D or $|f(z)| < M$ in D. This completes the proof.

Corollary 3.7.1. If $f(z)$ is analytic in a bounded domain D and continuous in $\bar{D}$, then $|f(z)|$ takes on its maximum on the boundary of D.

Proof: Since $|f(z)|$ is continuous in $\bar{D}$ it must take on a maximum in $\bar{D}$. It cannot take on its maximum in D, unless it is constant in D and hence in $\bar{D}$. Therefore, constant or not, the maximum modulus occurs on the boundary of D.

Corollary 3.7.2. If $f(z)$ is analytic in a bounded domain D, continuous in $\bar{D}$, and not zero in D, then $|f(z)|$ takes on its minimum on the boundary of D.

Proof: Since $|f(z)|$ is continuous in $\bar{D}$ it must take on a minimum in $\bar{D}$. If $f(z)$ is zero in $\bar{D}$, this zero must be on the boundary of D where

$$\min |f(z)| = 0.$$

If $f(z)$ is never zero in $\bar{D}$ then $1/f(z)$ is analytic in D and continuous in $\bar{D}$. Therefore, by Corollary 3.7.1, $\max 1/|f(z)| = \min |f(z)|$ occurs on the boundary of D.

Corollary 3.7.3. If $f(z)$ is analytic in a bounded domain D and continuous in $\bar{D}$, then $u = Re[f(z)]$ and $v = Im[f(z)]$ take on their maximum and minimum values on the boundary of D.

Proof: Consider $F(z) = e^{f(z)}$. $F(z)$ is analytic in D, continuous in $\bar{D}$, and is never zero in D. Therefore, $|F(z)| = e^u$ takes on its maximum and minimum on the boundary of D. But e^u is a monotone function of u. Therefore, u takes on its maximum and minimum on the boundary of D. To prove the same for v, we merely consider $G(z) = e^{-if(z)}$ and $|G(z)| = e^v$.

Theorem 3.7.2. *Schwarz's Lemma. If $f(z)$ is analytic for $|z| < R$, continuous for $|z| \leq R$, and $f(0) = 0$, then $|f(re^{i\theta})| \leq Mr/R$, where $M = max |f(z)|$ on $|z| = R$. Equality only occurs when $f(z) = zMe^{i\gamma}/R$, where γ is a real constant.*

Proof: Let $g(z) = f(z)/z$ for $0 < |z| \leq R$ and $g(0) = \lim_{z \to 0} \dfrac{f(z)}{z} = f'(0)$. Then $g(z)$ is analytic for $0 < |z| < R$ and continuous for $0 \leq |z| \leq R$.

Hence, max $|g(z)|$ occurs either at $z = 0$ or $|z| = R$. For $z = 0$, $|g(0)| = |f'(0)| \leq M/R$. For $|z| = R$, $|g(z)| = |f(z)/z| \leq M/R$. Hence,

$$|g(z)| \leq \frac{M}{R} \text{ for } |z| \leq R$$

and equality holds only when $g(z)$ is a constant $Me^{i\gamma}/R$. Therefore, $|f(re^{i\theta})| \leq Mr/R$ for $r \leq R$ and equality holds only when $f(z) = zMe^{i\gamma}/R$.

There are functions other than the modulus of an analytic function which satisfy a maximum principle. One such class of functions which have been widely studied are the subharmonic functions.

Definition 3.7.1. The real-valued function $\phi(x, y)$ is subharmonic in a domain D if and only if it is continuous in D and for each $(x_0, y_0) \in D$ there exists a ρ such that

$$\phi(x_0, y_0) \leq \frac{1}{2\pi} \int_0^{2\pi} \phi(x_0 + r\cos\theta, y_0 + r\sin\theta)\, d\theta,$$

for all r satisfying $0 \leq r \leq \rho$.

The reason why $\phi(x, y)$ is called subharmonic is the following. Suppose $u(x, y)$ is harmonic for $|z - z_0| < r$, continuous for $|z - z_0| \leq r$, and takes on the values $\phi(x_0 + r\cos\theta, y_0 + r\sin\theta)$ for $|z - z_0| = r$. Then $u(x, y)$ is the real part of an analytic function $f(z)$ and

$$f(z_0) = \frac{1}{2\pi} \int_0^{2\pi} f(z_0 + re^{i\theta})\, d\theta,$$

$$u(x_0, y_0) = \frac{1}{2\pi} \int_0^{2\pi} u(x_0 + r\cos\theta, y_0 + r\sin\theta)\, d\theta$$

$$= \frac{1}{2\pi} \int_0^{2\pi} \phi(x_0 + r\cos\theta, y_0 + r\sin\theta)\, d\theta.$$

Therefore, $\phi(x_0, y_0) \leq u(x_0, y_0)$. We shall show in Chapter 6 that the u can always be determined. Hence, in particular, harmonic functions are also subharmonic. Furthermore, the modulus of an analytic function is subharmonic. We shall now show that a subharmonic function satisfies the maximum principle.

Theorem 3.7.3. *Let $\phi(x, y)$ be subharmonic in a bounded domain D and continuous in $\overline{D}$. Then if $M = \max \phi(x, y)$ in $\overline{D}$, either $\phi \equiv M$ or $\phi(x, y) < M$ in D.*

Proof: Suppose $(x_0, y_0) \in D$ and $\phi(x_0, y_0) = M$. There exists a ρ such that $S_0 = \{z \mid |z - z_0| \leq \rho\} \subseteq D$ and

$$\phi(x_0, y_0) = M \leq \frac{1}{2\pi} \int_0^{2\pi} \phi(x_0 + r\cos\theta, y_0 + r\sin\theta)\, d\theta$$

for all r such that $0 \leq r \leq \rho$. But $\phi(x_0 + r \cos \theta, y_0 + r \sin \theta) \leq M$. Hence, just as in the proof of Theorem 3.7.1,

$$\phi(x_0 + r \cos \theta, y_0 + r \sin \theta) \equiv M$$

which implies that $\phi(x, y) \equiv M$ in S_0. Now, let (ξ, η) be any other point in D. Using the Heine-Borel theorem, just as we did in the proof of Theorem 3.7.1, we show that $\phi(\xi, \eta) = M$ and hence that $\phi(x, y) \equiv M$ in $\overline{D}$. Therefore, unless $\phi(x, y)$ is a constant, it is less than M throughout D and hence must take on its maximum on the boundary.

Theorem 3.7.4. *Let $\phi(x, y)$ be subharmonic in a bounded domain D and continuous in $\overline{D}$. If there exists a function $u(x, y)$ harmonic in D, continuous in $\overline{D}$, and taking on the same values as ϕ on the boundary of D,* then*

$$\phi(x, y) \leq u(x, y) \text{ in } \overline{D}.$$

Proof: Consider $v = \phi - u$, which is continuous in $\overline{D}$ and zero on the boundary of D. Now, v is subharmonic in D, since for $(x_0, y_0) \in D$, there exists a ρ, such that

$$\phi(x_0, y_0) \leq \frac{1}{2\pi} \int_0^{2\pi} \phi(x_0 + r \cos \theta, y_0 + r \sin \theta) \, d\theta,$$

$$u(x_0, y_0) = \frac{1}{2\pi} \int_0^{2\pi} u(x_0 + r \cos \theta, y_0 + r \sin \theta) \, d\theta,$$

$$v(x_0, y_0) = \phi(x_0, y_0) - u(x_0, y_0) \leq \frac{1}{2\pi} \int_0^{2\pi} [\phi(x_0 + r \cos \theta, y_0 + r \sin \theta)$$
$$- u(x_0 + r \cos \theta, y_0 + r \sin \theta)] \, d\theta,$$

$$v(x_0, y_0) \leq \int_0^{2\pi} v(x_0 + r \cos \theta, y_0 + r \sin \theta) \, d\theta,$$

for all r such that $0 \leq r \leq \rho$. This implies that v takes on its maximum on the boundary of D. Therefore, $v \leq 0$ or $\phi(x, y) \leq u(x, y)$ throughout D.

Theorem 3.7.5. *Let $\phi(x, y)$ have continuous second partial derivatives in a domain D. Then ϕ is subharmonic in D if and only if $\nabla^2 \phi \geq 0$ at every point in D.*

Proof: Recall the Green's identities of Exercise 3.1.4.

$$\iint\limits_R (v\nabla^2 u + \nabla u \cdot \nabla v) \, dx \, dy = \int_C v\nabla u \cdot \mathbf{N} \, ds,$$

$$\iint\limits_R (u\nabla^2 v - v\nabla^2 u) \, dx \, dy = \int_C (u\nabla v \cdot \mathbf{N} - v\nabla u \cdot \mathbf{N}) \, ds.$$

---

* We shall show in Chapter 6 that for a large class of domains such harmonic functions exist.

These identities hold for an annulus $\rho \leq |z - z_0| \leq r$, if this is a region where u and v have continuous second partial derivatives and C is the whole boundary, the normal on the outer circle pointing outward and the normal on the inner circle pointing inward. Let $v = \ln |z - z_0|$. Then $\nabla^2 v = 0$ in the annulus, and if u has continuous second partial derivatives for $|z - z_0| \leq r$

$$-\iint\limits_R \ln |z - z_0| \nabla^2 u \, dx \, dy = -\int_0^{2\pi} u(x_0 + \rho \cos \theta, y_0 + \rho \sin \theta) \, d\theta$$

$$+ \int_0^{2\pi} u(x_0 + r \cos \theta, y_0 + r \sin \theta) \, d\theta$$

$$+ \int_0^{2\pi} \rho \ln \rho (\nabla u \cdot \mathbf{N})_\rho \, d\theta$$

$$- \int_0^{2\pi} r \ln r (\nabla u \cdot \mathbf{N})_r \, d\theta.$$

Now,

$$2\pi u(x_0, y_0) = \int_0^{2\pi} u(x_0, y_0) \, d\theta,$$

and by continuity

$$\int_0^{2\pi} [u(x_0, y_0) - u(x_0 + \rho \cos \theta, y_0 + \rho \sin \theta)] \, d\theta \to 0$$

as $\rho \to 0$. Also since u_x and u_y are bounded

$$\int_0^{2\pi} \rho \ln \rho (\nabla u \cdot \mathbf{N})_\rho \, d\theta \to 0$$

as $\rho \to 0$. Hence,

$$u(x_0, y_0) = \frac{1}{2\pi} \int_0^{2\pi} u(x_0 + r \cos \theta, y_0 + r \sin \theta) \, d\theta$$

$$+ \frac{1}{2\pi} \iint\limits_{\tilde{R}} \ln |z - z_0| \nabla^2 u \, dx \, dy$$

$$- \frac{1}{2\pi} \int_0^{2\pi} r \ln r (\nabla u \cdot \mathbf{N})_r \, d\theta,$$

where $\tilde{R} = \{z \mid |z - z_0| \leq r\}$. It is not hard to show that the second integral on the right-hand side exists. Next, let $v = 1$. Then $\nabla v = 0$ and

$$\iint\limits_{\tilde{R}} \nabla^2 u \, dx \, dy = \int_0^{2\pi} r(\nabla u \cdot \mathbf{N})_r \, d\theta,$$

verifying that

$$u(x_0, y_0) = \frac{1}{2\pi} \int_0^{2\pi} u(x_0 + r \cos \theta, y_0 + r \sin \theta) \, d\theta$$

$$+ \frac{1}{2\pi} \iint_{\tilde{R}} [\ln |z - z_0| - \ln r] \nabla^2 u \, dx \, dy.$$

Let $u(x, y) = \phi(x, y)$, subharmonic, and assume $\nabla^2 \phi < 0$ at (x_0, y_0). By continuity, there exists an ϵ-neighborhood throughout which $\nabla^2 \phi < 0$. Since

$$\ln |z - z_0| - \ln r \leq 0 \text{ for } |z - z_0| \leq r,$$

$$\phi(x_0, y_0) > \frac{1}{2\pi} \int_0^{2\pi} \phi(x_0 + r \cos \theta, y_0 + r \sin \theta) \, d\theta \text{ for } r < \epsilon,$$

contradicting the assumption that ϕ is subharmonic at (x_0, y_0). Hence, if ϕ is subharmonic in D, $\nabla^2 \phi \geq 0$ throughout D.

Conversely, if $\nabla^2 \phi \geq 0$ throughout D, for every $z_0 \in D$ and every r such that

$$\tilde{R} = \{z \mid |z - z_0| \leq r\} \subseteq D,$$

$$\phi(x_0, y_0) \leq \frac{1}{2\pi} \int_0^{2\pi} \phi(x_0 + r \cos \theta, y_0 + r \sin \theta) \, d\theta,$$

which proves that ϕ is subharmonic in D.

Exercises 3.7

1. Let $f(z)$ be analytic at z_0. Prove that there exists a point z_1 such that $|f(z_0)| < |f(z_1)|$ unless $f(z)$ is constant.

2. Let $f(z)$ be analytic at z_0 and not constant. Prove that if $f(z_0) \neq 0$ there exists a point z_2 such that $|f(z_2)| < |f(z_0)|$.

3. Let $f(z)$ be analytic for $|z - z_0| \leq R$ and $|f(z_0 + Re^{i\theta})| \geq m > 0$. Prove that if $|f(z_0)| < m$, then $f(z)$ has at least one zero for $|z - z_0| < R$. Use this result to prove the Fundamental Theorem of Algebra.

4. Let u be harmonic in a domain D and continuous in $\overline{D}$. Prove that if v is harmonic in D continuous in $\overline{D}$ and takes on the same values as u on the boundary of D, then $u \equiv v$.

5. Let $f(z)$ be an entire function, not constant. Prove that $M(r) = \max |f(re^{i\theta})|$ is a strictly increasing function; that is, if $r_1 < r_2$, then $M(r_1) < M(r_2)$.

6. Let $f(x)$ be a real-valued continuous function of x. Then $f(x)$ is *convex* in the open interval $a < x < b$, if and only if for every x_0 in the interval there exists a δ such that $f(x_0) \leq \frac{1}{2}[f(x_0 + \epsilon) + f(x_0 - \epsilon)]$ for every

$0 \leq \epsilon \leq \delta$. Show that the convex functions are the one-dimensional counterpart of subharmonic functions by proving the following analogues of Theorems 3.7.3–5:

(a) If $f(x)$ is convex for $a < x < b$ and continuous for $a \leq x \leq b$, and if $M = \max\limits_{a \leq x \leq b} f(x)$, then either $f(x) \equiv M$ or $f(x) < M$, for $a < x < b$.

(b) Let $f(x)$ be convex for $a < x < b$ and continuous for $a \leq x \leq b$. Then $f(x) \leq f(a) + \dfrac{f(b) - f(a)}{b - a} (x - a)$ for $a \leq x \leq b$.

(c) Let $f(x)$ have a continuous second derivative for $a < x < b$. Then $f(x)$ is convex in the interval if and only if $f''(x) \geq 0$, for each x in the interval.

7. Prove that the modulus of an analytic function is subharmonic.

CHAPTER 4

Sequences and Series

4.1. SEQUENCES OF COMPLEX NUMBERS

In this chapter we shall study general questions relating to the convergence of infinite processes, such as sequences, series, products, and improper integrals. In many cases we shall want to define functions as limits of infinite processes and will be able to infer the properties of these functions by studying the sequences involved and the nature of convergence of these sequences. We begin the discussion by considering infinite sequences of complex numbers.

Definition 4.1.1. An infinite sequence of complex numbers is a function defined on the positive integers into the unextended complex plane. Let $w_n = f(n), n = 1, 2, 3, \ldots$. Then $w_1, w_2, w_3, \ldots$ is a sequence of complex numbers, which we abbreviate by $\{w_n\}$.

Definition 4.1.2. The complex number w is a *limit* of the sequence $\{w_n\}$ if and only if for every $\epsilon > 0$ there exists an $N(\epsilon)$ such that $|w_n - w| < \epsilon$, for all $n > N(\epsilon)$. We write $\lim_{n \to \infty} w_n = w$ and say that the sequence *converges* to w. If a sequence does not have a limit, we say that it *diverges*.

EXAMPLE 4.1.1. Let $w_n = 1/n + (n - 1)i/n$. Show that i is a limit of the sequence. We write

$$|w_n - w| = \left| \frac{1}{n} + \frac{n-1}{n} i - i \right| = \left| \frac{1}{n} - \frac{1}{n} i \right| \leq \frac{2}{n}.$$

Now, let $N(\epsilon) = [2/\epsilon]^* + 1$; then when $n > N(\epsilon)$, $n > 2/\epsilon$ and $2/n < \epsilon$. Therefore, $|w_n - w| < \epsilon$, for all $n > N(\epsilon)$.

In the example, we could have written $w_n = 1/n + (n - 1)i/n = u_n + iv_n$; that is $u_n = 1/n$, $v_n = (n - 1)/n$, where $\{u_n\}$ and $\{v_n\}$ are real sequences. Clearly, $\lim_{n \to \infty} u_n = u = 0$ and $\lim_{n \to \infty} v_n = v = 1$, and $w = u + iv$.

*[x] is the notation for the *greatest integer in x*.

Theorem 4.1.1. *Let* $\{w_n\}$ *be a sequence of complex numbers with* $w_n = u_n + iv_n$, u_n *and* v_n *real. Then* $\lim\limits_{n\to\infty} w_n = w = u + iv$ *if and only if* $\lim\limits_{n\to\infty} u_n = u$ *and* $\lim\limits_{n\to\infty} v_n = v$.

Proof: Let $\lim\limits_{n\to\infty} w_n = w = u + iv$. Then given $\epsilon > 0$, there exists $N(\epsilon)$ such that $|u_n + iv_n - (u + iv)| < \epsilon$, for all $n > N(\epsilon)$. Now $|u_n - u| \le |w_n - w| < \epsilon$ and $|v_n - v| \le |w_n - w| < \epsilon$. Hence, $\lim\limits_{n\to\infty} u_n = u$ and $\lim\limits_{n\to\infty} v_n = v$.

Let $\lim\limits_{n\to\infty} u_n = u$ and $\lim\limits_{n\to\infty} v_n = v$. Then given $\epsilon > 0$, there exists $N(\epsilon)$ such that $|u_n - u| < \epsilon/2$, $|v_n - v| < \epsilon/2$, for all $n > N(\epsilon)$. Then

$$|w_n - w| \le |u_n - u| + |v_n - v| < \epsilon.$$

Theorem 4.1.2. *Let* $\{w_n\}$ *be a sequence with a limit w. Then* $\{w_n\}$ *cannot have another limit w^* different from w.*

Proof: Since $w \ne w^*$, $|w - w^*| = \delta \ne 0$. If w and w^* are both limits of $\{w_n\}$, there exists $N(\delta)$ such that $|w_n - w| < \delta/4$ and $|w_n - w^*| < \delta/4$ for all $n > N(\delta)$. Hence,

$$|w - w^*| = |w - w_n + w_n - w^*| \le |w - w_n| + |w^* - w_n| < \frac{\delta}{2}.$$

This contradicts $|w - w^*| = \delta$. Therefore, w^* cannot be different from w.

Theorem 4.1.3. *Let* $\{z_n\}$ *be a sequence with limit z and* $\{\zeta_n\}$ *be a sequence with limit ζ. Then*

1. $\lim\limits_{n\to\infty} (z_n + \zeta_n) = z + \zeta$,
2. $\lim\limits_{n\to\infty} (z_n - \zeta_n) = z - \zeta$,
3. $\lim\limits_{n\to\infty} z_n\zeta_n = z\zeta$,
4. $\lim\limits_{n\to\infty} \dfrac{z_n}{\zeta_n} = \dfrac{z}{\zeta}$ provided $\zeta \ne 0$.

Proof: We shall prove 4 and leave the proofs of 1–3 to the reader. If $\zeta \ne 0$, then $|\zeta| = \delta \ne 0$. Since $\lim\limits_{n\to\infty} \zeta_n = \zeta$, there exists $M(\delta)$ such that $|\zeta_n - \zeta| < \delta/2$ for all $n > M(\delta)$. Now, $|\zeta_n| = |\zeta_n - \zeta + \zeta| \ge ||\zeta| - |\zeta_n - \zeta|| > \delta/2$. Therefore,

$$\left|\frac{z_n}{\zeta_n} - \frac{z}{\zeta}\right| = \left|\frac{z_n\zeta - z\zeta_n}{\zeta_n\zeta}\right| = \left|\frac{\zeta(z_n - z) - z(\zeta_n - \zeta)}{\zeta_n\zeta}\right|$$
$$\le \frac{|\zeta||z_n - z| + |z||\zeta_n - \zeta|}{|\zeta||\zeta_n|}$$
$$< \frac{\delta|z_n - z| + \delta'|\zeta_n - \zeta|}{\delta^2/2}.$$

Let $\delta^* = \max[\delta, \delta']$. Then

$$\left|\frac{z_n}{\zeta_n} - \frac{z}{\zeta}\right| < \frac{2\delta^*}{\delta^2}(|z_n - z| + |\zeta_n - \zeta|).$$

Now given $\epsilon > 0$, there exists N such that $|\zeta_n - \zeta| < \delta/2$, $|z_n - z| < \epsilon\delta^2/4\delta^*$, and $|\zeta_n - \zeta| < \epsilon\delta^2/4\delta^*$, for all $n > N$. Then

$$\left|\frac{z_n}{\zeta_n} - \frac{z}{\zeta}\right| < \frac{2\delta^*}{\delta^2}\left(\frac{\epsilon\delta^2}{4\delta^*} + \frac{\epsilon\delta^2}{4\delta^*}\right) = \epsilon.$$

Geometrically, the meaning of $\lim_{n\to\infty} w_n = w$ is that every ϵ-neighborhood of w contains all but a finite number of members of the sequence. Roughly speaking, this means that for large enough n all the members of the sequence must be close together. This statement can be made rigorous as follows: if $\{w_n\}$ is a sequence with a limit, then for every $\epsilon > 0$, there exists an $N(\epsilon)$ such that $|w_{n+p} - w_n| < \epsilon$, for all $n > N(\epsilon)$ and all positive integers p. Actually, this is also a sufficient condition for a sequence $\{w_n\}$ to have a limit. We can state and prove the following theorem:

Theorem 4.1.4. *Cauchy criterion for convergence. Let $\{w_n\}$ be a sequence. Then $\{w_n\}$ has a limit if and only if for every $\epsilon > 0$, there exists $N(\epsilon)$ such that $|w_{n+p} - w_n| < \epsilon$, for all $n > N(\epsilon)$ and all positive integers p.*

Proof: First, assume that $\{w_n\}$ has a limit w. Then given $\epsilon > 0$, there exists $N(\epsilon)$ such that $|w_n - w| < \epsilon/2$ and $|w_{n+p} - w| < \epsilon/2$, for all $n > N(\epsilon)$ and all positive integers p. Hence,

$$|w_{n+p} - w_n| = |(w_{n+p} - w) + (w - w_n)| \le |w_{n+p} - w| + |w - w_n| < \epsilon.$$

To prove the sufficiency, we consider the sequence as a point set in the unextended plane. This could be a finite set or an infinite set. If finite, then there must be at least one subsequence which is constant for all n sufficiently large. Such a subsequence obviously has a limit. If the point set is infinite it must, nevertheless, be bounded, since given $\epsilon > 0$ there exists $N(\epsilon)$ such that $|w_{n+p} - w_n| < \epsilon$, for all $n > N(\epsilon)$ and all positive integers p. Therefore,

$$|w_{N+p+1} - w_{N+1}| < \epsilon$$

and

$$|w_{N+p+1}| = |w_{N+p+1} - w_{N+1} + w_{N+1}| \le |w_{N+1}| + |w_{N+p+1} - w_{N+1}|$$
$$\le |w_{N+1}| + \epsilon,$$

for all positive integers p. Consequently,

$$|w_n| \le \max[|w_1|, |w_2|, \ldots, |w_{N+1}|, |w_{N+1}| + \epsilon],$$

for all n. We have shown that when $\{w_n\}$ represents an infinite point set, it is a bounded infinite set. By the Bolzano-Weierstrass Theorem (Theorem 1.4.1) there is at least one subsequence which has a limit. Finally, we shall show that if the given sequence has a convergent subsequence, then the limit of the subsequence is the limit of the sequence. Let $\{w_{n_k}\}$ be a convergent subsequence with limit w. Then, given $\epsilon > 0$ there exists an $M(\epsilon)$ such that $|w_{n_k} - w| < \epsilon/2$, for all $n_k > M$. Now,

$$|w_n - w| = |w_n - w_{n+p_k} + w_{n+p_k} - w_{n_k} + w_{n_k} - w|$$
$$\leq |w_n - w_{n+p_k}| + |w_{n+p_k} - w_{n_k}| + |w_{n_k} - w|.$$

Let $N(\epsilon)$ be an integer such that $|w_n - w_{n+p}| < \epsilon/2$, for all $n > N(\epsilon)$ and all positive integers p. Let n be any integer greater than $N(\epsilon)$. There exists an $n_k > M$ with $n_k > n$. Define $p_k = n_k - n$, which is a positive integer. Then $|w_{n+p_k} - w_{n_k}| = 0$ and

$$|w_n - w| \leq |w_n - w_{n+p_k}| + |w_{n_k} - w| < \epsilon.$$

This completes the proof.

The advantage of the Cauchy criterion for convergence is that we can determine if a sequence converges from the sequence itself without prior knowledge of the limit, which in many cases may be very difficult to determine.

EXAMPLE 4.1.2.   Let $\{w_n\}$ be a sequence with the property that $|w_{n+1} - w_n| \leq n^{-a}$, $a > 1$. Prove that the sequence converges. Here we have no hope of determining the limit since we do not know the sequence. However,

$$|w_{n+p} - w_n|$$
$$= |w_{n+p} - w_{n+p-1} + w_{n+p-1} - w_{n+p-2} + \cdots + w_{n+1} - w_n|$$
$$\leq |w_{n+p} - w_{n+p-1}| + |w_{n+p-1} - w_{n+p-2}| + \cdots + |w_{n+1} - w_n|$$
$$\leq (n + p - 1)^{-a} + (n + p - 2)^{-a} + \cdots + n^{-a}.$$

Now, it is well known that the series of real constants $\sum_{k=1}^{\infty} k^{-a}$ converges, for all $a > 1$. Hence, the sequence $\{S_n\}$, where $S_n = \sum_{k=1}^{n} k^{-a}$ converges. By the Cauchy criterion

$$|S_{n+p-1} - S_{n-1}| = (n + p - 1)^{-a} + (n + p - 2)^{-a} + \cdots + n^{-a}$$

approaches zero as n approaches infinity for all p. This proves that $\{w_n\}$ has a limit without determining the limit itself.

Exercises 4.1

1. Consider the members of a sequence of complex numbers as a point set in the plane. Is the point set necessarily infinite? If the sequence converges, is the limit a limit point of the point set? If the point set has a limit point,

is this a limit of the sequence? If the point set has a single limit point, is this the limit of the sequence? Is the point set bounded?

  2. Let $\{a_n\}$ be a bounded nondecreasing sequence of real numbers such that the sequence $\{w_n\}$ satisfies $|w_{n+1} - w_n| \le a_{n+1} - a_n$, for all n. Prove that $\{w_n\}$ converges.

  3. If $\lim_{n\to\infty} w_n = w$, prove that $\lim_{n\to\infty} \frac{1}{n} \sum_{k=1}^{n} w_k = w$.

  4. If $\lim_{n\to\infty} w_n = w$ and $\lim_{n\to\infty} w_n' = w'$, prove that $\lim_{n\to\infty} w_n w_n' = ww'$ and that $\lim_{n\to\infty} \frac{1}{n} \sum_{k=1}^{n-1} w_k w_{n-k}' = ww'$.

  5. Let $\{w_n\}$ be a convergent sequence with limit w. Prove that $\lim_{n\to\infty} \overline{w}_n = \overline{w}$ and that $\lim_{n\to\infty} |w_n| = |w|$. What can you say about $\lim_{n\to\infty} \arg w_n$?

4.2. SEQUENCES OF COMPLEX FUNCTIONS

If a sequence of functions $\{f_n(z)\}$ is defined for all z in a point set S, then, for $z_0 \in S$, $\{f_n(z_0)\}$ is a sequence of complex numbers. If $\{f_n(z_0)\}$ converges, then $f(z_0) = \lim_{n\to\infty} f_n(z_0)$ is uniquely defined. If the sequence converges for every $z \in S$, then we have a uniquely defined function $f(z) = \lim_{n\to\infty} f_n(z)$; that is, the sequence of functions $\{f_n(z)\}$, $z \in S$, defines a function $f(z)$ in S if and only if for every $\epsilon > 0$ and every $z \in S$, there exists $N(\epsilon, z)$ such that $|f(z) - f_n(z)| < \epsilon$, for all $n > N(\epsilon, z)$.

In general, it will not be possible to find a single N, for a given ϵ, which will work for all z in S. However, if this is possible we say we have *uniform convergence* in S.

Definition 4.2.1. A sequence of functions $\{f_n(z)\}$ converges uniformly in S to $f(z)$ if and only if for every $\epsilon > 0$ there exists an $N(\epsilon)$, independent of z, such that $|f_n(z) - f(z)| < \epsilon$, for all $n > N$.

Theorem 4.2.1. *A sequence of functions $\{f_n(z)\}$ converges uniformly in S if and only if for every $\epsilon > 0$ there exists an $N(\epsilon)$, independent of z, such that $|f_{n+p}(z) - f_n(z)| < \epsilon$, for all $n > N(\epsilon)$ and all positive integers p.*

Proof: The proof follows directly from Theorem 4.1.4.

Generally, S will have to be compact, that is, bounded and closed, in order to establish uniform convergence. Therefore, we shall make a special definition to cover this situation.

Definition 4.2.2. A sequence of functions $\{f_n(z)\}$ converges *normally* in S if and only if it converges uniformly in every compact subset of S.

EXAMPLE 4.2.1.   Prove that $f_n(z) = 1 + z + z^2 + \cdots + z^n = (1 - z^{n+1})/(1 - z)$ converges normally for $|z| < 1$. Let $|z| \leq r < 1$. Then

$$\left| f_n(z) - \frac{1}{1 - z} \right| = \frac{|z|^{n+1}}{|1 - z|} \leq \frac{r^{n+1}}{1 - r} < \epsilon,$$

if $r^{n+1} < \epsilon(1 - r)$ or $(n + 1) \ln r < \ln \epsilon(1 - r)$ or $n > \dfrac{\ln \epsilon(1 - r) - \ln r}{\ln r}$.

We may take $N(\epsilon) = \left[\dfrac{\ln \epsilon(1 - r) - \ln r}{\ln r} \right]$. For a fixed r this is obviously independent of z in $|z| \leq r$, and every compact subset of $|z| < 1$ can be considered a subset of $|z| \leq r < 1$ for some fixed r. On the other hand, suppose for all ϵ we assert that N exists independent of z in $|z| < 1$ such that

$$\left| f_n(z) - \frac{1}{1 - z} \right| < \epsilon,$$

for all $n > N$. Let x be defined by

$$N + 1 = \frac{\ln \epsilon(1 - x) - \ln x}{\ln x} \quad \text{or} \quad \frac{x^{N+2}}{1 - x} = \epsilon.$$

For each positive N, $x^{N+2}/(1 - x)$ is a positive, increasing, continuous function for $0 \leq x < 1$ which takes on all values between 0 and $+\infty$. Hence, for every $\epsilon > 0$ and N there are z's in $|z| < 1$ contradicting

$$\left| f_n(z) - \frac{1}{1 - z} \right| < \epsilon,$$

for some $n > N$. Therefore, we do not have uniform convergence for $|z| < 1$.

EXAMPLE 4.2.2.   Prove that $f_n(z) = n^{-z} = e^{-z \log n}$ converges normally in the right half-plane $\operatorname{Re}(z) > 0$. Let $\operatorname{Re}(z) \geq a > 0$. Then $|f_n(z)| = |e^{-z \log n}| = e^{-z \log n} = n^{-x} \leq n^{-a} < \epsilon$ if $n > \epsilon^{-1/a}$. Therefore, we can take $N = [\epsilon^{-1/a}]$ which is independent of z in $\operatorname{Re}(z) \geq a > 0$. We have shown that $\lim_{n \to \infty} f_n(z) = 0$ uniformly in any such closed right half-plane. Any compact set in $\operatorname{Re}(z) > 0$ can be made a subset of a closed right half-plane. Now consider $z = 0$. Then $f_n(0) = e^0 = 1$ and $\lim_{n \to \infty} f_n(0) = 1$. Therefore, we have convergence in $S = \{z \mid |\arg z| < \pi/2 \text{ or } z = 0\}$. However, we do not have uniform convergence in S because the limit function $\{f(z) = 0, \operatorname{Re}(z) > 0; f(0) = 1\}$ is not continuous in S. Uniform convergence in S would contradict the following theorem.

Theorem 4.2.1. *Let $f_n(z)$ be continuous in S for every n $= 1, 2, 3, \ldots$ and let $\{f_n(z)\}$ be uniformly convergent to $f(z)$ in S. Then $f(z)$ is continuous in S.*

Proof: Let z_0 be *any* point in S. Then given $\epsilon > 0$ there exists an $N(\epsilon)$ and a $\delta(\epsilon, z_0)$ such that

$$|f_{N+1}(z_0) - f(z_0)| < \frac{\epsilon}{3},$$

$$|f_{N+1}(z) - f(z)| < \frac{\epsilon}{3},$$

$$|f_{N+1}(z) - f_{N+1}(z_0)| < \frac{\epsilon}{3},$$

for all $z \in S$ such that $|z - z_0| < \delta$. Now,

$$
\begin{aligned}
|f(z) - f(z_0)| \\
&= |f(z) - f_{N+1}(z) + f_{N+1}(z) - f_{N+1}(z_0) + f_{N+1}(z_0) - f(z_0)| \\
&\leq |f(z) - f_{N+1}(z)| + |f_{N+1}(z) - f_{N+1}(z_0)| + |f_{N+1}(z_0) - f(z_0)| \\
&< \epsilon,
\end{aligned}
$$

for all $z \in S$ such that $|z - z_0| < \delta$. This proves that $f(z)$ is continuous at z_0. But z_0 was any point in S, which establishes the theorem.

Corollary 4.2.1. A normally convergent sequence of continuous functions in S converges to a function uniformly continuous in every compact subset of S.

Proof: This is just Theorem 4.2.1 with the added conclusion of uniform continuity of the limit function following from Theorem 2.1.3.

If $f_n(z)$ is continuous on a simple contour C we may integrate it along C, thus defining a sequence of complex numbers

$$I_n = \int_C f_n(z)\, dz.$$

Furthermore, if $\{f_n(z)\}$ is uniformly convergent on C to $f(z)$, then

$$I = \int_C f(z)\, dz.$$

The next theorem asserts that in these circumstances

$$I = \lim_{n \to \infty} I_n = \lim_{n \to \infty} \int_C f_n(z)\, dz = \int_C \lim_{n \to \infty} f_n(z)\, dz = \int_C f(z)\, dz,$$

that is, that we may take the limit under the integral sign.

Theorem 4.2.2. *Let* $\{f_n(z)\}$ *be a sequence of functions continuous on a simple contour C where it is uniformly convergent to* $f(z)$. *Then*

$$\int_C f(z)\, dz = \int_C \lim_{n\to\infty} f_n(z)\, dz = \lim_{n\to\infty} \int_C f_n(z)\, dz.$$

Proof: Given $\epsilon > 0$ there exists an $N(\epsilon)$ such that $|f(z) - f_n(z)| < \epsilon/L$, for all $n > N$ and for all $z \in C$, where L is the length of C. Then

$$\left| \int_C f(z)\, dz - \int_C f_n(z)\, dz \right| = \left| \int_C [f(z) - f_n(z)]\, dz \right|$$

$$\leq \int_C |f(z) - f_n(z)|\, |dz|$$

$$< \frac{\epsilon}{L} \cdot L = \epsilon.$$

Next, we turn to the study of the derivative of the limit of a sequence of analytic functions. Here the notion of normal convergence will play an important role.

Theorem 4.2.3. *Let* $\{f_n(z)\}$ *be a sequence of functions analytic in a domain D which converges normally to* $f(z)$ *in D. Then* $f(z)$ *is analytic in D, where* $f'(z) = \lim_{n\to\infty} f_n'(z)$ *and the convergence is normal in D.*

Proof: Let z_0 be *any* point in D. Then there is a $\rho > 0$ such that $S = \{z \mid |z - z_0| \leq \rho\} \subseteq D$. S is compact. Also

$$f_n(z_0) = \frac{1}{2\pi i} \int_{C_+} \frac{f_n(z)}{z - z_0}\, dz,$$

where C is the circle $|z - z_0| = \rho$. By Theorem 4.2.2

$$f(z_0) = \lim_{n\to\infty} f_n(z_0) = \frac{1}{2\pi i} \int_{C_+} \frac{\lim_{n\to\infty} f_n(z)}{z - z_0}\, dz = \frac{1}{2\pi i} \int_{C_+} \frac{f(z)}{z - z_0}\, dz.$$

This proves that $f'(z_0)$ exists and that

$$f'(z_0) = \frac{1}{2\pi i} \int_{C_+} \frac{f(z)}{(z - z_0)^2}\, dz = \frac{1}{2\pi i} \int_{C_+} \frac{\lim_{n\to\infty} f_n(z)}{(z - z_0)^2}\, dz$$

$$= \lim_{n\to\infty} \frac{1}{2\pi i} \int_{C_+} \frac{f_n(z)}{(z - z_0)^2}\, dz = \lim_{n\to\infty} f_n'(z_0).$$

To determine the nature of the convergence of $\{f_n'(z)\}$, consider

$$\tilde{S} = \{z \mid |z - z_0| \leq \tfrac{1}{2}\rho\}.$$

If $\zeta \in \tilde{S}$, then

$$f_n'(\zeta) = \frac{1}{2\pi i} \int_{C_+} \frac{f_n(z)}{(z - \zeta)^2}\, dz,$$

$$f'(\zeta) = \frac{1}{2\pi i} \int_{C_+} \frac{f(z)}{(z - \zeta)^2}\, dz,$$

$$|f_n'(\zeta) - f'(\zeta)| = \frac{1}{2\pi} \left| \int_{C_+} \frac{f_n(z) - f(z)}{(z - \zeta)^2}\, dz \right|,$$

$$\leq \frac{1}{2\pi} \frac{\epsilon}{\left(\dfrac{\rho}{2}\right)^2} 2\pi\rho = \frac{4\epsilon}{\rho},$$

for all $n > N(\epsilon)$, since we have uniform convergence of $\{f_n(z)\}$ on C. This proves uniform convergence of $\{f_n'(z)\}$ in $\tilde{S}$. But every compact subset of D can be covered by a finite collection of disks $S_i = \{z \mid |z - z_i| \leq r_i\}$ $i = 1, 2, \ldots, m$, where for a given $\epsilon > 0$ there exists $N_i(\epsilon)$ such that $|f_n'(z) - f'(z)| < \epsilon$ for all $n > N_i$ and all $z \in S_i$. Then $N = \max\limits_{i = 1, 2, \ldots, m} N_i$ will suffice for the whole compact subset. Similarly, we can prove normal convergence in D for $k = 2, 3, \ldots$ of

$$\lim_{n \to \infty} f_n^{(k)}(z) = f^{(k)}(z).$$

Definition 4.2.2. A collection of functions, each analytic in a domain D, is a *normal family* if and only if every sequence of functions from the family contains a normally convergent subsequence.

Definition 4.2.3. A collection of functions is locally uniformly bounded in a domain D, if and only if for each $z_0 \in D$ there exists a δ and an M such that $|f(z)| \leq M$ for all z satisfying $|z - z_0| < \delta$ and all functions in the family.

Theorem 4.2.4. *A family of analytic functions locally uniformly bounded in a domain D is a normal family.*

Proof: Consider the sequence of points $z_1, z_2, \ldots$, in D with rational coordinates. This set of points is denumerable and dense in D, that is, every point in D is a limit point of a set of points with rational coordinates. Take any sequence $f_1(z), f_2(z), \ldots$, from the family. Since the family is locally uniformly bounded, there exists an M such that $|f_n(z_1)| \leq M$. Now, the sequence $w_n = f_n(z_1)$ is bounded in the w plane, and therefore contains a subsequence $f_{n_1}(z_1), f_{n_2}(z_1), \ldots$, which converges. We have, consequently, a sequence of functions $f_{n_1}(z), f_{n_2}(z), \ldots$, which converges at z_1. Similarly, the numbers $f_{n_1}(z_2), f_{n_2}(z_2), \ldots$, are bounded in the w plane. Therefore, there exists a subsequence $f_{p_1}(z_2), f_{p_2}(z_2), \ldots$, which converges and the functions

$f_{p_1}(z)$, $f_{p_1}(z)$, ..., converge at z_1 and z_2. In this way, we construct the following array of functions

$$f_{n_1}(z), f_{n_2}(z), f_{n_3}(z), \ldots \quad \text{converges at } z_1,$$
$$f_{p_1}(z), f_{p_2}(z), f_{p_3}(z), \ldots \quad \text{converges at } z_1 \text{ and } z_2,$$
$$f_{q_1}(z), f_{q_2}(z), f_{q_3}(z), \ldots \quad \text{converges at } z_1, z_2, \text{ and } z_3,$$

and so on. Now, the diagonal sequence $f_{n_1}(z), f_{p_2}(z), f_{q_3}(z), \ldots$, converges at $z_1, z_2, z_3, \ldots$. From the original sequence we have obtained a subsequence which converges at every one of a dense set of points in D.

Next, we note that if $f(z)$ is locally uniformly bounded then so is $f'(z)$. Let $z_0 \in D$. Then there exists δ and M such that for all z satisfying

$$|z - z_0| < \delta/2, |f(z)| \leq M, z \in D, \text{ and } f'(z) = \frac{1}{2\pi i} \int_{C_+} \frac{f(\zeta)}{(\zeta - z)^2} \, d\zeta$$

where C is the circle $|\zeta - z_0| = \delta$. Then

$$|f'(z)| \leq \frac{1}{2\pi} \frac{M}{(\delta/2)^2} 2\pi\delta = \frac{4M}{\delta}.$$

Let S be a compact subset of D. For every point z_0 of S there is a disk $|z - z_0| < \delta(z_0)$, $\delta > 0$, such that $|f'(z)| \leq M(z_0)$ for all $f(z)$ in the family. By the Heine-Borel theorem, the set S is covered by a finite number of such disks. Therefore, there is an M independent of $z \in S$ and $f(z)$ in the family such that $|f'(z)| \leq M$. If z_1 and z_2 in D are close enough together they can be joined by a straight line segment in D. Then

$$|f(z_1) - f(z_2)| = \left| \int_{z_1}^{z_2} f'(z) \, dz \right| \leq M|z_2 - z_1|$$

independently of $f(z)$ in the family.*

Again, since S is compact, with the choice of any sequence $\{f_n(z)\}$ from the locally uniformly bounded family, we can find a finite collection of disks $|z - \zeta_j| < \delta_j, j = 1, 2, \ldots, J$ such that (a) S is covered by the disks, (b) $|f_n(z') - f_n(z'')| < \epsilon$ for z' and z'' in a given disk and for all n, (c) each disk contains a point with rational coordinates, (d) there is a subsequence $\{f_{n_k}(z)\}$ which converges on the set of points z_j in D with rational coordinates. Therefore,

$$|f_{n_k}(z) - f_{m_k}(z)| \leq |f_{n_k}(z) - f_{n_k}(z_j)| + |f_{n_k}(z_j) - f_{m_k}(z_j)| + |f_{m_k}(z_j) - f_{m_k}(z)|$$
$$< 3\epsilon,$$

for m_k and n_k sufficiently large. The first and third terms are small by the equicontinuity and the second term is small by the convergence of the subsequence on the dense sense of points. This completes the proof.

* This property is known as *equicontinuity*.

If a family of analytic functions is uniformly bounded in a domain D, that is, $|f(z)| \leq M$ for all f and for all $z \in D$, then it is obviously locally uniformly bounded and, therefore, a normal family. From every sequence $\{f_n(z)\}$ there is a subsequence $\{f_{n_k}(z)\}$ converging normally in D. Furthermore, the limit function is analytic in D and uniformly bounded. We have the following theorem:

Theorem 4.2.5. *Let $\mathfrak{F}$ be a family of functions analytic in a domain D where it is uniformly bounded. Then every sequence of functions in $\mathfrak{F}$ contains a subsequence which converges to a function in the family.*

Notice the similarity between this theorem and the Bolzano-Weierstrass theorem, Theorem 1.4.1. In fact, Theorem 4.2.5 asserts that a uniformly bounded family of analytic functions is compact in the same sense that a point set is compact.

Exercises 4.2

1. Prove that the sequence $(1 + z/n)^n$; $n = 1, 2, 3, \ldots$, converges uniformly in $|z| \leq R < \infty$, for every R. What is the limit?

2. Let $f_n(z) = z^n$. For what values of z does $\{f_n(z)\}$ converge? diverge? converge uniformly?

3. Let $f_n(z) = n^z$. For what values of z does $\{f_n(z)\}$ converge? diverge? converge uniformly?

4. Let $f_n(z) = (1/n) \sin nz$. For what values of z does $\{f_n(z)\}$ converge? diverge? converge uniformly?

5. Let $\{f_n(z)\}$ be a sequence of continuous functions uniformly convergent to $f(z)$ on a simple contour C. Let $g(z)$ be continuous on C. Prove that

$$\lim_{n \to \infty} \int_C f_n(z) g(z) \, dz = \int_C f(z) g(z) \, dz.$$

4.3. INFINITE SERIES

On many occasions it will be necessary for us to work with infinite sums. Therefore, we must carefully define what we mean by the sum of an infinite series. Now that we have studied sequences, we can give our definition in terms of the notion of convergence of infinite sequences. Consider the infinite series of complex numbers $\sum_{k=1}^{\infty} w_k$. Let

$$S_n = \sum_{k=1}^{n} w_k$$

be the sequence of partial sums. Then we give the following definition.

Definition 4.3.1. The infinite series $\sum\limits_{k=1}^{\infty} w_k$ converges to the sum S if and only if the sequence of partial sums $S_n = \sum\limits_{k=1}^{n} w_k$ has the limit S. If the series does not converge, it is said to diverge.

The following theorems follow without further proof from our work on sequences.

Theorem 4.3.1. *The infinite series $\sum\limits_{k=1}^{\infty} w_k$ converges if and only if for every $\epsilon > 0$ there is an $N(\epsilon)$ such that $|S_{n+p} - S_n| = |w_{n+p} + w_{n+p-1} + \cdots + w_{n+1}| < \epsilon$, for all $n > N$ and all positive integers p.*

Theorem 4.3.2. *The infinite series $\sum\limits_{k=1}^{\infty} w_k$, with $w_k = u_k + iv_k$, u_k and v_k real, converges to $S = U + iV$ if and only if $\sum\limits_{k=1}^{\infty} u_k$ converges to U and $\sum\limits_{k=1}^{\infty} v_k$ converges to V.*

Theorem 4.3.3. *A necessary condition for convergence of the infinite series $\sum\limits_{k=1}^{\infty} w_k$ is that*

$$\lim_{n \to \infty} |w_n| = \lim_{n \to \infty} |S_n - S_{n-1}| = 0.$$

Note that Theorem 4.3.3 does not give a sufficient condition for convergence, for the series $\sum\limits_{k=1}^{\infty} 1/k$ diverges in spite of the fact that $\lim\limits_{n \to \infty} (1/n) = 0$.

EXAMPLE 4.3.1. Prove that $\sum\limits_{k=1}^{\infty} z^{k-1}$ converges for $|z| < 1$ and diverges for $|z| \geq 1$. In this case we can find a formula for S_n.

$$S_n = 1 + z + z^2 + \cdots + z^{n-1},$$
$$zS_n = z + z^2 + z^3 + \cdots + z^n,$$
$$(1 - z)S_n = 1 - z^n,$$
$$S_n = \frac{1 - z^n}{1 - z}.$$

If $S = \dfrac{1}{1 - z}$, then $|S - S_n| = \dfrac{|z|^n}{|1 - z|} < \epsilon$, if $n \ln |z| < \ln \epsilon |1 - z|$. Now, if $0 < |z| < 1$, $\ln |z| < 0$ and $|S - S_n| < \epsilon$, if $n > \dfrac{\ln \epsilon |1 - z|}{\ln |z|}$ and we may take $N = \left[\dfrac{\ln \epsilon |1 - z|}{\ln |z|} \right]$. If $|z| \geq 1$ then $|w_n| = |z|^n \geq 1$ and $\lim\limits_{n \to \infty} |w_n| \neq 0$. This proves the divergence for $|z| \geq 1$.

We next give several tests for convergence.

Definition 4.3.2. The infinite series $\sum\limits_{k=1}^{\infty} w_k$ *converges absolutely* if and only if the series $\sum\limits_{k=1}^{\infty} |w_k|$ of nonnegative real numbers converges.

Theorem 4.3.4. *The infinite series* $\sum\limits_{k=1}^{\infty} w_k$ *converges if it converges absolutely.*

Proof: Let $S_n = \sum\limits_{k=1}^{n} w_k$ and $\sigma_n = \sum\limits_{k=1}^{n} |w_k|$. Then

$$|S_{n+p} - S_n| = |w_{n+p} + w_{n+p-1} + \cdots + w_{n+1}|,$$
$$|S_{n+p} - S_n| \leq |w_{n+p}| + |w_{n+p-1}| + \cdots + |w_{n+1}|,$$
$$|S_{n+p} - S_n| \leq \sigma_{n+p} - \sigma_n < \epsilon,$$

for all $n > N(\epsilon)$. The existence of $N(\epsilon)$ follows from the convergence of $\sum\limits_{k=1}^{\infty} |w_k|$.

EXAMPLE 4.3.2. Prove that $\sum\limits_{k=1}^{\infty} k^{-z}$ converges for $\text{Re}(z) > 1$. Here, $|w_k| = |k^{-z}| = k^{-x}$. Now, if $x > 1$,

$$\sigma_n = \sum_{k=1}^{n} k^{-x} < 1 + \int_1^2 t^{-x}\,dt + \int_2^3 t^{-x}\,dt + \cdots + \int_{n-1}^n t^{-x}\,dt,$$

$$\sigma_n < 1 + \int_1^n t^{-x}\,dt = 1 + \frac{n^{1-x}}{1-x} - \frac{1}{1-x},$$

$$\sigma_n < \frac{x - n^{1-x}}{x-1} < \frac{x}{x-1}.$$

Therefore, $\{\sigma_n\}$ is a bounded monotonic sequence, which must, therefore, have a limit. This means that $\sum\limits_{k=1}^{\infty} k^{-z}$ converges absolutely for $\text{Re}(z) > 1$. The given series diverges for $\text{Re}(z) \leq 0$, since $|w_n| = n^{-x}$ does not go to zero as n approaches infinity for $x \leq 0$. The present tests give no information in the case $0 < x \leq 1$.

A sequence may not have a limit and yet have many, even an infinite number, of subsequences which have limits. In connection with real sequences we make the following definitions:

Definition 4.3.3. If the real sequence $\{x_n\}$ is bounded above and has at least one convergent subsequence, then the limit superior ($\overline{\lim}\, x_n$) is the least upper bound of the limits of all convergent subsequences of $\{x_n\}$.

Definition 4.3.4. If the real sequence $\{x_n\}$ is bounded below and has at least one convergent subsequence, then the limit inferior ($\underline{\lim}\, x_n$) is the greatest lower bound of the limits of all convergent subsequences of $\{x_n\}$.

Theorem 4.3.5. *Ratio Test. Let*

$$M = \overline{\lim} \left| \frac{w_{n+1}}{w_n} \right| \quad \text{and} \quad m = \underline{\lim} \left| \frac{w_{n+1}}{w_n} \right|.$$

Then if $M < 1$, $\sum_{k=1}^{\infty} w_k$ *converges; if* $1 < m$, $\sum_{k=1}^{\infty} w_k$ *diverges; if* $m \leq 1 \leq M$
no conclusion can be reached concerning the convergence.

Proof: If $M < 1$ then there exists an N and an r such that $|w_{n+1}/w_n| \leq r < 1$, for all $n \geq N$. Therefore, $|w_{N+1}| \leq r|w_N|$, $|w_{N+2}| \leq r|w_{N+1}| \leq r^2|w_N|$, $\ldots$, $|w_{N+j}| \leq r^j|w_N|$. Then

$$\sum_{k=1}^{\infty} |w_k| \leq |w_1| + |w_2| + \cdots + |w_N|(1 + r + r^2 + \cdots),$$

$$\sum_{k=1}^{\infty} |w_k| \leq \sum_{k=1}^{N-1} |w_k| + |w_N| \frac{1}{1-r}.$$

Hence, $\sum_{k=1}^{\infty} w_k$ converges absolutely.

If $1 < m$ then there exists an N and an r such that $|w_{n+1}/w_n| \geq r > 1$, for all $n \geq N$. Therefore, $|w_{N+1}| \geq r|w_N|$, $|w_{N+2}| \geq r|w_{N+1}| \geq r^2|w_N|$, $\ldots$, $|w_{N+j}| \geq r^j|w_N|$. Hence, $|w_{N+j}| \geq r^j|w_N| \to \infty$ as $j \to \infty$, showing that the series diverges.

To show that the test fails in the third case we consider two examples, $\sum_{k=1}^{\infty} 1/k$ and $\sum_{k=1}^{\infty} 1/k^2$. In the first case the series diverges even though $M = \lim_{n \to \infty} \frac{n}{n+1} = 1$. In the second case the series converges even though $m = \lim_{n \to \infty} \frac{n^2}{(n+1)^2} = 1$.

Theorem 4.3.6. *Root Test. Let* $M = \overline{\lim}|w_n|^{1/n}$. *If* $M < 1$ *the series* $\sum_{k=1}^{\infty} w_k$ *converges; if* $M > 1$, $\sum_{k=1}^{\infty} w_k$ *diverges; if* $M = 1$ *then no conclusion can be reached about the convergence of* $\sum_{k=1}^{\infty} w_k$.

Proof: If $M < 1$ there exists N and r such that $|w_n|^{1/n} \leq r < 1$, for all $n \geq N$. Then $|w_n| \leq r^n$, for all $n \geq N$ and

$$\sum_{k=1}^{\infty} |w_k| \leq \sum_{k=1}^{N-1} |w_k| + r^N \sum_{j=1}^{\infty} r^{j-1}$$

$$\sum_{k=1}^{\infty} |w_k| \leq \sum_{k=1}^{N-1} |w_k| + r^N \frac{1}{1-r},$$

and the series converges absolutely. If $M > 1$ there is a subsequence $\{w_{k_n}\}$ such that $|w_{k_n}| \geq r^{k_n} > 1$, for all k_n. In this case, $\lim_{n \to \infty} |w_n| \neq 0$ and the

series diverges. If $M = 1$ we can have either convergence or divergence.
Consider $\sum_{k=1}^{\infty} k^{-z}$, where $|w_n|^{1/n} = n^{-x/n} = (n^{1/n})^{-x}$ which approaches one
for every x. Yet we know that the series converges for $x > 1$ and diverges
for $x \leq 0$.

Theorem 4.3.7. *The series* $\sum_{k=1}^{\infty} a_k b_k$ *converges if* $\sum_{k=1}^{\infty} |a_k - a_{k+1}|$ *converges,*
$\lim_{n \to \infty} a_n = 0$ *and* $\left| \sum_{k=1}^{n} b_k \right|$ *is bounded for all n.*

Proof: Let $B_n = \sum_{k=1}^{n} b_k$. Then

$$
\begin{aligned}
a_1 b_1 + a_2 b_2 + \cdots + a_n b_n = \; & a_n(b_n + b_{n-1} + \cdots + b_1) \\
& - a_n(b_{n-1} + b_{n-2} + \cdots + b_1) \\
& + a_{n-1}(b_{n-1} + b_{n-2} + \cdots + b_1) \\
& - a_{n-1}(b_{n-2} + b_{n-3} + \cdots + b_1) \\
& \cdots\cdots\cdots\cdots\cdots\cdots\cdots\cdots\cdots \\
& + a_2(b_2 + b_1) - a_2 b_1 + a_1 b_1
\end{aligned}
$$

$$
\sum_{k=1}^{n} a_k b_k = a_n B_n - \sum_{k=1}^{n-1} (a_{k+1} - a_k) B_k{}^*.
$$

Let $|B_n| \leq B$ and $S_n = \sum_{k=1}^{n} a_k b_k$. Then

$$
|S_{n+p} - S_n| \leq B|a_{n+p}| + B|a_n| + B \sum_{k=n}^{n+p-1} |a_{k+1} - a_k|.
$$

Under the hypotheses $|a_n|$, $|a_{n+p}|$, $\sum_{k=n}^{n+p-1} |a_{k+1} - a_k|$ each approach zero
as n approaches infinity for all positive integers p.

EXAMPLE 4.3.3.   Let $w_{2n-1} = (i/3)^{2n-1}$, $w_{2n} = (2i/3)^{2n}$, $n = 1$, 2,
3, ... Prove that $\sum_{k=1}^{\infty} w_k$ converges. Using the root test

$$
|w_{2n-1}|^{1/(2n-1)} = \tfrac{1}{3}, \qquad |w_{2n}|^{1/2n} = \tfrac{2}{3}.
$$

Then $\overline{\lim}_{k \to \infty} |w_k|^{1/k} = 2/3 < 1$. On the other hand, the ratio test gives us no
information since

$$
\left| \frac{w_{2n}}{w_{2n-1}} \right| = \frac{2^{2n}}{3} \to \infty, \qquad \left| \frac{w_{2n+1}}{w_{2n}} \right| = \frac{1}{3 \cdot 2^{2n}} \to 0,
$$

as $n \to \infty$.

---

* This is called the summation by parts formula. Note its similarity to the integration
by parts formula.

EXAMPLE 4.3.4. Let $w_n = z^n/n!$. Prove that $\sum\limits_{k=0}^{\infty} w_k$ converges for all z.
Using the ratio test

$$\left|\frac{w_{n+1}}{w_n}\right| = \frac{|z|^{n+1}n!}{|z|^n(n+1)!} = \frac{|z|}{n+1} \to 0,$$

as $n \to \infty$. On the other hand,

$$|w_n|^{1/n} = \frac{|z|}{(n!)^{1/n}}$$

and it is not so obvious that $\varlimsup\limits_{n\to\infty} |w_n|^{1/n} < 1$.

EXAMPLE 4.3.5. Prove that $\sum\limits_{k=1}^{\infty} \frac{z^k}{k}$ converges for $|z| \leq 1$ except at $z = 1$.
Using the ratio test,

$$\lim_{n\to\infty} \left|\frac{w_{n+1}}{w_n}\right| = \lim_{n\to\infty} \frac{n}{n+1} |z| = |z|.$$

Hence, the series converges for $|z| < 1$ and diverges for $|z| > 1$. For $z = 1$

$$\sum_{k=1}^{\infty} \frac{1}{k} = 1 + \tfrac{1}{2} + \tfrac{1}{3} + \cdots$$

is the harmonic series which diverges. For $|z| = 1$, $z \neq 1$, we use Theorem
4.3.7 with $a_k = 1/k$ and $b_k = z^k$. Now, $\sum\limits_{k=1}^{\infty} |a_k - a_{k+1}| = 1$ and for
$|z| \leq 1$, $|1 - z| > \delta$,

$$|B_n| = \left|\sum_{k=1}^{n} z^k\right| = \left|\frac{1 - z^{n+1}}{1 - z} - 1\right| = \left|\frac{z(1 - z^n)}{1 - z}\right| < \frac{2}{\delta}.$$

Definition 4.3.4. The series $\sum\limits_{k=1}^{\infty} w_k(z)$ converges uniformly to $S(z)$ on a
point set T, if and only if for every $\epsilon > 0$ there is an $N(\epsilon)$, independent of z
in T, such that $|S_n(z) - S(z)| < \epsilon$, for all $n > N$ and all $z \in T$, where
$S_n(z) = \sum\limits_{k=1}^{n} w_k(z)$.

The following theorems follow from corresponding theorems about sequences, taking $\{S_n(z)\}$ as the sequence in question.

Theorem 4.3.8. *The series $\sum\limits_{k=1}^{\infty} w_k(z)$ converges uniformly on a point set T,
if and only if for every $\epsilon > 0$ there is an $N(\epsilon)$, independent of z in T, such that
$|S_{n+p}(z) - S_n(z)| < \epsilon$, for all $n > N$, all $p \geq 1$, and all $z \in T$.*

Theorem 4.3.9. *If $w_k(z)$ is continuous in a point set T for all k and $\sum\limits_{k=1}^{\infty} w_k(z)$ converges to $S(z)$ uniformly in T, then $S(z)$ is continuous in T.*

Theorem 4.3.10. *If $w_k(z)$ is continuous on a simple contour C for all k and $\sum\limits_{k=1}^{\infty} w_k(z)$ converges to $S(z)$ uniformly on C, then*

$$\int_C S(z)\, dz = \sum_{k=1}^{\infty} \int_C w_k(z)\, dz.$$

Theorem 4.3.11. *If $w_k(z)$ is analytic in a domain D and $\sum\limits_{k=1}^{\infty} w_k(z)$ converges to $S(z)$ normally in D, that is, uniformly in every compact subset of D, then $S(z)$ is analytic in D, where*

$$S'(z) = \sum_{k=1}^{\infty} w_k'(z).$$

Finally, we give a test for uniform convergence known as the Weierstrass M-test.

Theorem 4.3.12. *Let $w_k(z)$ be defined for all $z \in T$ such that $|w_k(z)| \leq M_k$, where M_k is independent of $z \in T$. If $\sum\limits_{k=1}^{\infty} M_k$ converges, then $\sum\limits_{k=1}^{\infty} w_k(z)$ converges absolutely and uniformly in T.*

Proof: Since $\sum\limits_{k=1}^{\infty} M_k$ converges, there exists $N(\epsilon)$ such that

$$\left| \sum_{k=1}^{n+p} M_k - \sum_{k=1}^{n} M_k \right| = M_{n+p} + M_{n+p-1} + \cdots + M_{n+1} < \epsilon,$$

for all $n > N$ and all $p \geq 1$. Now,

$$\left| \sum_{k=1}^{n+p} w_k(z) - \sum_{k=1}^{n} w_k(z) \right| \leq \sum_{k=n+1}^{n+p} |w_k(z)| \leq \sum_{k=n+1}^{n+p} M_k < \epsilon,$$

for all $n > N$, all $p \geq 1$, and all $z \in T$.

EXAMPLE 4.3.6. Prove that $\zeta(z) = \sum\limits_{k=1}^{\infty} k^{-z}$ is an analytic function for $\mathrm{Re}(z) > 1$. This is called the *Riemann zeta function*. Let $\mathrm{Re}(z) \geq \rho > 1$. Then $|k^{-z}| = k^{-x} \leq k^{-\rho} = M_k$. Now, $\sum\limits_{k=1}^{\infty} M_k$ converges (see Example 4.3.2). This proves uniform convergence for $\mathrm{Re}(z) \geq \rho > 1$ and normal convergence for $\text{Re}(z) > 1$. By Theorem 4.3.11, $\zeta(z)$ is analytic for $\text{Re}(z) > 1$, where

$$\zeta'(z) = -\sum_{k=1}^{\infty} (\log k)k^{-z}.$$

Exercises 4.3

1. Prove that $\sum_{k=1}^{\infty} a_k b_k$ is convergent if the series $\sum_{k=1}^{\infty} a_k$ converges and the series $\sum_{k=1}^{\infty} (b_k - b_{k+1})$ is absolutely convergent.

2. Prove that the series $\sum_{k=1}^{\infty} a_k b_k(z)$ converges uniformly in S, if $\sum_{k=1}^{\infty} (a_k - a_{k+1})$ converges absolutely, $\lim_{n \to \infty} a_n = 0$ and the partial sums of $\sum_{k=1}^{\infty} b_k(z)$ are uniformly bounded in S.

3. Prove that $\sum_{k=1}^{\infty} a_k k^{-z}$ is analytic for $\text{Re}(z) > 0$ if $\sum_{k=1}^{\infty} a_k$ is absolutely convergent.

4. Let $\sum_{k=1}^{\infty} a_k$ and $\sum_{k=1}^{\infty} b_k$ be convergent with sums A and B. Let $c_n = \sum_{k=1}^{n} a_k b_{n-k+1}$, and $C_n = \sum_{k=1}^{n} c_k$. Prove that

$$\lim_{n \to \infty} \frac{1}{n} \sum_{k=1}^{n} C_k = AB.$$

Prove that if $\sum_{k=1}^{\infty} c_k$ is convergent, its sum is AB.

5. What are the regions of uniform convergence of

$$\sum_{k=1}^{\infty} \frac{1}{z^2 + k^2}, \quad \sum_{k=1}^{\infty} \frac{1}{z^2 - k^2}, \quad \text{and} \quad \sum_{k=1}^{\infty} \left(\frac{z+1}{z-1}\right)^k?$$

4.4. POWER SERIES

Among all series of functions, those which play the most important role in analytic function theory are the series in powers of z. The series $\sum_{k=0}^{\infty} a_k(z - z_0)^k$ will be called a power series. A trivial translation $\zeta = z - z_0$ transforms the series to the form $\sum_{k=0}^{\infty} a_k \zeta^k$. Therefore, without loss of generality we may study series of the form $\sum_{k=0}^{\infty} a_k z^k$.

Applying the root test, we have $\overline{\lim} |a_n|^{1/n}|z| = (1/R)|z|$ provided the limit superior exists. Then the series converges absolutely for $|z| < R$ and diverges for $|z| > R$, where

$$R = \frac{1}{\overline{\lim} |a_n|^{1/n}}.$$

If $\overline{\lim} |a_n|^{1/n} = 0$, then $R = \infty$ and if $\overline{\lim} |a_n|^{1/n} = \infty$, then $R = 0$. R is called the radius of convergence of the power series. The following examples show that all situations are possible.

EXAMPLE 4.4.1. Find the radius of convergence of $\sum_{k=0}^{\infty} z^k$. Here, $a_n = 1$ and $R = 1/\overline{\lim} |a_n|^{1/n} = 1$. The series converges for $|z| < 1$ and diverges for $|z| > 1$. In this case, the series diverges at all points on the unit circle $|z| = 1$, since the nth term of the series does not go to zero.

EXAMPLE 4.4.2. Find the radius of convergence of $\sum_{k=0}^{\infty} \dfrac{z^k}{k!}$. In this case, it is more convenient to use the ratio test. Here

$$\lim_{n\to\infty} \left| \frac{a_{n+1} z^{n+1}}{a_n z^n} \right| = \lim_{n\to\infty} \frac{|z|}{n+1} = 0 \text{ for } |z| \leq r < \infty,$$

for all r. Therefore, $R = \infty$.

EXAMPLE 4.4.3. Find the radius of convergence of $\sum_{k=0}^{\infty} (k+1)^k z^k$. Here, $\lim_{n\to\infty} |a_n|^{1/n} = \lim_{n\to\infty} (n+1) = \infty$, and $R = 0$. The series converges at one point $z = 0$. In fact, all power series have at least one point of convergence. However, those with a zero radius of convergence have no interesting properties and we shall not study them.

EXAMPLE 4.4.4. Find the radius of convergence of $\sum_{k=1}^{\infty} \dfrac{z^k}{k}$. The ratio test gives us $\lim_{n\to\infty} \left| \dfrac{a_{n+1}}{a_n} \right| |z| = \lim_{n\to\infty} \dfrac{n}{n+1} |z| = |z|$. Therefore, the series converges for $|z| < 1$ and diverges for $|z| > 1$. If $|z| = 1$, we have divergence for $z = 1$ and convergence for $0 < \arg z < 2\pi$. See Exercise 4.3.4.

EXAMPLE 4.4.5. Find the radius of convergence of $\sum_{k=1}^{\infty} \dfrac{z^k}{k^2}$. The ratio test gives us $\lim_{n\to\infty} \left| \dfrac{a_{n+1}}{a_n} \right| |z| = \lim_{n\to\infty} \dfrac{n^2}{(n+1)^2} |z| = |z|$. Therefore, the series converges for $|z| < 1$ and diverges for $|z| > 1$. If $|z| = 1$, the series converges absolutely since $\sum_{k=1}^{\infty} \dfrac{1}{k^2}$ converges.

We see in Examples 4.4.1, 4.4.4, and 4.4.5 that on the circle of convergence we may have convergence everywhere, divergence everywhere, or convergence at some points and divergence at others, so that the nature of convergence on the circle of convergence is really an open question. We shall return to this point later on.

If the radius of convergence R of a power series is positive then the interior of the region of convergence, $|z| < R$, is a domain. We now show that in this domain the power series converges normally and hence represents an analytic function.

Theorem 4.4.1. *If the radius of convergence R of a power series is positive, the series converges normally for $|z| < R$ where it represents an analytic function $f(z)$. The series may be differentiated as many times as we please and all the differentiated series have the radius of convergence R. The coefficients of the original series are given by $a_n = f^{(n)}(0)/n!$*

Proof: Let $|z| \le r < R$. Then $|a_n z^n| = |a_n| \dfrac{|z|^n}{r^n} r^n \le |a_n| r^n$. In the Weierstrass M-test for uniform convergence, we let $M_n = |a_n| r^n$. By definition of the radius of convergence, $\sum_{k=0}^{\infty} |a_k| r^k$ converges for $r < R$. Therefore, we have uniform convergence for $|z| \le r < R$, and since any compact subset of $|z| < R$ is contained in $|z| \le r$ for some $r < R$, we have normal convergence. Hence,

$$f(z) = \sum_{k=0}^{\infty} a_k z^k$$

is analytic for $|z| < R$, and by repeated application of Theorem 4.3.11,

$$f^{(n)}(z) = \sum_{k=n}^{\infty} k(k-1)\cdots(k-n+1)a_k z^{k-n},$$

$$f^{(n)}(0) = n(n-1)\cdots 2 \cdot 1 a_n,$$

$$a_n = \frac{f^{(n)}(0)}{n!}.$$

We have shown that each of the derived series converges for $|z| < R$. We still have to show that they diverge for $|z| > R$. By the root test $\overline{\lim} |k(k-1)\cdots(k-n+1)a_k|^{1/k} = \overline{\lim} |a_k|^{1/k} = 1/R$. This completes the proof.

A possible converse to this theorem would be that if $f(z)$ is analytic in a disk $|z| < R$, then it has a power series representation in the disk. If the function does have a power series representation, then by the previous theorem the coefficients would have to be $a_n = f^{(n)}(0)/n!$ We already know that if $f(z)$ is analytic at the origin, then it has all derivatives at the origin so that $f^{(n)}(0)/n!$ is defined for all n. Hence, there is some hope for a power series representation of the function. As a matter of fact, we have the following theorem:

Theorem 4.4.2. *If $f(z)$ is analytic in some ϵ-neighborhood of the origin, then the power series $\sum_{k=0}^{\infty} \dfrac{f^{(k)}(0)}{k!} z^k$ converges to $f(z)$ for $|z| < R$, where R is the minimum distance from the origin to the points where $f(z)$ fails to be analytic.*

Proof: Let R be the distance from the origin to the nearest point where $f(z)$ fails to be analytic. R is positive, since $f(z)$ is analytic in some ϵ-neighborhood of the origin. Let $C = \{z \mid |z| = r < R\}$. If $|z| < r$, then

$$f(z) = \frac{1}{2\pi i} \int_{C_+} \frac{f(\zeta)}{\zeta - z} \, d\zeta$$

$$= \frac{1}{2\pi i} \int_{C_+} \frac{f(\zeta)}{\zeta} \frac{1}{1 - \dfrac{z}{\zeta}} \, d\zeta$$

$$= \frac{1}{2\pi i} \int_{C_+} \frac{f(\zeta)}{\zeta} \sum_{k=0}^{\infty} \left(\frac{z}{\zeta}\right)^k d\zeta$$

$$= \sum_{k=0}^{\infty} \frac{z^k}{2\pi i} \int_{C_+} \frac{f(\zeta)}{\zeta^{k+1}} \, d\zeta$$

$$= \sum_{k=0}^{\infty} \frac{f^{(k)}(0)}{k!} z^k.$$

The interchange of summation and integration is justified by the uniform convergence of $\sum_{k=0}^{\infty} \left(\dfrac{z}{\zeta}\right)^k$, for $|z| < |\zeta| = r$, that is, uniformity in ζ on C for a fixed z inside C. This proves the theorem. We have obtained the *Maclaurin expansion* of the function.

So far we have shown that the series converges for $|z| < R$. To show that R is the radius of convergence, we would have to show that the series diverges for $|z| > R$. Suppose that the series converges for $z = z_0$, $|z_0| > R$. Then since $\lim_{n \to \infty} \dfrac{f^{(n)}(0)}{n!} z_0^n = 0$, $\dfrac{f^{(n)}(0)}{n!} z_0^n$ is bounded, and for $|z| < |z_0|$,

$$\left| \frac{f^{(n)}(0)}{n!} z^n \right| = \left| \frac{f^{(n)}(0)}{n!} z_0^n \right| \left| \frac{z}{z_0} \right|^n \leq M\rho^n$$

with $\rho < 1$. Hence, the series converges absolutely for $|z| < |z_0|$. In this case, the series represents an analytic function inside a larger circle identical with $f(z)$ for $|z| < R < |z_0|$. Now, it could be that $f(z)$ was not analytic on $|z| = R$ simply because it was not defined there. But if R is not the radius of convergence of the Maclaurin series, there exists a function identical with $f(z)$ for $|z| < R$, but analytic inside a larger circle. This raises some interesting questions such as: When can an analytic function's definition be continued into a larger domain? Is this process unique? How far can the continuation be carried out? We shall consider these questions in the next section.

Suppose that a function $f(z)$ is analytic in some ϵ-neighborhood of z_0. Then if $\zeta = z - z_0$, $z = \zeta + z_0$, $f(\zeta + z_0)$ is analytic in an ϵ-neighborhood of $\zeta = 0$, and

$$f(z) = f(\zeta + z_0) = \sum_{k=0}^{\infty} \frac{f^{(k)}(z_0)}{k!} \zeta^k = \sum_{k=0}^{\infty} \frac{f^{(k)}(z_0)}{k!} (z - z_0)^k.$$

This is the *Taylor Series expansion* of $f(z)$ about z_0. It converges for $|z - z_0| < R$, where R is the radius of the largest circle inside of which $f(z)$ is analytic.

EXAMPLE 4.4.6. Find the Maclaurin expansion of $1/(1 + z^2)$. Here,

$$f(z) = \frac{1}{1 + z^2}, f'(z) = \frac{-2z}{(1 + z^2)^2}, f''(z) = \frac{6z^2 - 2}{(1 + z^2)^3},$$

and so on. Then $f(0) = 1$, $f'(0) = 0$, $f''(0) = -2$, and so on. $f(z) = 1 - z^2 + \cdots$. We see that the differentiation can quickly get out of hand. However, we know that $\frac{1}{1 - w} = \sum_{k=0}^{\infty} w^k$, for $|w| < 1$. Letting $w = -z^2$ we get $\frac{1}{1 + z^2} = \sum_{k=0}^{\infty} (-1)^k z^{2k}$, for $|z| < 1$. The radius of convergence is one, which is the distance from the origin to the singularities at $\pm i$. We note from Theorem 4.4.1 that the power series representation of the function is unique, so that we can be sure that the second method does yield the Maclaurin series.

EXAMPLE 4.4.7. Find the Taylor series for $1/z$ about $z = 1$. We can write

$$\frac{1}{z} = \frac{1}{1 + (z - 1)} = \frac{1}{1 - [-(z - 1)]} = \sum_{k=0}^{\infty} (-1)^k (z - 1)^k$$

converging for $|z - 1| < 1$.

EXAMPLE 4.4.8. Find the Maclaurin expansion of e^z. Here, $f^{(n)}(z) = e^z$ and $f^{(n)}(0) = 1$ and $e^z = \sum_{k=0}^{\infty} \frac{z^k}{k!}$, converging for $|z| < \infty$. Note that entire functions have Maclaurin expansions which converge everywhere in the unextended plane.

With the Taylor series as a tool, we are now able to discuss the zeros of analytic functions. Suppose $f(z)$ is analytic in some ϵ-neighborhood of z_0 and $f(z_0) = 0$. Then

$$f(z) = \frac{f^{(m)}(z_0)}{m!}(z - z_0)^m + \frac{f^{(m+1)}(z_0)}{(m+1)!}(z - z_0)^{m+1} + \cdots,$$

where $m \geq 1$. If $f^{(m)}(z_0) = 0$, for all m, then $f(z) \equiv 0$ in some ϵ-neighborhood of z_0.

Definition 4.4.1. If $f(z)$ is analytic and not identically zero in some ϵ-neighborhood of z_0, then $f(z)$ has a zero of order m at z_0 if and only if $f(z_0) = f'(z_0) = \cdots = f^{m-1}(z_0) = 0$ and $f^{(m)}(z_0) \neq 0$.

Theorem 4.4.3. *If $f(z)$ is analytic in some ϵ-neighborhood of z_0 where it has a zero of order m, then*

$$f(z) = (z - z_0)^m g(z),$$

where $g(z)$ is analytic at z_0 and $g(z) \neq 0$ in some neighborhhood $|z - z_0| < \delta$.

Proof: By definition of a zero of order m

$$f(z) = \frac{f^{(m)}(z_0)}{m!}(z - z_0)^m + \frac{f^{(m+1)}(z_0)}{(m+1)!}(z - z_0)^{m+1} + \cdots$$
$$= (z - z_0)^m g(z),$$

where

$$g(z) = \frac{f^{(m)}(z_0)}{m!} + \frac{f^{(m+1)}(z_0)}{(m+1)!}(z - z_0) + \cdots \text{ and } g(z_0) = \frac{f^{(m)}(z_0)}{m!} \neq 0.$$

Since $g(z)$ is a power series with a positive radius of convergence, $g(z)$ is analytic at z_0. Let $|g(z_0)| = a > 0$. By continuity there exists a δ such that $|g(z) - g(z_0)| < a/2$ for all z such that $|z - z_0| < \delta$. Then $|g(z)| \geq ||g(z_0)| - |g(z) - g(z_0)|| > a/2$. This completes the proof.

Theorem 4.4.4. *If $f(z)$ is analytic in a domain D and vanishes on a sequence of distinct points with a limit in D, then $f(z) \equiv 0$ in D.*

Proof: Let $\{z_n\}$ be a sequence of zeros of $f(z)$ with a limit ζ in D. By continuity $f(\zeta) = \lim_{n \to \infty} f(z_n) = 0$. Therefore, ζ is a zero of $f(z)$. But every neighborhood of ζ contains zeros of the function other than ζ contradicting the previous theorem. Therefore, $f(z) \equiv 0$ in some δ-neighborhood of ζ, that is, there exists a $\delta(\zeta)$ such that $f(z) \equiv 0$ for all z in the disk $|z - \zeta| < \delta$.

Let ζ' be any point in D, not equal to ζ. There exists a Jordan curve C joining ζ and ζ' in D and a finite covering of C by disks, with $|z - \zeta| < \delta$ one of the disks. Let ζ'' be the center of a disk in the covering which overlaps the disk $|z - \zeta| < \delta$. The point $(\zeta + \zeta'')/2$ is in both disks. If we expand the function in a Taylor series about $(\zeta + \zeta'')/2$ we show that $f(z) \equiv 0$ in a disk which includes ζ'' or at least is closer to ζ'' than the original disk. After a finite number of steps we show that $f(\zeta'') = 0$. In this way we can move along C to eventually show that $f(\zeta') = 0$. This proves that $f(z) \equiv 0$ in D.

Corollary 4.4.1. If $f(z)$ and $g(z)$ are analytic in a domain D and are equal on any set of points with a limit point in D, then $f(z) = g(z)$ throughout D.

Proof: Let $h(z) = f(z) - g(z)$. Now, $h(z)$ is analytic in D, and there exists a sequence of distinct points in D, with a limit in D, on which $h(z)$ vanishes. Therefore, $h(z) = f(z) - g(z) \equiv 0$ in D.

This corollary has far reaching consequences which we shall explore more fully in the next section, but for now, we shall make just one observation. Suppose a real-valued function of x has a Taylor expansion,

$$f(x) = \sum_{k=0}^{\infty} \frac{f^{(k)}(x_0)}{k!} (x - x_0)^k,$$

convergent for $|x - x_0| < R$. A natural way to continue this function into the complex plane would be to define

$$f(z) = \sum_{k=0}^{\infty} \frac{f^{(k)}(x_0)}{k!} (z - x_0)^k$$

which converges for $|z - x_0| < R$ and is analytic in the same domain. It obviously is equal to $f(x)$ when $z = x$, $|x - x_0| < R$. Now, the following question should be raised: Is there another way to continue analytically $f(x)$ into the disk $|z - x_0| < R$ which will agree with $f(x)$ on the real axis? The answer to this is no. Suppose $g(z)$ is analytic in the disk and agrees with $f(x)$ on the real axis. Then by the corollary $g(z) \equiv f(z)$ in the disk.

EXAMPLE 4.4.8. Let $f(x) = \sum_{k=0}^{\infty} (-1)^k x^{2k}$, $|x| < 1$. Continue the function analytically into the complex plane. By the above remark the unique continuation is

$$f(z) = \sum_{k=0}^{\infty} (-1)^k z^{2k} = \frac{1}{1 + z^2}, |z| < 1.$$

This result is not surprising since the original function represents $1/(1 + x^2)$ in the given interval. However, this example illustrates an interesting fact about Taylor series on the real axis. The function $1/(1 + x^2)$ is well behaved for all values of x. From the point of view of the real variable x, there is no reason why the Maclaurin series should diverge for $|x| > 1$. The point is that the complex function $1/(1 + z^2)$ has singularities at $\pm i$ and therefore the radius of convergence of the Maclaurin series is one. For this reason the series must diverge on the real axis for $|x| > 1$. We see that to understand the convergence problem we must look at the series in the complex plane.

We conclude this section with a brief discussion of operations between power series. The following theorem covers most situations:

Theorem 4.4.5. *If the power series* $\sum\limits_{k=0}^{\infty} a_k z^k$ *and* $\sum\limits_{k=0}^{\infty} b_k z^k$ *converge for* $|z| < R$, *then the sum* $\sum\limits_{k=0}^{\infty} (a_k + b_k) z^k$, *the difference* $\sum\limits_{k=0}^{\infty} (a_k - b_k) z^k$, *and the Cauchy product* $\sum\limits_{k=0}^{\infty} c_k z^k$, *where* $c_n = \sum\limits_{k=0}^{n} a_k b_{n-k}$, *all converge for* $|z| < R$. *If* $g(z) = \sum\limits_{k=0}^{\infty} b_k z^k$ *never vanishes in* $|z| < R$ *then the series* $\sum\limits_{k=0}^{\infty} b_k z^k$ *can be divided into the series* $\sum\limits_{k=0}^{\infty} a_k z^k$, *where the quotient defined by ordinary long division will converge for* $|z| < R$.

Proof: Let $f(z) = \sum\limits_{k=0}^{\infty} a_k z^k$ and $g(z) = \sum\limits_{k=0}^{\infty} b_k z^k$ both be analytic for $|z| < R$. Then $f(z) + g(z)$, $f(z) - g(z)$, and $f(z)g(z)$ are analytic in the same domain and each has a convergent power series representation for $|z| < R$.

$$f(z) + g(z) = \sum_{k=0}^{\infty} \left[\frac{f^{(k)}(0)}{k!} + \frac{g^{(k)}(0)}{k!} \right] z^k = \sum_{k=0}^{\infty} (a_k + b_k) z^k,$$

$$f(z) - g(z) = \sum_{k=0}^{\infty} \left[\frac{f^{(k)}(0)}{k!} - \frac{g^{(k)}(0)}{k!} \right] z^k = \sum_{k=0}^{\infty} (a_k - b_k) z^k,$$

$$f(z)g(z) = \sum_{k=0}^{\infty} \frac{d^k[f(z)g(z)]}{k! \, dz^k} \bigg|_{z=0} z^k = \sum_{k=0}^{\infty} c_k z^k.$$

Now, by a well-known formula

$$\frac{d^n[f(z)g(z)]}{dz^n} = \sum_{k=0}^{n} \frac{n!}{k!(n-k)!} f^{(k)}(z) g^{(n-k)}(z).$$

Hence,

$$c_n = \frac{1}{n!} \sum_{k=0}^{n} \frac{n!}{k!(n-k)!} f^{(k)}(0) g^{(n-k)}(0) = \sum_{k=0}^{n} a_k b_{n-k}.$$

If $g(z)$ does not vanish, then $f(z)/g(z)$ is analytic for $|z| < R$, and has a Maclaurin expansion the coefficients of which can be computed by differentiating the quotient. For example,

$$\frac{f(z)}{g(z)} = \sum_{k=0}^{\infty} d_k z^k,$$

where

$$d_0 = \frac{f(0)}{g(0)} = \frac{a_0}{b_0}, d_1 = \frac{g(0)f'(0) - f(0)g'(0)}{[g(0)]^2} = \frac{b_0a_1 - a_0b_1}{b_0^2},$$

$$d_2 = \frac{2f(g')^2 - fgg'' - 2f'gg' + f''g^2}{2g^3}\bigg|_{z=0}$$

$$= \frac{a_0b_1^2 - a_0b_0b_2 - a_1b_0b_1 + a_2b_0^2}{b_0^3},$$

and so on. These coefficients are just what one obtains by long division.

Somewhat in the same vein, because of the method of proof we use, is the following theorem:

Theorem 4.4.6. *If $w = f(z)$ is analytic at z_0 and $f'(z_0) \neq 0$, then there exists an ϵ-neighborhood of $w_0 = f(z_0)$ in which the inverse function $z = g(w)$ exists and is analytic. Also, $g'(w_0) = 1/f'(z_0)$.*

Proof: Since $f(z)$ is analytic at z_0, there exists an R such that

$$w - w_0 = a_1(z - z_0) + a_2(z - z_0)^2 + \cdots$$

converges for $|z - z_0| \leq R$. Now, $a_1 = f'(z_0) \neq 0$, so we can make the change of variables $\omega = (w - w_0)/a_1, \zeta = z - z_0$. Then we have $\omega = \zeta + c_1\zeta^2 + c_2\zeta^3 + \cdots = \zeta g(\zeta)$, where $g(\zeta) = 1 + c_1\zeta + c_2\zeta^2 + \cdots$ is analytic for $|\zeta| \leq R$. By Cauchy's inequalities, we have $|c_n| \leq M/R^n$, where $M = \max_{|\zeta|=R} |g(\zeta)|$. We look for a series $\zeta = b_1\omega + b_2\omega^2 + b_3\omega^3 + \cdots$. Then,

$$\zeta = \omega - c_1\zeta^2 - c_2\zeta^3 - \cdots,$$
$$b_1\omega + b_2\omega^2 + b_3\omega^3 + \cdots = \omega - c_1\omega^2(b_1 + b_2\omega + b_3\omega^2 + \cdots)^2$$
$$- c_2\omega^3(b_1 + b_2\omega + b_3\omega^2 + \cdots)^3 - \cdots.$$

Equating coefficients of like powers of ω, we have $b_1 = 1$, $b_2 = -c_1$, $b_3 = -2c_1b_2 - c_2$, and so on. This shows that, at least formally, we can determine the coefficients of the power series. To complete the proof we have to show that the formal power series has a positive radius of convergence. To this purpose, we use the *Cauchy method of majorants*.

A power series $b_0 + b_1z + b_2z^2 + \cdots$ is said to be *majorized* by a second power series $B_0 + B_1z + B_2z^2 + \cdots$, if, for all n, $|b_n| \leq B_n$. Clearly, if the *majorant* $B_0 + B_1z + B_2z^2 + \cdots$ converges for $|z| \leq r$, then the original series converges absolutely in the same region. Now, consider the series

$$z = W + C_1z^2 + C_2z^3 + \cdots,$$

where $|c_n| \leq C_n$, for $n = 1, 2, 3, \ldots$. This determines a formal power series $z = B_1 W + B_2 W^2 + B_3 W^3 + \cdots$ which is a majorant of $\zeta = b_1\omega + b_2\omega^2 + b_3\omega^3 + \cdots$. Clearly, $B_1 = 1$, $B_2 = C_1$, $B_3 = 2C_1B_2 + C_2$, and so on. Then $b_1 = B_1$, $|b_2| \leq |c_1| \leq C_1 = B_2$, $|b_3| \leq 2|c_1| |b_2| + |c_2| \leq 2C_1B_2 + C_2 = B_3$, and so on. Therefore, to complete the proof all we have to do is obtain a majorant of $\zeta = b_1\omega + b_2\omega^2 + b_3\omega^3 + \cdots$. Let

$$z = W + \frac{M}{R}z^2 + \frac{M}{R^2}z^3 + \cdots,$$

$$z = W + \frac{M}{R}z^2 \left(1 + \frac{z}{R} + \frac{z^2}{R^2} + \cdots\right),$$

$$z = W + \frac{M}{R}z^2 \frac{1}{1 - \dfrac{z}{R}}, \qquad |z| < R,$$

$$z = W + \frac{Mz^2}{R - z},$$

or $(M + 1)z^2 - (R + W)z + WR = 0$. The root of this equation which goes to zero as W goes to zero is

$$z = \frac{R + W - \sqrt{(R + W)^2 - 4WR(M + 1)}}{2(M + 1)}.$$

Therefore, z is an algebraic function which has branch points where $(R + W)^2 - 4WR(M + 1) = 0$. Solving this equation for W, we have

$$W = R(1 + 2M \pm\sqrt{(1 + 2M)^2 - 1}).$$

Of the two branch points (both real) the one nearest the origin is the one with the minus sign. We have obtained an explicit inverse, which is analytic for $|W| \leq R(1 + 2M - \sqrt{(1 + 2M)^2 - 1})$, and therefore has a power series expansion $z = B_1 W + B_2 W^2 + B_3 W^3 + \cdots$ convergent in the same region. By the uniqueness of the Maclaurin expansion, the coefficients must be B's determined formally from $C_n = M/R^n$. Therefore, $\zeta = b_1\omega + b_2\omega^2 + b_3\omega^3 + \cdots$ must converge for

$$|\omega| < R(1 + 2M - \sqrt{(1 + 2M)^2 - 1})$$

and the inverse of $w = w_0 + a_1(z - z_0) + a_2(z - z_0)^2 + \cdots$ must exist for

$$|w - w_0| \leq |a_1|R(1 + 2M - \sqrt{(1 + 2M)^2 - 1}).$$

Finally,

$$\left(\frac{dz}{dw}\right)_{w=w_0} = \frac{1}{a_1}\left(\frac{d\zeta}{d\omega}\right)_{\omega=0} = \frac{1}{a_1} = \frac{1}{f'(z_0)}.$$

Another proof of this theorem, based on contour integration, will be given in Chapter 5. For a still different proof see E. Hille, *Analytic Function*

Theory, I. Boston: Ginn and Company, 1959, pp. 86–90. The method of majorants is important in ordinary and partial differential equations for proving existence and uniqueness theorems. One such proof will be given in Chapter 7.

Exercises 4.4

1. Find the radius of convergence of the power series with each of the following coefficients: (a) $a_k = k$; (b) $a_k = k!$; (c) $a_k = 2^k$; (d) $a_k = e^{-k}$; (e) $a_k = \cos k$.

2. Find the radius of convergence of the power series $\sum_{k=0}^{\infty} z^{2^k}$ and $\sum_{k=0}^{\infty} z^{k!}$.

3. Sum each of the following power series: (a) $\sum_{k=0}^{\infty} (-1)^k z^{2k}$, (b) $\sum_{k=1}^{\infty} (-1)^k k z^k$, (c) $\sum_{k=1}^{\infty} k^2 z^k$.

4. Find the Maclaurin series representation of (a) e^z, (b) e^{z^2}, (c) $\cos z$, (d) $\sinh z$, (e) $\dfrac{z+1}{z-1}$.

5. Find the radius of convergence of the binomial series

$$1 + \frac{\alpha z}{1!} + \frac{\alpha(\alpha-1)}{2!} z^2 + \frac{\alpha(\alpha-1)(\alpha-2)}{3!} z^3 + \cdots.$$

Discuss the convergence on the circle of convergence for various values of α.

6. Let $\sum_{k=0}^{\infty} a_k z^k$ converge at z_0, $|z_0| = R > 0$, where R is the radius of convergence. Prove that for $0 \le r \le 1$,

$$\lim_{r \to 1} \sum_{k=0}^{\infty} a_k z_0^k r^k = \sum_{k=0}^{\infty} a_k z_0^k.$$

7. Let a_n be real and $a_n \ge a_{n+1} \to 0$ as $n \to \infty$. Prove that $\sum_{k=0}^{\infty} a_k z^k$ converges for $|z| < 1$ and converges for $|z| = 1$, $z \neq 1$.

8. Let $\sum_{k=0}^{\infty} a_k = A$. Prove that $\lim_{z \to 1} \sum_{k=0}^{\infty} a_k z^k = A$ provided z approaches one in the sector $|z| < 1$, $\pi/2 < \arg(z-1) < 3\pi/2$.

9. Find the Taylor series for $\operatorname{Log} z$ in powers of $z - 1$ by integrating the appropriate series for $(d/dz) \operatorname{Log} z = 1/z$.

10. Prove that $e^{z_1} e^{z_2} = e^{z_1+z_2}$ using the power series for e^z.

11. Let $g(z) = b_0 + b_1 z + b_2 z^2 + \cdots$, $|z| \le R$, $b_0 \neq 0$. Find the power series representation of $f(z) = 1/g(z)$ and prove that it converges in some ϵ-neighborhood of the origin using the method of majorants. Find a lower bound for the radius of convergence.

4.5. ANALYTIC CONTINUATION

In the last section, we raised the question of the possibility of extending the definition of an analytic function to a larger domain. This will be the subject of the present section. Consider the following example.

EXAMPLE 4.5.1. The power series $\sum_{k=0}^{\infty} z^k$ converges for $|z| < 1$, where it represents an analytic function. The series diverges for $|z| \geq 1$ and so the domain $|z| < 1$ is the largest possible domain of convergence of the series. However, the series

$$\frac{1}{2} \sum_{k=0}^{\infty} \left(\frac{z+1}{2}\right)^k$$

converges for $|z + 1| < 2$ and agrees with the first series in the common domain $|z| < 1$. In a very real sense the second series extends the definition of the analytic function represented by the first to a larger domain. Now, consider the function $1/(1 - z)$, $z \neq 1$, which is analytic in the extended plane except at $z = 1$. Where they converge, both series represent $1/(1 - z)$. Hence, they both represent parts of the function $1/(1 - z)$ which encompasses all possible extensions of the original series as an analytic function. There is no extension which is analytic at $z = 1$ which agrees with $1/(1 - z)$ in a deleted ϵ-neighborhood of $z = 1$. This is because $|1/(1 - z)| \to \infty$ as $z \to 1$. We make these ideas more precise in the following discussion.

Definition 4.5.1. An analytic function element (f, D) is an analytic function $f(z)$ along with its domain of definition D. A function element (f_2, D_2) is a *direct analytic continuation* of another element (f_1, D_1) if $D_1 \cap D_2 \neq \emptyset$ and if $f_1 \equiv f_2$ *in* $D_1 \cap D_2$.

Theorem 4.5.1. *If* (f_2, D_2) *and* (f_3, D_3) *are both direct analytic continuations of* (f_1, D_1) *for which* $D_2 \cap D_3$ *is connected and not empty and* $D_1 \cap D_2 \cap D_3 \neq \emptyset$, *then* $f_2 \equiv f_3$ *in* $D_2 \cap D_3$.

Proof: $D_2 \cap D_3$ is a domain, since it is open and connected. In $D_1 \cap D_2 \cap D_3, f_1 \equiv f_2 \equiv f_3$. Therefore, $f_2 \equiv f_3$ in $D_2 \cap D_3$.

Corollary 4.5.1. If $D_1 \cap D_2 \neq \emptyset$ then a direct analytic continuation of (f_1, D_1) into D_2 is unique if it exists.

Proof: This follows from Theorem 4.5.1 with $D_2 = D_3$.

It may be that no continuation of a given function element exists. Consider the following example.

EXAMPLE 4.5.2. Show that $\sum_{k=0}^{\infty} z^{2^k}$ has no analytic continuation. The radius of convergence is one, so that the domain of definition is $|z| < 1$. If there existed an analytic continuation by a function element analytic in an

ϵ-neighborhood of $z = 1$, then $f(x) = \sum\limits_{k=0}^{\infty} x^{2^k}$, $-1 < x < 1$, would have
to approach a finite value as x approaches one. But this is not the case.
Let M be an arbitrary positive number. Then there are numbers N and a
such that $Na^{2^N} > M$, $0 < a < 1$. As a matter of fact, all that is required
is that $N > M$ and $\ln a > (\ln M - \ln N)/2^N$. Then

$$f(a) = \sum_{k=0}^{N} a^{2^k} + \sum_{k=N+1}^{\infty} a^{2^k} > M \qquad \text{and} \qquad f(x) > M,$$

for all x such that $a \leq x < 1$. Hence, there can be no analytic continuation
through $z = 1$. Next, we note that $f(z) = z + f(z^2)$, which means that
there can be no continuation through points $z^2 = 1$, that is, $z = \pm 1$.
Similarly, $f(z) = z + z^2 + f(z^4)$. Hence, there can be no continuation
through $z^4 = 1$ or $z = 1, -1, i, -i$. Likewise, there can be no continuation
through $z^n = 1$ for any positive integer n. But the roots of this polynomial
are evenly spaced around the unit circle and can be made arbitrarily close
to one another by taking n large. There can be no continuation anywhere
across the unit circle, which in this case is called a *natural boundary*.

Up to this point, when we have referred to an analytic function, we have
been talking about a single-valued function defined and differentiable in some
domain, that is, open connected set. Therefore, we have really been talking
about analytic function elements. If an element (f_1, D_1) has a direct continuation (f_2, D_2), there is nothing to keep us from defining a function $f(z)$
in $D_1 \cup D_2$ as follows: $f(z) = f_1(z)$ in D_1, $f(z) = f_2(z)$ in D_2. Also, there
is no reason to stop with one continuation, if others exist. In fact, starting
with the element (f_1, D_1) we can include all elements which are direct continuations of this element, then continue all these elements, and so on. By
this process, we obtain the definition of a *complete analytic function*. We
should be prepared for the possibility of this process leading to a multiple-
valued function. Actually, this is not so bad, for it gives us the possibility
of defining functions such as $z^{1/2}$, $\log z$, and so on, and gives us a better
understanding of the Riemann surface for multiple-valued functions.

Definition 4.5.2. A *complete analytic function* is the collection of all
possible analytic function elements starting with a given element, taking all
possible direct analytic continuations of this element, all possible direct
analytic continuations of these, and so on. A *singularity* of a complete
analytic function is a point which is a limit point of a domain of one or
more elements but which is not itself in the domain of any element.

EXAMPLE 4.5.3. Consider

$$\frac{1}{z} = \frac{1}{1 + (z - 1)} = \sum_{k=0}^{\infty} (-1)^k (z - 1)^k,$$

The series converges uniformly for $|z - 1| \leq \rho < 1$. Therefore, we may integrate term-by-term along a straight line path from one to z for any z such that $|z - 1| < 1$. Thus,

$$\log z = \int_1^z \frac{1}{\zeta} \, d\zeta = \sum_{k=0}^{\infty} \frac{(-1)^k (z - 1)^{k+1}}{k + 1}, |z - 1| < 1,$$

where $\log z$ is a branch which takes on the value zero for $z = 1$. To obtain a direct analytic continuation of this power series we can obtain the Taylor expansion about the point $z_0 = 1/\sqrt{2} + i/\sqrt{2} = e^{i\pi/4}$. We have

$$f(z_0) = \int_1^{e^{i\pi/4}} \frac{1}{\zeta} \, d\zeta = \int_0^{\pi/4} \frac{ie^{i\theta} \, d\theta}{e^{i\theta}} = \frac{i\pi}{4}, \frac{f^{(n)}(z_0)}{n!} = \frac{(-1)^{n+1}}{n} e^{-in\pi/4}.$$

Hence, continuing in this manner

$$f_1(z) = \sum_{k=1}^{\infty} \frac{(-1)^{k+1}(z - 1)^k}{k}, |z - 1| < 1,$$

$$f_2(z) = \frac{i\pi}{4} + \sum_{k=1}^{\infty} \frac{(-1)^{k+1}(z - e^{i\pi/4})^k}{ke^{ik\pi/4}}, |z - e^{i\pi/4}| < 1$$

$$f_3(z) = \frac{i\pi}{2} + \sum_{k=1}^{\infty} \frac{(-1)^{k+1}(z - e^{i\pi/2})^k}{ke^{ik\pi/2}}, |z - e^{i\pi/2}| < 1$$

$$\cdots \cdots \cdots \cdots \cdots \cdots \cdots \cdots \cdots$$

$$f_8(z) = \frac{7i\pi}{4} + \sum_{k=1}^{\infty} \frac{(-1)^{k+1}(z - e^{7i\pi/4})^k}{ke^{7ik\pi/4}}, |z - e^{7i\pi/4}| < 1$$

$$f_9(z) = 2\pi i + \sum_{k=1}^{\infty} \frac{(-1)^{k+1}(z - 1)^k}{k}, |z - 1| < 1.$$

By a process of direct analytic continuation by power series from one element to the next, we have arrived back at the original domain $|z - 1| < 1$. However, the process does not lead to the same values as the original element. This illustrates the fact that the complete analytic function may be many-valued since if $z_0 \in D$, the process of analytic continuation may lead to many elements $(f_1, D), (f_2, D), (f_3, D), \ldots$, such that $f_1(z_0), f_2(z_0), f_3(z_0), \ldots$, are all different.

The construction of a complete analytic function, which is many-valued, leads in a natural way to the construction of the Riemann surface for the function. We consider the elements as patches of the Riemann surface. Elements which are direct continuations of one another are attached over their common domain. As we continue the surface grows. If the continuation around a closed path (as in Example 4.5.3) leads back to a different element at the starting point, then we have two patches over the same domain, but the corresponding elements are not the same, so that the patches do not coalesce but form parts of two sheets of the Riemann surface. If the continuation around every closed path leads back to the original element, the Riemann surface has but one sheet. Otherwise, we end up with two or more sheets.

We now show that if we continue an analytic function around a simple closed Jordan curve we cannot come back to the starting point with a different element unless there is a singularity of the complete analytic function inside the curve. Or what is the same thing, if we start with a given element at z_1, and continue to z_2 along two different paths using power series elements, where there is no singularity of the complete analytic between the paths, then we must arrive at z_2, in either case, with the same value of the complete analytic function. This result is contained in the *monodromy theorem*.

Theorem 4.5.2. *Let z_1 and z_2 be two points in the unextended complex plane. Let C_1 and C_2 be two simple Jordan curves joining z_1 and z_2 such that $C_1 \cup C_2$ is a simple closed Jordan curve. Starting with an analytic function element $f(z)$ defined in D containing z_1, suppose we can continue $f(z)$ along C_1 and C_2 using power series elements with centers on the curves. If there is no singularity of the complete analytic function defined by continuation of $f(z)$ in the interior of $C_1 \cup C_2$, then the continuations along C_1 and C_2 must lead to the same value at z_2.*

Proof: We first prove the theorem in the case where $C_1 \cup C_2$ is a triangle, that is, where C_1 is a straight line segment joining z_1 and z_2 and C_2 is a broken line consisting of line segments joining z_1, a third point z_3, and z_2. See Figure 4.5.1. Let z_{12}, z_{13}, and z_{23} be the midpoints of the segments $[z_1, z_2]$, $[z_1, z_3]$, and $[z_2, z_3]$, respectively. We show that if the analytic continuations along C_1 and C_2 lead to different values at z_2, then the same

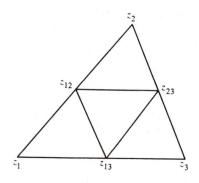

Figure 4.5.1.

property must be true of at least one of the smaller triangles. To do this we introduce the following notation. Let $C[a, b]$ denote the continuation by power series elements along the straight line segment joining a and b. Let $C[a, b, c]$ denote the continuation by power series elements along the broken line path joining the three points. We say that $C[a, b] = C[a, c, b]$ if and only if the two continuations from the same function element at a results in the same value at b.

In the present case, we wish to prove that if there is no singularity in the interior of the triangle, then $C[z_1, z_2] = C[z_1, z_3, z_2]$. Assume that

$C[z_1, z_2] \neq C[z_1, z_3, z_2]$. Now, if $C[z_1, z_{12}] = C[z_1, z_{13}, z_{12}], C[z_{12}, z_2] = C[z_{12}, z_{23}, z_2]$, $C[z_{13}, z_{12}, z_{23}] = C[z_{13}, z_{23}]$, and $C[z_{13}, z_{23}] = C[z_{13}, z_3, z_{23}]$, then

$$C[z_1, z_2] = C[z_1, z_{13}, z_{12}, z_{23}, z_2] = C[z_1, z_{13}, z_{23}, z_2]$$
$$= C[z_1, z_{13}, z_3, z_{23}, z_2] = C[z_1, z_3, z_2].$$

Therefore, if $C[z_1, z_2] \neq C[z_1, z_3, z_2]$ then the continuation around at least one of the smaller triangles must result in a different value. Let us say that $C[z_1, z_{12}] \neq C[z_1, z_{13}, z_{12}]$. Then we subject this triangle to same treatment as the original, and we arrive at a still smaller triangle with a continuation around it which results in a different element. Continuing the process, we can obtain a triangle as small as we please, lying in the closure of the interior of the original triangle, and such that the analytic continuation around it results in different elements. This is, however, impossible if there is no singularity in the interior of $C_1 \cup C_2$, for if the triangle is small enough, it lies entirely inside the circle of convergence of one of the power series elements used to define the complete analytic function. Inside the circle of convergence of a power series element the function is obviously single-valued.

Next, we proceed as in the proof of Cauchy's theorem (Theorem 3.3.1). We prove that the theorem is true where C_1 and C_2 are polygonal paths. We triangulate the closed polygon $C_1 \cup C_2$ by adding interior diagonals as we did in the proof of Cauchy's theorem. If the analytic continuation along C_1 and C_2 lead to different values at z_2, then there must be at least one triangle in the triangulation of the polygon which has the same property. But this is clearly impossible if there is no singularity in the interior of $C_1 \cup C_2$. Finally, in the general case, we observe that there is a finite chain of overlapping disks with centers on C_1, representing the power series elements in the analytic continuation, such that both the curve C_1 and the polygonal path joining the centers of the disks lie in the union of the disks. The same is true of C_2 and an appropriate polygonal path with vertices on C_2. If the continuations along C_1 and C_2 lead to different elements at z_2, then the same is true for the corresponding polygonal paths. But this is clearly impossible if there is no singularity in the interior of $C_1 \cup C_2$. This completes the proof.

We complete this section by giving one of the few, more or less, general methods of continuing an analytic function element, other than by power series.

Theorem 4.5.3. *Let D_1 and D_2 be two domains such that $D_1 \cap D_2 = \emptyset$, while $\overline{D}_1 \cap \overline{D}_2 = \Gamma$, a simple contour. Let (f_1, D_1) and (f_2, D_2) be analytic function elements which are continuous in $D_1 \cup \Gamma$ and $D_2 \cup \Gamma$, respectively. If for all $z \in \Gamma, f_1(z) = f_2(z)$, then*

$$F(z) = \begin{cases} f_1(z), & z \in D_1, \\ f_2(z), & z \in D_2, \\ f_1(z) = f_2(z), & z \in \Gamma, \end{cases}$$

is analytic in $D_1 \cup D_2 \cup \Gamma^*$, where Γ^* is Γ without its endpoints.

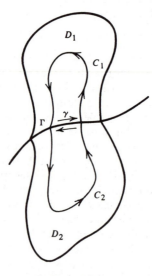

Figure 4.5.2.

Proof: Referring to Figure 4.5.2, let z_0 be in the interior of $C_1 \cup \gamma$. Then

$$f_1(z_0) = \frac{1}{2\pi i} \int_{(C_1 \cup \gamma)_+} \frac{f_1(z)\, dz}{z - z_0},$$

$$0 = \frac{1}{2\pi i} \int_{(C_2 \cup \gamma)_+} \frac{f_2(z)\, dz}{z - z_0}.$$

Adding, we have, since the integration along γ cancels,

$$f_1(z_0) = \frac{1}{2\pi i} \int_{(C_1 \cup C_2)_+} \frac{F(z)\, dz}{z - z_0}, \qquad z_0 \in D_1.$$

Let z_0 be in the interior of $C_2 \cup \gamma$. Then

$$f_2(z_0) = \frac{1}{2\pi i} \int_{(C_2 \cup \gamma)_+} \frac{f_2(z)\, dz}{z - z_0},$$

$$0 = \frac{1}{2\pi i} \int_{(C_1 \cup \gamma)_+} \frac{f_1(z)\, dz}{z - z_0}.$$

Adding, we have

$$f_2(z_0) = \frac{1}{2\pi i} \int_{(C_1 \cup C_2)_+} \frac{F(z)\, dz}{z - z_0}, \quad z_0 \in D_2.$$

Now, if $z_0 \in D_1 \cup D_2 \cup \Gamma^*$, there exists C_1, C_2, γ in $D_1 \cup D_2 \cup \Gamma^*$ such that z_0 is in the interior of $C_1 \cup C_2$. Since $F(z)$ is continuous on $C_1 \cup C_2$, the function

$$g(z_0) = \frac{1}{2\pi i} \int_{(C_1 \cup C_2)_+} \frac{F(z)\, dz}{z - z_0}$$

is analytic inside $C_1 \cup C_2$. From the above discussion

$$g(z_0) = \begin{cases} f_1(z_0), & z_0 \in D_1, \\ f_2(z_0), & z_0 \in D_2, \\ f_1(z_0) = f_2(z_0), & z_0 \in \Gamma^*. \end{cases}$$

Therefore, $g(z) \equiv F(z)$, which completes the proof.

Corollary 4.5.2. Let $f(z)$ be analytic in a domain D_1 in the upper half-plane which is bounded in part by the interval $I = \{x \mid a < x < b\}$ of the x axis. If $f(z)$ is continuous in $D_1 \cup I$ and $f(x)$ is real, then $\overline{f}(\bar{z})$ is the analytic continuation of $f(z)$ to D_2, which is the reflection of D_1 in the x axis.

Proof: To reduce this to Theorem 4.5.3 we have to show that $\overline{f}(\bar{z})$ is analytic in D_2 and that $\lim_{\bar{z} \to I} \overline{f}(\bar{z}) = \overline{f}(x) = f(x)$. The latter statement follows from the continuity of $f(z)$ in $D_1 \cup I$ and the fact that $f(x)$ is real. Let $f(z) = u(x, y) + iv(x, y)$. Then $\overline{f}(\bar{z}) = u(x, -y) - iv(x, -y) = U(x, y) + iV(x, y)$, and

$$\frac{\partial U}{\partial x} = \frac{\partial u}{\partial x} = \frac{\partial v}{\partial y} = \frac{\partial V}{\partial y}, \quad -\frac{\partial U}{\partial y} = \frac{\partial u}{\partial y} = -\frac{\partial v}{\partial x} = \frac{\partial V}{\partial x}.$$

Therefore, the Cauchy-Riemann equations are satisfied by the real and imaginary parts of $\overline{f}(\bar{z})$ in D_2. Then, by Theorem 4.5.3,

$$F(z) = \begin{cases} f(z), & z \in D_1, \\ \overline{f}(\bar{z}), & z \in D_2, \\ f(x), & z = x \in I, \end{cases}$$

is analytic in $D_1 \cup D_2 \cup I$.

Extensions of this result can be obtained by conformal mapping, which will be discussed in Chapter 6. The idea is to map D_1 into $\tilde{D}_1$ in the w plane, bounded in part by Γ, which is the image of the real axis. Then under the hypotheses of Corollary 4.5.2, the image of the function $f(z)$ will be continuous in $\tilde{D}_1 \cup \gamma$, where γ is the image of I in the w plane. Then the image of $\bar{f}(\bar{z})$ will provide the analytic continuation to the image of D_2. (See Exercise 4.5.5.)

Exercises 4.5

1. Show that $\sum\limits_{k=1}^{\infty} z^{k!}$ has a natural boundary at $|z| = 1$. Hint: Consider the point $z = e^{2\pi i p/q}$ where p and q are integers.

2. Prove that if $\sum\limits_{k=0}^{\infty} a_k(z - z_0)^k$ is a power series element of a complete analytic function $f(z)$ with a radius of convergence R, then $f(z)$ has at least one singularity on the circle of convergence. Hint: Prove that if the power series can be continued through every point on the circle of convergence, then necessarily the power series must converge within a larger circle.

3. If (f_2, D_2) is a direct analytic continuation of (f_1, D_1), $z_1 \in D_1$ and $z_2 \in D_2$, then there exists a simple Jordan Curve C joining z_1 and z_2 in $D_1 \cup D_2$, and a finite chain of power series elements with centers on C, which continues $f_1(z)$ to z_2 and gives the value $f_2(z_2)$ at z_2. In other words, every direct analytic continuation is equivalent to a finite chain of power series elements.

4. Starting with

$$\sqrt{x} = (1 + x - 1)^{1/2} = 1 + \tfrac{1}{2}(x - 1) + \tfrac{1}{2}(\tfrac{1}{2} - 1) \frac{(x - 1)^2}{2!} + \cdots,$$

which converges for $0 \le |x - 1| < 1$, continue to the complex plane as

$$\sqrt{z} = 1 + \tfrac{1}{2}(z - 1) + \tfrac{1}{2}(\tfrac{1}{2} - 1) \frac{(z - 1)^2}{2!} + \cdots.$$

Then continue this element by power series elements with centers on the unit circle. Show that after one circuit of the branch point at the origin, the value of the complete analytic function at $z = 1$ is -1, and after two circuits the value at $z = 1$ is 1.

5. Prove that $w = (z - i)/(-z - i)$ maps the upper-half z plane onto the interior of the unit circle in the w plane with the real axis going into the unit circle and the interval $I = \{x \mid |x| < 1\}$ going into the right semicircle between $-i$ and i. If D_1 is a domain in the upper-half z plane bounded in part by I, then $\tilde{D}_1$, the image of D_1, is a domain inside the unit circle in the w plane bounded in part by the right semicircle γ. Let $g(w)$ be analytic in $\tilde{D}_1$ and continuous in $\tilde{D}_1 \cup \gamma$ and $g[(x - i)/(-x - i)]$ be real. Find the analytic continuation $g(w)$ into $\tilde{D}_2$, which is the image of D_2, the reflection of D_1 in the real axis.

4.6. LAURENT SERIES

Let us consider the function $f(z) = 1/z(1 - z)$ which is analytic except at $z = 0$ and $z = 1$. Since $z = 0$ is a singularity, we cannot hope to obtain a Maclaurin expansion. However, we may do the following

$$f(z) = \frac{1}{z}\frac{1}{1 - z} = \frac{1}{z}\sum_{k=0}^{\infty} z^k = \sum_{k=0}^{\infty} z^{k-1},$$

which is valid for $0 < |z| < 1$. We have obtained an expansion in positive and negative powers of z which is valid in an annulus instead of inside a circle. As a matter of fact, we can obtain a general result, which we now prove in the following theorem.

Theorem 4.6.1. *Let $f(z)$ be analytic for $R_1 < |z - z_0| < R_2$. Then for $R_1 < |z - z_0| < R_2$*

$$f(z) = \sum_{k=-\infty}^{\infty} a_k(z - z_0)^k,$$

where

$$a_k = \frac{1}{2\pi i}\int_{C_+} \frac{f(\zeta)}{(\zeta - z_0)^{k+1}}\, d\zeta$$

and C is any simple closed contour lying in the given annulus and having z_0 in its interior, and this representation is unique.

Proof: Let $R_1 < r_1 < r_2 < R_2$ and $C_1 = \{z \mid |z - z_0| = r_1\}$ and $C_2 = \{z \mid |z - z_0| = r_2\}$. Now, if $r_1 < |z| < r_2$,

$$f(z) = \frac{1}{2\pi i}\int_{C_{2+}} \frac{f(\zeta)}{\zeta - z}\, d\zeta - \frac{1}{2\pi i}\int_{C_{1+}} \frac{f(\zeta)}{\zeta - z}\, d\zeta.$$

On C_1,

$$\frac{1}{\zeta - z} = \frac{-1}{(z - z_0) - (\zeta - z_0)} = \frac{-1}{(z - z_0)\left(1 - \dfrac{\zeta - z_0}{z - z_0}\right)}$$

$$= -\frac{1}{z - z_0}\sum_{k=0}^{\infty}\left(\frac{\zeta - z_0}{z - z_0}\right)^k = -\sum_{k=-\infty}^{0} \frac{(z - z_0)^{k-1}}{(\zeta - z_0)^k}$$

$$= -\sum_{k=-\infty}^{-1} \frac{(z - z_0)^k}{(\zeta - z_0)^{k+1}},$$

and the convergence is uniform in ζ for a fixed z. On C_2,

$$\frac{1}{\zeta - z} = \frac{1}{(\zeta - z_0) - (z - z_0)} = \frac{1}{(\zeta - z_0)\left(1 - \dfrac{z - z_0}{\zeta - z_0}\right)}$$

$$= \sum_{k=0}^{\infty} \frac{(z - z_0)^k}{(\zeta - z_0)^{k+1}},$$

where the convergence is uniform in ζ for a fixed z. Integrating term by term, we have

$$f(z) = \sum_{k=0}^{\infty} \frac{(z - z_0)^k}{2\pi i} \int_{C_{2+}} \frac{f(\zeta)}{(\zeta - z_0)^{k+1}} \, d\zeta$$

$$+ \sum_{k=-\infty}^{-1} \frac{(z - z_0)^k}{2\pi i} \int_{C_{1+}} \frac{f(\zeta)}{(\zeta - z_0)^{k+1}} \, d\zeta.$$

Since $f(\zeta)/(\zeta - z_0)^{k+1}$ is analytic in the annulus, we can integrate over any simple closed curve C lying in the annulus and having z_0 in its interior. Therefore, we have

$$f(z) = \sum_{k=-\infty}^{\infty} a_k(z - z_0)^k,$$

where

$$a_k = \frac{1}{2\pi i} \int_{C_+} \frac{f(\zeta)}{(\zeta - z_0)^{k+1}} \, d\zeta.$$

To show that the representation is unique, let $f(z) = \sum_{k=-\infty}^{\infty} b_k(z - z_0)^k$, where the convergence is in $R_1 < |z - z_0| < R_2$. Let C be any simple closed contour in the annulus with z_0 in its interior. The series converges uniformly on C, so we may multiply by $1/2\pi i(z - z_0)^{n+1}$ and integrate around C. The only term which survives is the one where $n - k + 1 = 1$, that is, $n = k$. We obtain

$$a_n = \frac{1}{2\pi i} \int_{C_+} \frac{f(z)}{(z - z_0)^{n+1}} \, dz = \sum_{k=-\infty}^{\infty} b_k \frac{1}{2\pi i} \int_{C_+} \frac{1}{(z - z_0)^{n-k+1}} \, dz = b_n.$$

EXAMPLE 4.6.1. Consider the following examples of Laurent expansions:

(a) $\dfrac{1}{z^2 + 1} = \dfrac{1}{(z - i)(z + i)} = \dfrac{1}{(z - i)(2i + z - i)}$

$$= \frac{1}{2i(z - i)\left(1 + \dfrac{z - i}{2i}\right)}$$

$$= \frac{1}{2i} \sum_{k=0}^{\infty} \left(\frac{-1}{2i}\right)^k (z - i)^{k-1} = \sum_{k=-1}^{\infty} \frac{(-1)^{k+1}}{(2i)^{k+2}} (z - i)^k$$

valid for $0 < |z - i| < 2$

(b) $\dfrac{1}{z^2 + 1} = \dfrac{1}{(z + i)(z - i)} = \dfrac{1}{(z + i)(-2i + z + i)}$

$$= \dfrac{-1}{2i(z + i)\left(1 - \dfrac{z + i}{2i}\right)}$$

$$= -\dfrac{1}{2i} \sum_{k=0}^{\infty} \dfrac{(z + i)^{k-1}}{(2i)^k} = -\sum_{k=-1}^{\infty} \dfrac{(z + i)^k}{(2i)^{k+2}}$$

valid for $0 < |z + i| < 2$

(c) $\dfrac{1}{z^2 + 1} = \dfrac{1}{z^2\left(1 + \dfrac{1}{z^2}\right)} = \dfrac{1}{z^2} \sum_{k=0}^{\infty} (-1)^k z^{-2k} = \sum_{k=-\infty}^{-1} (-1)^{k+1} z^{2k}$

valid for $1 < |z|$

(d) $\dfrac{\sin z}{z^3} = \dfrac{1}{z^3} \sum_{k=0}^{\infty} \dfrac{(-1)^k z^{2k+1}}{(2k + 1)!} = \sum_{k=-1}^{\infty} \dfrac{(-1)^{k+1} z^{2k}}{(2k + 3)!}$

valid for $0 < |z|$

(e) $e^{1/z} = \sum_{k=0}^{\infty} \dfrac{1}{k! z^k} = \sum_{k=-\infty}^{0} \dfrac{z^k}{(-k)!}$

valid for $0 < |z|$.

In examples (a), (b), (d), and (e) we have expanded functions in annuli of the form $0 < |z - z_0| < R$. In other words, the singularity at z_0 is isolated in the sense that there is a positive R such that $f(z)$ is analytic for $|z - z_0| < R$ except at z_0. This situation is important enough that we should study the behavior of functions in the neighborhood of an isolated singularity.

Definition 4.6.1. The function $f(z)$ has an isolated singularity at z_0 if there exists a *deleted ϵ-neighborhood* of z_0, $0 < |z - z_0| < \epsilon$, throughout which $f(z)$ is analytic, but is not analytic at z_0.

Definition 4.6.2. If $f(z)$ has an isolated singularity at z_0 there exists an R such that $f(z)$ has a Laurent expansion valid for $0 < |z - z_0| < R$. The part of this expansion containing negative powers of $z - z_0$ we call the *principal part* of the function.

Definition 4.6.3. If $f(z)$ has an isolated singularity at z_0, we classify the singularity according to the number of terms in the principal part of the function. If the principal part has no terms, the function has a *removable singularity* at z_0. If the principal part has a finite number of terms, the function has a *pole* at z_0. If the principal part has an infinite number of terms, the function has an *essential singularity* at z_0.

Definition 4.6.4. The function $f(z)$ has a pole of order m at z_0 if the principal part has the form

$$\sum_{k=-1}^{-m} a_k(z - z_0)^k$$

with $a_{-m} \neq 0$.

Theorem 4.6.2. *If $f(z)$ has a removable singularity at z_0, then by defining $f(z_0)$ as a_0 the function is made analytic for $|z - z_0| < R$. Hence, $|f(z)|$ is bounded for $0 < |z - z_0| \leq R/2$.*

Proof: Since $f(z)$ has a removable isolated singularity at z_0 the Laurent expansion

$$f(z) = \sum_{k=0}^{\infty} a_k(z - z_0)^k$$

is valid for $0 < |z - z_0| < R$. Hence, the function

$$g(z) = \begin{cases} \displaystyle\sum_{k=0}^{\infty} a_k(z - z_0)^k, & 0 < |z - z_0| < R \\ a_0, & z = z_0 \end{cases}$$

has a power series expansion valid for $|z - z_0| < R$ and is therefore analytic $|z - z_0| < R$. Hence, $|g(z)|$ is bounded for $0 < |z - z_0| \leq R/2$ where it is identical with $|f(z)|$.

Theorem 4.6.3. *If $f(z)$ has a pole of order m at z_0, then $1/f(z)$ has a zero of order m at $z = z_0$. Also there exist positive constants ϵ, A, and B such that in $0 < |z - z_0| < \epsilon$,*

$$\frac{A}{|z - z_0|^m} \leq |f(z)| \leq \frac{B}{|z - z_0|^m}.$$

Proof: Since $f(z)$ has a pole of order m at z_0, we have the Laurent representation valid for $0 < |z - z_0| < R$

$$\begin{aligned}
f(z) &= \frac{a_{-m}}{(z - z_0)^m} + \frac{a_{-m+1}}{(z - z_0)^{m-1}} + \cdots + \frac{a_{-1}}{z - z_0} + \sum_{k=0}^{\infty} a_k(z - z_0)^k \\
&= \frac{1}{(z - z_0)^m}\left[a_{-m} + a_{-m+1}(z - z_0) + \cdots + a_{-1}(z - z_0)^{m-1} \right. \\
&\qquad\qquad\qquad\left. + \sum_{k=0}^{\infty} a_k(z - z_0)^{k+m} \right] \\
&= \frac{g(z)}{(z - z_0)^m},
\end{aligned}$$

where $g(z_0) = a_{-m} \neq 0$. Now, $g(z)$ is analytic at z_0 and, therefore, $1/f(z) = (z - z_0)^m/g(z)$ is analytic at z_0 and has a zero of order m. Furthermore, $g(z) = a_{-m} + h(z)$, where $h(z_0) = 0$. There exists an ϵ such that for $|z - z_0| < \epsilon$, $|h(z)| \leq |a_{-m}|/2$, and $|g(z)| \geq |a_{-m}| - |h(z)| \geq |a_{-m}|/2 = A$. Also in the same ϵ-neighborhood $|g(z)| \leq B$. Consequently,

$$\frac{A}{|z - z_0|^m} \leq \frac{|g(z)|}{|z - z_0|^m} = |f(z)| \leq \frac{B}{|z - z_0|^m},$$

for $0 < |z - z_0| < \epsilon$. This inequality shows that there is an ϵ-neighborhood of z_0 in which $|f(z)|$ grows to infinity as a constant divided by $|z - z_0|^m$.

Theorem 4.6.4. *If $f(z)$ has an essential singularity at z_0, and c is any complex number, then $|f(z) - c|$ can be made arbitrarily small in every ϵ-neighborhood of z_0.*

Proof: Suppose $|f(z) - c| > \delta > 0$, for all z satisfying $0 < |z - z_0| < \epsilon < R$. Then $g(z) = 1/[f(z) - c]$ is analytic and bounded in the annulus $0 < |z - z_0| < \epsilon$, and hence has a removable singularity at z_0. If $\lim_{z \to z_0} g(z) \neq 0$, then $f(z) = c + 1/g(z)$ is analytic at z_0. If $\lim_{z \to z_0} g(z) = 0$, then $f(z) = c + 1/g(z)$ has a pole at z_0. These conclusions are impossible. Therefore, for arbitrary $\delta > 0$ there must exist at least one z satisfying $0 < |z - z_0| < \epsilon$ such that $|f(z) - c| < \delta$.

EXAMPLE 4.6.2. Prove that $e^{1/z}$ takes on every value except zero in every ϵ-neighborhood of the origin. Let $e^{1/z} = e^{1/r(\cos\theta + i\sin\theta)} = c = e^{\log|c| + i\arg c}$, $c \neq 0$. If we can solve $(1/r)\cos\theta = \log|c|$, $(1/r)\sin\theta = -\arg c$ in every ϵ-neighborhood, the proof will be complete. This requires $\tan\theta = -\arg c/\log|c|$ and $(1/r^2) = (\log|c|)^2 + (\arg c)^2$. The first equation can always be solved since, in the range $0 \leq \theta \leq 2\pi$, $\tan\theta$ takes on all values between $-\infty$ and ∞. Likewise, the second equation can always be solved with $0 < r < \epsilon$, since we are free to take $\arg c$ as large as we please.

This example is a special case of a famous theorem of Picard which we state without proof.

Theorem 4.6.5. *If $f(z)$ has an essential singularity at z_0, then it takes on every value with one possible exception in every ϵ-neighborhood of z_0.*

Let $f(z) = P(z)/Q(z)$ be a rational function with $P(z)$ a polynomial of degree m and $Q(z)$ a polynomial of degree n. If $P(z)$ and $Q(z)$ both vanish at z_0, each has $(z - z_0)$ as a factor. If z_0 is a zero of order α of $P(z)$ and a zero of order β of $Q(z)$, then we have three cases to consider (1) $\alpha > \beta$, (2) $\alpha = \beta$, and (3) $\alpha < \beta$. We have the following situations: (1) $f(z)$ has a removable singularity with a zero of order $\alpha - \beta$ at z_0, (2) $f(z)$ has a removable singularity with no zero at z_0, and (3) $f(z)$ has a pole of order $\beta - \alpha$ at z_0. Therefore, we can assume without loss of generality that $P(z)/Q(z)$ have been reduced to lowest terms by dividing out all the common factors.

If $m \geq n$ we may first divide $Q(z)$ into $P(z)$ to obtain

$$f(z) = \frac{P(z)}{Q(z)} = a_0 + a_1 z + a_2 z^2 + \cdots + a_{m-n} z^{m-n} + \frac{R(z)}{Q(z)},$$

where the degree of $R(z)$ is at most $n - 1$. We now show how we can simplify $R(z)/Q(z)$.

Theorem 4.6.6. *Let $R(z)/Q(z)$ be a rational function where $Q(z)$ is a polynomial of degree $n \geq 1$ and $R(z)$ is a polynomial of degree less than n which has no common factor with $Q(z)$. If $Q(z) = z^n + q_{n-1} z^{n-1} + \cdots + q_0 = (z - r_1)^{\alpha_1}(z - r_2)^{\alpha_2} \cdots (z - r_k)^{\alpha_k}$, $\alpha_1 + \alpha_2 + \cdots + \alpha_k = n$, then*

$$\frac{R(z)}{Q(z)} = \frac{a_{11}}{(z - r_1)} + \frac{a_{12}}{(z - r_1)^2} + \cdots + \frac{a_{1\alpha_1}}{(z - r_1)^{\alpha_1}}$$

$$+ \cdots + \frac{a_{k1}}{(z - r_k)} + \frac{a_{k2}}{(z - r_k)^2}$$

$$+ \cdots + \frac{a_{k\alpha_k}}{(z - r_k)^{\alpha_k}},$$

where

$$a_{i,\alpha_1 - j} = \frac{1}{j!} \lim_{z \to r_i} \frac{d^j}{dz^j} \left[(z - r_i)^{\alpha_i} \frac{R(z)}{Q(z)} \right], \qquad j = 0, 1, 2, \ldots, \alpha_i - 1.$$

Proof: Let r_i be the ith distinct root of $Q(z)$, a zero of order α_i. Therefore,

$$\frac{R(z)}{Q(z)} = \frac{S(z)}{(z - r_i)^{\alpha_i}} = \frac{a_{i,\alpha_i}}{(z - r_i)^{\alpha_i}} + \frac{a_{i,\alpha_i - 1}}{(z - r_i)^{\alpha_i - 1}} + \cdots + \frac{a_{i1}}{(z - r_i)} + H(z),$$

where $H(z)$ is a power series in $z - r_i$ which is analytic at r_i. Now,

$$(z - r_i)^{\alpha_i} \frac{R(z)}{Q(z)} = a_{i,\alpha_i} + a_{i,\alpha_i - 1}(z - r_i) + \cdots + a_{i1}(z - r_i)^{\alpha_i - 1}$$

$$+ H(z)(z - r_i)^{\alpha_i}$$

$$a_{i,\alpha_i} = \lim_{z \to r_i} (z - r_i)^{\alpha_i} \frac{R(z)}{Q(z)},$$

$$a_{i,\alpha_i - 1} = \lim_{z \to r_i} \frac{d}{dz} \left[(z - r_i)^{\alpha_i} \frac{R(z)}{Q(z)} \right],$$

$$a_{i,\alpha_i - j} = \frac{1}{j!} \lim_{z \to r_i} \frac{d^j}{dz^j} \left[(z - r_i)^{\alpha_i} \frac{R(z)}{Q(z)} \right],$$

$$j = 0, 1, 2, \ldots, \alpha_i - 1.$$

We have obtained explicitly the principal part of $R(z)/Q(z)$ at r_i. To obtain the rest of the expansion we do the same thing for the other distinct roots of $Q(z)$. We shall, in effect, be breaking down $H(z)$, which is a rational function, into simpler rational functions. However, we do not have to calculate $H(z)$, because, in working with a different root, the part of the expansion already computed becomes part of a residual analytic function which drops out of the calculation just as $H(z)$ did in the above development. Clearly, when we have calculated the principal parts for all the distinct roots, we are done, because of the unique factorization of $Q(z)$.

In this theorem, we have obtained the *partial fraction expansion* of the rational function. Such expansions are very useful in integration and in the inversion of Fourier and Laplace transforms, which will be treated in Chapters 8 and 9.

Corollary 4.6.1. Let $P(x)/Q(x)$ be a rational function of the real variable x consisting of the ratio of two polynomials with real coefficients with the degree of $Q(x)$ equal to $n \geq 1$ and the degree of $P(x)$ less than n. If $Q(x)$ has only real roots $r_1, r_2, \ldots, r_k$ with multiplicities $\alpha_1, \alpha_2, \ldots, \alpha_k$, then

$$\frac{P(x)}{Q(x)} = \frac{a_{11}}{(x - r_1)} + \frac{a_{12}}{(x - r_1)^2} + \cdots + \frac{a_{1\alpha_1}}{(x - r_1)^{\alpha_1}}$$

$$+ \cdots + \frac{a_{k1}}{(x - r_k)} + \frac{a_{k2}}{(x - r_k)^2}$$

$$+ \cdots + \frac{a_{k\alpha_k}}{(x - r_k)^{\alpha_k}},$$

where

$$a_{i,\alpha_i-j} = \frac{1}{j!} \lim_{x \to r_i} \frac{d^j}{dx^j}\left[(x - r_i)^{\alpha_i} \frac{P(x)}{Q(x)}\right], \qquad j = 0, 1, 2, \ldots, \alpha_i - 1.$$

Proof: This result follows from Theorem 4.6.6 by merely observing that, if the roots of $Q(z)$ are all real, the limits and differentiation can be carried out on the x axis.

Corollary 4.6.2. Let $P(x)/Q(x)$ be a rational function of the real variable x consisting of the ratio of two polynomials with real coefficients with the degree of $Q(x)$ equal to $n \geq 1$ and the degree of $P(x)$ less than n. Let $Q(x)$ have a nonrepeated complex root $a + ib$. Then

$$\frac{P(x)}{Q(x)} = \frac{Ax + B}{(x - a)^2 + b^2} + R(x),$$

where A and B are real constants and $R(x)$ is a rational function with real coefficients. If $L = \lim_{z \to a+ib} [(z - a)^2 + b^2] P(z)/Q(z)$, then $A = (1/b)\text{Im}(L)$ and $B = \text{Re}(L) - (a/b)\text{Im}(L)$.

Proof: Since $Q(\bar{z}) = \bar{Q}(z)$, $a - ib$ is a nonrepeated root of $Q(z)$. Then by Theorem 4.6.6

$$\frac{P(z)}{Q(z)} = \frac{a_{11}}{z - a - ib} + \frac{a_{21}}{z - a + ib} + R(z)$$

$$= \frac{(a_{11} + a_{21})z + [-a(a_{11} + a_{21}) + ib(a_{11} - a_{21})]}{(z - a)^2 + b^2} + R(z)$$

$$= \frac{Az + B}{(z - a)^2 + b^2} + R(z).$$

Now, if $z = x$, then

$$\frac{P(x)}{Q(x)} = \frac{Ax + B}{(x - a)^2 + b^2} + R(x).$$

Since $P(x)/Q(x)$ is real A, B, and $R(x)$ must be real. Also,

$$L = \lim_{z \to a+ib} [(z - a)^2 + b^2]\frac{P(z)}{Q(z)} = 2iba_{11},$$

$$\bar{L} = \lim_{z \to a-ib} [(z - a)^2 + b^2]\frac{P(z)}{Q(z)} = -2iba_{21},$$

$$A = a_{11} + a_{21} = \frac{L - \bar{L}}{2ib} = \frac{\text{Im}(L)}{b},$$

$$B = -a(a_{11} + a_{21}) + ib(a_{11} - a_{21}) = \frac{-a\,\text{Im}(L)}{b} + \text{Re}(L).$$

The importance of this corollary is that in many cases we wish a decomposition of $P(x)/Q(x)$ with real coefficients, and in this form A and B are real, whereas a_{11} and a_{21} are not. There is a corresponding theorem to take care of the case of repeated complex roots, but the details begin to be tedious.

EXAMPLE 4.6.3. Obtain a partial fraction expansion of

$$\frac{1}{x(x - 1)^2(x^2 + 2x + 2)}.$$

We can write

$$\frac{1}{x(x - 1)^2(x^2 + 2x + 2)} = \frac{A}{x} + \frac{B}{x - 1} + \frac{C}{(x - 1)^2} + \frac{Dx + E}{(x + 1)^2 + 1}.$$

Then,

$$A = \lim_{x \to 0} \frac{1}{(x - 1)^2(x^2 + 2x + 2)} = \frac{1}{2},$$

$$C = \lim_{x \to 1} \frac{1}{x(x^2 + 2x + 2)} = \frac{1}{5},$$

$$B = \lim_{x \to 1} \frac{d}{dx} \frac{1}{x^3 + 2x^2 + 2x} = \lim_{x \to 1} \frac{-(3x^2 + 4x + 2)}{(x^3 + 2x^2 + 2x)^2} = -\frac{9}{25},$$

$$L = \lim_{z \to -1+i} \frac{1}{z(z - 1)^2} = \frac{1}{50} - \frac{7}{50}i,$$

$$D = \text{Im}(L) = -\tfrac{7}{50},$$

$$E = \text{Re}(L) + \text{Im}(L) = -\tfrac{3}{25}.$$

EXAMPLE 4.6.4. Integrate

$$\int_{C_+} \frac{dz}{(z^4 - 1)^2},$$

where $C = \{z \mid |z - i| = 1\}$. We may write

$$\frac{1}{(z^4 - 1)^2} = \frac{A}{(z - 1)} + \frac{B}{(z - 1)^2} + \frac{C}{(z + 1)} + \frac{D}{(z + 1)^2} + \frac{E}{(z + i)}$$
$$+ \frac{F}{(z + i)^2} + \frac{G}{(z - i)} + \frac{H}{(z - i)^2}.$$

Among all these terms, the only one which contributes is $G/(z - i)$. Therefore, we have

$$\int_{C_+} \frac{dz}{(z^4 - 1)^2} = 2\pi i G = 2\pi i \lim_{z \to i} \frac{d}{dz} \frac{(z - i)^2}{(z^4 - 1)^2}$$

$$= 2\pi i \lim_{z \to i} \frac{d}{dz} \frac{1}{(z^2 - 1)^2(z + i)^2}$$

$$= 2\pi i \lim_{z \to i} \left[\frac{-4z}{(z^2 - 1)^3(z + i)^2} + \frac{-2}{(z^2 - 1)^2(z + i)^3} \right]$$

$$= 2\pi i \left(\frac{-6i}{32} \right)$$

$$= \frac{3\pi}{8}.$$

Exercises 4.6

1. Find expansions of $1/(z + 1)(z - 1)$ in powers of $z + 1$, $z - 1$, and z valid, respectively, in $0 < |z + 1| < 2$, $0 < |z - 1| < 2$, $|z| > 1$.

2. Find the Laurent expansion for $\sin(1/z)$ in powers of z. Where is it valid?

3. Prove that if $f(z)$ is analytic at z_0 where it has a zero of order m, then $1/f(z)$ has a pole of order m at z_0.

4. Let $f(z)$ be analytic on the unit circle $C = \{z \mid |z| = 1\}$. Then

$$f(e^{i\theta}) = u(\theta) + iv(\theta) = \sum_{k=-\infty}^{\infty} a_k e^{ik\theta}, \quad \text{where} \quad a_k = \frac{1}{2\pi i} \int_{C_+} \frac{f(z)\, dz}{z^{k+1}}$$

$$= \frac{1}{2\pi} \int_0^{2\pi} [u(\theta) + iv(\theta)] e^{-ik\theta}\, d\theta.$$

Let $f(e^{i\theta})$ be real. Show that

$$u(\theta) = \frac{\alpha_0}{2} + \sum_{k=1}^{\infty} (\alpha_k \cos k\theta + \beta_k \sin k\theta),$$

where

$$\alpha_k = \frac{1}{\pi} \int_0^{2\pi} u(\theta) \cos k\theta\, d\theta \quad \text{and} \quad \beta_k = \frac{1}{\pi} \int_0^{2\pi} u(\theta) \sin k\theta\, d\theta.$$

5. Find the poles and principal parts of tan z.

6. Let $f(z)$ be analytic for $R_1 < r < R_2$. Prove that

$$\frac{1}{2\pi} \int_0^{2\pi} |f(re^{i\theta})|^2 \, d\theta = \sum_{k=-\infty}^{\infty} |a_k|^2 \, r^{2k}.$$

7. A function $f(z)$ has an isolated singularity at ∞ if and only if $f(1/z)$ has an isolated singularity at the origin. Prove that a polynomial of degree $n > 0$ has a pole of order n at ∞.

8. Prove that an entire transcendental function has an essential singularity at ∞.

9. Prove that a rational function has at most a finite number of poles in the extended plane.

10. Find the partial fraction expansion with real coefficients for

$$\frac{1}{(z - 1)(z + 1)^3(z^2 + 2z + 5)}.$$

11. Integrate

$$\int_{C_+} \frac{dz}{(z^3 + 1)^2},$$

when $C = \{z \mid |z - 5| = 5\}$.

4.7. DOUBLE SERIES

For many reasons, including the study of functions of two complex variables, we will need to know some things about double series. As in the case of single series, we will define convergence of double series in terms of the convergence of double sequences. Therefore, we begin our study with double sequences.

Definition 4.7.1. A double sequence of complex numbers is a function defined on the points of the form $m + in$, m and n all positive integers, into the unextended plane. We designate a double sequence by the notation $\{w_{mn}\}$, where m and n are understood to take on all positive integer values.

Definition 4.7.2. The double sequence $\{w_{mn}\}$ has a limit w, if and only if for every $\epsilon > 0$ there exists an $N(\epsilon)$ such that $|w - w_{mn}| < \epsilon$, for all $m > N$ and $n > N$. If $\{w_{mn}\}$ has a limit it converges. Otherwise, it diverges.

Theorem 4.7.1. *The sequence $\{w_{mn}\}$ converges if and only if for every $\epsilon > 0$ there is an $N(\epsilon)$ such that $|w_{m+p,n+q} - w_{mn}| < \epsilon$, for all $m > N$, $n > N$, and all positive integers p and q.*

Proof: If $\{w_{mn}\}$ has a limit w, then for $\epsilon > 0$ there exists $N(\epsilon)$ such that

$$|w_{m+p,n+q} - w_{mn}| \leq |w - w_{m+p,n+q}| + |w - w_{mn}| < \frac{\epsilon}{2} + \frac{\epsilon}{2} = \epsilon,$$

for all $m > N$ and $n > N$.

Conversely, if $|w_{m+p,n+q} - w_{mn}| < \frac{\epsilon}{2}$, for all $m > N_1, n > N_1$, and all positive integers p and q, then

$$|w_{m+p,m+p} - w_{mm}| < \frac{\epsilon}{2},$$

for all $m > N_1$ and all positive integers p. This implies that there is a w such that $\lim\limits_{m \to \infty} w_{mm} = w$. Also, there exists an N_2, such that

$$|w_{mn} - w_{m+p,n+q}| < \frac{\epsilon}{2},$$

for $m > N_2$ and $n > N_2$ and $s = m + p = n + q > N_1$. Hence,

$$|w - w_{mn}| \leq |w - w_{ss}| + |w_{mn} - w_{ss}| < \frac{\epsilon}{2} + \frac{\epsilon}{2} = \epsilon,$$

for $m > N_2$ and $n > N_2$. This shows that $\{w_{mn}\}$ has the limit w.

A double sequence can be arranged in rows and columns in an array as follows:

$$
\begin{array}{cccc}
w_{11} & w_{12} & w_{13} & \cdots \\
w_{21} & w_{22} & w_{23} & \cdots \\
w_{31} & w_{32} & w_{33} & \cdots \\
\cdots & \cdots & \cdots & \cdots
\end{array}
$$

Then each row and each column by itself is a simple sequence and we can examine the following limits: $\lim\limits_{n \to \infty} w_{mn}$, $\lim\limits_{m \to \infty} w_{mn}$, $\lim\limits_{m \to \infty} (\lim\limits_{n \to \infty} w_{mn})$, and $\lim\limits_{n \to \infty} (\lim\limits_{m \to \infty} w_{mn})$. We have the following useful theorem.

Theorem 4.7.2. *If $\{w_{mn}\}$ has a limit w and the iterated limits $\lim\limits_{m \to \infty} (\lim\limits_{n \to \infty} w_{mn})$ and $\lim\limits_{n \to \infty} (\lim\limits_{m \to \infty} w_{mn})$ exist, then*

$$\lim_{m \to \infty} (\lim_{n \to \infty} w_{mn}) = \lim_{n \to \infty} (\lim_{m \to \infty} w_{mn}) = w.$$

Proof: Let $r_m = \lim\limits_{n \to \infty} w_{mn}$ be the limit of the mth row. Since for all $\epsilon > 0$ there exists $N(\epsilon)$ such that $|w - w_{mn}| < \epsilon/2$, for all $m > N$ and $n > N$. This shows that $|w - r_m| \leq \epsilon/2 < \epsilon$, for all $m > N$, which proves that $\lim\limits_{m \to \infty} r_m = \lim\limits_{m \to \infty} (\lim\limits_{n \to \infty} w_{mn}) = w$. Similarly, $\lim\limits_{n \to \infty} (\lim\limits_{m \to \infty} w_{mn}) = w$.

Definition 4.7.3. Let $\{w_{mn}(z)\}$ be a sequence of complex functions, each defined on a set S. Then $\{w_{mn}(z)\}$ converges uniformly on S, if and only if for every $\epsilon > 0$ there exists an $N(\epsilon)$, independent of $z \in S$, such that $|w_{m+p,n+q}(z) - w_{mn}(z)| < \epsilon$, for all $m > N$ and $n > N$.

Of course, the conditions of the definition imply that there is a function $w(z)$ which is defined on S and is the limit at each point of $\{w_{mn}(z)\}$. If each $w_{mn}(z)$ is continuous in S and $\{w_{mn}\}$ converges uniformly in S to $w(z)$, then $w(z)$ is continuous in S. If S is a domain and each $w_{mn}(z)$ is analytic in S and $\{w_{mn}(z)\}$ converges uniformly to $w(z)$ in every compact subset of S, then $w(z)$ is analytic in S.

Now, we turn to the main subject of this section, double series. Let w_{jk} be the jkth term of a double series $\sum\limits_{j,\,k=1}^{\infty} w_{jk}$. Let $S_{mn} = \sum\limits_{j=1}^{m} \sum\limits_{k=1}^{n} w_{jk}$ be the mnth partial sum. We define convergence of the double series in terms of the convergence of the sequence of partial sums $\{S_{mn}\}$.

Definition 4.7.4. The double series $\sum\limits_{j,\,k=1}^{\infty} w_{jk}$ converges to the sum S if and only if the sequence of partial sums $S_{mn} = \sum\limits_{j=1}^{m} \sum\limits_{k=1}^{n} w_{jk}$ converges to S. If the series does not converge it is said to diverge.

A special way to sum a double series would be to *sum by rows*, that is,

$$\sum_{j=1}^{\infty} \left(\sum_{k=1}^{\infty} w_{jk} \right).$$

Another way would be to sum by columns, that is,

$$\sum_{k=1}^{\infty} \left(\sum_{j=1}^{\infty} w_{jk} \right).$$

Theorem 4.7.2 implies that if both the iterated sums by rows and by columns exist and the series is convergent, then the iterated sums converge to the sum of the series.

Theorem 4.7.3. *A convergent double series of nonnegative terms may be summed by rows or by columns.*

Proof: Let $s = \sum\limits_{j,\,k=1}^{\infty} w_{jk}$. Then for each j, $\sigma_{jn} = \sum\limits_{k=1}^{n} w_{jk}$ is a nondecreasing sequence of real numbers bounded by s. Hence, $s_j = \lim\limits_{n \to \infty} \sigma_{jn} = \sum\limits_{k=1}^{\infty} w_{jk}$ exists. Now, if $\sum\limits_{j=1}^{m} s_j$ is bounded for all m, then $\lim\limits_{m \to \infty} \sum\limits_{j=1}^{m} s_j = \sum\limits_{j=1}^{\infty} \left(\sum\limits_{k=1}^{\infty} w_{jk} \right)$ exists and by Theorem 4.7.2 this limit is equal to s. On the other hand, if $\sum\limits_{j=1}^{m} s_j$ is unbounded, then there exists an M such that $\sum\limits_{j=1}^{m} s_j > 2s$, for all $m > M$. We fix $m > M$ and select n so large that

$$s_j \geq \sum_{k=1}^{n} w_{jk} > s_j - \frac{s}{2m}.$$

Then

$$s \geq \sum_{j=1}^{m} \sum_{k=1}^{n} w_{jk} > \sum_{j=1}^{m} s_j - \frac{s}{2} > \frac{3}{2} s.$$

This is clearly impossible, which proves that $\sum_{j=1}^{m} s_j$ must be bounded.

The corresponding result for summation by columns can be obtained by a similar argument.

Corollary 4.7.1. If the summation by rows or by columns of a double series of nonnegative terms does not exist then the series is divergent.

Proof: If the series is convergent, then both the summation by rows and by columns exist.

Definition 4.7.5. The double series $\sum_{j,\,k=1}^{\infty} w_{jk}$ converges absolutely if and only if the double series $\sum_{j,\,k=1}^{\infty} |w_{jk}|$ converges.

Theorem 4.7.4. *An absolutely convergent double series converges.*

Proof: Let $s_{mn} = \sum_{j=1}^{m} \left(\sum_{k=1}^{n} w_{jk} \right)$ and $\sigma_{mn} = \sum_{j=1}^{m} \left(\sum_{k=1}^{n} |w_{jk}| \right)$. Then given $\epsilon > 0$ there is an $N(\epsilon)$ such that

$$|s_{m+p,n+q} - s_{mn}| \leq \sigma_{m+p,n+q} - \sigma_{mn} < \epsilon,$$

for all $m > N$, $n > N$, and all positive integers p and q.

Definition 4.7.6. The double series $\sum_{j,\,k=1}^{\infty} w_{jk}(z)$, where each $w_{jk}(z)$ is a function of z defined on a set S, converges uniformly on S, if and only if for every $\epsilon > 0$ there exists an $N(\epsilon)$, independent of $z \in S$, such that

$$|s_{m+p,n+q}(z) - s_{mn}(z)| < \epsilon,$$

for all $m > N$, $n > N$, and all positive integers p and q.

Theorem 4.7.5. *The double series $\sum_{j,\,k=1}^{\infty} w_{jk}(z)$ converges absolutely and uniformly in S if there exists a set of constants M_{jk} such that $|w_{jk}(z)| \leq M_{jk}$ for all $z \in S$ and the double series $\sum_{j,\,k=1}^{\infty} M_{jk}$ converges.*

Proof: Let $s_{mn}(z) = \sum_{j=1}^{m} \left(\sum_{k=1}^{n} w_{jk}(z) \right)$ and $\sigma_{mn} = \sum_{j=1}^{m} \left(\sum_{k=1}^{n} M_{jk} \right)$. Then given $\epsilon > 0$, there exists $N(\epsilon)$, independent of $z \in S$, such that

$$|s_{m+p,n+q}(z) - s_{mn}(z)| \leq \sigma_{m+p,n+q} - \sigma_{mn} < \epsilon,$$

for all $m > N$, $n > N$, and all positive integers p and q.

By a double power series we shall mean a double series of the form $\sum_{j,k=0}^{\infty} a_{jk}(z - z_0)^j(\zeta - \zeta_0)^k$. Without loss of generality, we can consider the series $\sum_{j,k=0}^{\infty} a_{jk}z^j\zeta^k$ for we can always replace $z - z_0$ by z and $\zeta - \zeta_0$ by ζ. Clearly, no matter what the a_{jk}, the double series converges if $z = 0$, or $\zeta = 0$, or both. On the other hand, suppose that for all j and k $|a_{jk}| \leq MR_1^{-j}R_2^{-k}$, where M, R_1, R_2 are positive constants. Then

$$|a_{jk}z^j\zeta^k| \leq M\left(\frac{|z|}{R_1}\right)^j\left(\frac{|\zeta|}{R_2}\right)^k$$

and

$$\sum_{j,k=0}^{\infty} M\left(\frac{|z|}{R_1}\right)^j\left(\frac{|\zeta|}{R_2}\right)^k = \frac{M}{\left(1 - \dfrac{|z|}{R_1}\right)\left(1 - \dfrac{|\zeta|}{R_2}\right)},$$

provided $|z| < R_1$ and $|\zeta| < R_2$. Hence, $\sum_{j,k=0}^{\infty} a_{jk}z^j\zeta^k$ converges absolutely and uniformly for $|z| \leq r_1 < R_1$ and $|\zeta| \leq r_2 < R_2$. Let

$$f(z, \zeta) = \sum_{j,k=0}^{\infty} a_{jk}z^j\zeta^k.$$

Then for each ζ, $f(z, \zeta)$ is analytic in z for $|z| < R_1$, and for each z, $f(z, \zeta)$ is analytic in ζ for $|\zeta| < R_2$. We say that $f(z, \zeta)$ is an *analytic function of two complex variables* in $D = \{(z, \zeta) \mid |z| < R_1, |\zeta| < R_2\}$. Furthermore,

$$\frac{\partial f}{\partial z} = \sum_{j,k=0}^{\infty} ja_{jk}z^{j-1}\zeta^k,$$

$$\frac{\partial f}{\partial \zeta} = \sum_{j,k=0}^{\infty} ka_{jk}z^j\zeta^{k-1},$$

and these partial derivatives exist in D. We shall now "prove" the following.

Theorem 4.7.6. *Let $f(z, \zeta)$ be an analytic function of z and ζ in $D = \{(z, \zeta) \mid |z| < R_1, |\zeta| < R_2\}$. Then*

$$f(z, \zeta) = -\frac{1}{4\pi} \int_{|s|=\rho_1} \int_{|t|=\rho_2} \frac{f(s, t)}{(s - z)(t - \zeta)} \, ds \, dt,$$

where $\rho_1 < R_1$ and $\rho_2 < R_2$. All partial derivatives of f exist in D, where

$$\frac{\partial^{m+n}f}{\partial z^m\, \partial\zeta^n} = -\frac{m!n!}{4\pi} \int_{|s|=\rho_1} \int_{|t|=\rho_2} \frac{f(s, t)}{(s - z)^{m+1}(t - \zeta)^{n+1}} \, ds \, dt.$$

Furthermore, $f(z, \zeta)$ has a double power series expansion

$$f(z, \zeta) = \sum_{m,n=0}^{\infty} a_{mn}z^m\zeta^n,$$

where

$$a_{mn} = \frac{1}{m!n!} \frac{\partial^{m+n} f}{\partial z^m \partial \zeta^n}\bigg|_{z=\zeta=0}.$$

Proof: Just as with functions of two real variables, the existence of the two first partial derivatives at a point does not obviously imply the continuity of the function f at the point. However, it has been proved by Hartogs that under the present hypotheses f is continuous in D. We shall assume this and proceed. For a fixed ζ, $f(z, \zeta)$ is analytic for $|z| < R_1$. Therefore, if $\rho_1 < R_1$

$$f(z, \zeta) = \frac{1}{2\pi i} \int_{|s|=\rho_1} \frac{f(s, \zeta)}{(s - z)} \, ds.$$

For each z, $f(z, \zeta)$ is analytic for $|\zeta| < R_2$. Therefore, if $\rho_2 < R_2$

$$f(z, \zeta) = -\frac{1}{4\pi} \int_{|t|=\rho_2} \frac{1}{t - \zeta} \int_{|s|=\rho_1} \frac{f(s, t)}{(s - z)} \, ds \, dt$$

$$= -\frac{1}{4\pi} \int_{|s|=\rho_1} \int_{|t|=\rho_2} \frac{f(s, t)}{(s - z)(t - \zeta)} \, ds \, dt.$$

We can write the iterated integral as a double integral because of the continuity of $f(z, \zeta)$. As long as $|z| < \rho_1$, $|\zeta| < \rho_2$ we can differentiate under the integral signs as many times as we please. Hence,

$$\frac{\partial^{m+n} f}{\partial z^m \partial \zeta^n} = -\frac{m!n!}{4\pi} \int_{|s|=\rho_1} \int_{|t|=\rho_2} \frac{f(s, t)}{(s - z)^{m+1}(t - \zeta)^{n+1}} \, ds \, dt.$$

To obtain the power series expansion, we proceed as in the case of one complex variable.

$$\frac{1}{(s - z)(t - \zeta)} = \frac{1}{st} \frac{1}{\left(1 - \dfrac{z}{s}\right)} \frac{1}{\left(1 - \dfrac{\zeta}{t}\right)} = \sum_{m,n=0}^{\infty} \frac{z^m \zeta^n}{s^{m+1} t^{n+1}},$$

where the double series converges uniformly in s and t for $|z| < |s| = \rho_1$, $|\zeta| < |t| = \rho_2$. Therefore, we may insert the series in the Cauchy integral formula and interchange the operations of integration and summation. Hence,

$$f(z, \zeta) = -\frac{1}{4\pi} \int_{|s|=\rho_1} \int_{|t|=\rho_2} f(s, t) \sum_{m,n=0}^{\infty} \frac{z^m \zeta^n}{s^{m+1} t^{n+1}} \, ds \, dt$$

$$= \sum_{m,n=0}^{\infty} \left(-\frac{1}{4\pi} \int_{|s|=\rho_1} \int_{|t|=\rho_2} \frac{f(s, t)}{s^{m+1} t^{n+1}}\right) z^m \zeta^n$$

$$= \sum_{m,n=0}^{\infty} a_{mn} z^m \zeta^n,$$

where

$$a_{mn} = -\frac{1}{4\pi} \int_{|s|=\rho_1} \int_{|t|=\rho_2} \frac{f(s, t)}{s^{m+1} t^{n+1}} \, ds \, dt = \frac{1}{m!n!} \frac{\partial^{m+n} f}{\partial z^m \partial \zeta^n}\bigg|_{z=\zeta=0}.$$

Since for each z and ζ such that $|z| < R_1$ and $|\zeta| < R_2$ we can always find a ρ_1 and ρ_2 such that $|z| < \rho_1 < R_1$ and $|\zeta| < \rho_2 < R_2$, the double power series converges for $|z| < R_1$, $|\zeta| < R_2$.

Exercises 4.7

1. Show that $\left\{ \dfrac{mn}{(m + n)^2} \right\}$ does not converge although

$$\lim_{m \to \infty} (\lim_{n \to \infty} w_{mn}) = \lim_{n \to \infty} (\lim_{m \to \infty} w_{mn}).$$

2. Show that $\lim\limits_{m \to \infty} (\lim\limits_{n \to \infty} w_{mn})$ and $\lim\limits_{n \to \infty} (\lim\limits_{m \to \infty} w_{mn})$ do not exist if $w_{mn} = (-1)^{m+n} \dfrac{m + n}{mn}$, but that the sequence $\{w_{mn}\}$ converges.

3. Prove that if $w_{mn}(z)$ is continuous on a simple contour C for each m and n, and $w_{mn}(z)$ converges uniformly on C, then

$$\int_C \lim_{m,n \to \infty} w_{mn}(z) \, dz = \lim_{m,n \to \infty} \int_C w_{mn}(z) \, dz.$$

4. Let $\sum\limits_{j,k=1}^{\infty} a_{jk}$ be absolutely convergent to S. Prove that

$$S = \sum_{n=1}^{\infty} \left(\sum_{j=1}^{n} a_{jn-j+1} \right).$$

5. Let $\sum\limits_{j=1}^{\infty} a_j = A$ and $\sum\limits_{j=1}^{\infty} b_j = B$, where both series are absolutely convergent. Prove that

$$AB = \sum_{n=1}^{\infty} \left(\sum_{j=1}^{n} a_j b_{n-j+1} \right).$$

6. Let $\sum\limits_{j=1}^{\infty} a_j = A$ and $\sum\limits_{k=1}^{\infty} b_k = B$, where both series are absolutely convergent. Prove that

$$AB = \sum_{n=1}^{\infty} \left(\sum_{jk=n} a_j b_k \right).$$

7. Prove that $\sum\limits_{j,k=1}^{\infty} j^{-s} k^{-t}$ converges if $s > 1$ and $t > 1$.

8. Let $|a_{jk}| \leq M j^{-s} k^{-t}$, for all j and k. Prove that $\sum\limits_{j,k=1}^{\infty} a_{jk} z^j \zeta^k$ is absolutely convergent for $|z| \leq 1$ and $|\zeta| \leq 1$, if $s > 1$ and $t > 1$.

9. Suppose $f(z) = \sum\limits_{k=0}^{\infty} f_k(z)$ converges normally for $|z| < R$ where each $f_k(z)$ is analytic. Then each $f_k(z)$ has a Maclaurin expansion $\sum\limits_{j=0}^{\infty} a_{jk}z^j$ valid for $|z| < R$. Prove that

$$f(z) = \sum_{k=0}^{\infty}\left(\sum_{j=0}^{\infty} a_{jk}z^j\right) = \sum_{j=0}^{\infty}\left(\sum_{k=0}^{\infty} a_{jk}\right) z^j.$$

Hint: $f(z)$ must have a Maclaurin expansion

$$\sum_{j=0}^{\infty} \frac{f^{(j)}(0)}{j!}\, z^j,$$

so the problem is to show that

$$\frac{f^{(j)}(0)}{j!} = \sum_{k=0}^{\infty} a_{jk}.$$

10. Let $f(z, \zeta)$ be analytic at (z_0, ζ_0). If $z = x + iy$, $\zeta = \xi + i\eta$ and

$$f(z, \zeta) = u(x, y, \xi, \eta) + iv(x, y, \xi, \eta),$$

prove that $u_x = v_y$, $u_y = -v_x$, $u_\xi = v_\eta$, and $u_\eta = -v_\xi$ at (z_0, ζ_0).

11. Let $D = \{(z, \zeta) \,|\, |z| < R_1, |\zeta| < R_2\}$ and let $f(z, \zeta)$ be analytic in D. Let $f(z, \zeta) \equiv 0$ in $D_\epsilon = \{(z, \zeta) \,|\, |z - z_0| < \epsilon, |\zeta - \zeta_0| < \epsilon\}$, where $(z_0, \zeta_0) \in D$, $\epsilon < R_1 - |z_0|$ and $\epsilon < R_2 - |\zeta_0|$. Prove that $f(z, \zeta) \equiv 0$ in D.

12. Let $f(z, \zeta)$ be analytic in $D = \{(z, \zeta) \,|\, |z| < R_1, |\zeta| < R_2\}$ and let $R = \{(z, \zeta) \,|\, |z| \le r_1 < R_1, |\zeta| \le r_2 < R_2\}$. Prove that, if $M = \max\limits_{(z, \zeta) \in R} |f(z, \zeta)|$, then $|f(z, \zeta)| < M$ for $|z| < r_1$ and $|\zeta| < r_2$ unless $f(z, \zeta)$ is constant in D.

4.8. INFINITE PRODUCTS

Just as we can give meaning to the concept of an infinite series, we can also introduce the concept of an infinite product. Let $\{w_n\}$ be a sequence of complex numbers. Then we can form the product

$$\prod_{k=1}^{\infty} (1 + w_k) = (1 + w_1)(1 + w_2)(1 + w_3) \cdots .$$

We shall define convergence of an infinite product in terms of the convergence of the sequence of partial products, and shall use this definition to obtain theorems about the behavior of the w's. However, if $1 + w_N = 0$ for some N, then for N, $N + 1$, $N + 2, \ldots$ the partial products will all be zero regardless of the behavior of the w's. Therefore, to get something definitive we shall have to require that $1 + w_k \neq 0$ for all k.* We can rule out the possibility of the vanishing of any term by requiring that the limit of the partial product be different from zero. However, as the example

$$\prod_{k=1}^{\infty} \left(1 - \frac{1}{k+1}\right)$$

will show, the limit of the partial products could be zero without any single term vanishing. But for other reasons, which shall become apparent, we shall rule out convergence of a partial product to zero.

Definition 4.8.1. The infinite product $\prod_{k=1}^{\infty} (1 + w_k)$ converges to $p \neq 0$, if and only if the sequence of partial products $p_n = \prod_{k=1}^{n} (1 + w_k)$ converges to p. If the infinite product does not converge, it is said to diverge.

Theorem 4.8.1. *A necessary condition for convergence of an infinite product* $\prod_{k=1}^{\infty} (1 + w_k)$ *is that* $\lim_{n \to \infty} w_n = 0$.

Proof: We have $p_n = p_{n-1}(1 + w_n)$ and since $\lim_{n \to \infty} p_n = \lim_{n \to \infty} p_{n-1} = p \neq 0$,

$$\lim_{n \to \infty} w_n = \lim_{n \to \infty} \left(\frac{p_n}{p_{n-1}} - 1\right) = 0.$$

Theorem 4.8.2. *The infinite product* $\prod_{k=1}^{\infty} (1 + w_k)$ *with partial products* $p_n = \prod_{k=1}^{n} (1 + w_k)$ *converges if and only if there exists a* $\delta > 0$ *such that* $|p_n| > \delta$, *for all n, and for every* $\epsilon > 0$, *there is an* $N(\epsilon)$ *such that* $|p_{n+q} - p_n| < \epsilon$, *for all* $n > N$ *and all positive integers q.*

Proof: This is just the Cauchy criterion for convergence of the sequence $\{p_n\}$ with the added condition to make sure the limit is not zero.

Theorem 4.8.3. *If* $w_k \geq 0$ *for all k, then* $\prod_{k=1}^{\infty} (1 + w_k)$ *converges if and only if* $\sum_{k=1}^{\infty} w_k$ *converges.*

Proof: Since $e^{w_k} = 1 + w_k + \frac{w_k^2}{2!} + \cdots \geq 1 + w_k$, we have

$$w_1 + w_2 + \cdots + w_n \leq (1 + w_1)(1 + w_2) \cdots (1 + w_n) \leq e^{w_1 + w_2 + \cdots + w_n}.$$

* Some authors merely require that $1 + w_k \neq 0$ for all but a finite number of k. However, in this case the product can be split into two parts, one a finite product and the other an infinite product with no zero terms.

If all the w_k are zero then both series and product converge. Otherwise, it is obvious from the above inequality that the series and product converge or diverge together.

Theorem 4.8.4. *The infinite product* $\prod\limits_{k=1}^{\infty} (1 + w_k)$ *converges if and only if* $\sum\limits_{k=1}^{\infty}$ Log $(1 + w_k)$ *converges.*

Proof: We have $1 + w_k = e^{\text{Log}(1+w_k)}$ and

$$p_n = \prod_{k=1}^{n} (1 + w_k) = \exp\left[\sum_{k=1}^{n} \text{Log} (1 + w_k)\right] = e^{s_n},$$

where $s_n = \sum\limits_{k=1}^{n}$ Log $(1 + w_k)$. Hence, if $\lim\limits_{n \to \infty} s_n = s$, then $\lim\limits_{n \to \infty} p_n = e^s \neq 0$.

On the other hand, $s_n = \text{Log } p_n + 2\pi i q_n$, where q_n is an integer. Let $a_k = \arg (1 + w_k)$, where $-\pi \leq \arg (1 + w_k) < \pi$, and let $b_n = \arg p_n$, where $-\pi \leq \arg p_n < \pi$. Then

$$\sum_{k=1}^{n} a_k = b_n + 2\pi q_n$$

$$a_n = b_n - b_{n-1} + 2\pi(q_n - q_{n-1}).$$

Now, if $p = \prod\limits_{k=1}^{\infty} (1 + w_k) \neq 0$, then $\lim\limits_{n \to \infty} a_n = \lim\limits_{n \to \infty} \arg(1 + w_n) = \arg 1 = 0$. Also, $\lim\limits_{n \to \infty} b_n = \lim\limits_{n \to \infty} \arg p_n = \arg p = b$. Therefore,

$$\lim_{n \to \infty} (q_n - q_{n-1}) = \frac{1}{2\pi} \lim (b_n - b_{n-1}) + \frac{1}{2\pi} \lim a_n = 0,$$

and since q_n is an integer $q_n = q$, an integer, for all n sufficiently large. Hence, $\lim\limits_{n \to \infty} s_n = \text{Log } p + 2\pi i q$.

Definition 4.8.2. The infinite product $\prod\limits_{k=1}^{\infty} (1 + w_k)$ converges absolutely, if and only if the infinite product $\prod\limits_{k=1}^{\infty} (1 + |w_k|)$ converges.

Theorem 4.8.5. *The infinite product* $\prod\limits_{k=1}^{\infty} (1 + w_k)$ *converges absolutely, if and only if* $\sum\limits_{k=1}^{\infty} |w_k|$ *converges.*

Proof: This follows directly from Theorem 4.8.3.

Theorem 4.8.6. *An absolutely convergent infinite product* $\prod\limits_{k=1}^{\infty} (1 + w_k)$ *converges, if* $w_k \neq -1$ *for every k.*

Proof: Let $p_n = \prod_{k=1}^{n} (1 + w_k)$ and $P_n = \prod_{k=1}^{n} (1 + |w_k|)$. Then,

$$p_n - p_{n-1} = (1 + w_1)(1 + w_2) \cdots (1 + w_{n-1})w_n,$$
$$P_n - P_{n-1} = (1 + |w_1|)(1 + |w_2|) \cdots (1 + |w_{n-1}|)|w_n|,$$

and, clearly, $|p_n - p_{n-1}| \leq P_n - P_{n-1}$. Also, $P_n = \sum_{k=1}^{n} (P_k - P_{k-1})$ and $p_n = \sum_{k=1}^{n} (p_k - p_{k-1})$, if we define $p_0 = P_0 = 0$. The first series is known to converge and hence, by comparison, the second converges. Therefore, p_n tends to a limit p. Finally, we have to show that $p \neq 0$.

Since $\sum_{k=1}^{\infty} |w_k|$ converges, $\lim_{n \to \infty} (1 + w_n) = 1$. This implies that for all sufficiently large k

$$\frac{|w_k|}{|1 + w_k|} \leq 2|w_k|.$$

By comparison,

$$\sum_{k=1}^{\infty} \left| \frac{w_k}{1 + w_k} \right|$$

converges, and this implies the convergence of

$$\prod_{k=1}^{\infty} \left(1 - \frac{w_k}{1 + w_k} \right).$$

The partial sums of this product are $1/p_n$. Therefore, the product converges to $1/p$, which shows that $p \neq 0$. This completes the proof.

Definition 4.8.3. Let $w_k(z)$ be a function defined on a set S for each positive integer k. The infinite product $\prod_{k=1}^{\infty} [1 + w_k(z)]$ converges uniformly on S, if and only if the sequence of partial products $p_n(z) = \prod_{k=1}^{n} [1 + w_k(z)]$ converges uniformly on S to a function $p(z)$ which never vanishes in S.

Theorem 4.8.7. *If for all k, $w_k(z) \neq -1$ in S, and if there exist constants M_k such that $|w_k(z)| \leq M_k$ and $\sum_{k=1}^{\infty} M_k$ converges, then $\prod_{k=1}^{\infty} [1 + w_k(z)]$ converges uniformly on S.*

Proof: Let $p_n(z) = \prod_{k=1}^{n} [1 + w_k(z)]$ and $P_n = \prod_{k=1}^{n} (1 + M_k)$. Then,

$$p_n - p_{n-1} = (1 + w_1)(1 + w_2) \cdots (1 + w_{n-1})w_n,$$
$$P_n - P_{n-1} = (1 + M_1)(1 + M_2) \cdots (1 + M_{n-1})M_n \geq 0,$$
$$|p_n - p_{n-1}| \leq P_n - P_{n-1}.$$

Now, $p_n(z) = \sum\limits_{k=1}^{n} (p_k - p_{k-1})$ and $P_n = \sum\limits_{k=1}^{n} (P_k - P_{k-1})$, if we define $p_0 = P_0 = 0$. Since $\sum\limits_{k=1}^{\infty} M_k$ converges, $\prod\limits_{k=1}^{\infty} (1 + M_k)$ converges. This means that $\sum\limits_{k=1}^{\infty} (P_k - P_{k-1})$ converges and that $\sum\limits_{k=1}^{\infty} (p_k(z) - p_{k-1}(z))$ converges uniformly. Also, by comparison, we have that $\prod\limits_{k=1}^{\infty} [1 + |w_k(z)|]$ converges, which implies that $\lim\limits_{n \to \infty} [1 + w_n(z)] = 1$. Since $w_k(z) \neq -1$ in S, there exists a $K > 0$ such that for all k and all $z \in S$, $|w_k(z)|/|1 + w_k(z)| \leq K|w_k(z)|$. This proves that

$$\sum_{k=1}^{\infty} \left| \frac{w_k(z)}{1 + w_k(z)} \right|$$

converges and that

$$\prod_{k=1}^{\infty} \left[1 - \frac{w_k(z)}{1 + w_k(z)} \right]$$

converges. But this product converges to $1/p(z)$. Hence, $p(z) \neq 0$ in S.

The following two theorems follow directly from our discussion of uniformly convergent sequences.

Theorem 4.8.8. *Let $w_k(z)$ be continuous in S for each k. If the infinite product $\prod\limits_{k=1}^{\infty} [1 + w_k(z)]$ converges uniformly in S to $p(z)$, then $p(z)$ is continuous in S.*

Theorem 4.8.9. *Let $w_k(z)$ be analytic in a domain D for each k. If the infinite product $\prod\limits_{k=1}^{\infty} [1 + w_k(z)]$ converges uniformly to $p(z)$ in every compact subset of D, then $p(z)$ is analytic in D.*

EXAMPLE 4.8.1. Prove that

$$z \prod_{k=1}^{\infty} \left(1 - \frac{z^2}{\pi^2 k^2} \right)$$

is an entire function of z with zeros at $n\pi$, $n = 0, \pm 1, \pm 2, \ldots$ Let $|z| < R$ and $\pi^2 N^2 > R^2$. Then,

$$f(z) = z \prod_{k=1}^{N-1} \left(1 - \frac{z^2}{\pi^2 k^2} \right) \prod_{k=N}^{\infty} \left(1 - \frac{z^2}{\pi^2 k^2} \right) = g(z)h(z),$$

where

$$g(z) = z \prod_{k=1}^{N-1} \left(1 - \frac{z^2}{\pi^2 k^2} \right)$$

is a polynomial with zeros at $0, \pm\pi, \pm2\pi, \ldots, \pm(N-1)\pi$, and

$$h(z) = \prod_{k=N}^{\infty}\left(1 - \frac{z^2}{\pi^2 k^2}\right)$$

is an infinite product with no zero term. Now, $\left|\dfrac{z^2}{\pi^2 k^2}\right| < \dfrac{R^2}{\pi^2 k^2}$ and the

$$\sum_{k=N}^{\infty} \frac{R^2}{\pi^2 k^2}$$

converges. Therefore, $h(z)$ converges uniformly in compact subsets of $|z| < R$, and is analytic in this domain. But R can be taken as large as we please, implying that $f(z)$ is entire.

EXAMPLE 4.8.2. Prove that

$$\prod_{k=1}^{\infty}\left(1 + \frac{1}{k}\tan\frac{z}{k}\right)$$

is analytic except for poles in the unextended plane. We note that $\dfrac{1}{k}\tan\dfrac{z}{k}$ is analytic for $|z| < \dfrac{k\pi}{2}$ and its Maclaurin expansion is

$$\frac{1}{k}\tan\frac{z}{k} = \frac{z}{k^2} + \frac{1}{3}\frac{z^3}{k^4} + \frac{2}{15}\frac{z^5}{k^6} + \cdots$$

$$= \frac{z}{k^2}\left(1 + \frac{1}{3}\frac{z^2}{k^2} + \frac{2}{15}\frac{z^4}{k^4} + \cdots\right)$$

$$= \frac{z}{k^2}g(z),$$

where $g(z)$ is analytic for $|z| < \dfrac{k\pi}{2}$. Let $|z| < R < \dfrac{N\pi}{4}$. Then

$$f(z) = \prod_{k=1}^{\infty}\left(1 + \frac{1}{k}\tan\frac{z}{k}\right) = \prod_{k=1}^{N-1}\left(1 + \frac{1}{k}\tan\frac{z}{k}\right)\prod_{k=N}^{\infty}\left(1 + \frac{1}{k}\tan\frac{z}{k}\right),$$

$$f(z) = p(z)h(z).$$

The function

$$p(z) = \prod_{k=1}^{N-1}\left(1 + \frac{1}{k}\tan\frac{z}{k}\right)$$

is analytic except where $\tan(z/k)$ has poles for $k = 1, 2, \ldots, N-1$. $\tan(z/k)$ has poles of order one at $(z/k) = \pm(2n-1)(\pi/2)$, $n = 1, 2, 3, \ldots$. The function

$$h(z) = \prod_{k=N}^{\infty}\left(1 + \frac{1}{k}\tan\frac{z}{k}\right)$$

is an infinite product which has no zero terms provided N is large enough that $\frac{|z|}{N^2} M < \frac{RM}{N^2} < 1$, where $M = \max_{|z| < N\pi/4} |g(z)|$. Furthermore,

$$\left| \frac{1}{k} \tan \frac{z}{k} \right| < \frac{R}{k^2} M,$$

and the series

$$\sum_{k=N}^{\infty} \frac{RM}{k^2}$$

converges. Therefore, $h(z)$ converges uniformly in compact subsets of $|z| < R$, where it is analytic. However, R can be taken as large as we please, and therefore, $f(z)$ is analytic everywhere in the unextended plane except at the poles of $\tan (z/k)$, that is, $z = \pm(2n - 1)k\pi/2$, $k = 1, 2, 3, \ldots, n = 1, 2, 3, \ldots$.

Exercises 4.8

1. Show that

$$\prod_{k=1}^{\infty} \left(1 + \frac{i}{k} \right)$$

diverges, whereas

$$\prod_{k=1}^{\infty} \left| 1 + \frac{i}{k} \right|$$

converges.

2. Given that

$$\sin z = z \prod_{k=1}^{\infty} \left(1 - \frac{z^2}{\pi^2 k^2} \right).$$

Prove that

$$\sin z = z \prod_{k=-\infty}^{\infty} \left[\left(1 - \frac{z}{\pi k} \right) e^{z/\pi k} \right]$$

where the prime denotes a product where the term $k = 0$ is omitted.

3. Prove that

$$\prod_{k=1}^{\infty} \left(1 + \sin \frac{z^2}{k^2} \right)$$

is an entire function.

4. Prove that

$$\prod_{k=1}^{\infty} \left(1 + \frac{1}{k^2} \cot \frac{z}{k} \right)$$

is analytic except at the poles of $\cot (z/k)$. Locate the poles.

5. For what values of p does $\prod_{k=1}^{\infty} (1 + k^{-p})$ converge?

4.9. IMPROPER INTEGRALS

Our purpose in this section will be to define functions by integration, where the integrand contains the variable as a parameter, and where the path of integration does not have finite length. However, before discussing improper integrals, we shall prove an improved form of Theorem 3.5.3. At that time, we did not have the concept of uniform convergence, and for that reason we had to assume the continuity of the derivative of the integrand with respect to the parameter. Now, using uniform convergence, we can prove the following theorem:

Theorem 4.9.1. *Let $f(z, \zeta)$ be continuous for z in a domain D and ζ on a simple contour C. Let $f(z, \zeta)$ be analytic in D for each ζ on C. Then $F(z) = \int_C f(z, \zeta)\, d\zeta$ is analytic in D, and*

$$F'(z) = \int_C \frac{\partial f}{\partial z}\, d\zeta.$$

Proof: By hypothesis, there exists a continuous parameterization of C, $\zeta(t) = \xi(t) + i\eta(t)$, $0 \leq t \leq 1$. We divide the interval in t into n equal parts $0 = t_0 < t_1 < t_2 < \cdots < t_{n-1} < t_n = 1$, with $\Delta t_k = t_k - t_{k-1} = (1/n)$. We define*

$$F_n(z) = \sum_{k=1}^{n} f[z, \zeta(t_k)][\xi(t_k) + i\eta(t_k)]\, \Delta t_k.$$

From the theory of integration, we know that

$$\lim_{n \to \infty} F_n(z) = F(z) = \int_C f(z, \zeta)\, d\zeta.$$

We can show that this limit is uniform in every compact subset of D. Since $f(z, \zeta)$ is continuous in $z \in D$ and $\zeta \in C$, $f[z, \zeta(t)][\xi(t) + i\eta(t)]$ is uniformly continuous for z in a compact subset S of D and $0 \leq t \leq 1$. Therefore, given $\epsilon > 0$ there exists an $N(\epsilon)$, independent of $z \in S$, such that

$$|f[z, \zeta(t)][\xi(t) + i\eta(t)] - f[z, \zeta(t')][\xi(t') + i\eta(t')]| < \epsilon,$$

when $|t - t'| < (1/N)$. Then,

$$|F_n(z) - F(z)| = \left| \sum_{k=1}^{n} \int_{t_{k-1}}^{t_k} \{f[z, \zeta(t_k)][\xi(t_k) + i\eta(t_k)]\right.$$
$$\left. - f[z, \zeta(t)][\xi(t) + i\eta(t)]\} \, dt \right|$$
$$< \epsilon,$$

---

* We should really break up C into a finite number of arcs on each of which $\xi(t)$ and $\eta(t)$ are continuous, but this just adds a few small complications which are not essential to the understanding of the proof.

provided $n > N$. Now,

$$F_n'(z) = \sum_{k=1}^{n} f_z[z, \zeta(t_k)][\dot{\xi}(t_k) + i\dot{\eta}(t_k)] \Delta t_k$$

exists in D for each n. Therefore, by Theorem 4.2.3, $F(z)$ is analytic in D, and

$$F'(z) = \lim_{n \to \infty} F_n'(z) = \int_C \frac{\partial f}{\partial z} d\zeta.$$

Definition 4.9.1. Let C be a curve originating at z_0 with a parameterization $z(t) = \xi(t) + i\eta(t)$, $0 \le t < \infty$, such that $z_0 = z(0)$, and for each finite T the part of the curve traced by $0 \le t \le T$ is a simple contour with length

$$L(T) = \int_0^T (\dot{\xi}^2 + \dot{\eta}^2)^{1/2} dt$$

such that $\lim_{T \to \infty} L(T) = \infty$; then C is said to be an infinite contour.

It must not be assumed that the contour is infinite just because the parameter t goes to ∞. For example, if $z(t) = (2/\pi) \tan^{-1} t, 0 \le t < \infty$, the contour is just a unit interval on the real axis. On the other hand, the contour may be infinite even though the parameter is finite. For example, $z(t) = \tan(\pi t/2), 0 \le t \le 1$, represents the positive real axis. At any rate, if the contour is infinite, there is always one parameter, namely arclength, which goes from 0 to ∞.

Definition 4.9.2. Let C be an infinite contour. Let $f(z)$ be continuous on C for all finite t, where $z(t) = \xi(t) + i\eta(t)$, $0 \le t < \infty$ is a parameterization of C. Let

$$F(T) = \int_0^T f[z(t)][\dot{\xi}(t) + i\dot{\eta}(t)] dt.$$

Then the improper integral $\int_C f(z)\, dz$ converges, if and only if $\lim_{T \to \infty} F(T)$ exists as a finite limit. In this case,

$$\int_C f(z)\, dz = \lim_{T \to \infty} F(T).$$

Definition 4.9.3. Let C be an infinite contour. Let $f(z, \zeta)$ be a continuous function for $z \in S$ and ζ on C for all finite t, where $\zeta(t) = \xi(t) + i\eta(t)$, $0 \le t < \infty$. Then $\int_C f(z, \zeta)\, d\zeta$ converges uniformly in S if and only if

$$\lim_{T \to \infty} F(z, T) = \lim_{T \to \infty} \int_0^T f[z, \zeta(t)][\dot{\xi}(t) + i\dot{\eta}(t)] dt$$

converges uniformly in S.

Theorem 4.9.2. *Let C be an infinite contour. Let $f(z, \zeta)$ be a continuous function for $z \in S$ and ζ on C. Let $|f(z, \zeta)| \, |\dot{\xi}(t) + i\dot{\eta}(t)| \leq M(t)$, where $M(t)$ is independent of $z \in S$ and $\int_0^\infty M(t)\, dt$ converges. Then $\int_C f(z, \zeta)\, d\zeta$ converges uniformly in S.*

Proof: Let $s_n = \int_0^n M(t)\, dt$. Then $\lim\limits_{n\to\infty} s_n$ exists and given $\epsilon > 0$, there exists $N(\epsilon)$ such that $|s_{n+p} - s_n| < \epsilon$, for all $n > N$ and all positive integers p. Let

$$\sigma_n(z) = \int_0^n f[z, \zeta(t)][\dot{\xi}(t) + i\dot{\eta}(t)]\, dt.$$

Then,

$$|\sigma_{n+p} - \sigma_n| = \left| \int_n^{n+p} f[z, \zeta(t)][\dot{\xi}(t) + i\dot{\eta}(t)]\, dt \right|,$$

$$|\sigma_{n+p} - \sigma_n| \leq \int_n^{n+p} M(t)\, dt = s_{n+p} - s_n < \epsilon.$$

Therefore, $\lim\limits_{n\to\infty} \sigma_n(z)$ exists uniformly in S. Finally, if $T > n > N$

$$|F(z, T) - \sigma_n(z)| \leq \int_n^T M(t)\, dt \leq \int_n^{n+p} M(t)\, dt = s_{n+p} - s_n < \epsilon,$$

provided p is so large that $n + p > T$. Therefore,

$$\lim_{T\to\infty} F(z, T) = \lim_{n\to\infty} \sigma_n(z)$$

converges uniformly in S.

Theorem 4.9.3. *Let C be an infinite contour. Let $f(z, \zeta)$ be continuous for z in S and ζ on C. Let $\int_C f(z, \zeta)\, d\zeta$ be uniformly convergent in S. Then $F(z) = \int_C f(z, \zeta)\, d\zeta$ is continuous in S.*

Proof: $F_n(z) = \int_0^n f[z, \zeta(t)][\dot{\xi}(t) + i\dot{\eta}(t)]\, dt$ is a uniformly convergent sequence of continuous functions in S converging to $F(z)$.

Theorem 4.9.4. *Let C be an infinite contour. Let $f(z, \zeta)$ be an analytic function of z in a domain D for each ζ on C for any finite value of the parameter. Let $\int_C f(z, \zeta)\, d\zeta$ be uniformly convergent in every compact subset of D. Then $F(z) = \int_C f(z, \zeta)\, d\zeta$ is analytic in D, where*

$$F'(z) = \int_C \frac{\partial f}{\partial z}\, d\zeta.$$

Proof: $F_n(z) = \int_0^n f[z, \zeta(t)][\dot{\xi}(t) + i\dot{\eta}(t)]\, dt$ is sequence of analytic functions in D uniformly convergent on every compact subset of D. Hence,

$F(z) = \lim\limits_{n \to \infty} F_n(z)$ is analytic in D. Also

$$F'_n(z) = \int_0^n f_z[z, \zeta(t)][\dot{\xi}(t) + i\dot{\eta}(t)] \, dt,$$

$$F'(z) = \lim_{n \to \infty} F'_n(z) = \int_C \frac{\partial f}{\partial z} \, d\zeta.$$

EXAMPLE 4.9.1. Let $f(t)$ be a complex-valued function of the real variable t, $0 \le t < \infty$, such that $|f(t)| \le Me^{at}$ for some positive M and some real a. Show that the *Laplace Transform* of $f(t)$

$$\int_0^\infty f(t)e^{-zt} \, dt$$

is analytic for $\text{Re}(z) > a$. Let $\text{Re}(z) \ge \rho > a$. Then $|f(t)e^{-zt}| \le Me^{(a-\rho)t}$. Also,

$$\lim_{T \to \infty} \int_0^T Me^{(a-\rho)t} \, dt = \lim_{T \to \infty} \frac{M}{\rho - a}[1 - e^{(a-\rho)T}] = \frac{M}{\rho - a}.$$

Therefore, $\int_0^\infty f(t)e^{-zt} \, dt$ is uniformly convergent in every compact subset of the right half-plane $\text{Re}(z) > a$. Hence, since $f(t)e^{-zt}$ is analytic in z for each finite t, the result is proved.

This example gives us some interesting examples of analytic continuation. Suppose $f(t) = 1, 0 \le t < \infty$. Then the Laplace transform is

$$\lim_{T \to \infty} \int_0^T e^{-zt} \, dt = \lim_{T \to \infty} \frac{1}{z}(1 - e^{-zt}) = \frac{1}{z},$$

provided $\text{Re}(z) > 0$. However, the function $(1/z)$ is analytic everywhere except $z = 0$. In the right half-plane, $1/z$ and $\int_0^\infty e^{-zt} \, dt$ agree. Therefore, $1/z$ is the analytic continuation of the integral to the rest of the plane.

EXAMPLE 4.9.2. Let $f(t)$ be a complex-valued function of the real variable t, $-\infty < t < \infty$, such that

$$|f(t)| \le Ke^{at}, \qquad 0 \le t < \infty,$$
$$|f(t)| \le Me^{bt}, \qquad -\infty < t \le 0.$$

Show that the *Fourier Transform* of $f(t)$

$$\frac{1}{\sqrt{2\pi}} \int_{-\infty}^{\infty} f(t)e^{izt} \, dt$$

is analytic for $a < \text{Im}(z) < b$. We can write

$$F(z) = \frac{1}{\sqrt{2\pi}} \int_{-\infty}^{\infty} f(t)e^{izt} \, dt = \frac{1}{\sqrt{2\pi}} \int_{-\infty}^{0} f(t)e^{izt} \, dt + \frac{1}{\sqrt{2\pi}} \int_0^{\infty} f(t)e^{izt} \, dt$$

$$= G(z) + H(z),$$

where

$$G(z) = \frac{1}{\sqrt{2\pi}} \int_{-\infty}^{0} f(t)e^{izt}\, dt = \frac{1}{\sqrt{2\pi}} \int_{0}^{\infty} f(-t)e^{-izt}\, dt,$$

$$H(z) = \frac{1}{\sqrt{2\pi}} \int_{0}^{\infty} f(t)e^{izt}\, dt.$$

Let $a < \rho_1 \leq \text{Im}(z) \leq \rho_2 < b$. Then,

$$|f(-t)e^{-izt}| \leq Me^{-bt}e^{\rho_2 t} = Me^{(\rho_2 - b)t}, \quad 0 \leq t < \infty,$$
$$|f(t)e^{izt}| \leq Ke^{at}e^{-\rho_1 t} = Ke^{(a-\rho_1)t}, \qquad 0 \leq t < \infty.$$

Furthermore,

$$\lim_{T\to\infty} \int_{0}^{T} Me^{(\rho_2 - b)t}\, dt = \lim_{T\to\infty} \frac{M}{b - \rho_2}[1 - e^{(\rho_2 - b)T}] = \frac{M}{b - \rho_2},$$

$$\lim_{T\to\infty} \int_{0}^{T} Ke^{(a-\rho_1)t}\, dt = \lim_{T\to\infty} \frac{K}{\rho_1 - a}[1 - e^{(a-\rho_1)T}] = \frac{K}{\rho_1 - a}.$$

Therefore, $G(z)$ and $H(z)$ converge uniformly in compact subsets of the strip $a < \text{Im}(z) < b$, and this is the strip of analyticity of $F(z)$.

EXAMPLE 4.9.2. Show that $\Gamma(z) = \int_{0}^{\infty} t^{z-1}e^{-t}\, dt$ is analytic for $\text{Re}(z) > 0$. We can write

$$\Gamma(z) = \int_{0}^{1} t^{z-1}e^{-t}\, dt + \int_{1}^{\infty} t^{z-1}e^{-t}\, dt$$
$$= G(z) + H(z).$$

Let $0 < \rho \leq \text{Re}(z) \leq R$. Then $|t^{z-1}|e^{-t} \leq t^{R-1}e^{-t} \leq ke^{t/2}e^{-t} = ke^{-t/2}$ for $1 \leq t < \infty$. Also, $\lim_{T\to\infty} \int_{1}^{T} e^{-t/2}\, dt = 2e^{-1/2}$. Therefore, $H(z)$ converges uniformly in compact subsets of the right half-plane, where it is analytic. In the first integral, we let $t = 1/s$, $dt = -(1/s^2)\, ds$. Then

$$G(z) = \int_{0}^{1} t^{z-1}e^{-t}\, dt = \int_{1}^{\infty} s^{-1-z}e^{-1/s}\, ds.$$

Now, $|s^{-1-z}e^{-1/s}| \leq s^{-1-\rho}$ for $1 \leq s < \infty$, and

$$\lim_{T\to\infty} \int_{1}^{T} s^{-1-\rho}\, ds = \lim_{T\to\infty} \frac{1}{\rho}[1 - T^{-\rho}] = \frac{1}{\rho}.$$

Therefore, $G(z)$ converges uniformly in compact subsets of the right half-plane $\text{Re}(z) > 0$, and $G(z)$ and $\Gamma(z)$ are analytic in the half-plane. $\Gamma(z)$ is the well-known gamma function which we shall discuss in the next section.

Exercises 4.9

1. Show that $H(z)$ of Example 4.9.2 is an entire function.

2. Prove that $\int_0^\pi \cos(n\theta - z\sin\theta)\,d\theta$ is an entire function which satisfies the differential equation $z^2 w'' + z w' + (z^2 - n^2)w = 0$, if n is an integer.

3. Prove that

$$\frac{1}{2\pi i}\int_C \frac{(\zeta^2 - 1)^n}{2^n(\zeta - z)^{n+1}}\,d\zeta,$$

where C is a simple closed contour with z in its interior, is a polynomial in z which satisfies the differential equation $(1 - z^2)w'' - 2zw' + n(n+1)w = 0$.

4. Prove that $\int_0^\infty t^{z-1}/(e^t - 1)\,dt$ is analytic for $\mathrm{Re}(z) > 1$. Prove that for $\mathrm{Re}(z) > 1$

$$\int_0^\infty \frac{t^{z-1}}{e^t - 1}\,dt = \Gamma(z)\zeta(z),$$

where $\Gamma(z)$ is the gamma function of Example 4.9.2 and $\zeta(z)$ is the Riemann zeta function. Hint: Justify the formula for $z = x > 2$ by expanding $(e^t - 1)^{-1}$ as an infinite series and integrating term-by-term. You will have to prove the following lemma: suppose $u_n(x) \geq 0$ for all n and all x and that

$$\int_0^T \left(\sum_{n=0}^\infty u_n(x)\right)dx = \sum_{n=0}^\infty \int_0^T u_n(x)\,dx,$$

for all finite T, then

$$\int_0^\infty \left(\sum_{n=0}^\infty u_n(x)\right)dx = \sum_{n=0}^\infty \int_0^\infty u_n(x)\,dx,$$

provided either series converges.

4.10. THE GAMMA FUNCTION

We conclude this chapter with a section on the famous Euler gamma function, not only because it is an important function to be studied in its own right, but also because this study will illustrate many of the concepts which we have considered in this chapter. The original integral representation for the gamma function given by Euler is

$$\Gamma(z) = \int_0^\infty t^{z-1}e^{-t}\,dt,$$

but as we have seen in Example 4.9.2 this representation only exists for $\mathrm{Re}(z) > 0$. Therefore, the first problem to be solved will be to develop, if possible, an analytic continuation to the rest of the plane. To this end we first prove the following formula

$$\Gamma(z + 1) = z\Gamma(z),$$

valid for $\text{Re}(z) > 0$. First we note that each side of the equation is analytic for $\text{Re}(z) > 0$. Hence, their difference is analytic for $\text{Re}(z) > 0$, and if we can show that this difference vanishes on some point set with a limit point in the right half-plane, then it vanishes in the whole half-plane. This would prove the identity.

Let $z = x > 1$. Then

$$\Gamma(x + 1) = \lim_{T \to \infty} \int_0^T t^x e^{-t} \, dt$$

$$= \lim_{T \to \infty} \left[-t^x e^{-t} \Big|_0^T + x \int_0^T t^{x-1} e^{-t} \, dt \right]$$

$$= x \int_0^\infty t^{x-1} e^{-t} \, dt = x\Gamma(x).$$

This completes the proof of the formula $\Gamma(z + 1) = z\Gamma(z)$ for $\text{Re}(z) > 0$. This method of proof, based on Theorem 4.4.4, is very common and quite elegant for its simplicity. Note that if n is a positive integer

$$n! = \Gamma(n + 1) = n\Gamma(n) = n(n - 1)\Gamma(n - 1) = n(n - 1) \cdots 2 \cdot 1\Gamma(1),$$

since $\Gamma(1) = \int_0^\infty e^{-t} \, dt = 1$. For this reason $\Gamma(x + 1)$ is considered a generalization of $n!$ for noninteger values of x.

Actually, $\Gamma(z + 1)$ is analytic for $\text{Re}(z) > -1$. Therefore, we may *define*

$$\Gamma(z) = \frac{1}{z} \Gamma(z + 1)$$

for $\text{Re}(z) > -1$, and the function $(1/z)\,\Gamma(z + 1)$ becomes the analytic continuation of the integral representation to the half-plane $\text{Re}(z) > -1$. Also, since $\Gamma(z + 1)$ is analytic at $z = 0$ and $\Gamma(1) \neq 0$, $\Gamma(z)$ has a pole of order one at $z = 0$. The relation $\Gamma(z + 2) = (z + 1)\Gamma(z + 1) = (z + 1)z\Gamma(z)$ is the basis for the continuation

$$\Gamma(z) = \frac{\Gamma(z + 2)}{z(z + 1)}$$

to the half-plane $\text{Re}(z) > -2$. More generally,

$$\Gamma(z) = \frac{\Gamma(z + n)}{z(z + 1) \cdots (z + n - 1)}$$

gives the analytic continuation to $\text{Re}(z) > -n$, n a positive integer. Since $\Gamma(-m + n) = (n - m - 1)! \neq 0$, $0 \leq m < n$, the gamma function has simple poles at $0, -1, -2, \ldots$.

Another way to obtain an analytic continuation of the original integral representation is to proceed as follows. Let

$$\Gamma(z) = \int_0^\infty t^{z-1} e^{-t}\, dt = \int_0^1 t^{z-1} e^{-t}\, dt + \int_1^\infty t^{z-1} e^{-t}\, dt$$

$$= G(z) + H(z).$$

As we saw in Example 4.9.2 and Exercise 4.9.1, the function

$$H(z) = \int_1^\infty t^{z-1} e^{-t}\, dt$$

is an entire function. The problem is to continue $G(z)$. To this end, we expand the integrand as an infinite series, that is,

$$t^{z-1} e^{-t} = t^{z-1} \sum_{k=0}^\infty \frac{(-1)^k t^k}{k!} = t^{z-1} + \sum_{k=1}^\infty \frac{(-1)^k t^{k+z-1}}{k!}.$$

Let $\text{Re}(z) \geq \rho > 0$. Then $|(-1)^k t^{k+z-1}/k!| \leq t^{k+\rho-1}/k!$ and

$$\sum_{k=1}^\infty \frac{t^{k+\rho-1}}{k!}$$

converges. Therefore,

$$\sum_{k=1}^\infty \frac{(-1)^k t^{k+z-1}}{k!}$$

converges uniformly for $\text{Re}(z) \geq \rho > 0$. We may integrate term-by-term and we have

$$G(z) = \int_0^1 t^{z-1}\, dt + \sum_{k=1}^\infty \int_0^1 \frac{(-1)^k t^{k+z-1}}{k!}\, dt$$

$$= \frac{1}{z} + \sum_{k=1}^\infty \frac{(-1)^k}{k!(z+k)} = \sum_{k=0}^\infty \frac{(-1)^k}{k!(z+k)}.$$

Therefore,

$$\Gamma(z) = G(z) + H(z) = \sum_{k=0}^\infty \frac{(-1)^k}{k!(z+k)} + \int_1^\infty t^{z-1} e^{-t}\, dt$$

for $\text{Re}(z) > 0$. But the series

$$\sum_{k=0}^\infty \frac{(-1)^k}{k!(z+k)}$$

converges uniformly in compact subsets of the unextended complex plane with the points $0, -1, -2, \ldots$ excluded. Therefore, $G(z)$ is in fact a function analytic in the unextended plane except at the origin and the negative integers where it has simple poles. We have thus continued $\Gamma(z)$ to the unextended plane where we have shown that the only singularities are simple poles at $z = 0, -1, -2, \ldots$. From this it follows that $1/\Gamma(z)$ is analytic in the unextended plane except possibly at zeros of $\Gamma(z)$. As a matter of fact, $1/\Gamma(z)$ is an entire function, and we shall show this by finding an infinite product expansion.

As a formal process it is quite obvious that

$$\lim_{n \to \infty} \int_0^n \left(1 - \frac{t}{n}\right)^n t^{z-1}\, dt = \int_0^\infty e^{-t} t^{z-1}\, dt = \Gamma(z).$$

We shall rigorously prove this formula. Let

$$\Gamma_n(z) = \int_0^n \left(1 - \frac{t}{n}\right)^n t^{z-1}\, dt$$

for $\mathrm{Re}(z) > 0$. Let $t = ns$. Then

$$\Gamma_n(z) = n^z \int_0^1 (1 - s)^n s^{z-1}\, ds.$$

This is clearly an analytic function of z for each positive integer n for $\mathrm{Re}(z) > 0$. Let $z = x > 1$. Then

$$n^x \int_0^1 (1 - s)^n s^{x-1}\, ds = n^x \frac{n}{x} \int_0^1 (1 - s)^{n-1} s^x\, ds$$

$$= n^x \frac{n(n-1)}{x(x+1)} \int_0^1 (1 - s)^{n-2} s^{x+1}\, ds$$

$$= n^x \frac{n(n-1)(n-2)\cdots 2 \cdot 1}{x(x+1)(x+2)\cdots(x+n-1)} \int_0^1 s^{x+n-1}\, ds$$

$$= \frac{n^x n!}{x(x+1)(x+2)\cdots(x+n)}.$$

Now, the function $\dfrac{n^z n!}{z(z+1)(z+2)\cdots(z+n)}$ is analytic for $\mathrm{Re}(z) > 0$ and is equal to $\Gamma_n(z)$ for z real and greater than 2. But this proves that

$$\Gamma_n(z) = \frac{n^z n!}{z(z+1)(z+2)\cdots(z+n)}.$$

for $\text{Re}(z) > 0$. The denominator of this expression suggests an infinite product. Therefore, we write

$$\frac{1}{\Gamma_n(z)} = n^{-z}z(1 + z)\left(1 + \frac{z}{2}\right)\cdots\left(1 + \frac{z}{n}\right)$$

$$= n^{-z}z \prod_{k=1}^{n}\left(1 + \frac{z}{k}\right).$$

Now, the infinite product

$$\prod_{k=1}^{\infty}\left(1 + \frac{z}{k}\right)$$

does not converge. Hence, we put in the convergence factors $e^{-z/k}$ and write

$$\frac{1}{\Gamma_n(z)} = e^{z(1+1/2+1/3+\cdots+1/n-\ln n)}z \prod_{k=1}^{n}\left[\left(1 + \frac{z}{k}\right)e^{-z/k}\right].$$

Let us consider the infinite product

$$\prod_{k=1}^{\infty}\left[\left(1 + \frac{z}{k}\right)e^{-z/k}\right].$$

The function $\left(1 + \frac{z}{k}\right)e^{-z/k}$ is analytic for each positive integer k and

$$\left(1 + \frac{z}{k}\right)e^{-z/k} = \left(1 + \frac{z}{k}\right)\left(1 - \frac{z}{k} + \frac{z^2}{2k^2} - \cdots\right)$$

$$= 1 - \frac{z^2}{2k^2} + \frac{z^3}{3k^3} - \cdots$$

$$= 1 - \frac{z^2}{k^2}\left(\frac{1}{2} - \frac{z}{3k} + \cdots\right) = 1 + w_k(z),$$

where $w_k(z) = (z^2/k^2)g(z)$ and $g(z)$ is an entire function. Let $|z| \leq R$. Then $|w_k(z)| \leq MR^2/k^2$, where $M = \max_{|z|=R}|g(z)|$. Since

$$\sum_{k=1}^{\infty}\frac{MR^2}{k^2}$$

converges, the infinite product

$$\prod_{k=1}^{\infty}\left(1 + \frac{z}{k}\right)e^{-z/k}$$

converges uniformly in $|z| \leq R$, and therefore represents an analytic function for $|z| < R$. But R is arbitrary, which proves that the infinite product is an entire function.

To show that $\{1/\Gamma_n(z)\}$ converges to an entire function we have to show that the sequence $\{1 + 1/2 + \cdots + 1/n - \ln n\}$ converges. Consider the function $\ln (1 - x)$. By Taylor's theorem we have for $|x| < 1$

$$\ln (1 - x) = -x - \frac{1}{(1 - \theta)^2} \frac{x^2}{2},$$

where $0 < \theta < 1$. Therefore, $\ln (1 - x) + x < 0$. Hence,

$$\left[1 + \frac{1}{2} + \cdots + \frac{1}{n + 1} - \ln (n + 1)\right] - \left[1 + \frac{1}{2} + \cdots + \frac{1}{n} - \ln n\right]$$

$$= \ln\left(1 - \frac{1}{n + 1}\right) + \frac{1}{n + 1} < 0$$

Also,

$$\ln n + \frac{1}{n} = \int_1^n \frac{1}{t}\, dt + \frac{1}{n} < 1 + \frac{1}{2} + \frac{1}{3} + \cdots + \frac{1}{n - 1} + \frac{1}{n},$$

$$1 + \ln n = 1 + \int_1^n \frac{1}{t}\, dt > 1 + \frac{1}{2} + \frac{1}{3} + \cdots + \frac{1}{n},$$

$$0 < \frac{1}{n} < 1 + \frac{1}{2} + \frac{1}{3} + \cdots + \frac{1}{n} - \ln n < 1.$$

The sequence is decreasing and lies between 0 and 1, and therefore converges to a number γ between 0 and 1. This number is called Euler's constant. Its value is approximately .5772.

We have proved finally that $\{1/\Gamma_n(z)\}$ converges to the entire function

$$ze^{\gamma z} \prod_{k=1}^{\infty} \left[\left(1 + \frac{z}{k}\right)e^{-z/k}\right]$$

which has zeros at $0, -1, -2, -3, \ldots$. This function is the sought after infinite product representation of $1/\Gamma(z)$. To show this we have to prove that $\lim_{n \to \infty} \Gamma_n(z) = \Gamma(z)$ on some set of points with a limit in the domain of analyticity of $\Gamma(z)$. Let $z = x > 2$. It is well known that for $n > t$, $(1 - t/n)^n \to e^{-t}$ monotonically from below. Therefore,

$$\Gamma_n(x) = \int_0^n t^{x-1}\left(1 - \frac{t}{n}\right)^n dt < \int_0^{n+1} t^{x-1}\left(1 - \frac{t}{n + 1}\right)^{n+1} dt = \Gamma_{n+1}(x),$$

$$\Gamma_{n+1}(x) = \int_0^{n+1} t^{x-1}\left(1 - \frac{t}{n + 1}\right)^{n+1} dt < \int_0^{n+1} t^{x-1}e^{-t}\, dt = \Gamma(x),$$

and, therefore, $\lim_{n \to \infty} \Gamma_n(x) \leq \Gamma(x)$. Also, for $0 < T < n$

$$\Gamma_n(x) > \int_0^T t^{x-1}\left(1 - \frac{t}{n}\right)^n dt.$$

Letting n approach infinity, we have

$$\lim_{n \to \infty} \Gamma_n(x) \geq \int_0^T t^{x-1} e^{-t}\, dt,$$

for every T. Hence, $\lim_{n \to \infty} \Gamma_n(x) \geq \int_0^\infty t^{x-1} e^{-t}\, dt = \Gamma(x)$. Therefore, $\lim_{n \to \infty} \Gamma_n(x) = \Gamma(x)$. This completes the proof of the formula

$$\frac{1}{\Gamma(z)} = z e^{\gamma z} \prod_{k=1}^{\infty} \left[\left(1 + \frac{z}{k}\right) e^{-z/k} \right],$$

for all finite z.

A very useful formula is the following:

$$\Gamma(z)\Gamma(1 - z) = \pi \operatorname{cosec} \pi z$$

This can be proved as follows. First, we note that

$$\frac{1}{\Gamma(1 - z)} = \frac{1}{-z\Gamma(-z)} = e^{-\gamma z} \prod_{k=1}^{\infty} \left[\left(1 - \frac{z}{k}\right) e^{z/k} \right].$$

Therefore,

$$\frac{1}{\Gamma(z)\Gamma(1 - z)} = z \prod_{k=1}^{\infty} \left[\left(1 + \frac{z}{k}\right) e^{-z/k} \right] \prod_{k=1}^{\infty} \left[\left(1 - \frac{z}{k}\right) e^{z/k} \right]$$

$$= z \prod_{k=1}^{\infty} \left(1 - \frac{z^2}{k^2}\right) = \frac{\sin \pi z}{\pi}.$$

Finally, we develop the Hankel contour integral representation of $1/\Gamma(z)$. Consider

$$\frac{1}{2\pi i} \int_C e^t t^{-z}\, dt,$$

where $t^{-z} = e^{-z \log t}$, $-\pi < \arg t \leq \pi$, and C consists of the following three parts,

1. $t = \tau e^{-i\pi}$, $\rho \leq \tau < \infty$,
2. $t = \rho e^{i\theta}$, $-\pi \leq \theta \leq \pi$,
3. $t = \tau e^{i\pi}$, $\rho \leq \tau < \infty$.

It is easy to show by Theorem 4.9.4 that each part of the integral is an entire function of z, and therefore the whole integral is entire. We now show that

$$\frac{1}{\Gamma(z)} = \frac{1}{2\pi i} \int_C e^t t^{-z}\, dt.$$

Let $z = x < 1$. Then

$$\frac{1}{2\pi i}\int_C e^t t^{-z}\, dt = \frac{1}{2\pi i}\int_\infty^\rho e^{-\tau}e^{-x(\ln \tau - i\pi)}(-d\tau)$$

$$+ \frac{1}{2\pi}\int_{-\pi}^\pi e^{\rho(\cos \theta + i \sin \theta)}\rho^{1-x}e^{i\theta(1-x)}\, d\theta$$

$$+ \frac{1}{2\pi i}\int_\rho^\infty e^{-\tau}e^{-x(\ln \tau + i\pi)}(-d\tau)$$

$$= \frac{1}{2\pi i}(e^{ix\pi} - e^{-ix\pi})\int_\rho^\infty e^{-\tau}\tau^{-x}\, d\tau$$

$$+ \frac{1}{2\pi}\int_{-\pi}^\pi e^{\rho(\cos \theta + i \sin \theta)}\rho^{1-x}e^{i\theta(1-x)}\, d\theta.$$

Now,

$$\left|\frac{1}{2\pi}\int_{-\pi}^\pi e^{\rho(\cos \theta + i \sin \theta)}\rho^{1-x}e^{i\theta(1-x)}\, d\theta\right| \leq e^\rho \rho^{(1-x)} \to 0,$$

as $\rho \to 0$. Clearly, the contour integral is independent of ρ. Therefore, we let ρ approach zero and

$$\frac{1}{2\pi i}\int_C e^t t^{-z}\, dt = \frac{1}{\pi}\sin \pi x \int_0^\infty e^{-\tau}\tau^{-x}\, d\tau$$

$$= \frac{1}{\pi}\sin \pi x \Gamma(1 - x) = \frac{1}{\Gamma(x)}.$$

Exercises 4.10

1. Show that $\Gamma(1/2) = \sqrt{\pi}$. Find $\Gamma(3/2)$, $\Gamma(-3/2)$, and $\Gamma(-1/2)$.

2. Prove that $(2n)! = \dfrac{2^{2n}n!}{\sqrt{\pi}}\Gamma(n + \tfrac{1}{2})$.

3. Prove that the Laplace transform of t^a, $a > -1$, is $\Gamma(a + 1)/z^{a+1}$.

4. Prove that

$$\zeta(z) = \frac{\Gamma(1 - z)}{2\pi i}\int_C \frac{t^{z-1}}{e^{-t} - 1}\, dt,$$

where C is the same contour as in the integral representation of $1/\Gamma(z)$. This formula provides the analytic continuation of the Riemann zeta function. Prove that $\zeta(z)$ has only one singularity, a simple pole at $z = 1$.

5. Prove Legendre's duplication formula $\sqrt{\pi}\,\Gamma(2z) = 2^{2z-1}\Gamma(z)\Gamma(z + \tfrac{1}{2})$.

6. Let $\psi(z)$ be the logarithmic derivative of $\Gamma(z)$, namely

$$\psi(z) = \frac{d}{dz}\log\Gamma(z) = \frac{\Gamma'(z)}{\Gamma(z)}\,.$$

Prove that $\psi(1 - z) - \psi(z) = \pi\cot\pi z$, and that

$$\psi(z) = -\gamma - \frac{1}{z} + \sum_{k=1}^{\infty}\left(\frac{1}{k} - \frac{1}{k+z}\right).$$

CHAPTER 5

Residue Calculus

5.1. THE RESIDUE THEOREM

The residue calculus is that part of analytic function theory based on the residue theorem. We shall prove the residue theorem in this section, and then develop several of its more important consequences in the remainder of this chapter.

Definition 5.1.1. If $f(z)$ has an isolated singularity at z_0, then there exists an R such that the Laurent expansion

$$f(z) = \sum_{k=-\infty}^{\infty} a_k(z - z_0)^k$$

converges for $0 < |z - z_0| < R$. The residue of $f(z)$ at z_0 is the coefficient of $1/(z - z_0)$ in this expansion and is given by

$$a_{-1} = \frac{1}{2\pi i} \int_{|z-z_0|=R/2} f(z)\, dz,$$

where the integration is in the positive sense.

Theorem 5.1.1. *Residue Theorem. Let $f(z)$ be analytic within and on a simple closed contour C except for a finite number of isolated singularities inside C. Suppose that the geometry of C and the singularities inside C is such that $\int_{C_+} f(z)\, dz$ is equal to the sum of integrals taken in the positive direction around small circles inside C, which do not intersect each other or C, and each of which contains in its interior precisely one singularity of $f(z)$. Then $\int_{C_+} f(z)\, dz$ is equal to $2\pi i$ times the sum of the residues at the singularities inside C.*

Proof: The proof is self-evident from the definition of the residue and the rather explicit way the hypotheses are stated. Verifying the hypotheses

200

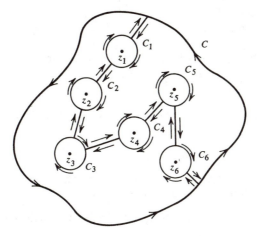

Figure 5.1.1.

depends on the following kind of consideration. See Figure 5.1.1. Let $z_1 - z_6$ be the isolated singularities. In this case, there are clearly two continuous closed paths of integration around which the integral of $f(z)$ is zero by Cauchy's theorem. When these integrals are added the parts over the paths connecting C with $C_1 - C_6$ and with each other drop out and we are left with

$$\int_{C_+} f(z)\,dz + \int_{C_{1-}} f(z)\,dz + \int_{C_{2-}} f(z)\,dz + \cdots + \int_{C_{6-}} f(z)\,dz = 0.$$

Hence,

$$\int_{C_+} f(z)\,dz = \int_{C_{1+}} f(z)\,dz + \int_{C_{2+}} f(z)\,dz + \cdots + \int_{C_{6+}} f(z)\,dz$$

$$= 2\pi i \sum_{i=1}^{6} K_i$$

where K_i is the residue at z_i. This of course does not prove that the hypotheses of the theorem hold for *any* C and *any* distribution of isolated singularities inside C. A complete discussion of this point would be far too complicated and boring. Furthermore, in all cases where we shall wish to apply the theorem the hypotheses will be obviously satisfied by considerations similar to the illustration we have just given.

In order to carry out explicit calculations, it will be helpful to consider the following cases where $f(z)$ has an isolated singularity at z_0.

1. The function $f(z)$ has a *removable singularity* at z_0, if and only if $\lim_{z \to z_0} f(z)$ exists and is finite. In this case, the residue at z_0 is zero.

2. The function $f(z)$ has a *pole of order m* at z_0 if and only if

$$\lim_{z \to z_0} (z - z_0)^m f(z)$$

exists and is not equal to zero or ∞. In this case, the residue at z_0 is equal

$$\frac{1}{(m-1)!} \lim_{z \to z_0} \frac{d^{m-1}}{dz^{m-1}} (z - z_0)^m f(z).$$

3. The function $f(z)$ has an *essential singularity* at z_0 if and only if $\lim_{z \to z_0} (z - z_0)^m f(z)$ exists for no positive integer m.

To facilitate the calculation of residues the following will be helpful.

L'Hospital's Rule. Let $f(z)$ and $g(z)$ be analytic at z_0, where $f(z)$ has a zero of order m and $g(z)$ has a zero of order n. Then if

1. $m > n$, $\lim_{z \to z_0} f(z)/g(z) = 0$,
2. $m = n$, $\lim_{z \to z_0} f(z)/g(z) = f^{(m)}(z_0)/g^{(m)}(z_0)$,
3. $m < n$, $\lim_{z \to z_0} f(z)/g(z) = \infty$.

Proof: Since $f(z)$ and $g(z)$ are analytic at z_0 there is an R such that

$$f(z) = \sum_{k=m}^{\infty} \frac{f^{(k)}(z_0)}{k!} (z - z_0)^k, \; g(z) = \sum_{k=n}^{\infty} \frac{g^{(k)}(z_0)}{k!} (z - z_0)^k$$

for $|z - z_0| < R$. Also, since the zeros of an analytic function are isolated, we can assume that $g(z) \neq 0$ for $0 < |z - z_0| < R$. In the same annulus, we have

$$\frac{f(z)}{g(z)} = \frac{\displaystyle\sum_{k=m}^{\infty} \frac{f^{(k)}(z_0)}{k!} (z - z_0)^k}{\displaystyle\sum_{k=n}^{\infty} \frac{g^{(k)}(z_0)}{k!} (z - z_0)^k}.$$

1. If $m > n$,

$$\frac{f(z)}{g(z)} = (z - z_0)^{m-n} \frac{\displaystyle\sum_{k=m}^{\infty} \frac{f^{(k)}(z_0)}{k!} (z - z_0)^{k-m}}{\displaystyle\sum_{k=n}^{\infty} \frac{g^{(k)}(z_0)}{k!} (z - z_0)^{k-n}}, \; z \neq z_0,$$

where $f^{(m)}(z_0) \neq 0$, $g^{(n)}(z_0) \neq 0$. Hence, $\lim_{z \to z_0} \dfrac{f(z)}{g(z)} = 0$.

2. If $m = n$,

$$\frac{f(z)}{g(z)} = \frac{\displaystyle\sum_{k=m}^{\infty} \frac{f^{(k)}(z_0)}{k!} (z - z_0)^{k-m}}{\displaystyle\sum_{k=m}^{\infty} \frac{g^{(k)}(z_0)}{k!} (z - z_0)^{k-m}}, \, z \neq z_0,$$

where $f^{(m)}(z_0) \neq 0$, $g^{(m)}(z_0) \neq 0$. Hence, $\displaystyle\lim_{z \to z_0} \frac{f(z)}{g(z)} = \frac{f^{(m)}(z_0)}{g^{(m)}(z_0)}$.

3. If $m < n$,

$$\frac{f(z)}{g(z)} = \frac{1}{(z - z_0)^{n-m}} \frac{\displaystyle\sum_{k=m}^{\infty} \frac{f^{(k)}(z_0)}{k!} (z - z_0)^{k-m}}{\displaystyle\sum_{k=n}^{\infty} \frac{g^{(k)}(z_0)}{k!} (z - z_0)^{k-n}}, \, z \neq z_0,$$

where $f^{(m)}(z_0) \neq 0$, $g^{(n)}(z_0) \neq 0$. Hence, $\displaystyle\lim_{z \to z_0} \frac{f(z)}{g(z)} = \infty$.

EXAMPLE 5.1.1. Compute the residue at $z = 0$ of $f(z) = (z \sin z)/(1 - e^z)^3$. Let $g(z) = z \sin z$ and $h(z) = (1 - e^z)^3$. Then $g'(z) = z \cos z + \sin z$, $g''(z) = 2 \cos z - z \sin z$, $h'(z) = -3(1 - e^z)^2 e^z$, $h''(z) = 6(1 - e^z)e^{2z} - 3(1 - e^z)^2 e^z$, $h'''(z) = -6e^{3z} + 18(1 - e^z)e^{2z} - 3(1 - e^z)^2 e^z$. Hence, $g(0) = g'(0) = 0$, $g''(0) = 2$, $h(0) = h'(0) = h''(0) = 0$, $h'''(0) = -6$. Therefore, $f(z)$ has a simple pole at the origin and if $g_1(z) = z^2 \sin z$, then

$$K = \lim_{z \to 0} \frac{z^2 \sin z}{(1 - e^z)^3} = \frac{g_1'''(0)}{h'''(0)} = \frac{3g''(0)}{-6} = -1.$$

EXAMPLE 5.1.2. Compute $\displaystyle\int_{C_+} \frac{z \sec z}{(1 - e^z)^2} \, dz$, where C is the circle of radius 2 with center at the origin. There are three isolated singularities inside C. They are $z = 0, \pi/2, -\pi/2$.

$$\lim_{z \to 0} \frac{z^2 \sec z}{(1 - e^z)^2} = 1, \quad \lim_{z \to \pi/2} \frac{(z - \pi/2)z \sec z}{(1 - e^z)^2} = -\frac{\pi}{2} \frac{1}{(e^{\pi/2} - 1)^2},$$

and

$$\lim_{z \to -\pi/2} \frac{(z + \pi/2)z \sec z}{(1 - e^z)^2} = -\frac{\pi}{2} \frac{1}{(e^{-\pi/2} - 1)^2}.$$

By the residue theorem,

$$\int_{C_+} \frac{z \sec z}{(1 - e^z)^2} \, dz = 2\pi i \left[1 - \frac{\pi}{2} \frac{1}{(e^{\pi/2} - 1)^2} - \frac{\pi}{2} \frac{1}{(e^{-\pi/2} - 1)^2} \right].$$

The residue theorem can be used to sum certain series. We illustrate this by the following example.

EXAMPLE 5.1.3. Sum the series $\sum_{k=1}^{\infty} \frac{1}{k^2}$. Consider the function $\pi \cot \pi z$
which has simple poles with residue one at all the positive and negative
integers. Let $R_n = (n + \frac{1}{2})$, $n = 1, 2, 3, \ldots$ and C_n be the square with
center at the origin and side $2R_n$. Then

$$\int_{C_{n+}} \frac{\pi \cot \pi z}{z^2} \, dz = 2\pi i \left[K_0 + \sum_{k=-n}^{n} \frac{1}{k^2} \right],$$

where K_0 is the residue of the integrand at the origin. Hence, $K_0 = -\pi^2/3$,
and

$$\sum_{k=1}^{\infty} \frac{1}{k^2} = \frac{1}{2} \sum_{k=-\infty}^{\infty} \frac{1}{k^2} = -\frac{K_0}{2} + \lim_{n \to \infty} \frac{1}{4\pi i} \int_{C_{n+}} \frac{\pi \cot \pi z}{z^2} \, dz.$$

We can show that $\lim_{n \to \infty} \int_{C_n} \frac{\pi \cot \pi z}{z^2} \, dz = 0$, and hence that $\sum_{k=1}^{\infty} \frac{1}{k^2} = \frac{\pi^2}{6}$.

On the horizontal sides of the square

$$\left| \int \frac{\pi \cot \pi z}{z^2} \, dz \right| \leq \frac{2\pi}{R_n} \frac{e^{\pi R_n} + e^{-\pi R_n}}{e^{\pi R_n} - e^{-\pi R_n}} \to 0, \quad \text{as} \quad R_n \to \infty.$$

On the vertical sides $z = \pm(n + 1/2) + iy$ and

$$|\pi \cot \pi z| = \pi|\tanh \pi y| \leq \pi|\tanh \pi(n + \tfrac{1}{2})|$$

$$\left| \int \frac{\pi \cot \pi z}{z^2} \, dz \right| \leq \frac{2\pi}{R_n} \left| \tanh \pi(n + \tfrac{1}{2}) \right| \to 0, \quad \text{as} \quad n \to \infty.$$

Exercises 5.1

1. Find the poles and residues of (a) $\csc z$; (b) $\operatorname{sech}^2 z$. (c) $(1/z)\tanh z$;
(d) $\tan z/(e^z - 1)$.

2. Integrate in the positive direction around a circle of radius 2 with
center at the origin:

(a) $\int_C z e^{1/z} \, dz$; (b) $\int_C \operatorname{sech}^2 z \, dz$; (c) $\int_C \frac{\tan z}{e^z - 1} \, dz$.

3. Sum the series $\sum_{k=0}^{\infty} \frac{1}{1 + k^2}$.

4. Sum the series $\sum_{k=0}^{\infty} \frac{(-1)^k}{k^2}$. Hint: The function $\pi \csc \pi z$ has simple
poles at the integers n with residue $(-1)^n$.

5. Let $f(z) = g(z)/h(z)$, where g and h are both analytic at z_0. If $g(z_0) \neq 0$ and z_0 is a zero of order 2 of h, prove that the residue of f at z_0 is

$$2\frac{g'(z_0)}{h''(z_0)} - \frac{2}{3}\frac{g(z_0)h'''(z_0)}{[h''(z_0)]^2}.$$

5.2. EVALUATION OF REAL INTEGRALS

One of the commonest applications of the residue theorem is the evaluation of real definite integrals. We shall consider examples of four basic types which lend themselves to this technique, (1) a real integral which by a proper transformation can be evaluated as a contour integral in the complex plane, (2) a real integral which is the contribution of the real axis as part of a contour integral, the other parts of which can be evaluated, (3) an integral which is the real or imaginary part of a contour integral which can be evaluated, and (4) a real integral which can be evaluated by integrating along a cut which is a branch cut for the integrand.

EXAMPLE 5.2.1. Evaluate $\int_0^{2\pi} (a + b\sin\theta)^{-1}d\theta$, where $|a| > |b|$. We let $z = e^{i\theta}$. Then $\sin\theta = (1/2i)(e^{i\theta} - e^{-i\theta}) = (1/2i)(z - z^{-1})$, $dz = ie^{i\theta}\,d\theta$, $d\theta = dz/iz$, and

$$\int_0^{2\pi} \frac{1}{a + b\sin\theta}\,d\theta = 2\int_{C_+} \frac{1}{(2ia + bz - bz^{-1})}\frac{dz}{z} = 2\int_{C_+} \frac{dz}{bz^2 + 2iaz - b}$$

and C is the unit circle. Clearly, the only singularities of the integrand are

$$z = -i\left(\frac{a}{b} \pm \sqrt{\left(\frac{a}{b}\right)^2 - 1}\right).$$

If $0 < b < a$, then $-i(a/b - \sqrt{(a/b)^2 - 1})$ is inside the unit circle while the other is outside. By the residue theorem,

$$\int_0^{2\pi} \frac{1}{a + b\sin\theta}\,d\theta$$

$$= \frac{2}{b}\int_{C_+} \frac{dz}{\left[z + i\left(\frac{a}{b} - \sqrt{\left(\frac{a}{b}\right)^2 - 1}\right)\right]\left[z + i\left(\frac{a}{b} + \sqrt{\left(\frac{a}{b}\right)^2 - 1}\right)\right]}$$

$$= 2\pi i\left[\frac{2}{b}\frac{1}{2i\sqrt{\left(\frac{a}{b}\right)^2 - 1}}\right] = \frac{2\pi}{\sqrt{a^2 - b^2}}.$$

If $a < b < 0$, then the value of the integral is $-2\pi/\sqrt{a^2 - b^2}$. If $b < 0 < a$, then the pole at $-i(a/b + \sqrt{(a/b)^2 - 1})$ is inside the unit circle while the other is outside. Then

$$\int_0^{2\pi} \frac{1}{a + b\sin\theta}\, d\theta = 2\pi i\left[\frac{2}{b}\, \frac{1}{-2i\sqrt{\left(\frac{a}{b}\right)^2 - 1}}\right]$$

$$= \frac{2\pi|b|}{-b}\, \frac{1}{\sqrt{a^2 - b^2}} = \frac{2\pi}{\sqrt{a^2 - b^2}}.$$

If $a < 0 < b$, then the value of the integral is $-2\pi/\sqrt{a^2 - b^2}$.

EXAMPLE 5.2.2. Evaluate $\int_{-\infty}^{\infty} (1 + x^4)^{-1}dx$. This is an improper integral which has the meaning

$$\int_{-\infty}^{\infty} \frac{dx}{1 + x^4} = \lim_{R\to\infty} \int_0^R \frac{dx}{1 + x^4} + \lim_{S\to\infty} \int_{-S}^0 \frac{dx}{1 + x^4},$$

provided these limits exist. That these limits exist is indicated by the following inequalities

$$\int_0^R \frac{dx}{1 + x^4} \leq \int_0^1 \frac{dx}{1 + x^4} + \int_1^R \frac{dx}{1 + x^2} \leq \int_0^1 \frac{dx}{1 + x^4} + \int_1^{\infty} \frac{dx}{1 + x^2}$$

$$\leq \int_0^1 \frac{dx}{1 + x^4} + \tan^{-1} x\,\Big|_1^{\infty} = \int_0^1 \frac{dx}{1 + x^4} + \frac{\pi}{4},$$

$$\int_{-S}^0 \frac{dx}{1 + x^4} \leq \int_{-1}^0 \frac{dx}{1 + x^4} + \int_{-\infty}^{-1} \frac{dx}{1 + x^2} = \int_{-1}^0 \frac{dx}{1 + x^4} + \frac{\pi}{4}.$$

Consider the contour C of Figure 5.2.1 consisting of the real axis from $-R$ to R and a semicircle of radius R in the upper half-plane. We integrate

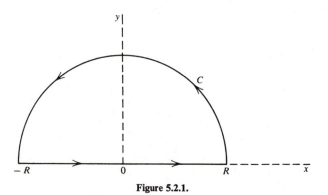

Figure 5.2.1.

$1/(1 + z^4)$ around C_+ using the residue theorem with $R > 1$. Letting R approach infinity gives us the value of the desired integral.

$$\int_{C_+} \frac{dz}{1 + z^4} = \int_{-R}^{R} \frac{dx}{1 + x^4} + \int_0^{\pi} \frac{iRe^{i\theta}\, d\theta}{1 + R^4 e^{4i\theta}}$$

$$= 2\pi i \left[\frac{1}{2i(\sqrt{2} + \sqrt{2}i)} + \frac{1}{2i(\sqrt{2} - \sqrt{2}i)} \right] = \frac{\pi}{\sqrt{2}}.$$

Now,

$$\left| \int_0^{\pi} \frac{iRe^{i\theta}\, d\theta}{1 + R^4 e^{4i\theta}} \right| \leq \frac{\pi R}{R^4 - 1} \rightarrow 0$$

as $R \rightarrow \infty$. Therefore,

$$\int_{-\infty}^{\infty} \frac{dx}{1 + x^4} = \lim_{R \to \infty} \int_{-R}^{R} \frac{dx}{1 + x^4} = \lim_{R \to \infty} \int_{C_+} \frac{dz}{1 + z^4} = \frac{\pi}{\sqrt{2}}.$$

EXAMPLE 5.2.3. Evaluate $\displaystyle\int_{-\infty}^{\infty} \frac{\cos x\, dx}{1 + x^2}.$ This improper integral converges since

$$\int_0^R \frac{|\cos x|\, dx}{1 + x^2} \leq \int_0^{\infty} \frac{1}{1 + x^2}\, dx = \frac{\pi}{2}.$$

It we use the same contour as in Example 5.2.2, it passes through the point iR and $\cos iR = \cosh R \rightarrow \infty$, as $R \rightarrow \infty$. Alternatively, we shall evaluate $\displaystyle\int_{-\infty}^{\infty} \frac{e^{ix}}{1 + x^2}\, dx$ and then

$$\int_{-\infty}^{\infty} \frac{\cos x\, dx}{1 + x^2} = \text{Re} \int_{-\infty}^{\infty} \frac{e^{ix}}{1 + x^2}\, dx.$$

Therefore, if $R > 1$

$$\int_{C_+} \frac{e^{iz}\, dz}{1 + z^2} = \int_{-R}^{R} \frac{e^{ix}\, dx}{1 + x^2} + \int_0^{\pi} \frac{ie^{i(R\cos\theta + iR\sin\theta)}Re^{i\theta}\, d\theta}{1 + R^2 e^{2i\theta}}$$

$$= 2\pi i \left(\frac{e^{-1}}{2i} \right) = \pi/e.$$

Furthermore,

$$\left| \int_0^{\pi} \frac{ie^{iR\cos\theta - R\sin\theta}Re^{i\theta}\, d\theta}{1 + R^2 e^{2i\theta}} \right| \leq \frac{R}{R^2 - 1} \rightarrow 0,$$

as $R \rightarrow \infty$. Hence,

$$\int_{-\infty}^{\infty} \frac{\cos x\, dx}{1 + x^2} = \text{Re} \int_{-\infty}^{\infty} \frac{e^{ix}\, dx}{1 + x^2} = \lim_{R \to \infty} \int_{C_+} \frac{e^{iz}\, dz}{1 + z^2} = \pi/e.$$

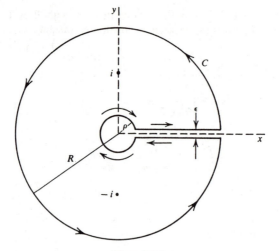

Figure 5.2.2.

EXAMPLE 5.2.4. Evaluate $\int_0^\infty \frac{\sqrt{x}}{1 + x^2}\, dx$. The improper integral exists since

$$\int_0^R \frac{\sqrt{x}\, dx}{1 + x^2} = \int_0^1 \frac{\sqrt{x}\, dx}{1 + x^2} + \int_1^R \frac{\sqrt{x}\, dx}{1 + x^2}$$

$$\leq \int_0^1 \frac{\sqrt{x}}{1 + x^2}\, dx + \int_1^\infty x^{-3/2}\, dx$$

$$\leq \int_0^1 \frac{\sqrt{x}}{1 + x^2}\, dx - 2x^{-1/2}\Big|_1^\infty = \int_0^1 \frac{\sqrt{x}}{1 + x^2}\, dx + 2.$$

The function $z^{1/2}/(1 + z^2)$ has a branch point at the origin. We pick a branch $z^{1/2} = |z|^{1/2}[\cos(1/2)\arg z + i\sin(1/2)\arg z]$, $0 \leq \arg z < 2\pi$, which has a branch cut on the positive real axis. We integrate around the contour shown in Figure 5.2.2.

$$\int_{C_+} \frac{z^{1/2}}{1 + z^2}\, dz = 2\pi i\left(\frac{e^{i\pi/4}}{2i} - \frac{e^{3i\pi/4}}{2i}\right) = \sqrt{2}\pi$$

$$= \int_\rho^R \frac{x^{1/2}}{1 + x^2}\, dx + \int_0^{2\pi - \epsilon} \frac{R^{1/2} e^{i\theta/2} R i e^{i\theta}\, d\theta}{1 + R^2 e^{2i\theta}}$$

$$+ \int_R^\rho \frac{r^{1/2} e^{i(2\pi - \epsilon)/2} e^{i(2\pi - \epsilon)}\, dr}{1 + r^2 e^{2i(2\pi - \epsilon)}}$$

$$+ \int_{2\pi - \epsilon}^0 \frac{\rho^{1/2} e^{i\theta/2} \rho i e^{i\theta}\, d\theta}{1 + \rho^2 e^{2i\theta}}.$$

The value of the integral is independent of $\rho \to 0$, $R \to \infty$, and $\epsilon \to 0$. Passing to the limits, the first and second integrals become

$$\int_0^\infty \frac{x^{1/2}}{1+x^2}\,dx,$$

while

$$\left| \int_0^{2\pi} \frac{R^{1/2}e^{i\theta/2}Rie^{i\theta}\,d\theta}{1+R^2e^{2i\theta}} \right| \le \frac{2\pi R^{3/2}}{R^2-1} \to 0, \text{ as } R \to \infty,$$

$$\left| \int_0^{2\pi} \frac{\rho^{1/2}e^{i\theta/2}\rho ie^{i\theta}\,d\theta}{1+\rho^2e^{2i\theta}} \right| \le \frac{2\pi\rho^{3/2}}{1-\rho^2} \to 0, \text{ as } \rho \to 0.$$

Therefore,

$$\int_0^\infty \frac{x^{1/2}}{1+x^2}\,dx = \pi/\sqrt{2}.$$

We illustrate one more type in which a singularity occurs on the path of integration.

EXAMPLE 5.2.5.　Evaluate $\int_0^\infty (1/x)\sin x\,dx$. This integral is proper at the origin since $(1/x)\sin x \to 1$, as $x \to 0$. To show that the integral exists we show the following

$$\int_1^R \frac{\sin x}{x}\,dx = -\left. \frac{\cos x}{x} \right|_1^R - \int_1^R \frac{\cos x}{x^2}\,dx$$

$$= -\frac{\cos R}{R} + \cos 1 - \int_1^R \frac{\cos x}{x^2}\,dx,$$

$$\lim_{R\to\infty} \int_1^R \frac{\sin x}{x}\,dx = \cos 1 - \int_1^\infty \frac{\cos x}{x^2}\,dx,$$

where the last integral quite obviously converges.

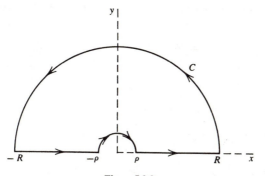

Figure 5.2.3.

We evaluate the given integral by integrating e^{iz}/z around the path shown in Figure 5.2.3

$$\int_{C_+} \frac{e^{iz}}{z}\, dz = \int_\rho^R \frac{e^{ix}}{x}\, dx + \int_0^\pi \frac{e^{iR(\cos\theta + i\sin\theta)} Rie^{i\theta}\, d\theta}{Re^{i\theta}}$$
$$+ \int_{-R}^{-\rho} \frac{e^{ix}}{x}\, dx + \int_\pi^0 \frac{e^{i\rho(\cos\theta + i\sin\theta)} \rho ie^{i\theta}\, d\theta}{\rho e^{i\theta}} = 0,$$

since there are no singularities inside C. In the third integral, we let $x = -t$. Then

$$\int_{-R}^{-\rho} \frac{e^{ix}}{x}\, dx = \int_\rho^R \frac{e^{it}}{t}\, dt.$$

In the second integral, we have

$$\left| \int_0^\pi e^{iR(\cos\theta + i\sin\theta)}\, d\theta \right| \le \int_0^\pi e^{-R\sin\theta}\, d\theta.$$

$$\int_0^\pi e^{-R\sin\theta}\, d\theta = \int_0^{\pi/2} e^{-R\sin\theta}\, d\theta + \int_0^{\pi/2} e^{-R\sin(\pi-\phi)}\, d\phi$$
$$= 2\int_0^{\pi/2} e^{-R\sin\theta}\, d\theta \le 2\int_0^{\pi/2} e^{-2R\theta/\pi}\, d\theta = \pi(1 - e^{-R})/R.$$

The last inequality follows from the fact that $2\theta/\pi \le \sin\theta$ for $0 \le \theta \le \pi/2$. Therefore, we have

$$\left| \int_0^\pi e^{iR(\cos\theta + i\sin\theta)}\, d\theta \right| \le \pi(1 - e^{-R})/R \to 0, \text{ as } R \to \infty.$$

Finally, we have $e^{iz}/z = 1/z + h(z)$, where $h(z)$ is continuous at the origin. Clearly, $\int h(z)\, dz$ around the small semicircle of radius ρ goes to zero, as $\rho \to 0$, and $\int (1/z)\, dz = -\pi i$ when we integrate around the small semicircle, independently of the value of ρ. Letting $\rho \to 0$ and $R \to \infty$, we have shown that

$$\int_0^\infty \frac{\sin x}{x}\, dx = \text{Im} \int_0^\infty \frac{e^{ix}}{x}\, dx = \pi/2.$$

Exercises 5.2

1. Evaluate $\int_0^{2\pi} \frac{1}{a + b\cos\theta}\, d\theta$, where $|a| > |b|$.

2. Evaluate $\int_0^\pi \frac{1}{(a + b\sin\theta)^2}\, d\theta$, where $|a| > |b|$.

3. Evaluate $\displaystyle\int_0^{2\pi} \frac{d\theta}{1 - 2a\cos\theta + a^2}$, where $0 \le a < 1$.

4. Evaluate $\displaystyle\int_{-\infty}^{\infty} \frac{dx}{(1 + x^2)^2}$.

5. Evaluate $\displaystyle\int_{-\infty}^{\infty} \frac{dx}{(a^2 + x^2)(b^2 + x^2)}$.

6. Evaluate $\displaystyle\int_0^{\infty} \frac{\cos ax}{1 + x^2}\, dx$.

7. Evaluate $\displaystyle\int_0^{\infty} \frac{x^{m-1}\, dx}{1 + x^n}$, where $0 < m < n$.

8. Evaluate $\displaystyle\int_0^{\infty} \frac{\sin ax}{x}\, dx$.

9. Evaluate $\displaystyle\int_0^{\infty} \frac{\sin^2 x}{x^2}\, dx$.

10. Prove that $\int_0^{\infty} e^{-x^2}\, dx = \frac{1}{2}\sqrt{\pi}$ by making the change of variable $t = x^2$. Evaluate $\int_0^{\infty} e^{-x^2} \cos 2mx\, dx$ by integrating around the contour shown in Figure 5.2.4 and letting $R \to \infty$.

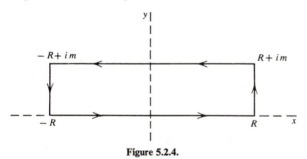

Figure 5.2.4.

5.3. THE PRINCIPLE OF THE ARGUMENT

Another application of the residue theorem is contained in the Principle of the Argument which is useful in locating the zeros and poles of analytic functions.

Theorem 5.3.1. *Principle of the Argument. Let $f(z)$ be analytic within and on a simple closed contour C except possibly for a finite number of poles inside C. If $f(z)$ does not vanish on C, then*

$$\frac{1}{2\pi} \Delta_{C_+} \arg f(z) = Z - P,$$

where $\Delta_{C_+} \arg f(z)$ *is the total change in the argument of* $f(z)$ *around* C_+, Z *is the number of zeros and* P *is the number of poles inside* C. *In counting the number of zeros and poles a zero of order m is counted m times and a pole of order n is counted n times.*

Proof: Consider the function $f'(z)/f(z)$. The only possible singularities of this function inside C are the zeros and poles of $f(z)$. If z_0 is a zero of order m of $f(z)$, then

$$f(z) = (z - z_0)^m g(z)$$

in some ϵ-neighborhood of z_0, where $g(z)$ is analytic and not zero. Then

$$\frac{f'(z)}{f(z)} = \frac{m}{z - z_0} + \frac{g'(z)}{g(z)},$$

which shows that $f'(z)/f(z)$ has a simple pole at z_0 with residue m. If z_p is a pole of order n of $f(z)$, then

$$f(z) = \frac{h(z)}{(z - z_p)^n},$$

for $0 < |z - z_p| < R$, for some $R > 0$, and $h(z)$ is analytic and not zero. Then

$$\frac{f'(z)}{f(z)} = \frac{-n}{z - z_p} + \frac{h'(z)}{h(z)},$$

which shows that $f'(z)/f(z)$ has a simple pole at z_p with residue $-n$. Since $f(z)$ does not vanish on C, $f'(z)/f(z)$ is analytic within and on C except for a finite number of simple poles inside C. By the residue theorem

$$\frac{1}{2\pi i} \int_{C_+} \frac{f'(z)}{f(z)} \, dz = Z - P,$$

where Z is the total number of zeros and P is the total number of poles counting multiplicities.

Now, consider the transformation $w = \log f(z)$. As z traces out C_+, w traces out some continuous curve Γ in the w plane (not necessarily simple). Also,

$$dw = \frac{f'(z)}{f(z)} \, dz$$

and

$$\frac{1}{2\pi i} \int_{C_+} \frac{f'(z)}{f(z)} \, dz = \frac{1}{2\pi i} \int_{\Gamma} dw = \frac{1}{2\pi i} \Delta_{\Gamma} w = \frac{1}{2\pi i} \Delta_{C_+} \log f(z)$$

$$= \frac{1}{2\pi i} \Delta_{C_+} \log |f(z)| + \frac{1}{2\pi} \Delta_{C_+} \arg f(z) = \frac{1}{2\pi} \Delta_{C_+} \arg f(z),$$

since $\log |f(z)|$ returns to its original value when C_+ is traversed. This completes the proof.

EXAMPLE 5.3.1. Locate the number of roots of $f(z) = z^4 + 4z^3 + 8z^2 + 8z + 4$ in each of the four quadrants. First, we note that $f(z)$ has no real roots since $f(x) = x^4 + 4x^3 + 8x^2 + 8x + 4$ has a minimum value of one at $x = -1$. Now, $f(iy) = y^4 - 8y^2 + 4 + i(8y - 4y^3) \neq 0$, since $y^4 - 8y^2 + 4 = 0$ implies $y = 4 \pm 2\sqrt{3}$, and $8y - 4y^3 = 0$ implies $y = 0, y = \sqrt{2}$, or $y = -\sqrt{2}$. We apply the principle of the argument to the contour consisting of the positive imaginary axis, the positive real axis and a large quarter circle. On the positive real axis $f(z)$ is real and positive, and there is no change in the argument. On the quarter circle $z = Re^{i\theta}, 0 \leq \theta \leq \pi/2$, and

$$f(Re^{i\theta}) = R^4 e^{4i\theta} \left(1 + \frac{4}{R} e^{-i\theta} + \frac{8}{R^2} e^{-2i\theta} + \frac{8}{R^3} e^{-3i\theta} + \frac{4}{R^4} e^{-4i\theta} \right).$$

If R is sufficiently large the second term is practically 1 and

$$\Delta \arg f(Re^{i\theta}) \to 2\pi \text{ as } R \to \infty.$$

When $z = iy$

$$\arg f(iy) = \tan^{-1} \frac{8y - 4y^3}{y^4 - 8y^2 + 4}.$$

The following table shows how the arg $f(iy)$ varies on the positive imaginary axis:

$y =$	∞	$4 + 2\sqrt{3}$	$\sqrt{2}$	$4 - 2\sqrt{3}$	0
$\dfrac{8y - 4y^3}{y^4 - 8y^2 + 4} =$	$0 \quad -$	∞	$+ \quad 0 \quad -$	∞	$+ \quad 0$
$\tan^{-1} \dfrac{8y - 4y^3}{y^4 - 8y^2 + 4} =$	0	$-\dfrac{\pi}{2}$	$-\pi$	$-\dfrac{3\pi}{2}$	-2π

Therefore, $\Delta_{C_+} \arg f(z) = 0$, which means that there are no zeros in the first quadrant. Also since the coefficients are real, if a root occurs in the second quadrant it also occurs in the third quadrant. Hence, there are two roots in the second quadrant and two roots in the third quadrant.

Theorem 5.3.2. *Rouché's Theorem. Let $f(z)$ and $g(z)$ be analytic within and on a simple closed contour C and satisfy the inequality $|g(z)| < |f(z)|$ on C, where $f(z)$ does not vanish. Then $f(z)$ and $f(z) + g(z)$ have the same number of zeros inside C.*

Proof: Clearly, $f(z)$ and $f(z) + g(z)$ are never zero on C. Therefore if Z and Z' are the number of zeros of $f(z)$ and $f(z) + g(z)$, respectively, then

$$Z = \frac{1}{2\pi} \Delta_{C_+} \arg f(z),$$

$$Z' = \frac{1}{2\pi} \Delta_{C_+} \arg (f + g) = \frac{1}{2\pi} \Delta_{C_+} \arg f + \frac{1}{2\pi} \Delta_{C_+} \arg (1 + g/f)$$

$$= Z + \frac{1}{2\pi} \Delta_{C_+} \arg (1 + g/f).$$

Let $w = 1 + g/f$. Then $|w - 1| = |g/f| < 1$, which implies that $-\pi/2 < \arg w < \pi/2$. Therefore, as z traces C_+ and returns to the starting point, w must return to its starting point without looping the origin in the w plane. This proves that $\Delta_{C_+} \arg (1 + g/f) = 0$, and hence that $Z = Z'$.

EXAMPLE 5.3.2. Prove the Fundamental Theorem of Algebra using Rouché's theorem. Let

$$f(z) = a_n z^n \quad \text{and} \quad g(z) = a_{n-1} z^{n-1} + a_{n-2} z^{n-2} + \cdots + a_1 z + a_0.$$

Let C be a circle with center at the origin and radius R. Then on C

$$|g(z)/f(z)| = \left| \frac{a_{n-1}}{a_n} \frac{1}{z} + \frac{a_{n-2}}{a_n} \frac{1}{z^2} + \cdots + \frac{a_1}{a_n} \frac{1}{z^{n-1}} + \frac{a_0}{a_n} \frac{1}{z^n} \right|$$

$$\leq \frac{1}{R} \left(\left| \frac{a_{n-1}}{a_n} \right| + \left| \frac{a_{n-2}}{a_n} \right| + \cdots + \left| \frac{a_1}{a_n} \right| + \left| \frac{a_0}{a_n} \right| \right) < 1,$$

provided $R > 1$ and $R > \left| \frac{a_{n-1}}{a_n} \right| + \left| \frac{a_{n-2}}{a_n} \right| + \cdots + \left| \frac{a_1}{a_n} \right| + \left| \frac{a_0}{a_n} \right|$. Now, $f(z)$ has n zeros inside $|z| = R$ and, therefore, so does

$$f(z) + g(z) = a_n z^n + a_{n-1} z^{n-1} + \cdots + a_1 z + a_0.$$

The Principle of the Argument can be used to give another proof of the inverse function theorem.

Theorem 5.3.3. *Let $f(z)$ be analytic at z_0 where $f'(z_0) = 0$. Then $w = f(z)$ has an analytic inverse $z = g(w)$ in some ϵ-neighborhood of $w_0 = f(z_0)$.*

Proof: Let $F(z) = f(z) - w_0$. Then $F(z_0) = 0$, and there is an R such that $F(z)$ is analytic for $|z - z_0| \leq R$ and $F(z) \neq 0$ for $0 < |z - z_0| \leq R$. Let $m = \min |F(Re^{i\theta})|$ and $|\omega| < m$. On the circle $C = \{z \mid |z - z_0| = R\}$, $|F(z)| \geq m > |\omega|$. By Rouché's theorem $F(z)$ and $F(z) - \omega$ have the same number of zeros inside C. But $F(z)$ has only a simple zero at z_0 inside C, since $F'(z_0) = f'(z_0) \neq 0$. Therefore, the equation $F(z) = \omega$ has precisely one solution for $|\omega| < m$, and this solution is given by

$$G(\omega) = \frac{1}{2\pi i} \int_{C_+} \frac{t F'(t)}{F(t) - \omega} \, dt = \frac{1}{2\pi i} \int_{C_+} \frac{t f'(t)}{f(t) - w_0 - \omega} \, dt.$$

Let $g(w_0 + \omega) = \dfrac{1}{2\pi i} \displaystyle\int_{C_+} \dfrac{t f'(t)}{f(t) - w_0 - \omega} \, dt$. Then if $w = w_0 + \omega$,

$$g(w) = \frac{1}{2\pi i} \int_{C_+} \frac{t f'(t)}{f(t) - w} \, dt$$

is the unique solution of $w = f(z)$ for $|w - w_0| = |\omega| < m$. This completes the proof.

By Theorem 4.9.1, $g(w)$ is analytic for $|w - w_0| < m$ and

$$g'(w) = \frac{1}{2\pi i} \int_{C_+} \frac{t f'(t)}{[f(t) - w]^2} \, dt$$

$$= \frac{1}{2\pi i} \int_{C_+} t \, \frac{d}{dt} [f(t) - w]^{-1} \, dt$$

$$= \frac{1}{2\pi i} \int_{C_+} \frac{dt}{f(t) - w}$$

$$= \frac{1}{2\pi i} \int_{C_+} \sum_{k=0}^{\infty} \frac{(w - w_0)^k}{[f(t) - w_0]^{k+1}} \, dt$$

$$= \sum_{k=0}^{\infty} (w - w_0)^k \frac{1}{2\pi i} \int_{C_+} \frac{dt}{[f(t) - w_0]^{k+1}} \, ,$$

$$z = g(w) = z_0 + \sum_{k=1}^{\infty} \frac{(w - w_0)^k}{k} \frac{1}{2\pi i} \int_{C_+} \frac{dt}{[f(t) - w_0]^k} \, .$$

We have obtained the power series

$$z - z_0 = \sum_{k=1}^{\infty} a_k (w - w_0)^k,$$

where

$$a_k = \frac{1}{2\pi i k} \int_{C_+} \frac{dt}{[f(t) - w_0]^k} \, .$$

Let $f(z) - w_0 = (z - z_0) h(z)$. Then if $H_k(t) = [h(t)]^{-k}$

$$a_k = \frac{1}{2\pi i k} \int_{C_+} \frac{dt}{(t - z_0)^k [h(t)]^k}$$

$$= \frac{1}{k!} H_k^{(k-1)}(z_0).$$

Hence,

$$z - z_0 = \sum_{k=1}^{\infty} H_k^{(k-1)}(z_0) \frac{(w - w_0)^k}{k!}.$$

EXAMPLE 5.3.3. Obtain a power series representation for the solution of $w = z^2 + 2z$ which approaches zero as w approaches zero. Here, $w = f(z) = z^2 + 2z$, $f(0) = 0$, $f'(0) = 2$,

$$m^2 = \min_{0 \le \theta \le 2\pi} |f(Re^{i\theta})|^2 = \min R^2(R^2 + 4 + 2R \cos \theta)$$

$$= R^2(R - 2)^2, R < 1.$$

Since $R^2(R - 2)^2$ has a maximum of one at $R = 1$, we can show that the power series will converge for $|w| < 1$. Specifically, $f(z) = z(z + 2)$, so that $h(z) = z + 2$ and

$$z = \sum_{k=1}^{\infty} H_k^{(k-1)}(0) \frac{w^k}{k!} = \sum_{k=1}^{\infty} \frac{(-1)^{k-1}(2k - 2)!}{(k - 1)! 2^{2k-1}} \frac{w^k}{k!}.$$

Note that we could have obtained the same result if we had solved for $z = -1 + \sqrt{1 + w}$ and used the binomial expansion. However, we would not have been so fortunate if $w = z^3 + 3z$.

Exercises 5.3

1. Let $g(z)$ be analytic within and on a simple closed contour C. Let $f(z)$ be analytic within and on C except for a finite number of zeros and/or poles inside C. Evaluate $\dfrac{1}{2\pi i} \displaystyle\int_{C_+} \dfrac{g(z)f'(z)}{f(z)}\, dz$.

2. Prove that if $f(z)$ is a polynomial of degree $n \ge 1$, then

$$\frac{1}{2\pi i} \int_{C_+} \frac{zf'(z)}{f(z)}\, dz$$

is equal to the sum of the roots if C is a very large circle.

3. Determine the number of roots of $z^4 + z^3 + 4z^2 + 2z + 3$ in each of the four quadrants.

4. The equation $z^3 + 3z = w$ has a root $z = g(w)$ which approaches zero as w approaches zero. Express $g(w)$ as a power series in w.

5. The equation $w = 2(z - t)/(z^2 - 1)$ has a solution which approaches t as w approaches zero. Express this solution as a power series in w. Prove that

$$(1 - 2tw + w^2)^{-1/2} = \sum_{k=0}^{\infty} w^k P_k(t),$$

where $P_k(t) = \dfrac{1}{2^k k!} \dfrac{d^k(t^2 - 1)^k}{dt^k}$.

6. Prove that $e^z = 3z^n$ has n roots inside the unit circle.

5.4. MEROMORPHIC FUNCTIONS

An important class of functions are those whose only singularities are poles.

Definition 5.4.1. A function $f(z)$ is meromorphic in D if and only if it is analytic in D except for poles.

Definition 5.4.2. A function $f(z)$ has a pole at infinity if and only if $f(1/z)$ has a pole at the origin.

Theorem 5.4.1. *The function $f(z)$ is meromorphic in the extended plane if and only if it is a rational function.*

Proof: Let $f(z) = P(z)/Q(z)$, where P and Q are polynomials with no common factors. If the degree of P is less than the degree of Q, then f has a finite number of poles and no pole at infinity. If the degree m of P is greater than the degree n of Q, then

$$
\begin{aligned}
f(z) &= \frac{a_m z^m + a_{m-1} z^{m-1} + \cdots + a_1 z + a_0}{b_n z^n + b_{n-1} z^{n-1} + \cdots + b_1 z + b_0} \\
&= C_0 + C_1 z + \cdots + C_{m-n-1} z^{m-n-1} + C_{m-n} z^{m-n} + \frac{R(z)}{Q(z)},
\end{aligned}
$$

which has a pole of order $m - n$ at infinity and a finite number of poles in the unextended plane.

On the other hand, if $f(z)$ is meromorphic in the extended plane it must have at most a finite number of poles since a pole, which is an isolated singularity, cannot be a limit point of poles. Let the poles be $z_1, z_2, \ldots, z_n$, ∞ with principal parts $\sum_{j=1}^{m_k} b_{jk}(z - z_k)^{-j}$, $k = 1, 2, \ldots, n$ and $\sum_{j=1}^{m} a_j z^j$. Now,

$$
f(z) - \sum_{k=1}^{n} \sum_{j=1}^{m_k} b_{jk}(z - z_k)^{-j} - \sum_{j=1}^{m} a_j z^j
$$

is analytic in the extended plane. By Liouville's theorem this function must be a constant a_0. Hence,

$$
f(z) = \sum_{k=1}^{n} \sum_{j=1}^{m_k} b_{jk}(z - z_k)^{-j} + \sum_{j=0}^{m} a_j z^j,
$$

which is clearly a rational function. This is incidentally the partial fraction expansion of the function.

A function like $\tan z$ is meromorphic in the unextended plane. It turns out that we can obtain an infinite series representation for $\tan z$ which is like a partial fraction expansion. Consider the following integral

$$
I_n = \frac{1}{2\pi i} \int_{C_{n+}} \frac{\tan \zeta}{\zeta(\zeta - z)} \, d\zeta,
$$

where C_n is a square with center at the origin and side $2n\pi$, $n = 1, 2, 3, \ldots$. The integrand has simple poles inside C_n at $\pm\pi/2$, $\pm 3\pi/2$, $\pm 5\pi/2, \ldots$, $\pm(2n - 1)\pi/2$, and at z, provided z is inside C_n and not an odd multiple of $\pi/2$. By the residue theorem, we have

$$I_n = \frac{\tan z}{z}$$

$$-\sum_{k=1}^{n}\left\{\frac{1}{\frac{(2k-1)\pi}{2}\left[\frac{(2k-1)\pi}{2}-z\right]} + \frac{1}{\frac{(2k-1)\pi}{2}\left[\frac{(2k-1)\pi}{2}+z\right]}\right\},$$

$$I_n = \frac{\tan z}{z} - 2\sum_{k=1}^{n}\frac{1}{\frac{(2k-1)^2\pi^2}{4} - z^2}.$$

To show that

$$\tan z = 2z\sum_{k=1}^{\infty}\frac{1}{\frac{(2k-1)^2\pi^2}{4} - z^2},$$

we have to show that $\lim_{n\to\infty} I_n = 0$. On the horizontal sides of the square C_n, $\zeta = \xi \pm n\pi i$, and $|\tan \zeta| \leq |\coth n\pi|$, which is bounded for $n = 1, 2, 3, \ldots$. Hence, on the horizontal sides

$$\left|\frac{1}{2\pi i}\int\frac{\tan \zeta}{\zeta(\zeta - z)}\,d\zeta\right| \leq \frac{M}{\pi}\frac{1}{n\pi - |z|} \to 0,$$

as $n \to \infty$. On the vertical sides $\zeta = \pm n\pi + i\eta$, and $|\tan \zeta| \leq |\tanh \eta| \leq \tanh n\pi$ which is bounded for $n = 1, 2, 3, \ldots$. Hence, on the vertical sides

$$\left|\frac{1}{2\pi i}\int\frac{\tan \zeta}{\zeta(\zeta - z)}\,d\zeta\right| \leq \frac{\tanh n\pi}{\pi}\frac{1}{n\pi - |z|} \to 0,$$

as $n \to \infty$. Evidently the convergence is normal in the unextended plane excluding the poles of $\tan z$.

EXAMPLE 5.4.1. Find an expansion for $\cot z$. This time the function has a simple pole at the origin and if we proceed as before we shall have to compute the residue at a double pole of the integrand at the origin. It is more convenient to subtract out the principal part of $\cot z$ at the origin before we expand. Therefore, we look for an expansion of $\cot z - 1/z$. Consider the integral

$$I_n = \frac{1}{2\pi i}\int_{C_{n+}}\frac{1}{\zeta(\zeta - z)}\left(\cot \zeta - \frac{1}{\zeta}\right)d\zeta,$$

where C_n is a square with center at the origin and side $(2n - 1)\pi$, $n = 1, 2, 3, \ldots$. If z is inside C_n and not a pole of $\cot z$, then the integrand has simple poles at z and at $\pm k\pi$, $k = 1, 2, 3, \ldots, n - 1$. Using the residue theorem to evaluate I_n, we have

$$I_n = \frac{1}{z}\left(\cot z - \frac{1}{z}\right) - \sum_{k=1}^{n-1}\left(\frac{1}{k\pi(z - k\pi)} - \frac{1}{k\pi(z + k\pi)}\right)$$

$$= \frac{1}{z}\left(\cot z - \frac{1}{z}\right) - 2\sum_{k=1}^{n-1}\frac{1}{z^2 - k^2\pi^2}.$$

To show that

$$\cot z = \frac{1}{z} + 2z\sum_{k=1}^{\infty}\frac{1}{z^2 - k^2\pi^2},$$

we have to show that $\lim_{n \to \infty} I_n = 0$. We can show that $\cot z - (1/z)$ is bounded on C_n. Hence,

$$|I_n| \leq \frac{M}{2\pi}\frac{4(2n - 1)\pi}{\frac{2n - 1}{2}\pi\left(\frac{2n - 1}{2}\pi - |z|\right)} \to 0,$$

as $n \to \infty$. To show that $\cot z - (1/z)$ is bounded on C_n, we first note that $1/z$ is bounded on the contours, while on the horizontal sides, where

$$z = x \pm \frac{2n - 1}{2}\pi i$$

$$|\cot z| \leq \left|\coth\frac{2n - 1}{2}\pi\right| \to 1,$$

as $n \to \infty$. On the vertical sides, where $z = \pm\dfrac{2n - 1}{2}\pi + iy$,

$$|\cot z| \leq \left|\tanh\frac{2n - 1}{2}\pi\right| \to 1,$$

as $n \to \infty$. This completes the proof.

More general situations than that of the previous example can be imagined. It may be that the given meromorphic function is not bounded on the sequence of contours selected. However, an expansion can still be obtained if $f(z)/z^m$ is bounded on the contours for some positive integer m. Then we take as the starting point the integral

$$I_n = \int_{C_n+}\frac{f(\zeta)}{\zeta^{m+1}(\zeta - z)}\,d\zeta.$$

If the length of C_n is an and the minimum distance from the origin to C_n is bn, then

$$|I_n| \leq \frac{aM}{b}\frac{1}{(bn - |z|)} \to 0,$$

as $n \to \infty$. It remains to compute the residues of $\dfrac{f(\zeta)}{\zeta^{m+1}(\zeta - z)}$ at the poles of $f(\zeta)$ and at z.

It is of some interest to obtain the most general representation of a function meromorphic in the unextended plane with a given set of poles and principal parts.

Theorem 5.4.2. *Let $f(z)$ be meromorphic in the unextended plane. Let $f(z)$ have an infinite number of poles $z_1, z_2, z_3, \ldots$, satisfying $\lim\limits_{k \to \infty} z_k = \infty$, with principal parts $P_k[(z - z_k)^{-1}]$. Then there is an entire function $g(z)$ and a set of polynomials $p_k(z)$ such that*

$$f(z) = \sum_{k=1}^{\infty} \left[P_k \left(\frac{1}{z - z_k} \right) - p_k(z) \right] + g(z).$$

Proof: We can assume that $z = 0$ is not a pole of $f(z)$. Otherwise, a translation would correct the situation. Now, $P_k[(z - z_k)^{-1}]$ is analytic for $|z| < |z_k|$. Therefore, it has a Maclaurin expansion. Let $p_k(z)$ be a partial sum of this expansion such that $|P_k[(z - z_k)^{-1}] - p_k(z)| < 2^{-k}$ for $|z| < |z_k|/2$. Let $|z| \leq R$ and $|z - z_k| \geq \delta$ for all k, then

$$\sum_{k=1}^{\infty} \left[P_k \left(\frac{1}{z - z_k} \right) - p_k(z) \right]$$

converges uniformly. To prove this let N be so large that $R < |z_N|/2$. Then

$$\sum_{k=1}^{\infty} \left[P_k \left(\frac{1}{z - z_k} \right) - p_k(z) \right] = \sum_{k=1}^{N-1} \left[P_k \left(\frac{1}{z - z_k} \right) - p_k(z) \right]$$

$$+ \sum_{k=N}^{\infty} \left[P_k \left(\frac{1}{z - z_k} \right) - p_k(z) \right].$$

The $\sum\limits_{k=1}^{N-1}$ is finite and $\sum\limits_{k=N}^{\infty}$ is absolutely and uniformly convergent by comparison with $\sum\limits_{k=N}^{\infty} 2^{-k}$. Therefore,

$$\sum_{k=1}^{\infty} \left[P_k \left(\frac{1}{z - z_k} \right) - p_k(z) \right]$$

is a meromorphic function with the same poles and principal parts as $f(z)$. Hence,

$$f(z) - \sum_{k=1}^{\infty} \left[P_k \left(\frac{1}{z - z_k} \right) - p_k(z) \right]$$

is an entire function $g(z)$. This completes the proof.

EXAMPLE 5.4.2. Using the result of Theorem 5.4.2, we obtain an expansion of $\pi \cot \pi z$. The given function has simple poles at $0, \pm 1, \pm 2, \ldots$ with residue 1. The series

$$\sum_{k=1}^{\infty} \frac{1}{z-k} \quad \text{and} \quad \sum_{k=-1}^{-\infty} \frac{1}{z-k}$$

do not converge. Therefore, we cannot write

$$\sum_{k=-\infty}^{\infty} \frac{1}{z-k}.$$

However,

$$\frac{1}{z-k} = -\frac{1}{k}\frac{1}{\left(1-\dfrac{z}{k}\right)} = -\frac{1}{k}\left(1 + \frac{z}{k} + \frac{z^2}{k^2} + \cdots\right).$$

Therefore,

$$\frac{1}{z-k} + \frac{1}{k} = -\frac{1}{k^2}\left(z + \frac{z^2}{k} + \cdots\right),$$

and

$$\left|\frac{1}{z-k} + \frac{1}{k}\right| \le \frac{M}{k^2} \text{ for } |z| \le \frac{k}{2},$$

where $M = \max\left|z + \dfrac{z^2}{k} + \cdots\right|$ on $|z| = \dfrac{k}{2}$. By comparison with

$$\sum_{k=1}^{\infty} \frac{1}{k^2}$$

we see that

$$\sum_{k=-\infty}^{\infty}{}' \left(\frac{1}{z-k} + \frac{1}{k}\right),$$

where $k = 0$ is omitted, converges uniformly in compact sets not including $z = \pm 1, \pm 2, \pm 3, \ldots$. Therefore,

$$\pi \cot \pi z = \frac{1}{z} + \sum_{k=-\infty}^{\infty}{}' \left(\frac{1}{z-k} + \frac{1}{k}\right) + g(z),$$

where $g(z)$ is an entire function. Differentiating term by term, we have

$$\pi^2 \csc^2 \pi z = \frac{1}{z^2} + \sum_{k=-\infty}^{\infty}{}' \frac{1}{(z-k)^2} - g'(z)$$

$$= \sum_{k=-\infty}^{\infty} \frac{1}{(z-k)^2} - g'(z).$$

Both

$$\csc^2 \pi z \quad \text{and} \quad \sum_{k=-\infty}^{\infty} \frac{1}{(z - k)^2}$$

are periodic with period 1 and for

$$0 \leq x \leq 1, |y| \geq 1, \lim_{|y| \to \infty} \csc^2 \pi z = \lim_{|y| \to \infty} \sum_{-\infty}^{\infty} \frac{1}{(z - k)^2} = 0.$$

This implies that $g'(z)$ is a bounded entire function which by Liouville's theorem, must be identically zero. In other words, $g(z)$ is identically constant. To evaluate g we merely evaluate $\pi \cot \pi z$ and

$$\frac{1}{z} + \sum_{k=-\infty}^{\infty}{}' \left(\frac{1}{z - k} + \frac{1}{k} \right)$$

at $z = 1/2$ and obtain $g = 0$. Combining the series for positive and negative k, we have

$$\pi \cot \pi z = \frac{1}{z} + \sum_{k=1}^{\infty} \frac{2z}{z^2 - k^2}.$$

We have incidentally proved that

$$\pi^2 \csc^2 \pi z = \sum_{k=-\infty}^{\infty} \frac{1}{(z - k)^2}.$$

Exercises 5.4

1. Find expansions for the following meromorphic functions

    (a) $\csc z$;   (b) $\sec z$;   (c) $\tanh z$;   (d) $\coth z$;

    (e) $\dfrac{1}{e^z - 1}$;   (f) $\sec^2 z$;   (g) $\mathrm{sech}\, z$;   (h) $\mathrm{csch}\, z$.

2. What is the simplest meromorphic function with double poles at all the integers n with principal parts $\dfrac{1}{(z - n)^2}$?

3. What is the simplest meromorphic function with simple poles at all the integers n with residues n?

5.5. ENTIRE FUNCTIONS

For meromorphic functions we have obtained representations in terms of their poles. For entire functions we seek representations in terms of their zeros.

Theorem 5.5.1. *If $f(z)$ is an entire function without zeros then there is an entire function $g(z)$ such that $f(z) = e^{g(z)}$.*

Proof: Consider $f'(z)/f(z)$. This function is entire since its only singularities would be zeros of $f(z)$. The function $\log f(z) = \int_{z_0}^{z} [f'(\zeta)/f(\zeta)] \, d\zeta + \log f(z_0)$ is also entire and

$$f(z) = e^{\log f(z)} = e^{g(z)},$$

where $g(z) = \int_{z_0}^{z} [f'(\zeta)/f(\zeta)] \, d\zeta + \log f(z_0)$.

Theorem 5.5.2. *Let $f(z)$ be an entire function with n zeros counting multiplicities. Then $f(z) = P(z)e^{g(z)}$, where $g(z)$ is entire and $P(z)$ is a polynomial of degree n.*

Proof: Let $z_1, z_2, \ldots, z_n$ be the zeros of $f(z)$ not necessarily distinct. Then $F(z) = f(z)/(z - z_1)(z - z_2) \cdots (z - z_n)$ is an entire function without zeros if we define $F(z_k) = \lim_{z \to z_k} f(z)/(z - z_1)(z - z_2) \cdots (z - z_n)$, $k = 1, 2, \ldots, n$. Therefore, $f(z) = (z - z_1)(z - z_2) \cdots (z - z_n)e^{g(z)}$, where $g(z)$ is entire.

Now, let us consider an entire function with an infinite number of zeros. Let $f(z) = \cos z$. Then

$$f'(z)/f(z) = -\tan z = -\sum_{k=1}^{\infty} \frac{2z}{(2k-1)^2 \frac{\pi^2}{4} - z^2},$$

where the convergence is uniform in compact sets not containing $z = \pm\pi/2$, $\pm3\pi/2, \ldots$. Therefore, we may integrate term by term along a finite path which does not pass through any of these singularities. Hence, if we integrate from 0 to z

$$\log \cos z = \sum_{k=1}^{\infty} \log \left(1 - \frac{z^2}{a^2}\right),$$

$a = (2k - 1)\pi/2$. There is some ambiguity in the definition of the logarithm, but this disappears in the next step when we take the exponential.

$$\cos z = \prod_{k=1}^{\infty} \left[1 - \frac{4z^2}{(2k-1)^2\pi^2}\right],$$

where the infinite product converges uniformly in any compact set. This expansion is analogous to the factored form of a polynomial. We are tempted to write

$$\prod_{k=-\infty}^{\infty} \left[1 - \frac{2z}{(2k-1)\pi}\right]$$

but the products

$$\prod_{k=1}^{\infty}\left[1 - \frac{2z}{(2k-1)\pi}\right] \quad \text{and} \quad \prod_{k=0}^{-\infty}\left[1 - \frac{2z}{(2k-1)\pi}\right]$$

do not by themselves converge. However, the products

$$\prod_{k=1}^{\infty}\left[1 - \frac{2z}{(2k-1)\pi}\right]e^{2z/(2k-1)\pi} \quad \text{and} \quad \prod_{k=0}^{-\infty}\left[1 - \frac{2z}{(2k-1)\pi}\right]e^{2z/(2k-1)\pi}$$

converge, and

$$\cos z = \prod_{k=1}^{\infty}\left[1 - \frac{4z^2}{(2k-1)^2\pi^2}\right] = \prod_{k=-\infty}^{\infty}\left[1 - \frac{2z}{(2k-1)\pi}\right]e^{2z/(2k-1)\pi}.$$

The latter product is called the *canonical product*.

EXAMPLE 5.5.1. Find an infinite product representation of sin z. Let $f(z) = \sin z/z$. Then

$$\frac{f'(z)}{f(z)} = \cot z - \frac{1}{z} = \sum_{k=1}^{\infty}\frac{2z}{z^2 - k^2\pi^2},$$

where the series converges uniformly in compact sets not containing $z = \pm\pi$, $\pm 2\pi, \pm 3\pi, \ldots$. Choosing a path from the origin to z, not containing these singularities, we may integrate term by term

$$\log f(z) - \log f(0) = \sum_{k=1}^{\infty}\log\left(1 - \frac{z^2}{k^2\pi^2}\right).$$

Taking exponentials, we have, since $f(0) = 1$,

$$\sin z = z\prod_{k=1}^{\infty}\left(1 - \frac{z^2}{k^2\pi^2}\right).$$

The canonical product for sin z is easily seen to be

$$\sin z = z\prod_{k=-\infty}^{\infty}{}'\left(1 - \frac{z}{k\pi}\right)e^{z/k\pi},$$

where the prime indicates that the term $k = 0$ is omitted.

It is of some interest to determine the most general entire function with a given set of zeros.

Theorem 5.5.3. *Let $f(z)$ be an entire function with an infinite number of zeros $z_1, z_2, z_3, \ldots$ such that $\lim_{k\to\infty} z_k = \infty$. Then there exists an entire function $g(z)$ and a set of polynomials $p_k(z/z_k)$ such that*

$$f(z) = z^m e^{g(z)}\prod_{k=1}^{\infty}\left(1 - \frac{z}{z_k}\right)e^{p_k(z/z_k)}.$$

Proof: Let $f(z)$ have a zero of order m at the origin. Then $f(z) = z^m h(z)$, where $h(z)$ is an entire function with zeros at $z_1, z_2, z_3, \ldots$. The Maclaurin expansion of $\log(1 - z)$ is

$$-\sum_{j=1}^{\infty} \frac{z^j}{j}$$

valid for $|z| < 1$. Therefore, for $|z| \leq 1/2$,

$$\left| \log(1 - z) + \sum_{j=1}^{N} \frac{z^j}{j} \right|$$

can be made as small as we please if N is sufficiently large. Hence, for $|z| \leq |z_k|/2$, there exists an N_k such that

$$\left| \log\left(1 - \frac{z}{z_k}\right) + \sum_{j=1}^{N_k} \frac{1}{j}\left(\frac{z}{z_k}\right)^j \right| \leq \frac{1}{k^2}.$$

Now, consider the product

$$\prod_{k=1}^{\infty} \left(1 - \frac{z}{z_k}\right) e^{p_k(z/z_k)}, \text{ where } p_k(z/z_k) = \sum_{j=1}^{N_k} \frac{1}{j}\left(\frac{z}{z_k}\right)^j.$$

Let K be so large that $|z| \leq R < |z_K|/2$, then

$$\prod_{k=1}^{\infty} \left(1 - \frac{z}{z_k}\right) e^{p_k(z/z_k)} = \prod_{k=1}^{K-1} \left(1 - \frac{z}{z_k}\right) e^{p_k(z/z_k)} \prod_{k=K}^{\infty} \left(1 - \frac{z}{z_k}\right) e^{p_k(z/z_k)}.$$

The $\prod_{k=1}^{K-1}$ is finite and $\prod_{k=K}^{\infty}$ is absolutely and uniformly convergent in $|z| \leq R$. To show this, let

$$\left(1 - \frac{z}{z_k}\right) e^{p_k(z/z_k)} = 1 + w_k(z).$$

Then

$$|w_k|/2 \leq |\log(1 + w_k(z))| = \left| \log\left(1 - \frac{z}{z_k}\right) + \sum_{j=1}^{N_k} \frac{1}{j}\left(\frac{z}{z_k}\right)^j \right| \leq \frac{1}{k^2}.$$

This shows that $\sum_{k=K}^{\infty} |w_k(z)|$ converges absolutely and uniformly, and that the product is analytic in the whole plane. It has the same zeros as $h(z)$. Hence,

$$\frac{h(z)}{\displaystyle\prod_{k=1}^{\infty} \left(1 - \frac{z}{z_k}\right) e^{p_k(z/z_k)}}$$

is an entire function without zeros. There is an entire function $g(z)$ such that

$$h(z) = e^{g(z)} \prod_{k=1}^{\infty} \left(1 - \frac{z}{z_k}\right) e^{p_k(z/z_k)},$$

$$f(z) = z^m e^{g(z)} \prod_{k=1}^{\infty} \left(1 - \frac{z}{z_k}\right) e^{p_k(z/z_k)}.$$

Theorem 5.5.4. *Let $f(z)$ be meromorphic in the extended or unextended plane. Then $f(z)$ can be written as the ratio of two entire functions.*

Proof: If $f(z)$ is meromorphic in the extended plane it is a rational function, the ratio of two polynomials, by Theorem 5.4.1. If it is meromorphic in the unextended plane and has a finite number of poles, then by Theorem 5.4.2 it is a rational function plus an entire function,

$$f(z) = \frac{P(z)}{Q(z)} + g(z),$$

where P and Q are polynomials and $g(z)$ is entire. Hence,

$$f(z) = \frac{P(z) + g(z)Q(z)}{Q(z)}.$$

If $f(z)$ is meromorphic in the unextended plane and has an infinite number of poles $z_1, z_2, z_3, \ldots$, then $\lim_{k \to \infty} z_k = \infty$, since a limit point of poles is not a pole. By Theorem 5.5.3 there is an entire function $h(z)$ with zeros at the poles of $f(z)$ and with the same multiplicities. Therefore, $f(z)h(z) = g(z)$ is entire. Hence, $f(z) = g(z)/h(z)$. This completes the proof.

Exercises 5.5

1. Find an infinite product expansion of
   (a) $\sinh z$;   (b) $\cosh z$;   (c) $e^{az} - e^{bz}$, a and b real.

2. What is the most general form for an entire function with zeros of order 2 at all the integers and no other zeros.

3. Let $f(z)$ be meromorphic and have double poles at all the integers and no other poles. Write $f(z)$ as the ratio of two entire functions.

PART TWO

Applications of Analytic Function Theory

CHAPTER 6

Potential Theory

6.1. LAPLACE'S EQUATION IN PHYSICS

Potential theory is, for the most part, the study of harmonic functions, that is, functions which satisfy Laplace's equation, $u_{xx} + u_{yy} + u_{zz} = 0$ in some region of three space, subject to certain conditions on the boundary of the region. If, for some reason, the function u is known to be independent of the third coordinate z, then we have the two-dimensional Laplace's equation $u_{xx} + u_{yy} = 0$. We have already seen that the real (or imaginary) part of an analytic function of a complex variable satisfies the two-dimensional Laplace's equation in the domain of analyticity. For this reason, analytic function theory becomes a powerful tool in the study of two-dimensional potential theory. In this chapter, we shall study potential theory from this point of view, and show how many typical boundary value problems can be solved. To set the stage for this study, we shall first consider some of the common physical situations which lead to harmonic functions.

Fluid dynamics. Consider a part of three space which is occupied by some homogeneous fluid. We shall assume that the fluid is flowing with a velocity $v(x, y, z, t)$ which measures the average velocity of all the molecules of the fluid very close to the point (x, y, z) at time t. Assume that at the point (x, y, z) at the time t, the fluid has a density $\rho(x, y, z, t)$, which is the mass per unit volume at that point at time t. Consider *any* fixed volume V with a smooth surface S which is occupied by the fluid. At time t the total mass of fluid in this volume is

$$M(t) = \iiint_V \rho(x, y, z, t) \, dx \, dy \, dz.$$

Since the volume is fixed, the change in mass per unit time is entirely due to the change in density with time; that is,

$$\frac{dM}{dt} = \iiint_V \frac{\partial \rho}{\partial t} \, dx \, dy \, dz.$$

229

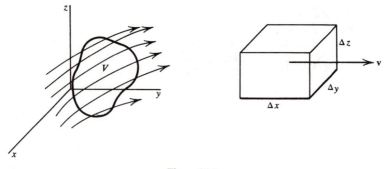

Figure 6.1.1.

We shall assume that there are no sources or sinks of fluid inside V; in other words, there is no way that fluid can enter or leave V except through the surface S.

Consider an element of space with edges Δx, Δy, Δz. See Figure 6.1.1. Assume that the velocity of all molecules inside this element is $(\Delta x/\Delta t, 0, 0)$ and that their total mass is Δm. In the time interval Δt, the mass Δm passes through the surface of area $(\Delta y)(\Delta z)$ which is one end of the given element. Clearly,

$$\frac{\Delta m}{\Delta t} = \frac{\Delta m}{\Delta V}\frac{\Delta x}{\Delta t}\,\Delta S$$

$$= \rho v_x\,\Delta S,$$

where $\Delta V = (\Delta x)(\Delta y)(\Delta z)$ and $\Delta S = (\Delta y)(\Delta z)$ and v_x is the component of velocity perpendicular to ΔS. Similarly,

$$\frac{dm}{dt} = \rho \mathbf{v} \cdot \mathbf{N}\,dS$$

is the change in mass in V per unit time due to the fluid with density ρ and velocity $\mathbf{v}$ at the surface passing through the element of area dS with inward pointing unit normal vector $\mathbf{N}$. Integrating over the whole surface of V, we have the rate of change of total mass due to the flow of fluid through the surface S,

$$\frac{dM}{dt} = \iint\limits_{S} \rho \mathbf{v} \cdot \mathbf{N}\,dS.$$

Since there are no sources or sinks inside V, we can equate the two expressions for dM/dt

$$\frac{dM}{dt} = \iiint\limits_{V} \frac{\partial \rho}{\partial t}\,dx\,dy\,dz = \iint\limits_{S} \rho \mathbf{v} \cdot \mathbf{N}\,dS.$$

Using the divergence theorem on the surface integral, remembering that it uses the outward pointing normal, we have

$$\iiint\limits_{V} \left[\frac{\partial \rho}{\partial t} + \nabla \cdot (\rho \mathbf{v}) \right] dx\, dy\, dz = 0.$$

Since V is arbitrary and we assume the continuity of the integrand, the partial differential equation

$$\frac{\partial \rho}{\partial t} + \nabla \cdot (\rho \mathbf{v}) = 0$$

is satisfied everywhere in the fluid where there are no sources or sinks. This is the *continuity equation* of fluid dynamics.

Let C be a simple closed smooth curve in three space. Let $\mathbf{T}$ be the unit tangent vector on C pointing in the direction of increasing arc length from some fixed point with a definite orientation. The *circulation* of fluid around C is defined as

$$\int_C \mathbf{v} \cdot \mathbf{T}\, ds = \int_C v_x\, dx + v_y\, dy + v_z\, dz,$$

where s is arc length. This is L times the average tangential velocity around C, where L is the length of C. Suppose a domain D occupied by the fluid is simply connected and the circulation around every simple closed smooth curve is zero.* Then

$$\int_C v_x\, dx + v_y\, dy + v_z\, dz = 0$$

for every admissible C in D. Then there is a function $\phi(x, y, z, t)$ such that $v_x = \partial\phi/\partial x$, $v_y = \partial\phi/\partial y$, $v_z = \partial\phi/\partial z$. ϕ is called the *velocity potential*. Suppose, further that the fluid is incompressible, that is, ρ does not change with time or with the space variables. Then $\partial\rho/\partial t = 0$ and the continuity equation becomes $\rho\nabla \cdot \mathbf{v} = 0$. Substituting $\mathbf{v} = \nabla\phi$, we have

$$\nabla \cdot \nabla\phi = \nabla^2\phi = 0.$$

If we assume a steady state situation, then $\mathbf{v}$ does not depend on time, and there exists a velocity potential ϕ which depends only on x, y, and z. Hence,

$$\nabla^2\phi = \phi_{xx} + \phi_{yy} + \phi_{zz} = 0$$

and ϕ satisfies Laplace's equation. Finally, if we assume a two-dimensional situation, that is, the velocity does not depend on the third variable z, then there exists a velocity potential $\phi(x, y)$ such that $\nabla^2\phi = 0$.

---

* In this case, we say that the fluid is *irrotational*.

In some domain D, of the xy plane, let ϕ be the real part of a complex analytic function $f(z)$. Then

$$f(z) = \phi(x, y) + i\psi(x, y)$$

and the curves $\phi = c_1$, $\psi = c_2$ are a mutually orthogonal set of curves if $f'(z) \neq 0$ in D. A normal vector to the curve $\psi = c_2$ is $(\partial\psi/\partial x, \partial\psi/\partial y)$, and it is easy to show that this vector is perpendicular to the velocity at that point; that is,

$$\mathbf{v} \cdot \nabla\psi = \frac{\partial\phi}{\partial x}\frac{\partial\psi}{\partial x} + \frac{\partial\phi}{\partial y}\frac{\partial\psi}{\partial y}$$

$$= -\frac{\partial\phi}{\partial x}\frac{\partial\phi}{\partial y} + \frac{\partial\phi}{\partial y}\frac{\partial\phi}{\partial x} = 0.$$

For this reason the ψ function is called the *stream* function because the curves $\psi = c_2$ are curves which are tangent to the flow. If a fluid is flowing past a rigid boundary then, since the fluid cannot penetrate the boundary, it must be a stream line of the flow. Hence, the problem reduces to finding a stream function for which the given boundary can be represented by the equation $\psi = c$. Once the stream function is found, the velocity can be determined as follows

$$v_x = \frac{\partial\phi}{\partial x} = \frac{\partial\psi}{\partial y},$$

$$v_y = \frac{\partial\phi}{\partial y} = -\frac{\partial\psi}{\partial x}.$$

The function $f(z)$ is called the *complex velocity potential*.

EXAMPLE 6.1.1. Show that the function $f(z) = U(z + 1/z)$ represents a uniform stream flowing past a circular cylinder of radius one, where U is a positive constant. Let $\phi(x, y) = \mathrm{Re}(f(z))$. Then

$$\phi(x, y) = U\left(x + \frac{x}{x^2 + y^2}\right).$$

For $x^2 + y^2$ very large, ϕ is approximately Ux and $v_x = U$, $v_y = 0$. Hence, at large distances from the origin, the flow is that of a uniform stream with velocity U to the right. Near the origin, the flow is not uniform, that is, the stream is disturbed by an obstacle. Consider the unit circle represented by $z = \cos\theta + i\sin\theta$. Here,

$$f(z) = U(\cos\theta + i\sin\theta + \cos\theta - i\sin\theta) = 2U\cos\theta.$$

Hence, the unit circle is the stream line $\psi = 0$. Since the flow is tangential to the unit circle, it can be considered as a rigid obstacle past which the stream is flowing. A few of the stream lines of this flow are shown in Figure 6.1.2.

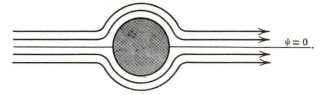

$\psi = 0$

Figure 6.1.2.

Heat transfer. Let a certain domain D of three space be occupied by a homogeneous solid which has a temperature $u(x, y, z, t)$ at the point (x, y, z) at time t. Experiments show that the quantity of heat contained in an element of volume $dx\, dy\, dz$ of density ρ and temperature u is

$$c\rho u\, dx\, dy\, dz,$$

where c is the specific heat of the material, which we assume is constant. If V is a fixed volume with a smooth surface contained in D, then

$$H(t) = \iiint_V c\rho u\, dx\, dy\, dz$$

is the total quantity of heat contained in V. We shall assume that ρ is a constant. Hence, as the temperature u changes with time

$$\frac{dH}{dt} = \iiint_V c\rho\, \frac{\partial u}{\partial t}\, dx\, dy\, dz$$

is the rate of change of heat in the volume V. Experiments also show that heat crosses a surface element dS at a rate

$$k\nabla u \cdot \mathbf{N}\, dS,$$

where k is a constant of thermal conductivity, ∇u is the gradient of temperature at the surface element, and $\mathbf{N}$ is a unit normal vector to the surface. If there are no sources or sinks of heat in V, then the total change of heat in V is due to heat crossing its surface S. Hence,

$$\frac{dH}{dt} = \iiint_V c\rho\, \frac{\partial u}{\partial t}\, dx\, dy\, dz = -\iint_S k\nabla u \cdot \mathbf{N}\, dS,$$

where $\mathbf{N}$ is the inward pointing normal. Using the divergence theorem on the surface integral, we have

$$\iiint_V \left[c\rho\, \frac{\partial u}{\partial t} - k\nabla^2 u \right] dx\, dy\, dz = 0.$$

Since this must hold for every volume V in D with a smooth surface, no matter how small, the integrand must vanish in D. Hence,

$$\nabla^2 u = \frac{c\rho}{k} \frac{\partial u}{\partial t}$$

must be satisfied by the temperature u. This is the *heat equation*.

If we assume that a certain heat conducting solid has reached a steady state temperature distribution because the losses and gains of heat through its surface are perfectly balanced, then u does not depend on t and $\nabla^2 u = 0$. Therefore, in the steady state the temperature is harmonic.

Suppose that in a given solid the steady state temperature is independent of the third variable z. Then we can express the temperature $u(x, y)$ as a harmonic function of two variables. Let u be the real part of an analytic function of the complex variable z. Then

$$f(z) = u(x, y) + iv(x, y).$$

The lines of constant temperature $u(x, y) = c_1$ are perpendicular to the gradient ∇u, which determines the direction of heat flow. Since the curves $v(x, y) = c_2$ are in the direction of ∇u, these are flux lines, or lines along which heat flows.

If heat is added to (or removed from) a solid at its boundary in such an amount so as to maintain that part of its boundary at a specified temperature, then we have a boundary condition, that is, that the temperature is specified on the boundary. If heat is added to (or removed from) a solid at its boundary at a specified rate, then this is the same as specifying the normal component of the gradient of temperature, $\nabla u \cdot \mathbf{N}$. This is another possible boundary condition. A third condition is realized when heat is allowed to radiate into a surrounding medium from the surface of a body. The law of radiation determines that

$$\nabla u \cdot \mathbf{N} = \alpha(u - u_0),$$

in other words, that the rate at which heat is lost is proportional to the difference between the surface temperature and the temperature of the surrounding medium u_0.

EXAMPLE 6.1.2. Find the steady state temperature of a thin metal plate in the shape of a semicircular disk with the curved edge perfectly insulated, while the straight edge is half maintained at $0°$ and the other half at $100°$. The top and bottom faces are insulated so that the flow of heat is strictly two-dimensional. See Figure 6.1.3 for the details. The obvious thing to look for is a function which is the real part of an analytic function which behaves properly on the boundary. Now, the arg z goes from 0 to π in the upper half-plane and it does not depend on $r = |z|$. Hence, $\partial u/\partial r = 0$ at $r = a$,

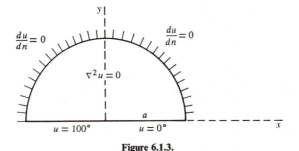

Figure 6.1.3.

if $u = c \arg z$, satisfying the boundary condition on the curved part of the boundary. The function

$$f(z) = \frac{100}{i\pi} \log z = \frac{100}{\pi} \arg z - \frac{100}{\pi} i \log |z|$$

has the desired property since $u(x, y) = (100/\pi) \arg z$ is the real part of an analytic function in the upper half-plane, if we specify $-\pi/2 \leq \arg z < 3\pi/2$, which takes on the value 0 on the positive real axis and the value 100 on the negative real axis. We shall show in the next section that if the boundary value problem has a solution, it is unique.

Electrostatics. Experiments show that if a fixed negative charge e is set up at the origin, then another charge e at the point (x, y, z) is repelled with a force with components $(Ke^2x/r^3, Ke^2y/r^3, Ke^2z/r^3)$, where K is a constant. The field of force is conservative since the line integral

$$\int_C \frac{Ke^2}{r^3} (x \, dx + y \, dy + z \, dz)$$

is independent of the path. There exists an electrostatic potential

$$\phi(x, y, z) = \frac{Ke^2}{r} = \frac{Ke^2}{\sqrt{x^2 + y^2 + z^2}}$$

such that $Ke^2x/r^3 = -\partial\phi/\partial x$, $Ke^2y/r^3 = -\partial\phi/\partial y$, $Ke^2z/r^3 = -\partial\phi/\partial z$. It is an easy exercise to prove that ϕ is harmonic except at the origin.

Consider an infinitely long wire along the z axis with a uniform charge of density ρ. Then a charge e at the point (x, y, z) will be repelled by a force **F** with components

$$F_x = \int_{-\infty}^{\infty} \frac{xKe\rho \, d\zeta}{[x^2 + y^2 + (\zeta - z)^2]^{3/2}} = \int_{-\infty}^{\infty} \frac{xKe\rho \, dt}{[x^2 + y^2 + t^2]^{3/2}} = \frac{2xKe\rho}{x^2 + y^2},$$

$$F_y = \int_{-\infty}^{\infty} \frac{yKe\rho \, d\zeta}{[x^2 + y^2 + (\zeta - z)^2]^{3/2}} = \int_{-\infty}^{\infty} \frac{yKe\rho \, dt}{[x^2 + y^2 + t^2]^{3/2}} = \frac{2yKe\rho}{x^2 + y^2},$$

$$F_z = \int_{-\infty}^{\infty} \frac{(z - \zeta)Ke\rho \, d\zeta}{[x^2 + y^2 + (\zeta - z)^2]^{3/2}} = -\int_{-\infty}^{\infty} \frac{tKe\rho \, dt}{[x^2 + y^2 + t^2]^{3/2}} = 0.$$

Hence, the force is independent of z. Nevertheless, the force field in two dimensions is still conservative since the line integral

$$\int_C \frac{2Ke\rho}{x^2 + y^2} (x \, dx + y \, dy)$$

is independent of the path and there is a potential function $\phi(x, y) = -Ke\rho \log r = -2Ke\rho \log \sqrt{x^2 + y^2}$. It is again easy to show that $\phi(x, y)$ is harmonic except at the origin.

Let S be the surface of an infinitely long cylinder with axis perpendicular to the xy plane and with intersection a simple closed curve C in the xy plane. Suppose charge is distributed over the surface S so that the electrostatic potential on S is specified as a function $g(s)$ of arc length on C, that is, the potential is independent of z. This is equivalent to a sheet of line charges with uniform density in the z direction. Hence, the potential inside S is independent of z and satisfies the two-dimensional Laplace's equation $\phi_{xx} + \phi_{yy} = 0$. The boundary condition is $\phi = g(s)$ on C. This boundary value problem is known as *Dirichlet's problem*. See Figure 6.1.4.

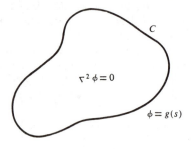

Figure 6.1.4.

EXAMPLE 6.1.3. The parallel plate condenser consists of two infinite conducting plates parallel to one another and a distance d apart. A battery is connected across the plates so as to maintain the potential 0 volts on one plate and 10 volts on the other. What is the potential between the plates? The problem is two-dimensional and consists of finding a function $\phi(x, y)$ which is harmonic for $0 < x < 1$, $-\infty < y < \infty$, is continuous for $0 \le x \le 1$, $-\infty < y < \infty$, and satisfies the boundary condition $\phi(0, y) = 0$, $\phi(1, y) = 10$. The obvious solution is the linear function $\phi = 10x$, which clearly satisfies all the conditions. Note that $10x = \mathrm{Re}(10z)$, where $10z$ is an obvious analytic function in the given strip.

There are many other places in physics where Laplace's equation is encountered. We shall not try to enumerate further. In addition, solutions of Laplace's equation become very important in the study of other partial differential equations such as the heat equation $\nabla^2 u = a^{-2} \, \partial u / \partial t$, the wave equation $\nabla^2 u = a^{-2} \partial^2 u/\partial t^2$, Poisson's equation $\nabla^2 u = f(x, y, z)$, and the equation $\nabla^2(\nabla^2 u) = 0$.

Exercises 6.1

1. Find the expression for $\nabla^2 u$ in polar coordinates by

(a) eliminating v from the polar coordinate form of the Cauchy-Riemann equations, and

(b) by evaluating $\dfrac{\partial^2 u}{\partial x^2} + \dfrac{\partial^2 u}{\partial y^2}$ using the transformation $x = r \cos \theta$, $y = r \sin \theta$.

2. Solve Example 6.1.3 using polar coordinates.

3. Find the potential inside a capacitor consisting of an infinitely long circular cylinder of radius b charged up to a potential of 10 volts and an infinitely long wire of radius a along the axis of the cylinder at a potential of zero. Hint: By symmetry, the solution must be independent of the polar angle θ.

4. Find the flow along a right angle barrier of a stream with initial velocity V parallel to one of the faces. Hint: The transformation $w = z^2$ opens the first quadrant up into the upper half-plane.

6.2. THE DIRICHLET PROBLEM

We shall consider several boundary value problems of potential theory, which, as we have indicated in the previous section, have applications in physics. Perhaps the most important of these is the Dirichlet problem.

Dirichlet Problem: Let D be a bounded domain bounded by a finite number of simple closed contours C_i which do not intersect. Let $u(x, y)$ be continuous in $D \cup C_1 \cup C_2 \cup \cdots \cup C_n$. Let u have continuous second partial derivatives in D and satisfy $u_{xx} + u_{yy} = 0$. Let n continuous functions g_i be defined on the n contours C_i. The problem is to find a u which takes on the values g_i on C_i.

Theorem 6.2.1. *If the Dirichlet problem has a solution for a given domain D and set of functions g_i, then the solution is unique.*

Proof: Consider two functions u_1 and u_2 which take on the same boundary values on the entire boundary of D. Then $w = u_1 - u_2$ is harmonic in D and is zero on the boundary. According to the maximum and minimum principle for the real part of an analytic function of Corollary 3.7.3, w takes on its maximum and minimum values on the boundary of D. This implies that $w \equiv 0$ and $u_1 \equiv u_2$ in D.

Theorem 6.2.2. *The Dirichlet problem is solvable for the disk $|z| < R$. If $g(\phi)$ is a continuous function for $0 \leq \phi \leq 2\pi$, such that $g(0) = g(2\pi)$, then*

$$u(r, \theta) = \frac{1}{2\pi} \int_0^{2\pi} \frac{(R^2 - r^2)g(\phi) \, d\phi}{R^2 + r^2 - 2Rr \cos(\theta - \phi)}$$

is harmonic for $0 \leq r < R$, $0 \leq \theta \leq 2\pi$, and $u(R, \theta) = \lim_{r \to R} u(r, \theta) = g(\theta)$.

Proof: By Theorem 3.4.2, Exercise 3.4.3, and Theorem 3.6.1, we have the result that a function $f(z)$ analytic for $|z| < R$, and continuous for $|z| \leq R$ can be represented by the Cauchy integral formula

$$f(z) = \frac{1}{2\pi i} \int_{C_+} \frac{f(\zeta)}{\zeta - z} \, d\zeta = \frac{1}{2\pi} \int_0^{2\pi} \frac{f(Re^{i\phi})Re^{i\phi} \, d\phi}{Re^{i\phi} - re^{i\theta}},$$

for $|z| = r < R$. The point $z' = \frac{R^2}{r} e^{i\theta}$ is outside the circle of radius R. Therefore,

$$0 = \frac{1}{2\pi i} \int_{C_+} \frac{f(\zeta)}{\zeta - z'} \, d\zeta = \frac{1}{2\pi} \int_0^{2\pi} \frac{f(Re^{i\phi})Re^{i\phi} \, d\phi}{Re^{i\phi} - \frac{R^2}{r} e^{i\theta}}$$

$$= \frac{1}{2\pi} \int_0^{2\pi} \frac{f(Re^{i\phi})re^{i\phi} \, d\phi}{re^{i\phi} - Re^{i\theta}}.$$

Now, $|Re^{i\phi} - re^{i\theta}|^2 = |re^{i\phi} - Re^{i\theta}|^2 = R^2 + r^2 - 2Rr \cos(\theta - \phi)$. Hence,

$$f(re^{i\theta}) = \frac{1}{2\pi} \int_0^{2\pi} \frac{f(Re^{i\phi})[R^2 - Rre^{i(\phi-\theta)}] \, d\phi}{R^2 + r^2 - 2Rr \cos(\theta - \phi)},$$

$$0 = \frac{1}{2\pi} \int_0^{2\pi} \frac{f(Re^{i\phi})[r^2 - Rre^{i(\phi-\theta)}] \, d\phi}{R^2 + r^2 - 2Rr \cos(\theta - \phi)}.$$

Subtracting, we have

$$f(re^{i\theta}) = \frac{1}{2\pi} \int_0^{2\pi} \frac{f(Re^{i\phi})(R^2 - r^2) \, d\phi}{R^2 + r^2 - 2Rr \cos(\theta - \phi)}.$$

Taking the real part,

$$u(r, \theta) = \frac{1}{2\pi} \int_0^{2\pi} \frac{u(R, \phi)(R^2 - r^2) \, d\phi}{R^2 + r^2 - 2Rr \cos(\theta - \phi)}.$$

This is *Poisson's integral formula* for the value of a harmonic function inside the circle of radius R, which is continuous in $|z| \leq R$, and takes on the values $u(R, \phi)$ on $|z| = R$.

Conversely, given a continuous function $g(\phi)$ we can form the integral

$$u(r, \theta) = \frac{1}{2\pi} \int_0^{2\pi} \frac{g(\phi)(R^2 - r^2)\, d\phi}{R^2 + r^2 - 2Rr \cos(\theta - \phi)},$$

and we can show that this function is the solution of the Dirichlet problem for the disk. To show that it is harmonic inside the circle, we merely differentiate under the integral sign using the Laplacian in polar coordinates. Thus for $r < R$,

$$\nabla^2 u = \int_0^{2\pi} g(\phi) \left[\frac{1}{r} \frac{\partial}{\partial r} \left(r \frac{\partial}{\partial r} \right) + \frac{1}{r^2} \frac{\partial^2}{\partial \theta^2} \right] \frac{R^2 - r^2}{R^2 + r^2 - 2Rr \cos(\theta - \phi)}\, d\phi.$$

It is an easy exercise to show that inside the circle of radius R

$$\nabla^2 \frac{R^2 - r^2}{R^2 + r^2 - 2Rr \cos(\theta - \phi)} = 0.$$

The differentiation under the integral sign is justified since, for $r < R$, $(R^2 - r^2)/[R^2 + r^2 - 2Rr \cos(\theta - \phi)]$ has continuous second partial derivatives.

We note that

$$1 = \frac{1}{2\pi} \int_0^{2\pi} \frac{R^2 - r^2}{R^2 + r^2 - 2Rr \cos(\theta - \phi)}\, d\phi.$$

This follows from the fact that $f(z) = 1$ for $|z| \leq R$ is analytic inside the circle of radius R and is continuous for $|z| \leq R$. Therefore,

$$u(r, \theta) - g(\theta) - \frac{1}{2\pi} \int_0^{2\pi} \frac{[g(\phi) - g(\theta)](R^2 - r^2)}{R^2 + r^2 - 2Rr \cos(\theta - \phi)}\, d\phi,$$

and to show that $\lim_{r \to R} u(r, \theta) = g(\theta)$, we have to show that

$$\lim_{r \to R} \int_0^{2\pi} \frac{[g(\phi) - g(\theta)](R^2 - r^2)}{R^2 + r^2 - 2Rr \cos(\theta - \phi)}\, d\phi = 0.$$

Because of the continuity of $g(\phi)$, for every $\epsilon > 0$ there is a δ such that $|g(\phi) - g(\theta)| < \epsilon$ for $|\phi - \theta| < 2\delta$. Then

$$\int_0^{2\pi} \frac{[g(\phi) - g(\theta)](R^2 - r^2)}{R^2 + r^2 - 2Rr \cos(\theta - \phi)}\, d\phi = \int_0^{\theta - \delta} \frac{[g(\phi) - g(\theta)](R^2 - r^2)}{R^2 + r^2 - 2Rr \cos(\theta - \phi)}\, d\phi$$

$$+ \int_{\theta - \delta}^{\theta + \delta} \frac{[g(\phi) - g(\theta)](R^2 - r^2)}{R^2 + r^2 - 2Rr \cos(\theta - \phi)}\, d\phi$$

$$+ \int_{\theta + \delta}^{2\pi} \frac{[g(\phi) - g(\theta)](R^2 - r^2)}{R^2 + r^2 - 2Rr \cos(\theta - \phi)}\, d\phi$$

$$= I_1 + I_2 + I_3.$$

Now,

$$|I_2| \leq \epsilon \int_0^{2\pi} \frac{R^2 - r^2}{R^2 + r^2 - 2Rr \cos (\theta - \phi)} \, d\phi = \epsilon,$$

and if $|g(\phi)| \leq M$.

$$|I_1| \leq M \int_0^{\theta - \delta} \frac{R^2 - r^2}{R^2 + r^2 - 2Rr \cos (\theta - \phi)} \, d\phi$$

$$\leq M \int_0^{\theta - \delta} \frac{R^2 - r^2}{R^2 + r^2 - 2Rr \cos \delta} \, d\phi$$

$$\leq 2\pi M \frac{R^2 - r^2}{[R - r \cos \delta]^2} \leq 2\pi M \frac{R^2 - r^2}{R^2(1 - \cos \delta)^2} < \epsilon,$$

for $|R - r|$ sufficiently small. Similarly, $|I_3| < \epsilon$. Therefore,

$$\left| \int_0^{2\pi} \frac{[g(\phi) - g(\theta)](R^2 - r^2)}{R^2 + r^2 - 2Rr \cos (\theta - \phi)} \right| \leq |I_1| + |I_2| + |I_3| < 3\epsilon.$$

Since ϵ is arbitrary, this proves that $\lim_{r \to R} [u(r, \theta) - g(\theta)] = 0$. This completes the proof.

Corollary 6.2.1. Let $g(\theta)$ be piecewise continuous for $0 \leq \theta \leq 2\pi$. Then

$$u(r, \theta) = \frac{1}{2\pi} \int_0^{2\pi} \frac{g(\phi)(R^2 - r^2)}{R^2 + r^2 - 2Rr \cos (\theta - \phi)} \, d\phi$$

is harmonic for $|z| < R$, and $\lim_{r \to R} u(r, \theta) = g(\theta)$ for all but a finite number of values of θ.

Proof: The same proof as for Theorem 6.2.2 shows that $u(r, \theta)$ is harmonic inside the circle of radius R, and that it takes on the value $g(\theta)$ at $r = R$ for any θ, where $g(\theta)$ is continuous.

EXAMPLE 6.2.1. Solve the Dirichlet problem in the region

$$B = \{z \mid |z| \leq 1, \text{Im}(z) \geq 0\}$$

such that $u(x, y) = 0$ when $y = 0$ and $-1 < x < 1$, $u(x, y) = 1$ when $x^2 + y^2 = 1$. We make use of the fact that an analytic function of an analytic function is analytic. Hence, if $u(x, y) = \text{Re}(f(z))$, where $f(z)$ is analytic in $D = \{z \mid |z| < 1, \text{Im}(z) > 0\}$ and if $\zeta = g(z)$ maps D onto the interior of a circle of radius R in the ζ plane so that the boundary of D maps onto the circle of radius R, and if $z = h(\zeta)$ is an analytic inverse of g, then $U(\xi, \eta) = \text{Re}(f[h(\zeta)])$ is harmonic for $|\zeta| < R$ and takes on given boundary values on $|\zeta| = R$. We can solve for $U(\xi, \eta)$ using the Poisson integral formula and then find $u(x, y)$ using the transformation $x = \text{Re}(h(\zeta))$, $y = \text{Im}(h(\zeta))$. We shall obtain $g(z)$ in three steps. First $\zeta_1 = z + 1/z$, maps D onto the lower half-plane $\text{Im}(\zeta_1) < 0$, such that $-1, i, 1$ map into $-2, 0, 2$, respectively. Second, $\zeta_2 = -\zeta_1/2$ maps the lower half-plane $\text{Im}(\zeta_1) < 0$ onto the upper half-plane $\text{Im}(\zeta_2) > 0$ so that $-2, 0, 2$ map into $1, 0, -1$, respectively. Third, $\zeta = (i - \zeta_2)/(i + \zeta_2)$ maps the upper half-plane $\text{Im}(\zeta_2) > 0$ onto the interior of the unit circle $|\zeta| < 1$, so that $-1, 0, 1$ map into $-i, 1, i$, respectively. Putting these together, we have

$$\zeta = \frac{2iz + z^2 + 1}{2iz - z^2 - 1}.$$

The inverse is

$$z = i\,\frac{\zeta - 1 + \sqrt{2(1 + \zeta^2)}}{1 + \zeta},$$

where a branch of the square root is chosen which is analytic inside the unit circle $|\zeta| = 1$. The boundary values satisfied by $U(\xi, \eta)$ are $U = 1$, $|\arg \zeta| \leq \pi/2$; $U = 0, \pi/2 < \arg < 3\pi/2$. Hence,

$$U(\xi, \eta) = \frac{1}{2\pi} \int_{-\pi/2}^{\pi/2} \frac{(1 - \rho^2)}{1 + \rho^2 - 2\rho \cos(\phi - \tau)}\, d\phi,$$

$\xi = \rho \cos \tau$, $\eta = \rho \sin \tau$. Suppose, for example, $z = i/2$, then $\zeta = 1/7$, $\rho = 1/7, \tau = 0$, and

$$u(0, \tfrac{1}{2}) = \frac{1}{2\pi} \int_{-\pi/2}^{\pi/2} \frac{1 - (\tfrac{1}{7})^2}{1 + (\tfrac{1}{7})^2 - (\tfrac{2}{7}) \cos \phi}\, d\phi.$$

Based on the method of the example, we have the following existence theorem.

Theorem 6.2.3. *Let D be a simply connected domain with boundary C. Let $\zeta = f(z)$ be analytic in D and map $D \cup C$ in a one-to-one fashion onto $|\zeta| \leq 1$, so that D maps onto $|\zeta| < 1$ and C maps onto $|\zeta| = 1$. Then*

$$U(\xi, \eta) = \frac{1}{2\pi} \int_0^{2\pi} \frac{(1 - \rho^2)g(\phi)}{1 + \rho^2 - 2\rho \cos(\phi - \tau)}\, d\phi$$

$\xi = \rho \cos \tau$, $\eta = \rho \sin \tau$, solves the Dirichlet problem in $D \cup C$ in the sense that $u(x, y) = U(\xi, \eta)$ at corresponding points under the mapping. (In Section 6.4, we shall prove the Riemann mapping theorem which asserts that every simply connected domain with more than two boundary points can be mapped in a one-to-one fashion onto the interior of the unit circle by an analytic mapping).

Before we complete the study of the Dirichlet problem let us derive the Poisson integral formula by a different method which will suggest an entirely different approach to the solution of boundary value problems in potential theory. Recall the Green's identity of Exercise 3.1.4

$$\iint_A (u\nabla^2 v - v\nabla^2 u)\, dx\, dy = \int_C (u\nabla v \cdot \mathbf{N} - v\nabla u \cdot \mathbf{N})\, ds,$$

where A is a bounded region bounded by the simple closed contour (or contours) C. The unit normal $\mathbf{N}$ is to point outward from the region A and u and v are to have continuous second partial derivatives in $A \cup C$. Let A be the annulus $\rho \leq |z| \leq R$, let u be harmonic for $|z| < R$ and take on the values $g(\theta)$ on $|z| = R$. Let v be harmonic in the interior of A and be zero on $|z| = R$. Then

$$0 = \iint_A (u\nabla^2 v - v\nabla^2 u)\, dx\, dy$$

$$= \int_0^{2\pi} g(\theta)\left(\frac{\partial v}{\partial r}\right)_{r=R} R\, d\theta - \int_0^{2\pi} u\left(\frac{\partial v}{\partial r}\right)_{r=\rho} \rho\, d\theta + \int_0^{2\pi} v\left(\frac{\partial u}{\partial r}\right)_{r=\rho} \rho\, d\theta.$$

We try to find a function v for which the second integral on the right will produce the value of u at the origin in the limit as $\rho \to 0$, while the third integral will vanish as $\rho \to 0$. To do this, $\lim_{\rho \to 0} \rho(\partial v/\partial r)_{r=\rho}$ must be $-1/2\pi$ while $\lim_{\rho \to 0} \rho v = 0$. A function which behaves this way and is harmonic except at the origin is $-(1/2\pi)\ln r$. Clearly,

$$\lim_{\rho \to 0} \int_0^{2\pi} u\, \frac{\partial}{\partial r}\left(\frac{1}{2\pi}\ln r\right)_{r=\rho} \rho\, d\theta = \lim_{\rho \to 0} \frac{1}{2\pi} \int_0^{2\pi} u\, d\theta = u(0,0),$$

$$\lim_{\rho \to 0} \int_0^{2\pi} v\left(\frac{\partial u}{\partial r}\right)_{r=\rho} \rho\, d\theta = \lim_{\rho \to 0} \frac{\rho \ln \rho}{2\pi} \int_0^{2\pi} \left(\frac{\partial u}{\partial r}\right)_{r=\rho} d\theta = 0,$$

since u and $\partial u/\partial r$ are continuous at the origin. Now, $-(1/2\pi)\ln r$ satisfies all the conditions except $v = 0$ on $|z| = R$. However, suppose we add $h(x, y)$ which is harmonic for $|z| < R$ and $h(R\cos\theta, R\sin\theta) = (1/2\pi)\ln R$. Then if $v = -(1/2\pi)\ln r + h(x, y)$, $\nabla^2 v = 0$ except at the origin, it behaves properly near the origin, and $v = 0$ on $|z| = R$. The result is

$$u(0,0) = -\int_0^{2\pi} g(\theta)\left(\frac{\partial v}{\partial r}\right)_{r=R} R\, d\theta.$$

Finally, it is clear that if we want the value of u at the point $z_0 = x_0 + iy_0$ inside the circle of radius R, we can proceed in the same way except with A replaced by a region bounded by $|z| = R$ and $|z - z_0| = \rho$. Using the function

$$v(x, y, x_0, y_0) = -\frac{1}{2\pi} \ln |z - z_0| + h(x, y, x_0, y_0),$$

where h is harmonic for $|z| < R$ and

$$h(R \cos \theta, R \sin \theta, x_0, y_0) = \frac{1}{2\pi} \ln |(R \cos \theta - x_0) + i(R \sin \theta - y_0)|.$$

The result is

$$u(x_0, y_0) = -\int_0^{2\pi} g(\theta) \left(\frac{\partial v}{\partial r}\right)_{r=R} R \, d\theta.$$

To help us find the proper $v(x, y, x_0, y_0)$ to express the value of the harmonic function $u(x, y)$ inside the circle of radius R in terms of its boundary values $g(\theta)$, we shall prove the following lemma.

Lemma 6.2.1. The circle $|z| = R$ is the locus of points the ratio of whose distances from z_0 and $R^2 z_0 / r_0^2$ is r_0 / R.

Proof:

$$(x - x_0)^2 + (y - y_0)^2 = \frac{r_0^2}{R^2} \left[\left(x - \frac{R^2 x_0}{r_0^2}\right)^2 + \left(y - \frac{R^2 y_0}{r_0^2}\right)^2 \right],$$

$$x^2 + y^2 - 2xx_0 - 2yy_0 + r_0^2 = \frac{r_0^2}{R^2} (x^2 + y^2) - 2xx_0 - 2yy_0 + R^2,$$

$$(x^2 + y^2)\left(1 - \frac{r_0^2}{R^2}\right) = R^2 - r_0^2,$$

$$x^2 + y^2 = R^2.$$

From the lemma, we see that

$$v(x, y, x_0, y_0) = -\frac{1}{2\pi} \ln \sqrt{(x - x_0)^2 + (y - y_0)^2}$$

$$+ \frac{1}{2\pi} \ln \sqrt{\left(x - \frac{R^2 x_0}{r_0^2}\right)^2 + \left(y - \frac{R^2 y_0}{r_0^2}\right)^2}$$

$$+ \frac{1}{2\pi} \ln (r_0/R)$$

^{i}s zero when $r = R$, has the proper singularity at z_0 inside the circle, and ^{i}s harmonic except at z_0. If we let $x = r \cos \theta$, $y = r \sin \theta$, $x_0 = r_0 \cos \phi$, $y_0 = r_0 \sin \phi$, then

$$v = -\frac{1}{4\pi} \ln [r^2 + r_0^2 - 2rr_0 \cos (\theta - \phi)]$$

$$+ \frac{1}{4\pi} \ln \left[\frac{r_0^2 r^2}{R^2} + R^2 - 2rr_0 \cos (\theta - \phi) \right],$$

$$\frac{\partial v}{\partial r} = -\frac{1}{2\pi} \frac{r - r_0 \cos (\theta - \phi)}{r^2 + r_0^2 - 2rr_0 \cos (\theta - \phi)}$$

$$+ \frac{1}{2\pi} \frac{\dfrac{r_0^2}{R^2} r - r_0 \cos (\theta - \phi)}{\dfrac{r_0^2 r^2}{R^2} + R^2 - 2rr_0 \cos (\theta - \phi)},$$

$$\left(\frac{\partial v}{\partial r} \right)_{r=R} = -\frac{1}{2\pi R} \frac{R^2 - r_0^2}{R^2 + r_0^2 - 2Rr_0 \cos (\theta - \phi)},$$

$$u(x_0, y_0) = \frac{1}{2\pi} \int_0^{2\pi} \frac{(R^2 - r_0^2)g(\theta)}{R^2 + r_0^2 - 2Rr_0 \cos (\theta - \phi)} \, d\theta,$$

which is the Poisson integral formula again. The importance of this derivation is that it starts from the Green's identity and does not make use of the analytic function theory. The function $v(x, y, x_0, y_0)$ plays a special role. This function is usually called a *Green's function*. The idea was to find an explicit representation for u in terms of the given boundary values and the Green's function v which is derived from a given set of properties. This idea will be explored further in the next section.

Exercises 6.2

1. Let D be a bounded simply connected domain bounded by a simple closed smooth curve C. Let $g(s)$ be a function continuous on C. Let $u(x, y)$ be the solution of Dirichlet's problem for $R = D \cup C$ such that $u = g$ on C. State the properties of a Green's function $G(x, y, x_0, y_0)$ in terms of which

$$u(x_0, y_0) = -\int_{C+} g(s) \nabla G \cdot \mathbf{N} \, ds.$$

2. Solve the following Dirichlet problem $\nabla^2 u = 0$ for $x < 0, 0 < y < \pi$; and $u = 1$ for $x = 0$, $0 < y < 1$; $u = 0$ for $y = 0$, $y = \pi$, $x < 0$. Hint: $w = e^z$ maps the given region onto a semidisk.

6.3. GREEN'S FUNCTIONS

In this section, we shall develop further the concept of the Green's function in the solution of boundary value problems.

Let D be a bounded simply connected domain bounded by a simple closed smooth contour C. Let $G(x, y, x_0, y_0)$ be a function satisfying the following properties:

1. $G(x, y, x_0, y_0) = -\frac{1}{2\pi} \ln |z - z_0| + H(x, y, x_0, y_0)$ where $H(x, y, x_0, y_0)$ is harmonic inside C.
2. G is continuous in $D \cup C$ except at (x_0, y_0) in D.
3. $G(x, y, x_0, y_0) = 0$ when (x, y) is on C.

If a function exists satisfying these properties, we call it a *Green's function* associated with the Dirichlet problem for the region $R = D \cup C$.

Theorem 6.3.1. *If a Green's function exists, it is unique.*

Proof: Let

$$G_1 = -\frac{1}{2\pi} \ln |z - z_0| + H_1(x, y, x_0, y_0)$$

and

$$G_2 = -\frac{1}{2\pi} \ln |z - z_0| + H_2(x, y, x_0, y_0).$$

Then $G_1 - G_2$ is harmonic in D, continuous in R, and zero on C. Hence, by Theorem 6.2.1, $G_1 - G_2 \equiv H_1 - H_2 \equiv 0$.

Theorem 6.3.2. *The Green's function for R is positive in D.*

Proof: Let D^* be the multiply connected domain bounded by C and a small circle of radius ρ centered at z_0. Now, G is zero on C and positive on the small circle for ρ sufficiently small. Hence, the minimum of G, which is taken on the boundary of D^*, is zero, and G cannot be zero in D^*. Every point of D, except z_0, is an interior point of D^* for ρ sufficiently small. This completes the proof.

Theorem 6.3.3. *If G is the Green's function for R, then $G(x_1, y_1, x_2, y_2) = G(x_2, y_2, x_1, y_1)$ for any distinct pair of points (x_1, y_1) and (x_2, y_2) in D.*

Proof: Let ρ_1 and ρ_2 be the radii of small circles with centers at (x_1, y_1) and (x_2, y_2), respectively, so small that both circles are in D and neither they, nor their interiors, intersect. Applying Green's theorem to D^*, the domain bounded by C and the two small circles, we have upon letting $G_1 = G(x, y, x_1, y_1)$ and $G_2 = G(x, y, x_2, y_2)$

$$0 = \iint_{D^*} (G_1 \nabla^2 G_2 - G_2 \nabla^2 G_1)\, dA = \int_{C_+} (G_1 \nabla G_2 \cdot \mathbf{N} - G_2 \nabla G_1 \cdot \mathbf{N})\, ds$$

$$+ \int_0^{2\pi} \left(G_2 \frac{\partial G_1}{\partial r} - G_1 \frac{\partial G_2}{\partial r} \right)_{r=\rho_1} \rho_1\, d\theta$$

$$+ \int_0^{2\pi} \left(G_2 \frac{\partial G_1}{\partial r} - G_1 \frac{\partial G_2}{\partial r} \right)_{r=\rho_2} \rho_2\, d\theta.$$

But since $G_1 = G_2 = 0$ on C, we have

$$\lim_{\rho_1 \to 0} \rho_1 \int_0^{2\pi} \left(G_2 \frac{\partial G_1}{\partial r} - G_1 \frac{\partial G_2}{\partial r} \right)_{r=\rho_1} d\theta$$

$$= \lim_{\rho_2 \to 0} \rho_2 \int_0^{2\pi} \left(G_1 \frac{\partial G_2}{\partial r} - G_2 \frac{\partial G_1}{\partial r} \right)_{r=\rho_2} d\theta,$$

$$G_2(x_1, y_1) = G_1(x_2, y_2),$$

using the properties of the Green's function. Hence,

$$G(x_1, y_1, x_2, y_2) = G(x_2, y_2, x_1, y_1).$$

Theorem 6.3.4. *If the Green's function exists for the region R, then for g(s) continuous on C*

$$u(x_0, y_0) = -\int_{C_+} g(s) \nabla G \cdot \mathbf{N}\, ds$$

solves the Dirichlet problem $\nabla u^2 = 0$ in D, $u = g$ on C.

We shall not give the proof of this theorem. It is based on the results of Theorems 6.3.1–6.3.3, and for those interested we refer them to R. Courant and D. Hilbert, *Methods of Mathematical Physics*, **II**. New York: Interscience, Inc., 1962, pp. 261–264. Rather, we shall take the attitude that the Dirichlet problem for the region R can be reduced to a corresponding problem for the unit disk $|z| \leq 1$ by introducing a mapping of R onto the disk. This approach will be explored further in the next section, but for now we can bring to light the following interesting connection between the Green's function and the mapping function.

Theorem 6.3.5. *Let $w = f(z)$ map the region R onto the unit disk $|z| \leq 1$ so that D, the interior of R, maps onto the open disk $|z| < 1$ and C, the boundary, maps onto the unit circle $|z| = 1$. Let $f'(z)$ exist and never vanish in D. Let z_0 map into the origin. Then $G(x, y, x_0, y_0) = -(1/2\pi)\ln |f(z)|$ is the Green's function for R.*

Proof: In the first place, G is harmonic in D except at z_0 since

$$G = \operatorname{Re}\left\{-\frac{1}{2\pi}\log f(z)\right\}.$$

Near z_0,

$$f(z) = f(z_0) + a_1(z - z_0) + a_2(z - z_0)^2 + \cdots$$
$$= a_1(z - z_0) + a_2(z - z_2)^2 + \cdots$$
$$= (z - z_0)g(z),$$

where $g(z_0) = f'(z_0) \neq 0$. Also,

$$-\frac{1}{2\pi}\ln|f(z)| = -\frac{1}{2\pi}\ln|z - z_0| - \frac{1}{2\pi}\ln|g(z)|$$

$$= -\frac{1}{2\pi}\ln\sqrt{(x - x_0)^2 + (y - y_0)^2} + H(x, y, x_0, y_0),$$

where H is harmonic at z_0. Finally, the boundary condition is satisfied on C, since

$$\lim_{z \to C} G = \lim_{z \to C}\left\{-\frac{1}{2\pi}\ln|f(z)|\right\}$$

$$= -\frac{1}{2\pi}\ln 1 = 0.$$

In the next section, we shall show that if D is a bounded simply connected domain bounded by a simple closed contour then a function $f(z)$ exists satisfying the conditions of Theorem 6.3.5. This will in turn show the existence of the Green's function associated with the Dirichlet problem for D.

We now turn to the study of some other boundary value problems of potential theory. The next most important boundary value problem after the Dirichlet problem is the *Neumann problem*. Let D be a bounded simply connected domain bounded by a simple closed smooth contour C. The Neumann problem is to determine a function u which is harmonic in D, has continuous first partial derivatives in $D \cup C$, and such that $\nabla u \cdot \mathbf{N} = g$, a given function defined on C. Clearly, a solution of the Neumann problem is not unique since, given u, u plus any constant will also satisfy the stated conditions. However, if we specify in addition that the average value of u on C shall be zero we make the solution unique.

Theorem 6.3.6. *The solution of the Neumann problem such that $\int_{C+} u\, ds = 0$ is unique.*

Proof: Let u_1 and u_2 both be harmonic in D and have the same normal derivative on C. Then $v = u_1 - u_2$ is harmonic in D and has vanishing normal derivative on C. Then

$$\iint_D \nabla v \cdot \nabla v \, dA = -\iint_D v \nabla^2 v \, dA + \int_{C+} v \nabla v \cdot \mathbf{N} \, ds$$

$$= 0.$$

Therefore, $\nabla v \equiv 0$ in D and $v = u_1 - u_2 \equiv k$, a constant. But

$$0 = \int_{C+} (u_1 - u_2) \, ds = \int_{C+} k \, ds = kL.$$

But $L \neq 0$, and hence $k = 0$ and $u_1 \equiv u_2$.

Theorem 6.3.7. *A necessary condition for the Neumann problem to have a solution is that $\int_{C+} g(s) \, ds = 0$.*

Proof: Assuming that there is a function u such that $\nabla^2 u = 0$ in D, u and its first partial derivatives are continuous in $D \cup C$ and $\nabla u \cdot \mathbf{N} = g(s)$ on C, then letting $v \equiv 1$, we have

$$0 = \iint_D (v \nabla^2 u - u \nabla^2 v) \, dA = \int_{C+} (v \nabla u \cdot \mathbf{N} - u \nabla v \cdot \mathbf{N}) \, ds$$

$$= \int_{C+} \nabla u \cdot \mathbf{N} \, ds = \int_{C+} g(s) \, ds.$$

Let us next see if we can define a Green's function associated with the Neumann problem. By analogy with the Dirichlet problem we shall assume that $G(x, y, x_0, y_0)$ is harmonic in D except at (x_0, y_0) where it has a singularity of the form $-(1/2\pi)\ln |z - z_0|$. Hence,

$$G(x, y, x_0, y_0) = -\frac{1}{2\pi} \ln |z - z_0| + H(x, y, x_0, y_0),$$

where H is harmonic in D. By analogy with the Dirichlet problem, we would like to say that $\nabla G \cdot \mathbf{N} = 0$ on C. However, this is impossible since

$$\int_{C+} \nabla H \cdot \mathbf{N} \, ds = 0 \quad \text{and} \quad \frac{1}{2\pi} \int_{C+} \nabla \ln |z - z_0| \cdot \mathbf{N} \, ds = 1.$$

Hence,

$$\int_{C+} \nabla G \cdot \mathbf{N} \, ds = -1.$$

We can still take $\nabla G \cdot \mathbf{N}$ as a constant on C, which will have to be $-1/L$, where L is the length of C. These conditions still do not make G unique, and for this reason we shall specify that the average value of G on C is zero; that is,

$$\int_{C+} G \, ds = 0$$

Theorem 6.3.8. *Let D be a bounded simply connected domain bounded by a simple closed smooth contour C. Let a Green's function G be defined as follows:*

1. $G(x, y, x_0, y_0) = -\dfrac{1}{2\pi} \ln |z - z_0| + H(x, y, x_0, y_0).$

2. H *is harmonic in* D.

3. G *and its first partial derivatives are continuous in* $D \cup C$.

4. $\nabla G \cdot \mathbf{N} = -1/L$ *when* (x, y) *is on* C.

5. $\int_{C+} G\, ds = 0.$

Then if G exists it is unique.

Proof: Let

$$G_1 = -\frac{1}{2\pi} \ln |z - z_0| + H_1 \quad \text{and} \quad G_2 = -\frac{1}{2\pi} \ln |z - z_0| + H_2.$$

Then $H_1 - H_2$ is harmonic in D, has zero normal derivative on C, and

$$\int_{C+} (G_1 - G_2)\, ds = \int_{C+} (H_1 - H_2)\, ds = 0.$$

Therefore, by Theorem 6.3.6, $H_1 - H_2 \equiv 0$, and $H_1 \equiv H_2$.

Theorem 6.3.9. *Let u be the solution of the Neumann problem for a domain D, satisfying the conditions of Theorems 6.3.6 and 6.3.7, and let G be the Green's function satisfying the conditions of Theorem 6.3.8. Then*

$$u(x_0, y_0) = \int_{C+} g(s) G\, ds.$$

Proof: We apply Green's theorem to u and G on the domain D^* bounded by C and a small circle of radius ρ centered at z_0. Then

$$0 = \iint_{D^*} (u\nabla^2 G - G\nabla^2 u)\, dA$$

$$= \int_{C+} (u\nabla G \cdot \mathbf{N} - G\nabla u \cdot \mathbf{N})\, ds$$

$$+ \int_0^{2\pi} \left(G\, \frac{\partial u}{\partial r} \right)_{r=\rho} \rho\, d\theta - \int_0^{2\pi} \left(u\, \frac{\partial G}{\partial r} \right)_{r=\rho} \rho\, d\theta.$$

Since,

$$\lim_{\rho \to 0} \int_0^{2\pi} \left(G\, \frac{\partial u}{\partial r} \right)_{r=\rho} \rho\, d\theta = 0,$$

we have

$$u(x_0, y_0) = \lim_{\rho \to 0} \int_0^{2\pi} \left(u \frac{\partial G}{\partial r} \right)_{r=\rho} d\theta = \int_{C+} gG \, ds + \frac{1}{L} \int_{C+} u \, ds$$

$$= \int_{C+} gG \, ds,$$

since $\int_{C+} u \, ds = 0$.

One way to prove existence of the solution of the Neumann problem is to first prove existence of the Green's function and then show that for $g(s)$ continuous on C and $\int_{C+} g(s) \, ds = 0$

$$u(x_0, y_0) = \int_{C+} g(s)G \, ds$$

is a solution of the Neumann problem. We shall do this for the special case where D is the open disk $|z| < R$, and C is the circle $|z| = R$.

It is not difficult to prove that

$$G(x, y, x_0, y_0) = -\frac{1}{2\pi} \ln r_1 - \frac{1}{2\pi} \ln r_2 + \frac{1}{2\pi} \ln R/\rho,$$

where $r_1 = |z - z_0|$ and $r_2 = |z - (R^2/\rho^2)z_0|$ and $\rho = \sqrt{x_0^2 + y_0^2}$, is the Green's function for the disk. We shall leave this as an exercise. In polar coordinates $z = re^{i\theta}$ and $z_0 = \rho e^{i\phi}$, and

$$G(r, \theta, \rho, \phi) = -\frac{1}{4\pi} \ln [r^2 + \rho^2 - 2r\rho \cos (\theta - \phi)]$$

$$- \frac{1}{4\pi} \ln \left[\frac{R^4}{\rho^2} + r^2 - 2r \frac{R^2}{\rho} \cos (\theta - \phi) \right] + \frac{1}{2\pi} \ln R/\rho.$$

Let $g(\theta)$ be continuous for $0 \leq \theta \leq 2\pi$, $g(0) = g(2\pi)$, and $\int_0^{2\pi} g(\theta) \, d\theta = 0$. Then

$$u(\rho, \phi) = \int_0^{2\pi} g(\theta)G(R, \theta, \rho, \phi)R \, d\theta$$

is the solution of the Neumann problem for the disk with boundary conditions $(\partial u/\partial \rho)_{\rho=R} = g(\theta)$ and $\int_0^{2\pi} u(R, \phi)d\phi = 0$. Clearly, for $\rho < R$ we may take the Laplacian under the integral sign. Hence, since $\nabla^2 G = 0$ for $\rho < R$

$$\nabla^2 u = \int_0^{2\pi} g(\theta)\nabla^2 GR \, d\theta = 0.$$

For $\rho < R$,

$$\frac{\partial u}{\partial \rho} = R \int_0^{2\pi} g(\theta) \frac{\partial G}{\partial \rho} \, d\theta$$

$$= \frac{R}{2\pi\rho} \int_0^{2\pi} g(\theta) \frac{R^2 - \rho^2}{R^2 + \rho^2 - 2R\rho \cos(\theta - \phi)} \, d\theta - \frac{R}{2\pi\rho} \int_0^{2\pi} g(\theta) \, d\theta$$

$$= \frac{R}{2\pi\rho} \int_0^{2\pi} g(\theta) \frac{R^2 - \rho^2}{R^2 + \rho^2 - 2R\rho \cos(\theta - \phi)} \, d\theta.$$

By a calculation similar to that used in Theorem 6.2.2, using the continuity of $g(\theta)$, we have

$$\lim_{\rho \to R} \frac{\partial u}{\partial \rho} = g(\phi).$$

Also,

$$\lim_{\rho \to R} u(\rho, \phi) = -\frac{R}{2\pi} \int_0^{2\pi} g(\theta) \ln 2R^2[1 - \cos(\theta - \phi)] \, d\theta$$

and it is not difficult to show that this integral exists. Finally, we can show, using Green's theorem applied to a domain between the circle of radius R and a small circle centered at (x_0, y_0) and the usual limiting argument as the radius of the small circle goes to zero, that

$$u(\rho, \phi) = \int_0^{2\pi} g(\theta) G(R, \theta, \rho, \phi) R \, d\theta - \frac{1}{2\pi} \int_0^{2\pi} u(R, \phi) \, d\phi.$$

Therefore,

$$\int_0^{2\pi} u(R, \phi) \, d\phi = 0.$$

Suppose D is a bounded simply connected domain bounded by a simple closed smooth contour C. Suppose $w = f(z)$ maps $D \cup C$ onto the closed disk $|w| \leq R$, and $f'(z) \neq 0$ in $D \cup C$. Under the mapping Laplace's equation is preserved. If the normal derivatives of a function are given on C, then they map as follows:

$$g = \nabla u \cdot \mathbf{N} = \frac{\partial u}{\partial x} n_x + \frac{\partial u}{\partial y} n_y$$

$$= \left(\frac{\partial u}{\partial \xi} \frac{\partial \xi}{\partial x} + \frac{\partial u}{\partial \eta} \frac{\partial \eta}{\partial x} \right) \cos(\alpha^* - \beta)$$

$$+ \left(\frac{\partial u}{\partial \xi} \frac{\partial \xi}{\partial y} + \frac{\partial u}{\partial \eta} \frac{\partial \eta}{\partial y} \right) \sin(\alpha^* - \beta),$$

where $f(z) = \xi + i\eta$, $\alpha^* = \arg \mathbf{N}^*$, and $\beta = \arg f'(z)$. Also $f'(z) = \xi_x + i\eta_x = \eta_y - i\xi_y = |f'(z)| (\cos \beta + i \sin \beta)$. Hence,

$$g = \frac{1}{|f'(z)|} \left[\frac{\partial u}{\partial \xi} \left(\frac{\partial \xi}{\partial x} \right)^2 \cos \alpha^* + \frac{\partial u}{\partial \xi} \left(\frac{\partial \eta}{\partial x} \right)^2 \cos \alpha^* \right.$$

$$\left. + \frac{\partial u}{\partial \eta} \left(\frac{\partial \eta}{\partial x} \right)^2 \sin \alpha^* + \frac{\partial u}{\partial \eta} \left(\frac{\partial \xi}{\partial x} \right)^2 \sin \alpha^* \right]$$

$$= |f'(z)| \left(\frac{\partial u}{\partial \xi} \cos \alpha^* + \frac{\partial u}{\partial \eta} \sin \alpha^* \right),$$

where $u_\xi \cos \alpha^* + u_\eta \sin \alpha^*$ is the normal derivative of u on C^* the image of C. Furthermore,

$$0 = \int_{C+} g(s)\, ds = \int g^*(s^*) |f'(z)|\, ds = \int g^*(s^*)\, ds^*.$$

Hence, the Neumann problem for D goes over into a Neumann problem for the disk $D^* = \{w \mid |w| \le R\}$. The conformal mapping technique for solving boundary value problems will be the subject of the next section.

The Green's function approach is also applicable to other types of boundary value problems of potential theory. We might, for example, prescribe the values of $\Delta u \cdot \mathbf{N} + ku$ for a harmonic function on the boundary of a domain D. Or we might prescribe the value of the harmonic function on part of the boundary and the value of its normal derivative on another part of the boundary. The reader can describe other boundary value problems with the appropriate conditions on a Green's function in terms of which the solution can be expressed.

Exercises 6.3

1. The Dirichlet problem for the upper half-plane is the following: $u(x, y)$ is harmonic in the open upper half-plane, continuous in the closed upper half-plane, $u(x, 0) = g(x)$, and u behaves for large values of r in such a way that

$$\lim_{R \to \infty} \int_0^\pi \left(u \frac{\partial u}{\partial r} \right)_{r=R} R\, d\theta = 0.$$

Show that the solution, if it exists, is unique.

2. Referring to Exercise 1, show that

$$G(x, y, x_0, y_0) = -\frac{1}{2\pi} \ln \sqrt{(x - x_0)^2 + (y - y_0)^2}$$

$$+ \frac{1}{2\pi} \ln \sqrt{(x - x_0)^2 + (y + y_0)^2}$$

is a Green's function for the half-plane problem in the sense that the solution, if it exists, can be expressed as

$$u(x_0, y_0) = \int_{-\infty}^{\infty} g(x) G_y(x, 0, x_0, y_0)\, dx$$

$$= \frac{y_0}{\pi} \int_{-\infty}^{\infty} \frac{g(x)}{(x - x_0)^2 + y_0^2}\, dx.$$

6.4. CONFORMAL MAPPING

We have already indicated how mappings by analytic functions can be useful in the solution of boundary value problems in potential theory. (See Examples 2.4.7, 6.1.2, 6.1.3, and 6.2.1.) However, to assume that under a mapping $w = f(z) = u(x, y) + iv(x, y)$ Laplace's equation is preserved at z_0, we need the further condition that $f'(z_0) \neq 0$. This is because

$$\left(\frac{\partial^2 \phi}{\partial x^2} + \frac{\partial^2 \phi}{\partial y^2}\right)_{z_0} = \left(\frac{\partial^2 \Phi}{\partial u^2} + \frac{\partial^2 \Phi}{\partial v^2}\right)_{w_0} |f'(z_0)|^2,$$

where $\Phi(u, v)$ is the image of $\phi(x, y)$, and $w_0 = f(z_0)$. (See Section 2.4.)

Recall also from Section 2.4 that if $w = f(z)$ has the property that it preserves angles both in magnitude and sense at z_0, then it is said to be *conformal*. We showed in Section 2.4 that if $f(z)$ is analytic at z_0 and $f'(z_0) \neq 0$, then the mapping is conformal at z_0. Furthermore, (see Theorem 4.4.6) if $w = f(z)$ is analytic at z_0 and $f'(z_0) \neq 0$, then at least locally it is invertable and the inverse is also conformal at w_0, that is, there exists some ϵ-neighborhood of z_0 throughout which the mapping is one-to-one, the inverse is analytic and its derivative does not vanish. However, for the purpose of solving boundary value problems local invertability is usually not enough. Even if we say that $f(z)$ is analytic and $f'(z) \neq 0$ in a domain D, this is not enough to guarantee that the transformation has a unique inverse throughout D. Consider, for example, $w = z^2$ in $D = \{z \mid 0 < |z| < 1\}$. Then $f'(z) = 2z$ does not vanish in D, but no single inverse exists in D since the function maps D onto $D' = \{w \mid 0 < |w| < 1\}$ twice.

For our purposes we shall be most interested in the following class of function:

Definition 6.4.1. A function $w = f(z)$ is said to be *simple* in a domain D if it is analytic and one-to-one in D; that is, if z_1 and z_2 are in D and $z_1 \neq z_2$ then $f(z_1) \neq f(z_2)$.

Theorem 6.4.1. *If $f(z)$ is simple in a domain D, then $f'(z) \neq 0$ in D.*

Proof: Assume that for some z_0 in D, $f'(z_0) = 0$. Then $f(z) - f(z_0)$ has a zero of order $n \geq 2$ at z_0. Since $f(z)$ is not constant there is a circle $|z - z_0| = \rho$ on which $f(z) - f(z_0)$ does not vanish and inside of which $f'(z)$ vanishes only at z_0. Let

$$m = \min_{|z-z_0|=\rho} |f(z) - f(z_0)|.$$

If $0 < |c| < m$, then by Rouche's theorem $f(z) - f(z_0) - c$ has n zeros inside the circle. But $f(z) - f(z_0) - c = 0$ cannot occur as a multiple zero inside the circle since $f'(z) = 0$ only at z_0. This implies that $f(z) = f(z_0) + c$ at two or more points, contradicting the one-to-one property of the function.

Corollary 6.4.1. If $f(z)$ is simple in a domain D, then it is conformal in D.

We see that simple functions have highly desirable properties as mappings. They preserve angles in magnitude and direction; they preserve Laplace's equation; they have a unique inverse which at every point is the same as the local inverse, and the inverse is a simple function. We also have the following desirable property of composite simple functions.

Theorem 6.4.2. *If $f(z)$ has range D' and is simple in a domain D and $F(w)$ is simple in D', then $F[f(z)]$ is simple in D.*

Proof: The analyticity of $F[f(z)]$ is obvious. Let z_1 and z_2 be in D. If $F[f(z_1)] = F[f(z_2)]$, then $f(z_1) = f(z_2)$ since F is simple. But if $f(z_1) = f(z_2)$, then $z_1 = z_2$, since f is simple.

The following two theorems give sufficient conditions for a function to be simple in a domain D:

Theorem 6.4.3. *Let C be a simple closed contour with interior D. Let $f(z)$ be analytic within and on C and take no value more than once on C. Then $f(z)$ is simple in D.*

Proof: The function $w = f(z)$ maps C onto a simple closed contour C' in the w plane. Let w_0 be any point in the w plane not on C'. Then

$$n = \frac{1}{2\pi i} \int_{C_+} \frac{f'(z)}{f(z) - w_0}\, dz = \frac{1}{2\pi i} \int_{C'} \frac{dw}{w - w_0}.$$

Now, the last integral is zero if w_0 is outside C' and is ± 1 if w_0 is inside C' depending on whether the positive direction on C corresponds to the positive or negative direction on C'. However, n cannot be negative since the first integral gives the number of zeros of $f(z) - w_0$ inside C. Therefore, $n = 1$ if w_0 is inside C'. This proves that $f(z) = w_0$ has just one solution when w_0 is inside C', that $f(z)$ is simple in D, that D maps simply on D', the interior of C', and that the positive direction on C' corresponds to the positive direction on C.

Theorem 6.4.4. *Let $f_n(z)$ be simple in a domain D for $n = 1, 2, 3, \ldots$. Let $f_n(z)$ converge normally to $f(z)$ in D, that is, uniformly in compact subsets of D. Then $f(z)$ is either constant or simple in D.*

Proof: That $f(z)$ is analytic in D follows from Theorem 4.2.3. Assume that $f(z)$ is not constant and that z_1 and z_2 are any two distinct points in D. If $w_0 = f(z_1) = f(z_2)$, then there are disjoint disks $|z - z_1| \leq \rho_1$ and $|z - z_2| \leq \rho_2$ lying in D on whose boundaries $f(z) - w_0$ does not vanish. Let $m = \min |f(z) - w_0|$ on the boundaries of the disks and let n be so large that $|f(z) - f_n(z)| < m$ on the disks. By Rouché's theorem,

$$f_n(z) - w_0 = f(z) - w_0 + f_n(z) - f(z)$$

has as many zeros in the interior of the disks as does $f(z) - w_0$, namely, two. This implies that $f_n(z)$ is not simple in D since it takes on the value w_0 at least twice. Therefore, $f(z)$ is either simple or constant. That it can be constant is indicated by the example $f_n(z) = z/n$, where $\lim\limits_{n \to \infty} f_n(z) = 0$.

As we indicated in Sections 6.2 and 6.3, one of the central problems in the theory of conformal mapping is to find a simple function $w = f(z)$ which maps a given simply connected domain onto the unit disk $|w| < 1$. We shall prove (Riemann mapping theorem) that such a mapping exists under rather general hypotheses. In preparation for this theorem we shall need the following theorems:

Theorem 6.4.5. *Let $w = f(z)$ be simple in the unit disk $|z| \leq 1$, which it maps onto the unit disk $|w| \leq 1$ so that the origin maps into the origin and a given direction at the origin is preserved. Then $f(z)$ is the identity z.*

Proof: We have $|f(z)| = 1$, when $|z| = 1$ and $f(0) = 0$. By Schwarz's lemma $|w| = |f(z)| \leq |z|$. Applying the same argument to the inverse, we have $|z| \leq |w|$. Hence, $|f(z)| = |z|$ or $f(z) = e^{i\alpha}z$. Then $\arg w = \arg z + \alpha$, and since a given direction at the origin must be preserved $\alpha = 0$. Therefore, $w = f(z) = z$.

Theorem 6.4.6. *Let $w = f(z)$ be simple in the unit disk $|z| \leq 1$ which it maps onto the unit disk $|w| \leq 1$. Then $f(z)$ is a linear fractional transformation (see Section 2.4).*

Proof: In Exercise 2.4.2, we saw that

$$w = e^{i\alpha} \frac{z - \beta}{\bar{\beta}z - 1}$$

is a simple mapping of the unit disk onto the unit disk which takes $z = 0$ into $\beta e^{i\alpha}$, $|\beta| < 1$. Hence, the inverse takes $\beta e^{i\alpha}$ into the origin. Now,

$w = f(z)$ maps the unit disk onto the unit disk with $w_0 = f(0) = \beta e^{i\alpha}$. Also,

$$\frac{w - \beta e^{i\alpha}}{\bar{\beta}w - e^{i\alpha}}$$

maps w_0 into the origin. Hence,

$$\frac{f(z) - \beta e^{i\alpha}}{\bar{\beta}f(z) - e^{i\alpha}}$$

maps the origin into the origin. By the previous theorem

$$\frac{f(z) - \beta e^{i\alpha}}{\bar{\beta}f(z) - e^{i\alpha}} = e^{i\gamma}z.$$

Therefore,

$$w = f(z) = e^{i\alpha}\frac{e^{i\gamma}z - \beta}{\bar{\beta}e^{i\gamma}z - 1}$$

which is a linear fractional transformation.

The Riemann mapping theorem deals with the question of existence of a simple mapping which maps a simply connected domain D onto a simply connected domain D'. Actually, it is enough to prove that a given simply connected domain D can be mapped simply onto the unit disk $|w| < 1$. This is because if D' is in the ζ plane and there exists a simple mapping of D' onto the unit disk by $w = g(\zeta)$ and a simple mapping of D onto the unit disk by $w = f(z)$, then

$$\zeta = g^{-1}[f(z)]$$

is a simple mapping of D onto D'. Therefore, in the actual proof we will take D' to be the unit disk $|w| < 1$.

Obviously, beside the simple connectivity of D there must be some restriction on the boundary of D. Suppose, for example that D has a single boundary point z_0. We may as well assume that $z_0 = \infty$, because a preliminary transformation $\zeta = 1/(z - z_0)$ will make it so. Therefore, if the theorem were true for a domain D with a single boundary point at infinity, then $w = f(z)$ is analytic for all z in the unextended plane and $|w| = |f(z)| < 1$. However, by Liouville's theorem $f(z)$ is constant and therefore not simple. Actually, we shall prove the theorem if D is simply connected and has at least two boundary points.

Theorem 6.4.7. *Riemann Mapping Theorem. Let D be a simply connected domain with at least two boundary points. Then there exists a simple function $w = f(z)$ which maps D onto the unit disk $|w| < 1$. If we specify that a given point z_0 in D maps into the origin and a given direction at z_0 is mapped into a given direction at the origin, then the mapping is unique.*

Proof: In the proof we shall make use of our work on normal families of functions of Section 4.2. Essentially, we shall construct a compact family of functions and shall find the desired mapping function as the limit of a certain sequence of functions in this family.

Let G be a family of functions $g(z)$ which are simple in D and such that $|g(z)| \leq 1$ for all z in D. We shall first have to show that G is not empty. Let a and b be boundary points of D and form the function

$$h(z) = \sqrt{\frac{z-a}{z-b}}.$$

This can be done by starting from some analytic element of a given branch of the square root inside D and then by continuing it throughout D by analytic continuation. This process must define a single-valued analytic function in D by the monodromy theorem and the fact that D is simply connected. Now, $h(z)$ is simple, for suppose $h(z_1) = h(z_2)$, then $[h(z_1)]^2 = [h(z_2)]^2$ and

$$\frac{z_1 - a}{z_1 - b} = \frac{z_2 - a}{z_2 - b},$$

which implies that $z_1 = z_2$. Let D^* be the image of D under $h(z)$. If β is in D^* then $-\beta$ is not in D^*. If it were

$$\beta = \sqrt{\frac{z_1 - a}{z_1 - b}}, \qquad -\beta = \sqrt{\frac{z_2 - a}{z_2 - b}},$$

and

$$\beta^2 = \frac{z_1 - a}{z_1 - b} = \frac{z_2 - a}{z_2 - b},$$

which implies that $z_1 = z_2$. Let w_0 be in D^*. Since D^* is a domain it contains an ϵ-neighborhood $|w - w_0| < \epsilon$. Also, since w_0 and $-w_0$ cannot both be in D^* the ϵ-neighborhood $|w + w_0| < \epsilon$ is not in D^*. Therefore, $|h(z) + w_0| \geq \epsilon$, for all z in D. Now, the function

$$g_1(z) = \frac{\epsilon}{h(z) + w_0}$$

is simple in D and $|g_1(z)| \leq 1$ for all z in D. Therefore, g_1 is in G. This shows that G is not empty.

Now, G is a uniformly bounded family and therefore is a normal family of analytic functions. However, G is not compact because it contains constant functions which are not simple but which may be limits of sequences of simple functions. Let G' be the subfamily of functions in G such that $|g'(z_0)| \geq |g_1'(z_0)|$, where z_0 is in D. G' is not empty since, obviously, $g_1(z)$ is in G'. Also, G' is compact since we have excluded constant functions, since the derivative of a constant function vanishes and $|g_1'(z_0)| > 0$ since $g_1(z)$ is simple.

We now assert that $|g'(z_0)|$ attains a maximum value M in G', that is, there exists a function $f(z)$ in G' such that $|f'(z_0)| = M \geq |g'(z_0)|$ for all other $g(z)$ in G'. This can be shown as follows. Let $F[g] = |g'(z_0)|$. (F is called a functional of g.) Then F is continuous in the following sense: if $\{g_n(z)\}$ is a sequence of functions in G' with a limit $g(z)$, then $\lim_{n \to \infty} F[g_n] = F[g]$. The collection of positive real numbers $F[g]$ for all g in G' is a set of real numbers with a least upper bound M (which for all we know may be ∞). Therefore, there is a sequence $\{g_n(z)\}$ such that $\lim_{n \to \infty} F[g_n] = M$. However, the family G' is normal and compact. Therefore, $\{g_n(z)\}$ has a subsequence $\{g_{n_k}(z)\}$ with a limit $f(z)$ in G', and hence

$$F[f] = \lim_{n_k \to \infty} F[g_{n_k}] = M.$$

But $F[f]$ must be finite and therefore M is finite and $M \geq F[g]$ for all g in G'.

Finally, we assert that $w = f(z)$ maps D onto the disk $|w| < 1$. First, we show that $f(z_0) = 0$, since $|f(z_0)| < 1$, because $|f(z)|$ cannot attain its maximum at an interior point of D, and

$$f_1(z) = \frac{f(z) - f(z_0)}{\overline{f(z_0)}f(z) - 1}$$

is also in G'. However,

$$|f_1'(z_0)| = \frac{|f'(z_0)|}{1 - |f(z_0)|^2} \geq |f'(z_0)|$$

and this implies that $|f_1'(z_0)| > M$ unless $f(z_0) = 0$. Suppose ω, such that $|\omega| < 1$, is a value not taken by $f(z)$ in D. Then

$$p(z) = \sqrt{\frac{\omega - f(z)}{1 - \overline{\omega}f(z)}}$$

is simple in D, for if $p(z_1) = p(z_2)$ for z_1 and z_2 in D

$$\frac{\omega - f(z_1)}{1 - \overline{\omega}f(z_1)} = \frac{\omega - f(z_2)}{1 - \overline{\omega}f(z_2)}$$

implying that $f(z_1) = f(z_2)$ and $z_1 = z_2$, since f is simple in D. Also, $|p(z)|^2 = |\omega - f(z)|/|1 - \overline{\omega}f(z)| \leq 1$ in D. Therefore, $p(z)$ is in G. Furthermore,

$$g(z) = \frac{p(z) - p(z_0)}{1 - \overline{p(z_0)}p(z)}$$

is also in G. However,

$$g'(z_0) = -\frac{1 + |\omega|}{2\sqrt{\omega}} f'(z_0),$$

which implies that $|g'(z_0)| > |f'(z_0)|$, since

$$1 + |\omega| = 2\sqrt{|\omega|} + (1 - \sqrt{|\omega|})^2 > 2\sqrt{|\omega|}.$$

This contradicts the fact that $|f'(z_0)| = M \geq |g'(z_0)|$, proving that $w = f(z)$ takes on all values in D'.

The uniqueness is proved as follows. Suppose $f_1(z)$ and $f_2(z)$ map D onto the disk $|w| < 1$ so that z_0 maps into the origin and a given direction at z_0 maps into the same direction at the origin. Then $f_1[f_2^{-1}]$ maps the disk onto the disk with the origin going into the origin and a direction at the origin preserved. By Theorem 6.4.5, $f_1[f_2^{-1}]$ is the identity. Hence, $f_1 = f_2$. This completes the proof of Theorem 6.4.7.

For our purposes, the Riemann mapping theorem is inadequate from two points of view. In the first place, it is strictly an existence theorem which asserts the existence of a mapping function but does not tell us how to construct the mapping. There is, however, an iteration scheme† for constructing the desired mapping. Also, the Schwarz-Christoffel method, which we study in the next section, gives a more or less general scheme for mapping domains bounded by straight line segments onto a half-plane which can in turn be mapped onto the unit disk. In the second place, the Riemann mapping theorem does not adequately deal with the boundaries of the domains. For example, if $w = f(z)$ maps D conformally onto D^* in what way do the boundary points of D correspond to the boundary points of D^*? We state without proof the following theorem:

Theorem 6.4.8. *Let D and D^* be domains bounded by simple closed contours C and C^*. Then the conformal mcp of D onto D^* is continuous in $D \cup C$ and establishes a one-to-one correspondence between C and C^*.*

The proof of this theorem can be found in Nehari, Z., *Conformal Mapping*, New York: McGraw-Hill Book Company, Inc., 1952, pp. 179–181.

Exercises 6.4

1. Prove that the function $f(z) = z + a_2z^2 + a_3z^3 + \cdots$ is simple for $|z| < 1$ if $\sum_{k=2}^{\infty} k|a_k| \leq 1$.

2. Let $f(z)$ be simple within and on a simple closed contour C. Let the interior of C be D and let D map onto D^* and C onto C^*, the boundary of D^*.

(a) Prove that the length of C^* is $\int_C |f'(z)| \, |dz|$;

(b) Prove that the area of D^* is $\iint_D |f'(z)|^2 \, dx \, dy$;

(c) If C is a circle of radius R centered at the origin, prove that the area of $D^* \geq \pi|f'(0)|^2R^2$.

3. Prove that $w = 3z + z^2$ is simple on the disk $|z| < 1$.

† See Wilf, H. S., *Mathematics for the Physical Sciences*. New York: John Wiley and Sons, Inc., 1962, pp. 202–203.

6.5. THE SCHWARZ-CHRISTOFFEL TRANSFORMATION

The Riemann mapping theorem is an existence theorem which tells us that under quite general conditions there is a simple function which maps a given simply connected domain onto another simply connected domain, or what is equivalent, onto the unit disk. However, it does·not tell us how to find the desired mapping. In this section, we shall consider the problem of finding specific mappings which will have a variety of applications in the solution of boundary value problems in potential theory. We shall consider a number of examples leading up to the Schwarz-Christoffel transformation which maps domains with straight line or straight line segments as boundaries onto the upper half-plane, which can in turn be mapped onto the unit disk.

EXAMPLE 6.5.1. Find a simple mapping which maps the sector $0 \leq \arg z \leq \alpha \leq 2\pi$, $|z| \geq 0$, onto the upper half-plane $\text{Im}(w) \geq 0$ so that the origin maps into the origin. Clearly, $w = z^{\pi/\alpha}$ does the required job since $\arg w = (\pi/\alpha) \arg z$, and $|w| = |z|^{\pi/\alpha}$.

EXAMPLE 6.5.2. Find a simple mapping which maps the sector $0 \leq \arg (z - z_0) \leq \alpha \leq 2\pi$, $|z - z_0| \geq 0$, onto the upper half-plane $\text{Im}(w) \geq 0$, so that z_0 maps into w_1 (real). Clearly, $w = w_1 + (z - z_0)^{\pi/\alpha}$ does the required job since $\arg (w - w_1) = (\pi/\alpha) \arg (z - z_0)$ and $|w - w_0| = |z - z_0|^{\pi/\alpha}$.

EXAMPLE 6.5.3. Find a simple mapping which maps the sector $\beta \leq \arg (z - z_0) \leq \alpha + \beta \leq 2\pi + \beta$, $|z - z_0| \geq 0$, onto the upper half-plane $\text{Im}(w) \geq 0$, so that z_0 maps into w_1 (real). We can solve the problem by a translation $\zeta_1 = z - z_0$, a rotation $\zeta_2 = e^{-i\beta}\zeta_1 = e^{-i\beta}(z - z_0)$, and then one application of the result of Example 6.5.2, that is,

$$w = w_1 + e^{-i\beta\pi/\alpha}(z - z_0)^{\pi/\alpha}.$$

In Example 6.5.3, we have been able to map a sector in an arbitrary position onto the upper half-plane with the vertex mapping into any given point on the real w axis. Note that the inverse is given by

$$z = z_0 + e^{i\beta}(w - w_1)^{\alpha/\pi}$$

and

$$\frac{dz}{dw} = (\alpha/\pi)e^{i\beta}(w - w_1)^{(\alpha/\pi)-1}.$$

Let w be on the real axis and let us move the point to the right by an amount $dw > 0$ so that $\arg dw = 0$. Then in this small change

$$\arg dz = \beta + [(\alpha/\pi) - 1] \arg (w - w_1).$$

Now when w is to the left of w_1, $\arg (w - w_1) = \pi$ and $\arg dz = \beta + \alpha - \pi$. This shows that in this displacement z changes in such a way that it moves along a line which makes an angle of $\alpha + \beta - \pi$ with the x axis. This holds true until w reaches w_1. Then $\arg dz = \beta$ which means that as w passes through w_1, z undergoes an abrupt change of direction of $\pi - \alpha$. See Figure 6.5.1.

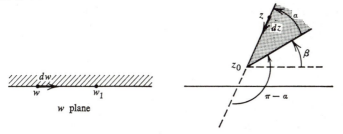

Figure 6.5.1.

Of course, we knew this already, but looking at the mapping in this new way lays the ground work for much more general situations.

EXAMPLE 6.5.4. Find a simple mapping which maps the half-strip $S = \{z \mid \text{Im}(z) \geq 0, -\pi/2 \leq \text{Re}(z) \leq \pi/2\}$ onto the upper half-plane so that $\pi/2$ maps into 1 and $-\pi/2$ maps into -1. Referring back to Example 6.5.3, we shall begin by writing an expression for dz/dw which will have the property that $\arg dz$ will be constant as w moves along the negative real axis from left to right, has an abrupt change of $\pi/2$ as w passes through -1, remains constant as w moves from -1 to 1, has an abrupt change of $\pi/2$ as w passes through 1, and remains constant as w moves from 1 to the right along the positive real axis. The following expression for dz/dw will have these properties, namely

$$\frac{dz}{dw} = A(w + 1)^{-1/2}(w - 1)^{-1/2} = A(w^2 - 1)^{-1/2}.$$

We see that

$$\arg dz = \arg dw + \arg A - \tfrac{1}{2}\arg(w + 1) - \tfrac{1}{2}\arg(w - 1),$$

verifying the stated properties. Here, A is a constant which we can use to rotate the entire strip if necessary and also expand or contract it. Upon integrating dz/dw we will introduce a constant of integration which can be used to translate the entire figure if necessary. This freedom to expand, contract, rotate, and translate is very important, since the built in properties at the corners do not, by any stretch of the imagination, determine precisely the desired mapping. We integrate to obtain

$$z = A \int_0^w (\omega^2 - 1)^{-1/2} \, d\omega + B.$$

Now, $(w^2 - 1)^{-1/2}$ is the derivative of $\sin^{-1} w$. Hence, we can write

$$z = A \sin^{-1} w + B,$$
$$w = \sin [(1/A)z - B].$$

Finally, we require that $\pm\pi/2$ map into ± 1, respectively. Hence,

$$\sin\left[(1/A)\frac{\pi}{2} - B\right] = 1,$$

$$\sin\left[(1/A)\frac{\pi}{2} + B\right] = 1.$$

Let $C = 1/A = c_1 + ic_2$, $B = b_1 + ib_2$. Then

$$\sin(c_1\pi/2 - b_1)\cosh(c_2\pi/2 - b_2) = 1,$$
$$\sinh(c_2\pi/2 - b_2)\cos(c_1\pi/2 - b_1) = 0,$$
$$\sin(c_1\pi/2 + b_1)\cosh(c_2\pi/2 + b_2) = 1,$$
$$\sinh(c_2\pi/2 + b_2)\cos(c_1\pi/2 + b_1) = 0.$$

A solution of these equations is not unique. However, one solution is $c_1 = 1, c_2 = b_1 = b_2 = 0$. This gives the result

$$w = \sin z,$$

which the reader can easily verify as a mapping with the required properties.

EXAMPLE 6.5.5. Find a simple mapping which maps the strip $S = \{z \mid 0 \leq \text{Im}(z) \leq \pi\}$ onto the upper half-plane, $\text{Im}(w) \geq 0$, so that the origin in the w plane corresponds to the left-hand end of the strip and the origin in the z plane maps into $w = 1$. We can treat this problem as a slight modification of the simple sector problem. We wish arg dz to remain constant as w moves from left to right along the negative real axis up to the origin, undergo an abrupt change of π as w passes through the origin, and then remain constant as w moves along the positive real axis. For this reason we write

$$\frac{dz}{dw} = Aw^{-1}.$$

Integrating this, we have

$$z = A \int_1^w \omega^{-1} d\omega + B = A \operatorname{Log} w + B,$$
$$w = e^{(1/A)(z-B)} = e^{Cz+D} = Ke^{Cz}.$$

Since $z = 0$ maps into $w = 1$, $K = 1$. Also, since w is to be real when z is real, C must be real. In order that

$$\lim_{x \to \infty} w = \lim_{x \to \infty} e^{Cx} = \infty,$$

C must be positive. When $z = x + i\pi$

$$w = e^{Cx} (\cos C\pi + i \sin C\pi)$$

must be real and negative. Therefore, C must be an odd integer. For definiteness, let us take $C = 1$. Then

$$w = e^z$$

The reader can easily verify that this mapping has all the desired properties.

We are now ready to discuss the Schwarz-Christoffel transformation. Suppose we wish to map the interior of an n-sided polygon with interior angles α_1, $\alpha_2, \ldots, \alpha_n$ onto the upper half-plane so that the vertices P_1, $P_2, \ldots, P_n$ map into the points $-\infty < u_1 < u_2 < \cdots < u_n < \infty$ on the real axis. We start with an expression for dz/dw with the property that $\Delta \arg dz = \pi - \alpha_k$ when w passes through u_k, that is,

$$\frac{dz}{dw} = A \prod_{k=1}^{n} (w - u_k)^{(\alpha_k/\pi - 1)}.$$

Integrating, we have

$$z = A \int_{w_0}^{w} \prod_{k=1}^{n} (\omega - u_k)^{(\alpha_k/\pi - 1)} \, d\omega + B.$$

At this point, we have to decide what constants we have to assign and what information we have to determine them. First, the α's are determined by the interior angles of the given polygon. The constant B can be used to translate the polygon to any location in the z plane. The constant A can be used to expand or contract and rotate the polygon. In other words, if without the A and the B we have mapped a similar polygon, then there is enough freedom of choice in the A and the B to map the actual polygon. In the case of a triangle, we have angle-angle-angle symmetry and therefore the images of the vertices, u_1, u_2, u_3 can be assigned arbitrarily. In the case of a quadrilateral, the interior angles are not enough to determine similar figures. In addition we must specify the ratio of the lengths of two adjacent sides. This means that only three of the four constants u_1, u_2, u_3, u_4 can be assigned arbitrarily. The fourth one must be picked so that the function

$$\int_{w_0}^{w} \prod_{k=1}^{4} (\omega - u)^{(\alpha_k/\pi - 1)} \, d\omega$$

maps a quadrilateral similar to the given one. In general, for an n-sided polygon, in addition to the interior angles, one must specify the ratios of $n - 3$ pairs of adjacent sides. This means that only three of the u_k's can be assigned arbitrarily. The other $n - 3$ must be determined to achieve symmetry.

Finally, we must settle the question of whether the polygon mapped is really closed. This is a matter of whether

$$\int_{w_0}^{-\infty} \prod_{k=1}^{n} (\omega - u_k)^{(\alpha_k/\pi - 1)} \, d\omega \quad \text{is equal to} \quad \int_{w_0}^{\infty} \prod_{k=1}^{n} (\omega - u_k)^{(\alpha_k/\pi - 1)} d\omega.$$

Let w_0 be a point in the upper half-plane. Then since the sum of the exterior angles of the polygon is 2π, $\sum_{k=1}^{n} (\pi - \alpha_k) = 2\pi$. Hence,

$$\sum_{k=1}^{n} \frac{\alpha_k - \pi}{\pi} = -2$$

and the integral behaves as $1/\omega^2$ for large $|\omega|$. Therefore both improper integrals exist and they have the same value since the integral around a large semicircle in the upper half-plane goes to zero as the radius of the semicircle goes to infinity.

EXAMPLE 6.5.6. Find a mapping which maps the interior of the equilateral triangle with vertices at $z = -1$, $z = 1$, and $z = \sqrt{3}i$ onto the upper half-plane so that the vertices map into $w = -1, 0, 1$, respectively. By the above discussion, we have

$$\frac{dz}{dw} = A(w + 1)^{-2/3} w^{-2/3} (w - 1)^{-2/3},$$

$$\frac{dz}{dw} = A[w(w^2 - 1)]^{-2/3},$$

$$z = A \int_{i}^{w} \frac{1}{[\omega(\omega^2 - 1)]^{2/3}} \, d\omega + B.$$

Now, A and B can be determined so that the images of the given vertices are properly fixed. For example,

$$-1 = A \int_{i}^{-1} \frac{1}{[\omega(\omega^2 - 1)]^{2/3}} \, d\omega + B,$$

$$1 = A \int_{i}^{0} \frac{1}{[\omega(\omega^2 - 1)]^{2/3}} \, d\omega + B,$$

$$\sqrt{3}\, i = A \int_{i}^{1} \frac{1}{[\omega(\omega^2 - 1)]^{2/3}} \, d\omega + B.$$

The three integrals are improper but they are no worse than something of the form $\int_{0}^{1} x^{-2/3} \, dx$ and therefore they exist as finite complex numbers.

The situation is somewhat different if we specify that the point at infinity be the image of one of the vertices. In this case, we have one less u_k on the real axis as an image point of a vertex. Hence, if the nth vertex with interior angle α_n is the image of ∞

$$\frac{dz}{dw} = A \prod_{k=1}^{n-1} (w - u_k)^{(\alpha_k/\pi - 1)}.$$

Now,

$$\sum_{k=1}^{n-1} \frac{\alpha_k - \pi}{\pi} = -2 - \left(\frac{\alpha_n - \pi}{\pi}\right) = -1 - \frac{\alpha_n}{\pi}$$

and since $0 < \alpha_n < \pi$

$$\int_{w_0}^{\infty} \prod_{k=1}^{n-1} (\omega - u_k)^{(\alpha_k/\pi - 1)} \, d\omega = \int_{w_0}^{-\infty} \prod_{k=1}^{n-1} (\omega - u_k)^{(\alpha_k/\pi - 1)} \, d\omega.$$

Therefore, as before, the polygon closes on itself, that is, is joined at the nth vertex. Clearly, the change in the argument of dz from the positive real axis to the negative real axis in the w plane is $\pi - \alpha_n$. In this case, since we have already fixed the image of one of the vertices we can only assign arbitrarily two more images, which means that $n - 2$ of the $u_1, u_2, \ldots, u_{n-1}$ must be determined.

EXAMPLE 6.5.7. Find a mapping which maps the interior of the isosceles right triangle with vertices $z = -1, 1, i$ onto the upper half-plane so that the vertices map into $w = -1, 1$, and ∞, respectively. By the above discussion, we have

$$\frac{dz}{dw} = A(w + 1)^{-3/4}(w - 1)^{-3/4}$$

$$= A(w^2 - 1)^{-3/4},$$

$$z = \int_0^w \frac{1}{(\omega^2 - 1)^{3/4}} \, d\omega + B,$$

where A and B are determined by the equations

$$-1 = A \int_0^{-1} \frac{1}{(\omega^2 - 1)^{3/4}} \, d\omega + B,$$

$$1 = A \int_0^1 \frac{1}{(\omega^2 - 1)^{3/4}} \, d\omega + B,$$

$$i = A \int_0^{\infty} \frac{1}{(\omega^2 - 1)^{3/4}} \, d\omega + B.$$

All the integrals are improper, but they exist as finite complex numbers.

Exercises 6.5

1. Find a function which maps the strip $S = \{z \mid 0 < \text{Im}(z) < \pi\}$ into the interior of the unit circle. Write down a solution of the Dirichlet problem for this strip.

2. Find a function which maps the strip

$$S = \{z \mid 0 \le \text{Re}(z) \le \pi, \text{Im}(z) \ge 0\}$$

onto the upper half-plane so that the origin maps into the origin and π maps into 1.

3. Find a function which maps the upper half-plane D onto D' as shown in Figure 6.5.2. Find the velocity potential for a flow over the "falls" in the w plane.

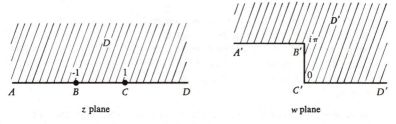

Figure 6.5.2.

4. Find a function which maps the upper half-plane D onto D' as shown in Figure 6.5.3.

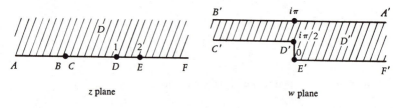

Figure 6.5.3.

6.6. FLOWS WITH SOURCES AND SINKS

The equations we derived in Section 6.1 for the flow of an incompressible, irrotational, homogeneous fluid were based on the assumption that there were no sources or sinks of fluid in the region of the flow. This meant that there were no singularities of the complex velocity potential in the region of the flow. In this section, we will assume a slightly more general situation, that is, that fluid may be introduced (a source) or removed (a sink) at some points on the boundary.

EXAMPLE 6.6.1.   Find the velocity potential of the flow in the upper half-plane if the real axis is considered a rigid boundary except at the origin where fluid is introduced at a constant rate through a slit. Consider the function $w = A \log z = A (\ln |z| + i \arg z)$, where $-\pi/2 < \arg z < 3\pi/2$, and A is real. This function is analytic in the upper half-plane. Let $w = \phi(x, y) + i\psi(x, y)$; then $\psi(x, y) = A \arg z$ is constant for constant $\arg z$ and therefore the stream lines are radial lines. The velocity is

$$v_x = \frac{\partial \phi}{\partial x} = \frac{Ax}{x^2 + y^2}, \qquad v_y = \frac{\partial \phi}{\partial y} = \frac{Ay}{x^2 + y^2}.$$

The fluid flowing across the semicircular arc, $|z| = R$, $0 \leq \arg z \leq \pi$, per unit time is

$$\int_0^\pi \rho \mathbf{v} \cdot \mathbf{N} \, ds = \rho A \int_0^\pi d\theta = \pi \rho A,$$

which is constant and is independent of R. This is precisely what should be the case if fluid is introduced at the origin at a constant rate and is allowed to flow out in all directions.

Definition 6.6.1.   The complex velocity potential of a source of strength A at the point z_0 is $w(z) = A \log (z - z_0)$ where A is a positive real constant. The complex velocity potential of a sink of strength A at the point z_0 is $w(z) = -A \log (z - z_0)$, where A is a positive real constant.

EXAMPLE 6.6.2.   Find the flow in a straight channel of width h due to a source of strength A at some finite point on its boundary. Without loss of generality we can assume that the channel is the strip $S = \{z \mid 0 < \text{Im}(z) < h\}$ and that the source is at the origin. In this case however, we cannot use the velocity potential of Example 6.6.1 because the fluid must stay in the channel and cannot flow outward in all directions. By symmetry, the fluid will split, half flowing down the channel to the right and the other half flowing to the left. If we map the strip into the upper half-plane by the transformation

$$\zeta = e^{\pi z/h},$$

we will have a half-plane problem but the source will have moved and we have to introduce a sink in an appropriate place to take into account the flow in the channel. The origin, $z = 0$, corresponds to $\zeta = 1$. Under the transformation a semicircle $|z| = R < h$, $0 \leq \arg z \leq \pi$ maps into an arc in the upper half of the ζ plane passing above $\zeta = 1$ from the positive real ζ axis to the right of $\zeta = 1$ to the left of $\zeta = 1$. It can be shown that the fluid crossing this curve per unit time is $\pi \rho A$ from a source of strength A in the ζ plane at $\zeta = 1$. Hence, a source of strength A at $\zeta = 1$ is mapped into a source of strength A at $z = 0$ and vice versa. Also, since the left-hand end of the channel maps into $\zeta = 0$ it is necessary to place a sink of strength $A/2$ at $\zeta = 0$ to provide for the loss of fluid which flows away from the origin down the left-hand side of the channel. Therefore, the mapped problem for the upper half ζ plane has a source of strength A at $\zeta = 1$ and a sink of strength $A/2$ at $\zeta = 0$. The required velocity potential is

$$
\begin{aligned}
w(\zeta) &= A \log (\zeta - 1) - A/2 \log \zeta \\
&= A \log (e^{\pi z/h} - 1) - A/2 \log e^{\pi z/h} \\
&= A \log (e^{\pi z/h} - 1) - (\pi A/2h)z.
\end{aligned}
$$

EXAMPLE 6.6.3. Find the flow in the channel with an offset of Exercise 6.5.4, if on the far left the fluid has a uniform velocity of U to the right. By the Schwarz-Christoffel transformation one finds a transformation $\zeta = f(z)$ which maps the channel onto the upper half-plane so that the left-hand end maps into the origin. Therefore, we have a half-plane problem to solve with a source of the appropriate strength at the origin. The fluid crossing a given vertical line in the left-hand channel per unit time is $\rho U \pi/2$. Therefore, we must put a source of strength $U/2$ at the origin in the ζ plane, and the required velocity potential is

$$
w(z) = \frac{U}{2} \log \zeta = \frac{U}{2} \log f(z).
$$

There is a special potential which is useful to know about. This is the potential of the *dipole* or *doublet*. Consider a source of strength A at $ae^{i\alpha}$ and a sink of strength A at the origin. The potential of this combination is

$$
\begin{aligned}
w &= A \log (z - ae^{i\alpha}) - A \log z \\
&= -ae^{i\alpha} A \frac{\log (z - ae^{i\alpha}) - \log z}{-ae^{i\alpha}} \\
&= -aAe^{i\alpha} \frac{\log (z + \Delta z) - \log z}{\Delta z},
\end{aligned}
$$

where $\Delta z = -ae^{i\alpha}$. Now, we let a approach zero and A approach infinity in such a way that aA is constant. If $aA = M$, then

$$
w = -e^{i\alpha} \lim_{\Delta z \to 0} \frac{\log (z + \Delta z) - \log z}{\Delta z} = \frac{-e^{i\alpha} M}{z}.
$$

This is the potential of a doublet of strength M in the direction α at the origin.

The potential of a doublet of strength M in the direction α at z_0 would be $w = -e^{i\alpha}M/(z - z_0)$. Consider the velocity potential, assuming that $\theta = \arg(z - z_0)$

$$\begin{aligned}
\phi(x, y) &= -\text{Re}[e^{i\alpha}M/(z - z_0)] \\
&= -\text{Re}\left[\frac{e^{i\alpha}Me^{-i\theta}}{|z - z_0|}\right] \\
&= \frac{-M\cos(\theta - \alpha)}{|z - z_0|} \\
&= \frac{-M|z - z_0|}{|z - z_0|^2}[\cos\theta\cos\alpha + \sin\theta\sin\alpha] \\
&= -M\frac{x - x_0}{|z - z_0|^2}\cos\alpha - M\frac{y - y_0}{|z - z_0|^2}\sin\alpha \\
&= -M\frac{d}{dn}\log|z - z_0|,
\end{aligned}$$

where d/dn stands for the directional derivative in the direction α.

EXAMPLE 6.6.4. Show that the potential of a flow past a unit cylinder is that of a uniform flow of velocity U plus a doublet at the origin of strength U in the direction π. We have already seen (Example 6.1.1) that the potential is

$$w = U\left(z + \frac{1}{z}\right) = Uz - \frac{e^{i\pi}U}{z}.$$

The potentials of sources and sinks and of doublets have their counterparts in electrostatics, magnetostatics, heat transfer, and so on. We have already seen (Section 6.1) that the potential of a uniform line charge perpendicular to the xy plane through the point z_0 is of the form

$$A\ln|z - z_0| = \text{Re}[A\log(z - z_0)].$$

A point heat source from which heat is transported radially in all directions will have the same potential. A magnetic dipole at z_0 will have the potential $w = -\text{Re}[e^{i\alpha}M/(z - z_0)]$. In the next section we shall consider continuous distributions of point charges, point sources, and dipoles.

Exercises 6.6

1. Draw a few of the stream lines for the flows of Examples 6.6.2 and 6.6.3.

2. Draw a few of the stream lines of a doublet.

3. Prove that under a conformal transformation a source of strength A is mapped into a source of strength A, that is, if $2\pi\rho A$ is the amount of fluid crossing a circle of radius R centered on the source in the z plane then this is the amount crossing the image of the circle in the w plane.

4. Prove that under a conformal transformation $w = f(z)$ a doublet of strength M at z_0 is mapped into a doublet of strength $M|f'(z_0)|$ at $w_0 = f(z_0)$. Is the direction of the doublet changed?

6.7. VOLUME AND SURFACE DISTRIBUTIONS

As we have pointed out in the last section, the complex potential of a line source or sink is $w = A \log (z - z_0)$, where A is real, positive for a source and negative for a sink. If we have a distribution of sources or sinks on a sheet perpendicular to the xy plane and intersecting it in the curve C we have the potential

$$w(z) = \int_C \omega(\xi, \eta) \log (z - z_0) \, d \, |z_0|,$$

where $\omega(\xi, \eta)$ is the density at $z_0 = \xi + i\eta$ on the curve. The velocity potential is

$$\phi(x, y) = \text{Re}(w) = \int_C \omega(\xi, \eta) \log \sqrt{(x - \xi)^2 + (y - \eta)^2} \, ds.$$

We shall call this the potential of a *single layer*.

If we have a distribution of doublets on such a sheet in the direction of the normal to the curve C, the potential is

$$w(z) = -\int_C \frac{\omega(\xi, \eta)e^{i\alpha(\xi, \eta)}}{z - z_0} \, d \, |z_0|,$$

and the velocity potential is

$$\phi(x, y) = \text{Re}(w) = -\int_C \omega(\xi, \eta) \frac{d}{dn} \log \sqrt{(x - \xi)^2 + (y - \eta)^2} \, ds,$$

where d/dn refers to the directional derivative in the direction of the normal to C. We shall call this the potential of a *double layer*.

If we have a distribution of line sources or sinks over an area R in the xy plane with strength $\rho(\xi, \eta) \, d\xi \, d\eta$ proportional to the element of area $d\xi \, d\eta$ then the complex potential is

$$w(z) = \iint_R \rho(\xi, \eta) \log (z - z_0) \, d\xi \, d\eta,$$

and the velocity potential is

$$\phi(x, y) = \text{Re}(w) = \iint_R \rho(\xi, \eta) \log \sqrt{(x - \xi)^2 + (y - \eta)^2} \, d\xi \, d\eta.$$

We shall call this the potential of a *volume distribution*.

Clearly, if the densities $\omega(\xi, \eta)$ and $\rho(\xi, \eta)$ are continuous then these three potentials are harmonic at points not occupied by sources, sinks, or doublets. In general, it will require stronger conditions on the densities, to prove existence of these potentials at points occupied by sources or sinks or doublets. We shall not consider this general problem here. The reader should see a book like O. D. Kellogg, *Foundations of Potential Theory*. New York: Dover Publications, Inc., 1953.

There is one case where we can obtain a definitive answer by considering an integral of the Cauchy type. Let C be a simple closed contour and let $f(\zeta)$ be a real-valued function defined on C satisfying a Lipschitz condition. Then by Theorem 3.5.4,

$$\frac{1}{\pi i} \int_C \frac{f(\zeta)}{\zeta - z_0} \, d\zeta$$

exists for z_0 on C, in the Cauchy principal value sense, and satisfies

$$\frac{1}{\pi i} \int \frac{f(\zeta)}{\zeta - z_0} \, d\zeta = F(z_0^-) - f(z_0) = F(z_0^+) + f(z_0),$$

where $F(z_0^-)$ and $F(z_0^+)$ are equal to the limits of

$$F(z) = \frac{1}{\pi i} \int_C \frac{f(\zeta)}{\zeta - z} \, d\zeta,$$

as z approaches z_0 from the inside and the outside of C, respectively. Now, let $\zeta - z = re^{i\theta}$. Then $d\zeta = dre^{i\theta} + ire^{i\theta} \, d\theta$ and

$$F(z) = \frac{1}{\pi i} \int_C \frac{f(\zeta)}{re^{i\theta}} (dre^{i\theta} + ire^{i\theta} \, d\theta)$$

$$= \frac{1}{\pi} \int_C f(\zeta) \, d\theta + \frac{1}{\pi i} \int_C \frac{f(\zeta)}{r} \, dr,$$

$$\mathrm{Re}\,[F(z)] = \frac{1}{\pi} \int_C f(\zeta) \, d\theta = \frac{1}{\pi} \int_C f(\zeta) \frac{d\theta}{ds} \, ds$$

$$= -\frac{1}{\pi} \int_C f(\zeta) \frac{d}{dn} \ln |\zeta - z| \, ds.$$

Therefore, we see that the real part of $F(z)$ is the potential of a double layer. Consequently, we have the following theorem about double layers:

Theorem 6.7.1. *Let C be a simple closed contour and let $\omega(\xi, \eta)$ be a real-valued function defined on C and satisfying a Lipschitz condition. Then the potential of the double layer*

$$\phi(x, y) = -\int_C \omega(\xi, \eta) \frac{d}{dn} \ln |\zeta - z| \, ds$$

is defined at $z_0 = x_0 + iy_0$ on C in the Cauchy principal value sense and satisfies

$$\phi(x_0, y_0) = \Phi(z_0^-) - \pi\omega(x_0, y_0) = \Phi(z_0^+) + \pi\omega(x_0, y_0),$$

where $\Phi(z_0^-)$ and $\Phi(z_0^+)$ are the limits of $\phi(x, y)$ as z approaches z_0 on C from the inside and the outside respectively.

Theorem 6.7.2. Let $u(x, y)$ be any real-valued function with continuous second partial derivatives within and on a simple closed contour C. Then u can be expressed inside C as the sum of a single layer, a double layer, and a volume distribution.

Proof: Let (x_0, y_0) be a point inside C. Let $v = (-1/2\pi) \ln |z - z_0|$ and apply Green's lemma to the region R between C and a circle with center at z_0 and small radius ρ. Then,

$$\iint\limits_R (u\nabla^2 v - v\nabla^2 u) \, dx \, dy = \int_C \left(u \frac{dv}{dn} - v \frac{du}{dn}\right) ds$$

$$- \int_0^{2\pi} \left(u \frac{\partial v}{\partial r} - v \frac{\partial u}{\partial r}\right)_{r=\rho} \rho \, d\theta,$$

where $r = |z - z_0|$. In R, $\nabla^2 v = 0$, and therefore letting $\rho \to 0$ we have

$$\frac{1}{2\pi} \iint\limits_R (\nabla^2 u) \log r \, dx \, dy$$

$$= -\frac{1}{2\pi} \int_C u \frac{d}{dn} \log r \, ds + \frac{1}{2\pi} \int_C \left(\frac{du}{dn}\right) \log r \, ds$$

$$+ \lim_{\rho \to 0} \frac{1}{2\pi} \int_0^{2\pi} u \, d\theta - \lim_{\rho \to 0} \frac{1}{2\pi} \int_0^{2\pi} \left(\frac{\partial u}{\partial r}\right)_{r=\rho} \rho \log \rho \, d\theta$$

$$u(x_0, y_0) = \iint\limits_R \left(\frac{\nabla^2 u}{2\pi}\right) \log r \, dx \, dy + \int_C \left(-\frac{1}{2\pi} \frac{du}{dn}\right) \log r \, ds$$

$$- \int_C \left(-\frac{1}{2\pi} u\right) \frac{d}{dn} \log r \, ds,$$

and we see that $u(x_0, y_0)$ can be expressed as a volume distribution with density $\nabla^2 u / 2\pi$, a single layer with density $(-1/2\pi) \, du/dn$, and a double layer with density $-u/2\pi$.

Corollary 6.7.1. Let u satisfy the conditions of Theorem 6.7.2 and be harmonic inside the simple closed contour C. Then u can be expressed at points inside C in terms of a single layer and a double layer on C, that is,

$$u(x_0, y_0) = \int_C \left(-\frac{1}{2\pi} \frac{du}{dn}\right) \log r \, ds - \int_C \left(-\frac{u}{2\pi}\right) \frac{d}{dn} \log r \, ds.$$

This corollary does not assert that this is the only way to express u in terms of a single layer and a double layer on C. As a matter of fact, in the next section we shall prove the existence of a double layer in terms of which the Dirichlet problem can be solved.

Corollary 6.7.1. If u is harmonic in a domain D, then

$$u(x_0, y_0) = \frac{1}{2\pi} \int_0^{2\pi} u(x, y) \, d\theta,$$

$x = x_0 + R \cos \theta, y = y_0 + R \sin \theta$, where R is the radius of a circle with center at (x_0, y_0) such that the disk $|z - z_0| \leq R$ is in D. In other words, the value of u at the center of the disk is equal to the mean value on the circumference.

Proof: This follows from Theorem 6.7.2 applied to the disk where C is the circle $|z - z_0| = R$. Then

$$u(x_0, y_0) = \frac{1}{2\pi R} \int_0^{2\pi} u(x, y) R \, d\theta - \frac{1}{2\pi} \log R \int_0^{2\pi} \left(\frac{\partial u}{\partial r}\right)_{r=R} R \, d\theta.$$

But

$$\int_0^{2\pi} \left(\frac{\partial u}{\partial r}\right)_{r=R} R \, d\theta = \iint_{|z-z_0|<R} \nabla^2 u \, dx \, dy = 0.$$

Therefore,

$$u(x_0, y_0) = \frac{1}{2\pi} \int_0^{2\pi} u(x, y) \, d\theta.$$

Exercises 6.7

1. In three dimensions the potentials of a volume distribution, a single layer, and a double layer are respectively

$$\iiint_V \frac{\rho(\xi, \eta, \zeta)}{r} \, d\xi \, d\eta \, d\zeta, \quad \iint_S \frac{\omega(\xi, \eta)}{r} \, dS, \quad \text{and} \quad -\iint_S \omega(\xi, \eta) \frac{d}{dn} \frac{1}{r} \, dS,$$

where $r = \sqrt{(x - \xi)^2 + (y - \eta)^2 + (z - \zeta)^2}$. Prove the analogs of Theorem 6.7.2 and Corollaries 6.7.1 and 6.7.2 in three dimensions.

2. Prove that if u is harmonic within and on the circle $|z - z_0| = R$, then

$$u(x_0, y_0) = \frac{1}{\pi R^2} \iint_D u \, dx \, dy,$$

where the integration is over the disk.

6.8. SINGULAR INTEGRAL EQUATIONS

We have seen in the last section that there is a close connection between integrals of the Cauchy type and potential theory. In this section, we shall expand on this idea a bit further and show how certain singular integral equations involving Cauchy type integrals can be solved. This type of integral equation has wide application in the field of elasticity. Finally, we shall show that the Dirichlet problem can be reformulated in terms of integral equations and another existence theorem can be proved.

EXAMPLE 6.8.1. Let C be a simple closed smooth contour. Let D_- denote the interior of C and D_+ denote the exterior of C. Find a piecewise analytic function $F(z)$ analytic in D_- and D_+ such that $\lim_{|z| \to \infty} |F(z)| = 0$, and

$$F(z_0^-) - F(z_0^+) = f(z_0),$$

where z_0 is on C, $f(z)$ is a function defined on C and satisfying a Lipschitz condition, and $F(z_0^-)$ and $F(z_0^+)$ are limits as z approaches z_0 from D_- and D_+ respectively. This problem can easily be solved using Theorem 3.5.4. As a matter of fact,

$$F(z) = \frac{1}{2\pi i} \int_{C+} \frac{f(\zeta)}{\zeta - z} \, d\zeta$$

solves the problem, because according to Theorem 3.5.4, if z_0 is on C

$$\frac{1}{2\pi i} \int_{C+} \frac{f(\zeta)}{\zeta - z_0} \, d\zeta = F(z_0^-) - \tfrac{1}{2} f(z_0) = F(z_0^+) + \tfrac{1}{2} f(z_0),$$

where the integral is defined in the Cauchy principal value sense. Therefore, it follows that

$$F(z_0^-) - F(z_0^+) = f(z_0).$$

The other conditions are clearly satisfied. Actually, the solution of this problem is unique, since the difference G of two solutions would satisfy

$$G(z_0^-) - G(z_0^+) = 0.$$

Therefore, $G(z)$ would be analytic in the unextended plane and approach zero, as z approaches ∞. Hence, $G(z) \equiv 0$.

EXAMPLE 6.8.2. Solve the following singular integral equation

$$\frac{1}{\pi i} \int_{C+} \frac{f(\zeta)}{\zeta - z_0} \, d\zeta = g(z_0)$$

for $f(z)$, given $g(z_0)$ on C, a simple closed smooth contour, satisfying a Lipschitz condition. We again obtain a solution using Theorem 3.5.4. Introduce the piecewise analytic function

$$F(z) = \frac{1}{2\pi i} \int_{C+} \frac{f(\zeta)\, d\zeta}{\zeta - z}.$$

Then

$$F(z_0^-) + F(z_0^+) = g(z_0).$$

Consider

$$G(z) = \begin{cases} F(z) & \text{in } D_- \\ -F(z) & \text{in } D_+. \end{cases}$$

Then

$$G(z_0^-) - G(z_0^+) = g(z_0)$$

and by the previous example

$$G(z) = \frac{1}{2\pi i} \int_{C+} \frac{g(\zeta)}{\zeta - z}\, d\zeta.$$

Therefore,

$$f(z_0) = F(z_0^-) - F(z_0^+) = G(z_0^-) + G(z_0^+)$$

$$= \frac{1}{\pi i} \int_{C+} \frac{g(\zeta)\, d\zeta}{\zeta - z_0}.$$

It can be shown that $f(z)$ satisfies a Lipschitz condition on C when defined this way.* Therefore, the formulas for $f(z)$ and $g(z)$ are completely symmetric. This implies that the solution of the integral equation is unique.

We can also use Theorem 3.5.4 to derive an important formula which is very useful in the solution of singular integral equations of the Cauchy type. Let

$$g(z_0) = \frac{1}{2\pi i} \int_{C+} \frac{f(\zeta)}{\zeta - z_0}\, d\zeta,$$

$$h(z_0) = \frac{1}{2\pi i} \int_{C+} \frac{g(\zeta)}{\zeta - z_0}\, d\zeta,$$

where $f(z)$ and hence, $g(z)$ and $h(z)$ satisfy Lipschitz conditions on C, a simple closed smooth contour. The problem is to express $h(z)$ in terms of

---

* See N. I. Muskhelishvili, *Singular Integral Equations*. Groningen, Holland: P. Noordhoff, N. V., 1953, pp. 46–48.

$f(z)$, if possible. We define the following piecewise analytic functions

$$G(z) = \frac{1}{2\pi i} \int_{C+} \frac{f(\zeta)}{\zeta - z} \, d\zeta,$$

$$H(z) = \frac{1}{2\pi i} \int_{C+} \frac{g(\zeta)}{\zeta - z} \, d\zeta.$$

Then

$$G(z_0^-) = \tfrac{1}{2}f(z_0) + \frac{1}{2\pi i} \int_{C+} \frac{f(\zeta)}{\zeta - z_0} \, d\zeta = \tfrac{1}{2}f(z_0) + g(z_0),$$

$$H(z_0^-) = \tfrac{1}{2}g(z_0) + \frac{1}{2\pi i} \int_{C+} \frac{g(\zeta)}{\zeta - z_0} \, d\zeta = \tfrac{1}{2}g(z_0) + h(z_0).$$

Substituting in the definition for $H(z)$, we have

$$H(z) = \frac{1}{2\pi i} \int_{C+} \frac{G(\zeta_-)}{\zeta - z} \, d\zeta - \frac{1}{4\pi i} \int_{C+} \frac{f(\zeta)}{\zeta - z} \, d\zeta$$

$$= G(z) - \tfrac{1}{2}G(z) = \tfrac{1}{2}G(z).$$

Therefore, $H(z_0^-) = \tfrac{1}{2}G(z_0^-)$, and

$$h(z_0) = \tfrac{1}{2}G(z_0^-) - \tfrac{1}{2}[G(z_0^-) - \tfrac{1}{2}f(z_0)] = \tfrac{1}{4}f(z_0)$$

or

$$\tfrac{1}{4}f(z_0) = \frac{1}{2\pi i} \int_{C+} \frac{g(\tau)}{\tau - z_0} \, d\tau = \frac{1}{(2\pi i)^2} \int_{C+} \frac{1}{t - z_0} \int_{C+} \frac{f(\zeta)}{\zeta - \tau} \, d\zeta \, d\tau.$$

This last result is known as the *Poincaré-Bertrand formula.*

EXAMPLE 6.8.3. Solve the following integral equation

$$g(z_0) = af(z_0) + \frac{b}{\pi i} \int_{C+} \frac{f(\zeta) \, d\zeta}{\zeta - z_0},$$

where a and b are constants, $a \neq b$, C is a simple closed smooth contour, $g(z)$ satisfies a Lipschitz condition on C, and z_0 is on C. First we multiply by a

$$ag(z_0) = a^2 f(z_0) + \frac{ab}{\pi i} \int_{C+} \frac{f(\zeta) \, d\zeta}{\zeta - z_0}.$$

Then we multiply by $b/\pi i(z_0 - z)$ and integrate z_0 around C_+ using the Poincaré-Bertrand formula

$$\frac{b}{\pi i} \int_{C+} \frac{g(z_0)}{z_0 - z} \, dz_0 = \frac{ab}{\pi i} \int_{C+} \frac{f(z_0)}{z_0 - z} \, dz_0$$

$$+ \frac{b^2}{(\pi i)^2} \int_{C+} \left[\frac{1}{z_0 - z} \int_{C+} \frac{f(\zeta) \, d\zeta}{\zeta - z_0} \right] dz_0$$

$$= \frac{ab}{\pi i} \int_{C+} \frac{f(z_0) \, dz_0}{z_0 - z} + b^2 f(z).$$

Subtracting and multiplying by $1/(b^2 - a^2)$ we have

$$f(z) = \frac{a}{b^2 - a^2}\, g(z) - \frac{b}{\pi i(b^2 - a^2)} \int_{C+} \frac{g(z_0)\,dz_0}{z_0 - z}.$$

We now return to the Dirichlet problem. Recall that in the last section we found that we could express certain harmonic function on the interior of a simple closed smooth contour C as the potential of a double layer on C. Let us assume that C has continuous curvature. We wish to solve the Dirichlet problem for a function harmonic inside C continuous on C plus its interior and taking on given real boundary values $g(z_0)$ when z_0 is on C. We seek a solution in terms of the potential of a double layer. As we saw in the previous section, if $f(\zeta)$ is real and

$$F(z) = \frac{1}{2\pi i} \int_{C+} \frac{f(\zeta)}{\zeta - z}\, d\zeta,$$

$\operatorname{Re}[F(z)]$ is the potential of a double layer. If we let z approach z_0 on C from the inside

$$F(z_0^-) = \tfrac{1}{2}f(z_0) + \frac{1}{2\pi i} \int_{C+} \frac{f(\zeta)}{\zeta - z_0}\, d\zeta.$$

For $\operatorname{Re}[F(z)]$ to be the solution of the Dirichlet problem $F(z_0^-) = g(z_0)$ and

$$
\begin{aligned}
g(z_0) &= \tfrac{1}{2}f(z_0) + \frac{1}{2\pi} \operatorname{Im}\left[\int_{C+} \frac{f(\zeta)}{\zeta - z_0}\, d\zeta \right] \\
&= \tfrac{1}{2}f(z_0) + \frac{1}{2\pi} \int_{C+} f(\zeta) \operatorname{Im}\left(\frac{1}{\zeta - z_0} \right) d\zeta \\
&= \tfrac{1}{2}f(z_0) + \frac{1}{2\pi} \int_{C+} f(\zeta) \frac{d}{dn} \ln|\zeta - z_0|\, ds \\
&= \tfrac{1}{2}f(z_0) - \frac{1}{2\pi} \int_{C+} f(\zeta) \frac{\cos\theta}{|\zeta - z_0|}\, ds,
\end{aligned}
$$

where θ is the angle between the vector $\zeta - z_0$ and the outward normal to C at z_0. Using the continuity of the curvature of C it is not hard to show that $|\zeta - z_0|^{-1} \cos\theta$ is continuous on C. Therefore, we have the following *Fredholm integral equation* to solve

$$f(z_0) = 2g(z_0) + \frac{1}{\pi} \int_{C+} f(\zeta) K(\zeta, z_0)\, d\zeta,$$

where K is a continuous function of ζ and z_0 on C.

We state without proof the following result concerning the Fredholm equation: The Fredholm integral equation has a unique solution if and only if the corresponding homogeneous equation ($g \equiv 0$) has only the trivial

solution ($f \equiv 0$).* Using this result we can obtain an existence theorem for
for the Dirichlet problem by showing that the equation

$$f(z_0) = \frac{1}{\pi} \int_{C+} f(\zeta) K(\zeta, z_0)\, d\zeta$$

has only the trivial solution. Consider

$$H(z) = \frac{1}{2\pi i} \int_{C+} \frac{f(\zeta)}{\zeta - z}\, d\zeta.$$

For the homogeneous problem

$$Re[H(z_0^-)] = 0,$$

which implies that $Re[H(z)] \equiv 0$ inside C, which implies that $Im[H(z)] = k$,
a constant. Therefore,

$$H(z) = \frac{1}{2\pi i} \int_{C+} \frac{f(\zeta)}{\zeta - z}\, d\zeta = ik.$$

Now consider the integral when z is outside C.

$$G(z) = \frac{1}{2\pi i} \int_{C+} \frac{f(\zeta)}{\zeta - z}\, d\zeta, \; z \text{ outside } C.$$

By Theorem 3.5.4

$$H(z_0^-) - G(z_0^+) = f(z_0),$$
$$ik - G(z_0^+) = f(z_0).$$

Since $f(z_0)$ is real, $Im[G(z_0^+)] = k$. This implies that $Im[G(z)] = k$ and that
$G(z)$ is constant. But $\lim\limits_{z \to \infty} G(z) = 0$. Therefore, $k = 0$, $G(z_0^+) = 0$, and

$f(z_0) \equiv 0$. This proves that the nonhomogeneous integral equation has a
solution which establishes the existence of the solution of the Dirichlet problem. The actual solution of the integral equation can be carried out by a series
expansion technique.†

Exercises 6.8

1. Solve Example 6.8.2 using the Poincaré-Bertrand formula.

2. Show that the double integral in the Poincaré-Bertrand formula cannot
be evaluated by interchanging the order of integration.

3. Prove that the solution of the integral equation in Example 6.8.3 is
unique.

4. Prove that $d/dn \ln |z - z_0|$ is continuous on C, where z and z_0 are on
C and C has continuous curvature.

---

* The proof can be found in J. W. Dettman, *Mathematical Methods in Physics and
Engineering*. New York: McGraw-Hill Book Company, Inc., 1962. Compare this result
with the work on the Sturm-Liouville problem of Sections 7.7 and 7.8.
   † *Ibid*, pp. 246–258.

CHAPTER 7

Ordinary
Differential Equations

7.1. SEPARATION OF VARIABLES

As is well known by anyone with even a little experience in engineering or science, ordinary differential equations play a very important role. What is probably not so well known at this point is that ordinary differential equations play an important role in the solution of certain partial differential equations. We shall illustrate this role, in this section, by considering the important separation of variables technique in the study of partial differential equations.

In most applications, we seek a real-valued function of a real variable which satisfies a given differential equation plus certain boundary and/or initial conditions. For this reason, the problem is usually one of real variables. However, the additional insight gained by considering these problems in the complex plane justifies including this study in a course in complex variable theory.

Consider the Dirichlet problem for the disk of radius R: to determine a function $u(x, y)$ which is harmonic for $|z| < R$, continuous for $|z| \leq R$, and takes on given boundary values $g(\theta)$ on $|z| = R$. In polar coordinates Laplace's equation takes the form

$$\frac{\partial^2 u}{\partial r^2} + \frac{1}{r} \frac{\partial u}{\partial r} + \frac{1}{r^2} \frac{\partial^2 u}{\partial \theta^2} = 0.$$

Let us seek a solution which is a function of r times a function of θ, that is, $u(r, \theta) = v(r)w(\theta)$. Then

$$v''w + \frac{1}{r} v'w + \frac{1}{r^2} vw'' = 0,$$

$$\frac{w''}{w} = \frac{-r^2 v'' - rv'}{v}.$$

The function on the left is a function of θ only, while the function on the right is a function of r only. The equation must be satisfied identically in θ and r. Therefore,

$$\frac{w''}{w} = -\frac{r^2v'' + rv'}{v} = a,$$

where a is a constant. We have to find solutions of the two ordinary differential equations

$$w'' - aw = 0,$$
$$r^2v'' + rv' + av = 0.$$

Looking at the first equation, if a is positive then

$$w(\theta) = Ae^{\sqrt{a}\theta} + Be^{-\sqrt{a}\theta},$$

but this cannot be periodic in θ, as required to make w continuous, that is, $w(0) = w(2\pi)$, unless $A = B = 0$. Therefore, $a \le 0$. Let $a = -b^2$, then $w'' + b^2w = 0$, and

$$w(\theta) = A\cos b\theta + B\sin b\theta.$$

In order for w to be continuous and have continuous derivatives, $b = n$, $n = 0, 1, 2, 3, \ldots$. Therefore, $a = -n^2$, and the second equation becomes

$$r^2v'' + rv' - n^2v = 0.$$

This is Euler's differential equation, which has solutions r^n and r^{-n}. If $n \ne 0$, then r^{-n} has to be excluded because it is not finite at the origin. Finally, we have arrived at the following possible solutions of Laplace's equation,

$$u_0(r, \theta) = 1, u_{n_1}(r, \theta) = r^n\cos n\theta, u_{n_2}(r, \theta) = r^n\sin n\theta.$$

It is very unlikely that any of these takes on the boundary values $g(\theta)$ at $r = R$. However, because of the linearity of the partial differential equation, we may take a linear combination of a finite (or infinite) number of these solutions. Hence,

$$u(r, \theta) = \frac{a_0}{2} + \sum_{n=1}^{\infty}(a_nr^n\cos n\theta + b_nr^n\sin n\theta).$$

For $r = R$,

$$u(R, \theta) = g(\theta) = \frac{a_0}{2} + \sum_{n=1}^{\infty}(a_nR^n\cos n\theta + b_nR^n\sin n\theta).$$

This is the *Fourier series* (see Section 8.1) for $g(\theta)$, if

$$a_n = \frac{1}{\pi R^n}\int_0^{2\pi}g(\theta)\cos n\theta\, d\theta, \qquad b_n = \frac{1}{\pi R^n}\int_0^{2\pi}g(\theta)\sin n\theta\, d\theta.$$

We prove in Section 8.1 that if $g(\theta)$ is continuous, $g(0) = g(2\pi)$, and has a piecewise continuous derivative, then its Fourier series converges uniformly to it. Therefore, our solution is

$$u(r, \theta) = \frac{1}{2\pi} \int_0^{2\pi} g(\phi)\, d\phi$$

$$+ \sum_{n=1}^{\infty} \frac{r^n}{R^n} \frac{1}{\pi} \int_0^{2\pi} g(\phi)\, (\cos n\theta \cos n\phi + \sin n\theta \sin n\phi)\, d\phi$$

$$= \sum_{n=0}^{\infty} \frac{1}{2\pi} \int_0^{2\pi} g(\phi)\, \frac{2r^n}{R^n} \cos n(\theta - \phi)\, d\phi - \frac{1}{2\pi} \int_0^{2\pi} g(\phi)\, d\phi$$

$$= \frac{1}{2\pi} \int_0^{2\pi} g(\phi) \left[\sum_{n=0}^{\infty} \frac{2r^n}{R^n} \cos n(\theta - \phi) - 1 \right] d\phi.$$

Consider the function $\dfrac{\zeta + z}{\zeta - z}$, $\zeta = Re^{i\phi}$, $z = re^{i\theta}$, $r < R$.

$$\frac{\zeta + z}{\zeta - z} = 1 + \frac{2z}{\zeta - z} = 1 + 2\left(\frac{z}{\zeta}\right) \frac{1}{1 - \dfrac{z}{\zeta}} = 1 + 2 \sum_{n=1}^{\infty} \left(\frac{z}{\zeta}\right)^n$$

$$= \sum_{n=0}^{\infty} \frac{2r^n}{R^n} [\cos n(\theta - \phi) + i \sin n(\theta - \phi)] - 1.$$

Therefore,

$$u(r, \theta) = \frac{1}{2\pi} \int_0^{2\pi} g(\phi)\, \text{Re} \left[\frac{\zeta + z}{\zeta - z}\right] d\phi$$

$$= \text{Re}\, \frac{1}{2\pi} \int_0^{2\pi} g(\phi)\, \frac{\zeta + z}{\zeta - z}\, d\phi$$

$$= \text{Re}\, \frac{1}{2\pi} \int_0^{2\pi} g(\phi)\, \frac{(\zeta + z)(\bar{\zeta} - \bar{z})}{R^2 + r^2 - 2rR \cos (\theta - \phi)}\, d\phi$$

$$= \frac{1}{2\pi} \int_0^{2\pi} g(\phi)\, \frac{R^2 - r^2}{R^2 + r^2 - 2rR \cos (\theta - \phi)}\, d\phi,$$

which is Poisson's formula again.

This solution of the Dirichlet problem illustrates the separation of variables technique. As another example, consider *Helmholtz's equation* $\nabla^2 u + \lambda u = 0$ in polar coordinates, that is,

$$\frac{\partial^2 u}{\partial r^2} + \frac{1}{r} \frac{\partial u}{\partial r} + \frac{1}{r^2} \frac{\partial^2 u}{\partial \theta^2} + \lambda u = 0.$$

Assuming $u(r, \theta) = v(r)w(\theta)$, then

$$wv'' + \frac{1}{r}\, wv' + \frac{1}{r^2}\, vw'' + \lambda vw = 0,$$

$$\frac{w''}{w} = -\frac{r^2v'' + rv' + \lambda r^2}{v} = -n^2.$$

We are lead to the two ordinary differential equations

$$w'' + n^2 w = 0,$$
$$r^2 v'' + rv' + (\lambda r^2 - n^2)v = 0.$$

If in the second equation we make the change of independent variable $x = \sqrt{\lambda}r$, we arrive at

$$x^2 v'' + xv' + (x^2 - n^2)v = 0,$$

where the prime refers to differentiation with respect to x. This is *Bessel's differential equation.* We shall study its solution in some detail.

As a final example, consider Laplace's equation in spherical coordinates, that is,

$$\nabla^2 u = \frac{\partial^2 u}{\partial \rho^2} + \frac{2}{\rho}\, \frac{\partial u}{\partial \rho} + \frac{1}{\rho^2}\, \frac{\partial^2 u}{\partial \phi^2} + \frac{\cot \phi}{\rho^2}\, \frac{\partial u}{\partial \phi} + \frac{1}{\rho^2 \sin^2 \phi}\, \frac{\partial^2 u}{\partial \theta^2} = 0.$$

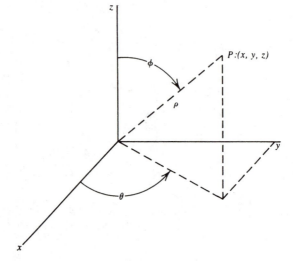

Figure 7.1.1.

We assume that $u(\rho, \theta, \phi) = f(\rho)g(\theta)h(\phi)$. Then

$$ghf'' + \frac{2}{\rho}\, ghf' + \frac{fg}{\rho^2}\, h'' + \frac{\cot \phi}{\rho^2}\, fgh' + \frac{1}{\rho^2 \sin^2 \phi}\, fhg'' = 0,$$

$$\frac{\rho^2 f'' + 2\rho f'}{f} = -\left[\frac{h''}{h} + \cot \phi \, \frac{h'}{h} + \frac{1}{\sin^2 \phi} \frac{g''}{g}\right] = a,$$

$$\rho^2 f'' + 2\rho f' - af = 0,$$

$$\frac{\sin^2 \phi h'' + \sin \phi \cos \phi h' + a \sin^2 \phi h}{h} = -\frac{g''}{g} = n^2,$$

$$g'' + n^2 g = 0,$$

$$\sin^2 \phi h'' + \sin \phi \cos \phi h' + a \sin^2 \phi h - n^2 h = 0.$$

In the last equation, we make the change of variable $x = \cos \phi$, and the equation becomes

$$(1 - x^2)h'' - 2xh' + \left(a - \frac{n^2}{1 - x^2}\right)h = 0.$$

This is the *associated Legendre equation*. If $n = 0$, we have

$$(1 - x^2)h'' - 2xh' + ah = 0,$$

which is the *Legendre equation*. We shall study both the Legendre equation and the associated Legendre equation which have solutions which are closely related.

Exercises 7.1

1. Consider the ordinary differential equation $x^2 y'' + axy' + by = 0$, where a and b are constants. By assuming solutions of the form $y = x^m$, show that the equation has solutions $y_1 = x^{m_1}$ and $y_2 = x^{m_2}$ provided m_1 and m_2 are roots of the indicial equation $m(m - 1) + am + b = 0$.

2. Consider the separation of the Helmholtz equation $\nabla^2 u + \lambda u = 0$ in spherical coordinates. Show that the radial part $f(\rho)$ satisfies the equation

$$\rho^2 f'' + 2\rho f' + \lambda \rho^2 f - af = 0,$$

where a is a separation constant. Let a be $n(n + 1)$, where n is an integer. By making the substitution $x = \lambda^{1/2}\rho$ and $y = \rho^{1/2}f$, show that y satisfies the Bessel equation

$$x^2 y'' + xy' + [x^2 - (n + \tfrac{1}{2})^2]y = 0.$$

3. Suppose that the Legendre equation

$$(1 - x^2)y'' - 2xy' + n(n + 1)y = 0,$$

where n is a nonnegative integer, has a solution $P_n(x)$. Show that the associated Legendre equation

$$(1 - x^2)u'' - 2xu' + \left[n(n + 1) - \frac{m^2}{1 - x^2}\right]u = 0$$

has a solution $u = (1 - x^2)^{m/2} \dfrac{d^m}{dx^m} P_n(x)$ if m is a nonnegative integer.

7.2. EXISTENCE AND UNIQUENESS THEOREMS

In dealing with a problem in ordinary differential equations, one of the first questions which should be asked is "Is the problem formulated in such a way that a solution exists and is unique?" We shall illustrate some common approaches to this question by proving some existence and uniqueness theorems.

Theorem 7.2.1. *Let $u(t, x)$ be a continuous real-valued function of the real variables t and x defined in the rectangle $R = \{(t, x)| |t - t_0| \leq a, |x - x_0| \leq b\}$. Let $u(t, x)$ satisfy a Lipschitz condition in R, that is, there exists a positive constant K such that for any two points (t, x_1) and (t, x_2) in R, $|u(t, x_1) - u(t, x_2)| \leq K|x_1 - x_2|$. Then there exists a unique function $x(t)$ such that $\dot{x}(t) = u[t, x(t)]$ for $|t - t_0| \leq c = \min [a, b/M]$, and $x(t_0) = x_0$, where $M = \max\limits_{(t,x)\in R} |u(t, x)|$.*

Proof: We seek a solution of the integral equation

$$x(t) = x_0 + \int_{t_0}^{t} u[\tau, x(\tau)] \, d\tau,$$

because, obviously, a solution of this equation will satisfy $\dot{x} = u(t, x)$ and $x(t_0) = x_0$. We solve the integral equation by iteration. As a first approximation we take $x_0(t) = x_0$. Then

$$x_1(t) = x_0 + \int_{t_0}^{t} u(\tau, x_0) \, d\tau,$$

$$x_2(t) = x_0 + \int_{t_0}^{t} u[\tau, x_1(\tau)] \, d\tau$$

$$\cdots\cdots\cdots\cdots\cdots\cdots\cdots\cdots\cdots$$

$$x_{n+1}(t) = x_0 + \int_{t_0}^{t} u[\tau, x_n(\tau)] \, d\tau.$$

We can show that each successive approximation lies in R provided

$$|t - t_0| \leq c.$$

For example,

$$|x_1(t) - x_0| = \left| \int_{t_0}^{t} u(\tau, x_0)\, d\tau \right| \leq M|t - t_0| \leq b.$$

Suppose $|x_k(t) - x_0| \leq b$. Then

$$|x_{k+1}(t) - x_0| = \left| \int_{t_0}^{t} u[\tau, x_k(t)]\, d\tau \right| \leq M|t - t_0| \leq b.$$

Hence, by induction $|x_n(t) - x_0| \leq b$ for all n.

Consider the series $x_0 + \sum_{n=1}^{\infty} [x_n(t) - x_{n-1}(t)]$ which has $x_N(t)$ as Nth partial sum. We can show that for $|t - t_0| \leq c$

$$|x_n(t) - x_{n-1}(t)| \leq \frac{MK^{n-1}|t - t_0|^n}{n!} \leq \frac{M}{K} \frac{(Kc)^n}{n!}.$$

We again proceed by induction. Let $n = 1$. Then

$$|x_1(t) - x_0| = \left| \int_{t_0}^{t} u(\tau, x_0)\, d\tau \right| \leq M|t - t_0|.$$

Assume that the inequality holds for $n = k$, that is,

$$|x_k(t) - x_{k-1}(t)| \leq \frac{MK^{k-1}|t - t_0|^k}{k!}.$$

Then

$$|x_{k+1}(t) - x_k(t)| = \left| \int_{t_0}^{t} \{u[\tau, x_k(\tau)] - u[\tau, x_{k-1}(\tau)]\}\, d\tau \right|$$

$$\leq \left| \int_{t_0}^{t} K|x_k(\tau) - x_{k-1}(\tau)|\, d\tau \right|$$

$$\leq \frac{MK^k}{k!} \left| \int_{t_0}^{t} |\tau - t_0|^k\, d\tau \right|$$

$$\leq \frac{MK^k}{(k + 1)!} |t - t_0|^{k+1}.$$

Therefore, the inequality holds for all n. Hence, by the Weierstrass M-test the series $x_0 + \sum_{n=1}^{\infty} [x_n(t) - x_{n-1}(t)]$ converges absolutely and uniformly in $|t - t_0| \leq c$. The Nth partial sum $x_N(t)$ converges uniformly to a continuous function $x(t)$. Let N be sufficiently large that $|x(t) - x_N(t)| < \epsilon$ for all t satisfying $|t - t_0| \le c$. Then

$$|u[t, x(t)] - u[t, x_N(t)]| \le K|x(t) - x_N(t)| < K\epsilon.$$

Since ϵ is arbitrary, this proves that $\lim_{N \to \infty} u[t, x_N(t)] = u[t, x(t)]$ uniformly in t. Then

$$x_N(t) = x_0 + \int_{t_0}^{t} u[\tau, x_{N-1}(\tau)] \, d\tau,$$

$$x(t) = \lim_{N \to \infty} x_N(t) = x_0 + \int_{t_0}^{t} \lim_{N \to \infty} u[\tau, x_{N-1}(\tau)] \, d\tau$$

$$= x_0 + \int_{t_0}^{t} u[\tau, x(\tau)] \, d\tau$$

which proves that $x(t)$ satisfies the integral equation.

It remains to prove that $x(t)$ is the only solution. Suppose that $x^*(t)$ is another solution. Then

$$x^*(t) = x_0 + \int_{t_0}^{t} u[\tau, x^*(\tau)] \, d\tau,$$

$$|x^*(t) - x_0| = \left| \int_{t_0}^{t} u[\tau, x^*(\tau)] \, d\tau \right| \le M|t - t_0| \le Mc,$$

$$|x^*(t) - x_1(t)| = \left| \int_{t_0}^{t} [u(\tau, x^*) - u(\tau, x_0)] \, d\tau \right|$$

$$\le K \left| \int_{t_0}^{t} |x^*(\tau) - x_0| \, d\tau \right|$$

$$\le MK \left| \int_{t_0}^{t} |\tau - t_0| \, d\tau \right| = \frac{MK|t - t_0|^2}{2} \le \frac{MKc^2}{2},$$

$$|x^*(t) - x_2(t)| = \left| \int_{t_0}^{t} [u(\tau, x^*) - u(\tau, x_1)] \, d\tau \right|$$

$$\le K \left| \int_{t_0}^{t} |x^*(\tau) - x_1(\tau)| \, d\tau \right|$$

$$\le \frac{MK^2}{2} \left| \int_{t_0}^{t} (\tau - t_0)^2 \, d\tau \right| = \frac{MK^2|t - t_0|^3}{3!} \le \frac{MK^2c^3}{3!}.$$

Inductively, we have

$$|x^*(t) - x_N(t)| \le \frac{MK^N c^{N+1}}{(N + 1)!} < \epsilon,$$

for N sufficiently large. Then

$$|x^*(t) - x(t)| \leq |x^*(t) - x_N(t)| + |x(t) - x_N(t)| < 2\epsilon.$$

Therefore, since ϵ is arbitrary, $x^*(t) \equiv x(t)$. This completes the proof.

There are similar theorems for systems of first order equations. We shall consider one of these which, for the sake of algebraic simplicity, involves just two first order equations. Consider the system

$$\dot{x} = u(t, x, y),$$
$$\dot{y} = v(t, x, y),$$

or

$$\dot{z} = \dot{x} + i\dot{y} = u(t, x, y) + iv(t, x, y) = f(t, z).$$

If there is a solution $z(t) = x(t) + iy(t)$ satisfying the initial conditions $x(t_0) = x_0, y(t_0) = y_0$, that is, $z(t_0) = x_0 + iy_0$, then it satisfies the integral equation

$$\dot{z}(t) = \dot{x}(t) + i\dot{y}(t) = x_0 + \int_{t_0}^{t} u(\tau, x, y) \, d\tau + iy_0 + i \int_{t_0}^{t} v(\tau, x, y) \, d\tau$$

$$= z_0 + \int_{t_0}^{t} f[\tau, z(\tau)] \, d\tau.$$

Theorem 7.2.2. *Let $f(t, z)$ be a continuous complex-valued function of the real variable t and the complex variable z in the region*

$$R = \{(t, z) \mid |t - t_0| \leq a, |z - z_0| \leq b\}.$$

Let $f(t, z)$ satisfy a Lipschitz condition in R, that is, there exists a positive constant K such that for any two points (t, z_1) and (t, z_2) in R, $|f(t, z_1) - f(t, z_2)| \leq K|z_1 - z_2|$. Then there exists a unique function $z(t)$ such that $\dot{z}(t) = f[t, z(t)]$ for $|t - t_0| \leq c = \min[a, b/M]$, and $z(t_0) = z_0$, where $M = \max\limits_{(t, z) \in R} |f(t, z)|$.

Proof: We shall not give the proof of this theorem since it follows exactly the proof of Theorem 7.2.1. The reader should check the details for himself.

EXAMPLE 7.2.1. Prove the existence and uniqueness of the solution of the second order differential equation $\ddot{x} - 3\dot{x} + 2x = 0$, subject to $x(0) = 1$, $\dot{x}(0) = 1$. Let $x = x(t)$, $\dot{x}(t) = y$. Then $\dot{y} = 3y - 2x$ and

$$\dot{z} = \dot{x} + i\dot{y} = y + 3iy - 2ix = f(z).$$

Now, $f(z)$ is continuous for all z, and

$$|f(z_1) - f(z_2)| = |y_1 - y_2 + 3i(y_1 - y_2) - 2i(x_1 - x_2)|$$
$$\leq 4|y_1 - y_2| + 2|x_1 - x_2| \leq 6|z_1 - z_2|.$$

Hence, $f(z)$ satisfies a Lipschitz condition for all t and all z. This establishes existence and uniqueness by Theorem 7.2.2. It is interesting to look at the iteration procedure.

$$z_0 = 1 + i,$$

$$z_1 = 1 + i + \int_0^t (1 + i)\, d\tau = 1 + t + i(1 + t),$$

$$z_2 = 1 + i + \int_0^t [1 + \tau + i(1 + \tau)]\, d\tau = \left(1 + t + \frac{t^2}{2}\right) + i\left(1 + t + \frac{t^2}{2}\right).$$

The iteration is generating the power series for $z(t) = e^t + ie^t$, which is the unique solution.

EXAMPLE 7.2.2. Prove existence and uniqueness of the solution of the second order linear differential equation $\ddot{x} + a(t)\dot{x} + b(t)x = 0$ in the interval $|t - t_0| \le c$, assuming that $a(t)$ and $b(t)$ are continuous in the same interval, and $x(t_0) = x_0$, $\dot{x}(t_0) = \dot{x}_0$. Let $x = x(t)$ and $y = \dot{x}(t)$. Then $\dot{y} = -[a(t)y + b(t)x]$ and $\dot{z}(t) = \dot{x} + i\dot{y} = y - ia(t)y - ib(t)x = f(t, z)$. Clearly, $f(t, z)$ is continuous for $|t - t_0| \le c$ and for all z. Furthermore,

$$\begin{aligned} |f(t, z_1) - f(t, z_2)| &= |y_1 - y_2 - ia(t)(y_1 - y_2) - ib(t)(x_1 - x_2)| \\ &\le |y_1 - y_2| + A|y_1 - y_2| + B|x_1 - x_2| \\ &\le (1 + A + B)|z_1 - z_2|, \end{aligned}$$

where $A = \max |a(t)|$ and $B = \max |b(t)|$ in $|t - t_0| \le c$. Therefore, $f(t, z)$ satisfies a Lipschitz condition and this establishes existence and uniqueness.

Another situation confronted in the complex plane is the following. Suppose $w' = f(z, w)$, $w(z_0) = w_0$. Under what circumstances will there exist a unique solution to the problem? First, we might expect a solution $w(z)$ to be differentiable in some domain D. Then in D

$$w'' = \frac{\partial f}{\partial z} + \frac{\partial f}{\partial w}\frac{dw}{dz}$$

should exist. Therefore, in D f_z should exist and there should be a domain of the w plane in which f_w exists. For this reason, it makes sense to require that $f(z, w)$ be an analytic function of the two complex variables z and w for $|z - z_0| \le a$, $|w - w_0| \le b$. We shall prove the following theorem:

Theorem 7.2.3. *Let $f(z, w)$ be an analytic function of the two complex variables z and w for $|z - z_0| \le a$, $|w - w_0| \le b$. Then there exists a unique analytic function $w(z)$ satisfying $w' = f(z, w)$ and $w_0 = w(z_0)$.*

First Proof: Without loss of generality we can assume that $w_0 = z_0 = 0$. If not we can make the change of variables $w - w_0 \to w$ and $z - z_0 \to z$.

By the analyticity of $f(z, w)$ for $|z| \leq a$, $|w| \leq b$, we can integrate $f(z, w)$ along the straight line path from 0 to z. Hence,

$$w(z) = \int_0^z f[\zeta, w(\zeta)] \, d\zeta = \int_0^1 f[zt, w(zt)]z \, dt.$$

From this point on, the proof follows pretty much the proof of Theorem 7.2.1. For example, we solve the integral equation by iteration taking $w_0 = 0$,

$$w_1(z) = \int_0^1 f(zt, 0)z \, dt,$$

$$w_2(z) = \int_0^1 f[zt, w_1(zt)]z \, dt,$$

$$\dots\dots\dots\dots\dots\dots\dots\dots$$

$$w_n(z) = \int_0^1 f[zt, w_{n-1}(zt)]z \, dt.$$

Then $|w_n(z)| = |\int_0^1 f[zt, w_{n-1}(zt)] z \, dt| \leq M|z| \leq b$, provided $|z| \leq c = \min [a, b/M]$, where $M = \max |f(z, w)|$. The unique solution is the sum of the series

$$\sum_{n=1}^{\infty} [w_n(z) - w_{n-1}(z)].$$

This series converges absolutely and uniformly by comparison with the series

$$\sum_{n=1}^{\infty} \frac{M}{K} (K|z|)^n.$$

As a matter of fact, we have

$$|w_1(z)| = \left| \int_0^1 f(zt, 0)z \, dt \right| \leq M|z|,$$

$$|w_2(z) - w_1(z)| = \left| \int_0^1 [f(zt, w_1) - f(zt, 0)]z \, dt \right|$$
$$\leq K|w_1| \, |z| \leq MK|z|^2,$$

and by induction $|w_n(z) - w_{n-1}(z)| \leq MK^{n-1}|z|^n = (M/K)(K|z|)^n$. We have used the fact that $f(z, w)$ satisfies a Lipschitz condition. This follows from the fact that

$$\left| \frac{f(z, w_1) - f(z, w_2)}{w_1 - w_2} \right|$$

is uniformly bounded in $R = \{(z, w) \mid |z| \leq a, |w| \leq b\}$, which follows from the analyticity of f. The comparison series converges uniformly for $K|z| \leq \rho < 1$. Also, $w_1(z), w_2(z), \ldots$ are analytic for $|z| < c$. Hence,

$$w(z) = \lim_{N \to \infty} \sum_{n=1}^{N} [w_n(z) - w_{n-1}(z)]$$

is the uniform limit of a sequence of analytic functions and is therefore analytic for $|z| < c^* = \min [a, b/M, 1/K]$. From here on the proof is identical with that for Theorem 7.2.1. The reader can supply the details.

Second Proof: Because of the importance of this method, we present a second proof of this theorem based on the *Cauchy majorant method*. See Theorem 4.4.6. Since $f(z, w)$ is an analytic function of the two complex variables it has a double power series representation

$$f(z, w) = a_{00} + a_{10}z + a_{01}w + a_{11}zw + \cdots$$

valid for $|z| \leq \rho_1 < a$, $|w| \leq \rho_2 < b$. We assume a solution

$$w(z) = c_1z + c_2z^2 + c_3z^3 + \cdots.$$

Formally substituting in the differential equation, we have

$$\begin{aligned} w'(z) &= c_1 + 2c_2z + 3c_3z^2 + \cdots \\ &= a_{00} + a_{10}z + a_{01}(c_1z + c_2z^2 + \cdots) \\ &\quad + a_{11}(c_1z^2 + c_2z^3 + \cdots) \\ &\quad + a_{20}z^2 + a_{02}(c_1z + c_2z^2 + \cdots)^2 + \cdots \end{aligned}$$

Equating coefficients of like powers of z, we have

$$\begin{aligned} c_1 &= a_{00}, \\ 2c_2 &= a_{10} + a_{01}c_1, \\ 3c_3 &= a_{01}c_2 + a_{11}c_1 + a_{20} + a_{02}c_1^2, \\ &\cdots\cdots\cdots\cdots\cdots\cdots\cdots\cdots\cdots\cdots \end{aligned}$$

Hence, we determine successively $c_1 = a_{00}$, $c_2 = (a_{10} + a_{01}a_{00})/2$, $c_3 = [a_{01}(a_{10} + a_{01}a_{00})/2 + a_{11}a_{00} + a_{20} + a_{02}a_{00}^2]/3$, and so on. Therefore, a formal series solution is determined uniquely. We shall have proved the theorem if we can prove that the formal power series converges in some ϵ-neighborhood of the origin.

Consider another differential equation

$$\frac{dW}{dz} = F(z, W) = A_{00} + A_{01}z + A_{10}W + A_{11}zW + \cdots$$

where the right-hand side is a majorant for $a_{00} + a_{10}z + a_{10}w + \cdots$, that is, $|a_{jk}| \leq A_{jk}$, for all j and k. If $W = C_1z + C_2z^2 + C_3z^3 + \cdots$, then the formal series is determined, as follows:

$$C_1 = A_{00},$$
$$C_2 = (A_{10} + A_{01}A_{00})/2,$$
$$C_3 = [A_{01}(A_{10} + A_{01}A_{00})/2 + A_{11}A_{00} + A_{20} + A_{02}A_{00}^2]/3,$$

and so on. Clearly, this series is a majorant for $c_1z + c_2z^2 + c_3z^3 + \cdots$, since

$$|c_1| = |a_{00}| \leq A_{00} = C_1,$$
$$|c_2| \leq (|a_{10}| + |a_{01}||a_{00}|)/2 \leq (A_{10} + A_{01}A_{00})/2,$$
$$|c_3| \leq [|a_{01}|(|a_{10}| + |a_{01}||a_{00}|)/2 + |a_{11}||a_{00}| + |a_{20}| + |a_{02}||a_{00}|^2]/3$$
$$\leq [A_{01}(A_{10} + A_{01}A_{00})/2 + A_{11}A_{00} + A_{20} + A_{02}A_{00}^2]/3$$
$$= C_3.$$

and so on. So all we have to do is find a function $F(z, W)$ with a power series representation which is a majorant for $f(z, w)$ and for which we can explicitly solve the differential equation. Then by comparison of $c_1z + c_2z^2 + c_3z^3 + \cdots$ with a known series $C_1z + C_2z^2 + C_3z^3 + \cdots$ we shall have established the convergence of the former and the existence of $w(z)$ analytic in some ϵ-neighborhood of the origin. Now

$$|a_{jk}| \leq M\rho_1^{-j}\rho_2^{-k},$$

where $M = \max |f(z, w)|$ in R. So we can take $A_{jk} = M\rho_1^{-j}\rho_2^{-k}$. Then

$$F(z, W) = M \sum_{j=0}^{\infty} \left(\frac{z}{\rho_1}\right)^j \sum_{k=0}^{\infty} \left(\frac{W}{\rho_2}\right)^k = \frac{M\rho_1\rho_2}{(\rho_1 - z)(\rho_2 - W)},$$

$$(\rho_2 - W)\frac{dW}{dz} = \frac{M\rho_1\rho_2}{\rho_1 - z},$$

$$\frac{(W - \rho_2)^2}{2} = M\rho_1\rho_2 \, \text{Log}\left(1 - \frac{z}{\rho_1}\right) + C,$$

where C is a constant of integration. Since $W(0) = 0$, $C = \rho_2^2/2$. Then

$$W(z) = \rho_2 - \rho_2 \sqrt{1 + 2M\frac{\rho_1}{\rho_2} \, \text{Log}\left(1 - \frac{z}{\rho_1}\right)},$$

where the square root is a branch which is analytic at $z = 0$ and takes on the value one. Therefore, $W(z)$ is a unique analytic solution of $W' = F(z, W)$ which has a power series $C_1z + C_2z^2 + \cdots$ which converges in some ϵ-neighborhood of $z = 0$. It is a majorant for $c_1z + c_2z^2 + \cdots$, and therefore by comparison we have proved that $c_1z + c_2z^2 + \cdots$ converges and is a unique analytic solution of $w' = f(z, w)$ satisfying $w(0) = 0$. This completes the proof.

EXAMPLE 7.2.3. Prove the existence and uniqueness of the solution of the first order linear differential equation $w' = f(z) + wg(z)$ satisfying $w(0) = 0$, where $f(z)$ and $g(z)$ are analytic for $|z| < r$. Let $|z| \leq \rho < r$. Then

$$f(z) = a_0 + a_1 z + a_2 z^2 + \cdots,$$
$$g(z) = b_0 + b_1 z + b_2 z^2 + \cdots,$$

with $|a_k| \leq M\rho^{-k}$, $|b_k| \leq M\rho^{-k}$, where $M = \max_{|z| = \rho} [|f(z)|, |g(z)|]$. Assuming a series solution $w(z) = c_1 z + c_2 z^2 + \cdots$ proceeding as in the second proof of the theorem, we can determine a unique formal power series solution. A majorant for this formal series is determined from the differential equation

$$\frac{dW}{dz} = F(z) + WG(z),$$

where

$$F(z) = G(z) = M\left[1 + \frac{z}{\rho} + \left(\frac{z}{\rho}\right)^2 + \cdots\right] = \frac{\rho M}{\rho - z}.$$

Then

$$\frac{dW}{1 + W} = \frac{\rho M \, dz}{\rho - z},$$

$$\text{Log}\,(1 + W) = -\rho M \,\text{Log}\left(1 - \frac{z}{\rho}\right) + C,$$

where C is a constant of integration. Since $W(0) = 0$, $C = 0$. Then

$$W(z) = \left(1 - \frac{z}{\rho}\right)^{-\rho M} - 1.$$

Now, $W(z)$ is analytic for $|z| < \rho$ and therefore remains a majorant for $w(z)$ for all $\rho < r$. Hence, it can be used to establish the convergence of the formal power series for $w(z)$ for $|z| < r$. We thus establish the existence of a unique analytic solution for $|z| < r$, where r is the distance from the origin to the nearest singularity of $f(z)$ or $g(z)$.

We can also treat systems of the form

$$w_1'(z) = f_1(z, w_1, w_2)$$
$$w_2'(z) = f_2(z, w_1, w_2)$$

subject to $w_1(z_0) = w_1^0$, $w_2(z_0) = w_2^0$, where f_1 and f_2 are analytic functions of three complex variables for $|z| \leq a$, $|w_1| \leq b$, $|w_2| \leq c$. The proper way to treat such systems is in vector notation. Let $\mathbf{w}$ be a vector with two components $w_1(z)$ and $w_2(z)$. Let $\mathbf{f}$ be a vector with two components f_1 and f_2. Then the system becomes

$$\mathbf{w}' = \mathbf{f}(z, \mathbf{w})$$

subset to $\mathbf{w}(z_0) = \mathbf{w}^0$, where $\mathbf{w}^0$ has two components w_1^0 and w_2^0. Equality in the vector sense means that the two vectors are equal, component by component. Let

$$|\mathbf{w}| = \sqrt{|w_1|^2 + |w_2|^2}.$$

Then

$$
\begin{aligned}
|\mathbf{u} + \mathbf{v}|^2 &= |u_1 + v_1|^2 + |u_2 + v_2|^2 \\
&\le |u_1|^2 + |u_2|^2 + |v_1|^2 + |v_2|^2 + 2|u_1||v_1| + 2|u_2||v_2| \\
&\le |\mathbf{u}|^2 + |\mathbf{v}|^2 + 2|\mathbf{u}||\mathbf{v}| \\
&\le (|\mathbf{u}| + |\mathbf{v}|)^2.
\end{aligned}
$$

We then have the triangle inequality $|\mathbf{u} + \mathbf{v}| \le |\mathbf{u}| + |\mathbf{v}|$. Armed with this inequality we can prove the existence and uniqueness theorem in much the same way as the first proof of Theorem 7.2.3. We shall not carry out this proof. Instead, we shall prove the following theorem about a system of linear first order equations:

Theorem 7.2.4. *Let $f_1(z)$, $f_2(z)$, $g_1(z)$, $g_2(z)$, $h_1(z)$, $h_2(z)$ be analytic for $|z| < r$. Then there exists a unique solution $w_1(z)$ and $w_2(z)$ satisfying simultaneously*

$$
\begin{aligned}
w_1' &= f_1(z) + w_1 g_1(z) + w_2 h_1(z), \\
w_2' &= f_2(z) + w_1 g_2(z) + w_2 h_2(z),
\end{aligned}
$$

and $w_1(0) = w_2(0) = 0$, and analytic for $|z| < r$.

Proof: We use the Cauchy Majorant method. The given functions have the following representations

$$
\begin{aligned}
f_1(z) &= a_{10} + a_{11}z + a_{12}z^2 + \cdots \\
f_2(z) &= a_{20} + a_{21}z + a_{22}z^2 + \cdots \\
g_1(z) &= b_{10} + b_{11}z + b_{12}z^2 + \cdots \\
g_2(z) &= b_{20} + b_{21}z + b_{22}z^2 + \cdots \\
h_1(z) &= c_{10} + c_{11}z + c_{12}z^2 + \cdots \\
h_2(z) &= c_{20} + c_{21}z + c_{22}z^2 + \cdots
\end{aligned}
$$

valid for $|z| \le \rho < r$. The following inequalities hold:

$$|a_{ij}| \le M\rho^{-j}, \ |b_{ij}| \le M\rho^{-j}, \ |c_{ij}| \le M\rho^{-j},$$

$i = 1, 2; j = 0, 1, 2, \ldots$. We assume solutions

$$
\begin{aligned}
w_1(z) &= d_{11}z + d_{12}z^2 + d_{13}z^3 + \cdots \\
w_2(z) &= d_{21}z + d_{22}z^2 + d_{23}z^3 + \cdots
\end{aligned}
$$

Then

$$d_{11} + 2d_{12}z + 3d_{13}z^3 + \cdots$$
$$= a_{10} + a_{11}z + a_{12}z^2 + \cdots$$
$$+ (d_{11}z + d_{12}z^2 + \cdots)(b_{10} + b_{11}z + b_{12}z^2 + \cdots)$$
$$+ (d_{21}z + d_{22}z^2 + \cdots)(c_{10} + c_{11}z + c_{12}z^2 + \cdots),$$

$$d_{21} + 2d_{22}z + 3d_{23}z^2 + \cdots$$
$$= a_{20} + a_{21}z + a_{22}z^2 + \cdots$$
$$+ (d_{11}z + d_{12}z^2 + \cdots)(b_{20} + b_{21}z + b_{22}z^2 + \cdots)$$
$$+ (d_{21}z + d_{22}z^2 + \cdots)(c_{20} + c_{21}z + c_{22}z^2 + \cdots).$$

Equating coefficients of like powers of z, we have

$$d_{11} = a_{10} \qquad\qquad\qquad d_{21} = a_{20}$$
$$2d_{12} = a_{11} + d_{11}b_{10} + d_{21}c_{10} \qquad 2d_{22} = a_{21} + d_{11}b_{20} + d_{21}c_{20}$$
$$3d_{13} = a_{12} + d_{11}b_{11} + d_{12}b_{10} \qquad 3d_{23} = a_{22} + d_{11}b_{21} + d_{12}b_{20}$$
$$+ d_{21}c_{11} + d_{22}c_{10} \qquad\qquad + d_{21}c_{21} + d_{22}c_{20}$$

$$\cdots\cdots\cdots\cdots\cdots\cdots \qquad\qquad \cdots\cdots\cdots\cdots\cdots\cdots$$

These equations determine unique formal power series. We next consider the related system of equations

$$W_1' = F_1(z) + W_1 G_1(z) + W_2 H_1(z),$$
$$W_2' = F_2(z) + W_1 G_2(z) + W_2 H_2(z),$$

where

$$F_1(z) = A_{10} + A_{11}z + A_{12}z^2 + \cdots,$$
$$F_2(z) = A_{20} + A_{21}z + A_{22}z^2 + \cdots,$$
$$G_1(z) = B_{10} + B_{11}z + B_{12}z^2 + \cdots,$$
$$G_2(z) = B_{20} + B_{21}z + B_{22}z^2 + \cdots,$$
$$H_1(z) = C_{10} + C_{11}z + C_{12}z^2 + \cdots,$$
$$H_2(z) = C_{20} + C_{21}z + C_{22}z^2 + \cdots,$$

are majorants for $f_1, f_2, g_1, g_2, h_1, h_2$, respectively. We determine the unique formal power series for $W_1(z)$ and $W_2(z)$, as above. Let

$$W_1(z) = D_{11}z + D_{12}z^2 + \cdots$$
$$W_2(z) = D_{21}z + D_{22}z^2 + \cdots$$

Then

$$D_{11} = A_{10} \qquad\qquad D_{21} = A_{20}$$
$$2D_{12} = A_{11} + D_{11}B_{10} \qquad 2D_{22} = A_{21} + D_{11}B_{20}$$
$$+ D_{21}C_{10} \qquad\qquad + D_{21}C_{20}$$
$$3D_{13} \qquad\qquad 3D_{23}$$
$$= A_{12} + D_{11}B_{11} + D_{12}B_{10} \qquad = A_{22} + D_{11}B_{21} + D_{12}B_{20}$$
$$+ D_{21}C_{11} + D_{22}C_{10} \qquad\qquad + D_{21}C_{21} + D_{22}C_{20}$$

. . . .

We prove that $W_1(z)$ and $W_2(z)$ are majorants for $w_1(z)$ and $w_2(z)$, respectively.

$$|d_{11}| = |a_{10}| \le A_{10} = D_{11} \qquad |d_{21}| = |a_{20}| \le A_{20} = D_{21}$$
$$2|d_{12}| \le |a_{11}| + |a_{10}||b_{10}| \qquad 2|d_{22}| \le |a_{21}| + |a_{10}||b_{20}|$$
$$+ |a_{20}||c_{10}| \qquad\qquad + |a_{20}||c_{20}|$$
$$\le A_{11} + A_{10}B_{10} \qquad\qquad \le A_{21} + A_{10}B_{20}$$
$$+ A_{20}C_{10} = 2D_{12} \qquad\qquad + A_{20}C_{20} = 2D_{22}$$

. . . .

We let

$$F_1(z) = F_2(z) = G_1(z) = G_2(z) = H_1(z) = H_2(z)$$
$$= M\left[1 + \frac{z}{\rho} + \left(\frac{z}{\rho}\right)^2 + \cdots\right] = \frac{\rho M}{\rho - z}.$$

Then

$$W_1' = \frac{\rho M}{\rho - z}\,(1 + W_1 + W_2),$$

$$W_2' = \frac{\rho M}{\rho - z}\,(1 + W_1 + W_2),$$

$$W_1' = W_2',$$

$$W_1' + W_2' = \frac{2\rho M}{\rho - z}\,(1 + W_1 + W_2).$$

Therefore, $W_1 = W_2 + C$, where C is a constant of integration. However, since $W_1(0) = W_2(0) = 0$, $C = 0$, and

$$W_1' = \frac{\rho M}{\rho - z}\,(1 + 2W_1),$$

$$\frac{dW_1}{1 + 2W_1} = \frac{\rho M}{\rho - z}\,dz,$$

$$\text{Log}\,(1 + 2W_1) = -2\rho M\,\text{Log}\left(1 - \frac{z}{\rho}\right) + C',$$

where C' is a constant of integration. But $W_1(0) = 0$, implying that $C' = 0$. Therefore,

$$W_1(z) = W_2(z) = \frac{1}{2}\left(1 - \frac{z}{\rho}\right)^{-2\rho M} - \frac{1}{2}.$$

These functions are analytic for $|z| < \rho$. Therefore, by comparison $w_1(z)$ and $w_2(z)$ have convergent power series expansions valid for $|z| < r$, since the above analysis holds for every $0 < \rho < r$. This completes the proof.

EXAMPLE 7.2.4.   Let $f(z)$, $g(z)$, and $h(z)$ be analytic for $|z| < r$. Prove that $w'' + f(z)w' + g(z)w = h(z)$, $w(0) = a$, $w'(0) = b$, has a unique analytic solution for $|z| < r$. Let $w_1(z) = w(z)$, $w_2(z) = w'(z)$, $w_1(0) = a$, $w_2(0) = b$. We then have

$$w_1' = w_2,$$
$$w_2' = h(z) - f(z)w_2 - g(z)w_1.$$

The system is linear, so the result follows from Theorem 7.2.4, after we make a preliminary change of variables $w_1 - a \to w_1$, $w_2 - b \to w_2$.

Exercises 7.2

1.   Consider the ordinary differential equation $\ddot{y}(t) = -f(t)$, where f is a real-valued continuous function of t for $t \geq 0$. Show that y satisfies the differential equation and the initial conditions $y(0) = a$, $\dot{y}(0) = b$ if and only if it is given by the integral

$$a + bt + \int_0^t (x - t)f(x)\,dx.$$

Show that

$$G(x, t) = x - t, 0 \leq x \leq t$$

satisfies the following conditions: $G_x = 1, 0 < x < t$, and $G(t, t) = 0$. $G(x, t)$ is called the *Green's function* for this initial value problem.

2.   Consider the following initial value problem $\ddot{y}(t) + y(t) = 0$, $y(0) = a$, $\dot{y}(0) = b$. Show that y solves this problem if and only if it satisfies the following *Volterra integral equation*:

$$y(t) = a + bt + \int_0^t G(x, t)y(x)\,dx$$

where G is the Green's function of Problem 1.

3.   Prove that there exists a unique solution of the initial value problem in Problem 2 using iteration as in the proof of Theorem 7.2.1.

4.   Generalize the results of problems 1-3 to the more general initial value problem $\ddot{y} + c_1\dot{y} + c_2 y = -f(t)$, $y(0) = a$, $\dot{y}(0) = b$, where c_1 and c_2 are constants and $f(t)$ is continuous for $t \geq 0$. Hint: find a Green's function which satisfies $G_x - c_1 G = 1$ and $G(t, t) = 0$.

7.3. SOLUTION OF A LINEAR SECOND-ORDER DIFFERENTIAL EQUATION NEAR AN ORDINARY POINT

The most general linear second-order ordinary differential equation is of the form

$$a_2(z)w'' + a_1(z)w' + a_0(z)w = b(z).$$

We shall consider the problem of solving this equation subject to the conditions $w(z_0) = w_0$, $w'(z_0) = w_0'$, assuming that the functions a_0, a_1, a_2, and b are analytic in some disk $|z - z_0| < R$. Without loss of generality, we can assume that $z_0 = 0$, for if not, the change of variable $z - z_0 \rightarrow z$ will achieve this.

We say that the origin is an *ordinary point* of the differential equation if $a_2(0) \neq 0$. Since $a_2(z)$ is analytic at the origin, there exists an ϵ-neighborhood of the origin throughout which $a_2(z) \neq 0$. Let us assume that $a_2(z) \neq 0$ for $|z| < R$. Then we can write

$$w'' + p(z)w' + q(z)w = r(z),$$

where $p(z) = a_1(z)/a_2(z)$, $q(z) = a_0(z)/a_2(z)$, and $r(z) = b(z)/a_2(z)$ are analytic for $|z| < R$.

Definition 7.3.1. Two functions $w_1(z)$ and $w_2(z)$ are linearly independent in $|z| < R$ if and only if there exist no constants c_1 and c_2, not both zero, such that $c_1w_1(z) + c_2w_2(z) \equiv 0$ in $|z| < R$. If $w_1(z)$ and $w_2(z)$ are not linearly independent they are said to be linearly dependent.

If w_1 and w_2 are linearly dependent then they are proportional, that is, either $w_1 \equiv kw_2$ or $w_2 \equiv kw_1$, since if $c_1 \neq 0$, then $w_1 \equiv (-c_2/c_1)w_2$, or $c_2 \neq 0$, then $w_2 \equiv (-c_1/c_2)w_1$. If $w_1 \equiv 0$, then w_1 and *any* other function are linearly dependent, since $c_1 = 1$, $c_2 = 0$ will cause $c_1w_1 + c_2w_2 \equiv 0$. If w_1 and w_2 are analytic and there exists a set of constants c_1 and c_2, not both zero, such that $c_1w_1 + c_2w_2 = 0$ at a set of points with a limit point in $|z| < R$, then w_1 and w_2 are linearly dependent. This follows because $c_1w_1 + c_2w_2$ is analytic in $|z| < R$ and vanishes on a set of points with a limit point in the given domain. Hence, it vanishes identically.

Definition 7.3.2. Let $w_1(z)$ and $w_2(z)$ be analytic for $|z| < R$. Then the function $W(z) = w_1w_2' - w_2w_1'$ is called the Wronskian of w_1 and w_2.

Theorem 7.3.1. *Let $w_1(z)$ and $w_2(z)$ be analytic for $|z| < R$. Then w_1 and w_2 are linearly independent if and only if the Wronskian of w_1 and w_2 is not zero at some point z_0 such that $|z_0| < R$.*

Proof: Assume that w_1 and w_2 are linearly independent. Then there exists no constants c_1 and c_2, not both zero, such that

$$c_1w_1 + c_2w_2 \equiv 0.$$

In other words, $c_1w_1 + c_2w_2 \equiv 0$ implies $c_1 = c_2 = 0$. If $c_1w_1 + c_2w_2 \equiv 0$, then $c_1w_1' + c_2w_2' \equiv 0$. Let $|z_0| < R$. Then

$$c_1w_1(z_0) + c_2w_2(z_0) = 0,$$
$$c_1w_1'(z_0) + c_2w_2'(z_0) = 0,$$

should have only the trivial solution $c_1 = c_2 = 0$. This implies that the determinant

$$\begin{vmatrix} w_1(z_0) & w_2(z_0) \\ w_1'(z_0) & w_2'(z_0) \end{vmatrix} = w_1(z_0)w_2'(z_0) - w_2(z_0)w_1'(z_0) = W(z_0) \neq 0.$$

Conversely, if w_1 and w_2 are linearly dependent, then either $w_1 \equiv kw_2$ or $w_2 \equiv kw_1$. In either case, $W(z) \equiv 0$.

Notice that the vanishing of the Wronskian at one point does not imply linear dependence, since $w_1 = z^2$, $w_2 = z^3$ are linearly independent even though their Wronskian vanishes at the origin. On the other hand the vanishing of the Wronskian on a set of points with a limit point in $|z| < R$ implies the identical vanishing of the Wronskian which implies linear dependence.

Theorem 7.3.2. *Let $w_1(z)$ and $w_2(z)$ be analytic solutions of $w'' + p(z)w' + q(z)w = 0$, where $p(z)$ and $q(z)$ are analytic for $|z| < R$. Then w_1 and w_2 are linearly independent if and only if their Wronskian never vanishes for $|z| < R$.*

Proof: Since both w_1 and w_2 satisfy the differential equation, we have

$$w_1'' + p(z)w_1' + q(z)w_1 = 0,$$
$$w_2'' + p(z)w_2' + q(z)w_2 = 0.$$

Multiplying the first equation by w_2 and the second by w_1 and subtracting, we have

$$w_1w_2'' - w_2w_1'' + p(z)(w_1w_2' - w_2w_1') = 0,$$
$$\frac{d}{dz}(w_1w_2' - w_2w_1') + p(z)(w_1w_2' - w_2w_1') = 0,$$
$$W' + p(z)W = 0.$$

Let $|z_0| < R$. Then there is a unique solution of $W' + p(z)W = 0$ satisfying $W(z_0) = W_0$, for $|z - z_0| < R - |z_0|$, which is given by

$$W(z) = W_0 \exp\left[-\int_{z_0}^{z} p(\varsigma)\, d\varsigma \right],$$

where the straight line path of integration can be taken. Clearly, if $W(z_0) = 0$, then $W(z) \equiv 0$ in $|z - z_0| < R - |z_0|$ and hence also vanishes for $|z| < R$. If $W(z_0) \neq 0$, then $W(z)$ never vanishes for $|z - z_0| < R - |z_0|$, since the function $\exp\left[-\int_{z_0}^{z} p(\zeta)\,d\zeta\right]$ never vanishes, and by analytic continuation $W(z)$ never vanishes for $|z| < R$. If w_1 and w_2 are linearly independent then there is at least one z_0, $|z_0| < R$, such that $W(z_0) \neq 0$ and, consequently, $W(z)$ never vanishes for $|z| < R$. Conversely, if $W(z_0) = 0$, $W(z) \equiv 0$ and this proves that w_1 and w_2 are linearly dependent. This completes the proof.

We now return to the solution of $w'' + p(z)w' + q(z)w = r(z)$ near an ordinary point, which we assume is the origin. By Example 7.2.4, this equation has a unique solution for $|z| < R$ satisfying $w(0) = a$ and $w'(0) = b$. If a and b are thought of as parameters, then we have a two parameter family of solutions. Another two parameter family of solutions is

$$w(z) = c_1 w_1(z) + c_2 w_2(z) + u(z),$$

where w_1 and w_2 are linearly independent solutions of the *homogeneous equation* $w'' + p(z)w' + q(z) = 0$, and u is any particular solution of the full equation. We can show that from this family of solutions we can always select the unique solution satisfying $w(0) = a$ and $w'(0) = b$. We first must show that for every c_1 and c_2 we have a solution. Clearly,

$$\begin{aligned} w'' + pw' + qw = {} & c_1[w_1'' + pw_1' + qw_1] \\ & + c_2[w_2'' + pw_2' + qw_2] \\ & + u'' + pu' + qu = r(z), \end{aligned}$$

for any c_1 and c_2. Finally,

$$\begin{aligned} w(0) &= c_1 w_1(0) + c_2 w_2(0) + u(0) = a \\ w'(0) &= c_1 w_1'(0) + c_2 w_2'(0) + u'(0) = b \end{aligned}$$

can always be solved uniquely for c_1 and c_2 since $W(0) = w_1(0)w_2'(0) - w_2(0)w_1'(0) \neq 0$. This is a well-known theorem from the theory of equations. We can summarize this result in the following theorem.

Theorem 7.3.3. *Let $w_1(z)$ and $w_2(z)$ be two linearly independent solutions of the homogeneous linear differential equation $w'' + p(z)w' + q(z) = 0$ for $|z| < R$, where $p(z)$ and $q(z)$ are analytic. Let $u(z)$ be a particular solution of the nonhomogeneous equation $w'' + p(z)w' + q(z)w = r(z)$, where $r(z)$ is analytic for $|z| < R$. Then $c_1 w_1 + c_2 w_2 + u$ is a general solution for arbitrary constants c_1 and c_2 which includes the unique solution of $w'' + p(z)w' + q(z)w = r(z)$, satisfying $w(0) = a$, $w'(0) = b$, valid for $|z| < R$.*

Therefore, for the solution of the second order linear differential equation near an ordinary point the problem reduces to finding the w_1, w_2, and u. Consider the homogeneous equation $w'' + p(z)w' + q(z) = 0$. We already know from Theorem 7.2.4 that there is a power series solution satisfying

$w(0) = c_0$, $w'(0) = c_1$, and converging for $|z| < R$. To determine this power series we assume that

$$w(z) = c_0 + c_1 z + c_2 z^2 + \cdots,$$
$$w'(z) = c_1 + 2c_2 z + 3c_3 z^2 + \cdots,$$
$$w''(z) = 2c_2 + 6c_3 z + 12c_4 z^2 + \cdots.$$

Let

$$p(z) = a_0 + a_1 z + a_2 z^2 + \cdots$$
$$q(z) = b_0 + b_1 z + b_2 z^2 + \cdots$$

valid for $|z| < R$. Substituting in the equation, we have

$$(2c_2 + 6c_3 z + 12c_4 z^2 + \cdots)$$
$$+ (a_0 + a_1 z + a_2 z^2 + \cdots)(c_1 + 2c_2 z + 3c_3 z^2 + \cdots)$$
$$+ (b_0 + b_1 z + b_2 z^2 + \cdots)(c_0 + c_1 z + c_2 z^2 + \cdots) = 0.$$

Equating coefficients of the various powers of z to zero, we have

$$2c_2 + a_0 c_1 + b_0 c_0 = 0,$$
$$6c_3 + 2a_0 c_2 + a_1 c_1 + b_1 c_0 + b_0 c_1 = 0,$$
$$12c_4 + a_2 c_1 + 2a_1 c_2 + 3a_0 c_3 + b_0 c_2 + b_1 c_1 + b_2 c_0 = 0,$$
$$c_2 = -(a_0 c_1 + b_0 c_0)/2,$$
$$c_3 = -(2a_0 c_2 + a_1 c_1 + b_1 c_0 + b_0 c_1)/6,$$
$$c_4 = -(a_2 c_1 + 2a_1 c_2 + 3a_0 c_3 + b_0 c_2 + b_1 c_1 + b_2 c_0)/12,$$

and so on. Hence, the coefficients are uniquely and recursively determined. Let $w_1(0) = 1$ and $w_1'(0) = 0$; then

$$w_1(z) = 1 - (b_0/2)z^2 + [(a_0 b_0 - b_1)/6]z^3 + \cdots.$$

Let $w_2(0) = 0$ and $w_2'(0) = 1$; then

$$w_2(z) = z - (a_0/2)z^2 + [(a_0^2 - a_1 - b_0)/6]z^3 + \cdots.$$

The Wronskian of w_1 and w_2 at the origin is $w_1(0)w_2'(0) - w_2(0)w_1'(0) = 1$. Therefore, w_1 and w_2 are linearly independent.

To determine the particular solution $u(z)$ we use a method known as *variation of parameters.* Assume a solution of the form

$$u(z) = A(z)w_1(z) + B(z)w_2(z).$$

Then

$$u'(z) = A(z)w_1' + B(z)w_2' + A'w_1(z) + B'w_2(z).$$

At this point we set $A'w_1 + B'w_2 = 0$ and differentiate again.

$$u''(z) = A(z)w_1'' + B(z)w_2'' + A'w_1' + B'w_2'.$$

Substituting in the differential equation, we have

$$A'w_1' + B'w_2' + A(z)[w_1'' + pw_1' + qw_1]$$
$$+ B(z)[w_2'' + pw_2' + qw_2] = r(z).$$

Hence, we have to solve the following system for A' and B'

$$A'w_1 + B'w_2 = 0,$$
$$A'w_1' + B'w_2' = r(z).$$

Solving for A' and B', we have

$$A' = -r(z)w_2(z)/W(z), \qquad B' = r(z)w_1(z)/W(z),$$

where $W(z)$ is the Wronskian of w_1 and w_2. If $|z| < R$, we can integrate along a straight line path from the origin to z. Hence,

$$A(z) = -\int_0^z \frac{r(\zeta)w_2(\zeta)}{W(\zeta)} d\zeta, \qquad B(z) = \int_0^z \frac{r(\zeta)w_1(\zeta)}{W(\zeta)} d\zeta,$$

$$u(z) = \int_0^z r(\zeta)G(z, \zeta) d\zeta,$$

where $G(z, \zeta) = (w_1(\zeta)w_2(z) - w_1(z)w_2(\zeta))/W(\zeta)$ is an analytic function of z and ζ for $|z| < R$ and $|\zeta| < R$. $G(z, \zeta)$ is a *Green's function* associated with the given differential equation.

Exercises 7.3

1. By making the transformation $z = 1/\zeta$ in the equation

$$w'' + p(z)w' + q(z)w = r(z)$$

state a criterion for the differential equation to have an ordinary point at infinity. Show that, under these circumstances, the substitution

$$w(z) = c_0 + \frac{c_1}{z} + \frac{c_2}{z^2} + \cdots$$

will determine a series solution valid for large values of $|z|$.

2. Find two linearly independent solutions of $(1 - z^2)w'' - 2zw' + n(n + 1)w = 0$ valid for $|z| < 1$. Show that one solution is a polynomial in the case where n is a nonnegative integer.

3. Find two linearly independent solutions of

$$w'' - z^2w + (2n + 1)w = 0,$$

where n is a nonnegative integer. In case $n = 0$, show that $e^{-z^2/2}$ is a solution. Make the transformation $w = e^{-z^2/2}u(z)$ and show that the resulting equation for u has polynomial solutions.

4. Find the general solution of $w'' + zw' + 2w = z$.

5. Let $p(z)$ be meromorphic in the unextended plane. Let $w_1(z)$ and $w_2(z)$ be linearly independent solutions of $w'' + pw' + qw = 0$ in $|z - z_0| < R$ which does not contain any singularities of p or q. Let w_1 and w_2 be analytically continued around a simple contour C to ω_1 and ω_2. Show that $W(\omega_1, \omega_2) = W(w_1, w_2)e^{-2\pi i S}$, where S is the sum of the residues at the poles of p inside C.

7.4. SOLUTION OF A LINEAR SECOND-ORDER DIFFERENTIAL EQUATION NEAR A REGULAR SINGULAR POINT

Recall that the most general linear second-order ordinary differential equation is of the form

$$a_2(z)w'' + a_1(z)w' + a_0(z)w = b(z),$$

where we shall again assume that a_0, a_1, a_2, and b are analytic for $|z - z_0| < R$. In the previous section, we assumed that $a_2(z)$ never vanished in this same disk. A case which comes up too frequently to be ignored is the case where $a_2(z)$ has a zero of order two at z_0 and $a_1(z)$ has a zero of order one at z_0. In this case, we say that z_0 is a *regular singular point* of the differential equation. Again we can assume that $z_0 = 0$ and hence that the differential equation can be written as

$$z^2 b_2(z)w'' + zb_1(z)w' + a_0(z)w = b(z)$$

or

$$z^2 w'' + zp(z)w' + q(z)w = r(z),$$

where $p(z) = b_1(z)/b_2(z)$, $q(z) = a_0(z)/b_2(z)$, and $r(z) = b(z)/b_2(z)$ are analytic for $|z| < R$. We shall prove that every solution of this equation in the annulus $0 < |z| < R$ can be written in the form $z^m u(z)$ or

$$z^m[u(z)\log z + z^n v(z)],$$

where $u(z)$ and $v(z)$ are analytic in $|z| < R$ and n is an integer. This condition is also a sufficient condition for $z = 0$ to be a regular singular point of the differential equation.* We shall not prove this. Instead, we shall use as our justification for treating this special case the fact that it occurs frequently in the applications.

---

* B. A. Fuchs and V. I. Levin, *Functions of a Complex Variable and some of their Applications*, II. Reading, Mass.: Addison-Wesley Publishing Company, Inc., 1961, p. 114.

We shall begin by considering the homogeneous differential equation

$$z^2 w'' + zp(z)w' + q(z)w = 0,$$

where $p(z)$ and $q(z)$ are analytic for $|z| < R$. A closely related differential equation is Euler's equation

$$z^2 w'' + zp_0 w' + q_0 w = 0,$$

where p_0 and q_0 are constants. It is well known that the assumption $w = z^m$ leads to a solution. Substituting, we have

$$[m(m - 1) + p_0 m + q_0]z^m = 0,$$

if m is a root of the *indicial equation*

$$m(m - 1) + p_0 m + q_0 = 0.$$

This equation has, in general, two complex roots, m_1 and m_2. Let $p(0) = p_0$ and $q(0) = q_0$. Then since $p(z)$ and $q(z)$ are near p_0 and q_0 for small $|z|$, it is reasonable to assume that a solution of $z^2 w'' + zp(z)w' + q(z)w = 0$ is near z^{m_1} or z^{m_2} for small values of $|z|$. For this reason we try a solution of the form

$$w(z) = z^{m_1}(c_0 + c_1 z + c_2 z^2 + \cdots) = z^{m_1} u(z),$$

where $\text{Re}(m_1) \geq \text{Re}(m_2)$. Substituting in the differential equation, we have

$$z^{m_1+2} u'' + [2m_1 + p(z)]z^{m_1+1} u'$$
$$+ [m_1(m_1 - 1) + m_1 p(z) + q(z)]z^{m_1} u = 0.$$

The function $m_1(m_1 - 1) + m_1 p(z) + q(z)$ is analytic for $|z| < R$ and vanishes at $z = 0$ since $m_1(m_1 - 1) + m_1 p_0 + q_0 = 0$. Therefore, the differential equation can be written as

$$zu'' + f(z)u' + g(z)u = 0,$$

where $f(z)$ and $g(z)$ are analytic for $|z| < R$ and

$$f(0) = 2m_1 + p_0 = 2m_1 + (1 - m_1 - m_2) = m_1 - m_2 + 1 \neq 0,$$

since $\text{Re}[f(0)] = \text{Re}(m_1 - m_2 + 1) \geq 1$. We shall prove that there exists a unique solution of this equation satisfying $u(0) = c_0$ and analytic for $|z| < R$.

Theorem 7.4.1. *Let* $f(z)$ *and* $g(z)$ *be analytic for* $|z| < R$ *with* $\text{Re}[f(0)] \geq 1$. *Then the differential equation* $zu'' + f(z)u' + g(z)u = 0$ *has a unique solution satisfying* $u(0) = c_0$ *and analytic for* $|z| < R$.

Proof: We assume a power series

$$u(z) = \sum_{k=0}^{\infty} c_k z^k,$$

$$u'(z) = \sum_{k=1}^{\infty} k c_k z^{k-1},$$

$$u''(z) = \sum_{k=2}^{\infty} k(k-1) c_k z^{k-2}.$$

Now, $f(z)$ and $g(z)$ have power series representations valid for $|z| < R$

$$f(z) = \sum_{k=0}^{\infty} a_k z^k, \qquad g(z) = \sum_{k=0}^{\infty} b_k z^k.$$

Substituting in the differential equation, we have

$$\sum_{k=2}^{\infty} k(k-1) c_k z^{k-1} + \left(\sum_{k=0}^{\infty} a_k z^k \right) \left(\sum_{k=1}^{\infty} k c_k z^{k-1} \right)$$

$$+ \left(\sum_{k=0}^{\infty} b_k z^k \right) \left(\sum_{k=0}^{\infty} c_k z^k \right) = 0,$$

$$\sum_{k=0}^{\infty} (k+1)(k+a_0) c_{k+1} z^k + \sum_{k=0}^{\infty} \sum_{j=1}^{k+1} (k+1-j) a_j c_{k+1-j} z^k$$

$$+ \sum_{k=0}^{\infty} \sum_{j=0}^{k} b_j c_{k-j} z^k = 0.$$

Setting the coefficients of the various powers of z equal to zero, we have

$$a_0 c_1 + b_0 c_0 = 0,$$
$$2(1+a_0) c_2 + a_1 c_1 + b_0 c_1 + b_1 c_0 = 0,$$
$$3(2+a_0) c_3 + 2a_1 c_2 + a_2 c_1 + b_0 c_2 + b_1 c_1 + b_2 c_0 = 0,$$

. . . .

$$(k+1)(k+a_0) c_{k+1} + \sum_{j=1}^{k+1} (k+1-j) a_j c_{k+1-j} + \sum_{j=0}^{k} b_j c_{k-j} = 0.$$

Therefore, the differential equation determines a unique formal power series. If we can prove that this formal series converges for $|z| < R$, we will have proved the existence of a unique solution analytic for $|z| < R$.

Let $|z| \le \rho < R$. Then $|a_k| \le A\rho^{-k}$, $|b_k| \le B\rho^{-k} = (B\rho)\rho^{-k-1}$, where $A = \max_{|z| = \rho} |f(z)|$ and $B = \max_{|z| = \rho} |g(z)|$. Let $M = \max [A, B\rho]$. Then

$$|a_k| \le M\rho^{-k}, \qquad |b_k| \le M\rho^{-k-1}.$$

Let $|c_0| = C_0$. Then $|a_0||c_1| \leq M\rho^{-1}C_0$. Let $|a_0|C_1 = M\rho^{-1}C_0$. Then, clearly, $|c_1| \leq C_1$. Also,

$$2|1 + a_0||c_2| \leq |a_1||c_1| + |b_0||c_1| + |b_1||c_0|$$
$$\leq 2M\rho^{-1}C_1 + M\rho^{-2}C_0.$$

Let $2|1 + a_0|C_2 = 2M\rho^{-1}C_1 + M\rho^{-2}C_0$. Then $|c_2| \leq C_2$, and

$$3|2 + a_0||c_3| \leq 2|a_1||c_2| + |b_0||c_2| + |a_2||c_1| + |b_1||c_1| + |b_2||c_0|$$
$$\leq 3M\rho^{-1}C_2 + 2M\rho^{-2}C_1 + M\rho^{-3}C_0.$$

We let

$$3|2 + a_0|C_3 = 3M\rho^{-1}C_2 + 2M\rho^{-2}C_1 + M\rho^{-3}C_0,$$

in which case $|c_3| \leq C_3$. We are constructing a majorant $C_0 + C_1 z + C_2 z^2 + C_3 z^3 + \cdots$ for our formal power series such that $|c_k| \leq C_k$. Finally, we have to show that the majorant converges. The C's satisfy the following recurrence relation

$$k|k - 1 + a_0|C_k = kM\rho^{-1}C_{k-1} + (k - 1)M\rho^{-2}C_{k-2} + \cdots + M\rho^{-k}C_0.$$

Then, clearly,

$$\rho(k + 1)|k + a_0|C_{k+1} = (k + 1)MC_k + kM\rho^{-1}C_{k-1} + \cdots$$
$$+ M\rho^{-k}C_0.$$

Subtracting we have

$$\rho(k + 1)|k + a_0|C_{k+1} - k|k - 1 + a_0|C_k = (k + 1)MC_k$$

or

$$\frac{C_{k+1}}{C_k} = \frac{k|k - 1 + a_0| + (k + 1)M}{\rho(k + 1)|k + a_0|}.$$

By the ratio test

$$\lim_{k \to \infty} \frac{C_{k+1}|z|}{C_k} = \frac{|z|}{\rho} < 1,$$

the majorant series converges for $|z| < \rho$. This implies that

$$u(z) = c_0 + c_1 z + c_2 z^2 + \cdots$$

converges for $|z| < \rho$. But ρ is any positive number less than R. Therefore, the series converges for $|z| < R$. This completes the proof.

Our first reaction to this result is that we should not have gotten a unique solution since we specified the initial value $u(0) = c_0$ but not the derivative $u'(0)$, whereas in Example 7.2.4 where the origin was an ordinary point, we had to specify the initial value of the solution and its derivative in order to get a unique solution. We suspect therefore that there must be another solution of

$$z^2 w'' + z p(z) w' + q(z) w = 0$$

valid near the origin. Let us make the substitution

$$w = w_1(z) v(z),$$

where $w_1(z) = z^{m_1} u(z)$, which satisfies the differential equation for $0 < |z| < R$. Then

$$w' = w_1' v + w_1 v',$$

$$w'' = w_1'' v + 2 w_1' v' + w_1 v'',$$

$$(z^2 w_1'' + z p w_1' + q w_1) v + z^2 w_1 v'' + (2 z^2 w_1' + z p w_1) v' = 0,$$

$$v'' + \left[\frac{2 w_1'}{w_1} + \frac{p(z)}{z} \right] v' = 0,$$

for $0 < |z| < R$. Then

$$\frac{v''}{v'} = - \left[\frac{2 w_1'}{w_1} + \frac{p(z)}{z} \right],$$

$$\log v' = -2 \log w_1 - \int_{z_0}^{z} \frac{p(\zeta)}{\zeta} \, d\zeta,$$

where z_0 is some fixed point in the annulus and the path of integration lies in the annulus. Taking exponentials we have

$$= \frac{1}{w_1^2} \exp \left[- \int_{z_0}^{z} \frac{p(\zeta)}{\zeta} \, d\zeta \right],$$

$$w = w_1 v = w_1(z) \int_{z_0}^{z} \frac{1}{w_1^2} \exp \left[- \int_{z_0}^{\zeta} \frac{p(t)}{t} \, dt \right] d\zeta.$$

Let us cut the annulus by restricting the arg z as follows: $-\pi < \arg z \le \pi$. The first solution $w_1(z)$ may be multiple-valued by virtue of z^{m_1}. However, if we select a definite branch of w_1, then

$$w_2 = w_1 \int_{z_0}^{z} \frac{1}{w_1^2} \exp \left[- \int_{z_0}^{\zeta} \frac{p(t)}{t} \, dt \right] d\zeta$$

is analytic in the cut annulus. In the cut annulus w_1 and w_2 are linearly independent, since

$$\begin{vmatrix} w_1 & w_2 \\ w_1' & w_2' \end{vmatrix} = w_1^2 v' = \exp \left[- \int_{z_0}^{z} \frac{p(\zeta)}{\zeta} \, d\zeta \right]$$

which never vanishes.

Let us examine w_2 more carefully. Let $p(z) = p_0 + p_1 z + p_2 z^2 + \cdots$. Then

$$\int_{z_0}^{z} \frac{p(\zeta)}{\zeta} \, d\zeta = p_0 \log z + p_1 z + p_2 \frac{z^2}{2} + \cdots + C,$$

where C is a constant of integration depending on z_0 and the branch of $\log z$ chosen. Let $P(z) = C + p_1 z + p_2 z^2/2 + \cdots$. Then

$$\exp\left[-\int_{z_0}^{z} \frac{p(\zeta)}{\zeta} \, d\zeta\right] = z^{-p_0} e^{-P(z)} = z^{-p_0}(\alpha_0 + \alpha_1 z + \alpha_2 z^2 + \cdots)$$

$$\frac{1}{w_1^2} \exp\left[-\int_{z_0}^{z} \frac{p(\zeta)}{\zeta} \, d\zeta\right] = z^{-2m_1 - p_0} \frac{1}{u^2} (\alpha_0 + \alpha_1 z + \alpha_2 z^2 + \cdots)$$

$$= z^{-2m_1 - p_0}(\beta_0 + \beta_1 z + \beta_2 z^2 + \cdots).$$

Since the indicial equation is $m^2 - (1 - p_0)m + q_0 = 0$, $m_1 + m_2 = 1 - p_0$ and $2m_1 + p_0 = m_1 - m_2 + 1$

$$w_2(z) = w_1 \int_{z_0}^{z} t^{-m_1 + m_2 - 1}(\beta_0 + \beta_1 t + \beta_2 t^2 + \cdots) \, dt.$$

Now let us consider three separate cases.

Case 1. $m_1 = m_2$

In this case

$$w_2(z) = z^{m_1} u(z) \int_{z_0}^{z} (\beta_0 t^{-1} + \beta_1 + \beta_2 t + \cdots) \, dt$$

$$= z^{m_1} u(z)[\beta_0 \log z + B(z)],$$

where $u(z)$ and $B(z)$ are analytic for $|z| < R$. The logarithm term is never missing since $\beta_0 = \alpha_0/c_0$ and $\alpha_0 = e^{-P(0)} \neq 0$ and $c_0 \neq 0$, if $w_1(z)$ is not identically zero.

Case 2. $m_1 - m_2 = n$ a positive integer

In this case

$$w_2(z) = z^{m_1} u(z) \int_{z_0}^{z} (\beta_0 t^{-n-1} + \beta_1 t^{-n} + \beta_2 t^{-n+1} + \cdots + \beta_n t^{-1} + \cdots) \, dt$$

$$= z^{m_1} u(z)\left[\frac{\beta_0 z^{-n}}{-n} + \frac{\beta_1 z^{-n+1}}{-n+1} + \cdots + \beta_n \log z + A(z)\right],$$

where $A(z)$ is analytic for $|z| < R$. In this case the logarithm term may be missing if $\beta_n = 0$. If this is the case, then the solution takes the form $w_2 = z^{m_2} \phi(z)$, where ϕ is analytic for $|z| < R$. This suggests that in certain cases the substitution $w = z^{m_2}\phi(z)$ will lead to the determination of a power series for $\phi(z)$ just as we were led to the above by the substitution $w = z^{m_1}u(z)$.

Case 3. $m_1 - m_2$ not an integer

In this case

$$w_2(z) = z^{m_1}u(z) \int_{z_0}^{z} (\beta_0 t^{-m_1+m_2-1} + \beta_1 t^{-m_1+m_2} + \cdots) \, dt$$

$$= z^{m_1}u(z) \left(\frac{\beta_0 z^{-m_1+m_2}}{-m_1 + m_2} + \frac{\beta_1 z^{-m_1+m_2+1}}{-m_1 + m_2 + 1} + \cdots + C \right),$$

where C is a constant of integration. Then

$$w_2(z) = z^{m_2}\psi(z),$$

where $\psi(z)$ is analytic for $|z| < R$. This tells us that whenever $m_1 - m_2$ is not an integer the substitution $w = z^{m_2}\psi(z)$ can always be used to obtain a power series for $\psi(z)$.

Now let us turn to the solution of the nonhomogeneous differential equation

$$z^2 w'' + z p(z) w' + q(z) w = r(z),$$

where p, q, and r are analytic in $|z| < R$. For definiteness let us look for the general solution in the disk $|z - R/2| < R/2$. Let $w_1(z)$ and $w_2(z)$ be particular solutions of the homogeneous equation obtained above, where definite branches of the functions have been selected which are analytic in the disk. We already know that w_1 and w_2 are linearly independent and, therefore, that the Wronskian

$$W(w_1, w_2) = w_1 w_2' - w_2 w_1'$$

$$= [w_1(R/2)w_2'(R/2) - w_2(R/2)w_1'(R/2)] \exp\left[-\int_{R/2}^{z} \frac{p(\zeta)}{\zeta} \, d\zeta \right]$$

is analytic and never vanishes in the disk. We look for a particular solution of the nonhomogeneous equation in the form

$$\tilde{w} = A(z)w_1 + B(z)w_2.$$

Then

$$\tilde{w}' = Aw_1' + Bw_2' + A'w_1 + B'w_2.$$

We set $A'w_1 + B'w_2 = 0$ and differentiate again

$$\tilde{w}'' = Aw_1'' + Bw_2'' + A'w_1' + B'w_2'$$

and then

$$z^2(A'w_1' + B'w_2') = r(z).$$

Solving for A' and B', we have

$$A' = \frac{-w_2 r}{z^2 W(w_1, w_2)}, \qquad B' = \frac{w_1 r}{z^2 W(w_1, w_2)},$$

$$\tilde{w} = \int_{R/2}^{z} r(\zeta) G(z, \zeta) \, d\zeta,$$

where

$$G(z, \zeta) = \frac{w_1(\zeta) w_2(z) - w_2(\zeta) w_1(z)}{\zeta^2 W(w_1, w_2)}.$$

$G(z, \zeta)$ serves as a Green's function for the problem and it is an analytic function of z and ζ in the disk.

We know that if we prescribe the initial conditions $w(R/2) = w_0$ and $w'(R/2) = w_0'$, then there exists a unique solution of the differential equation valid in the disk satisfying the given condition. This unique solution can always be found from the general solution

$$w(z) = c_1 w_1 + c_2 w_2 + \tilde{w},$$

by solving the equations

$$w_0 = c_1 w_1(R/2) + c_2 w_2(R/2) + \tilde{w}(R/2),$$
$$w_0' = c_1 w_1'(R/2) + c_2 w_2'(R/2) + \tilde{w}'(R/2),$$

for c_1 and c_2. These equations always have a unique solution because of the linear independence of w_1 and w_2.

Exercises 7.4

1. By making the transformation $z = 1/\zeta$ in $z^2 w'' + z p(z) w' + q(z) w = r(z)$, state a criterion for the differential equation to have a regular singular point at $z = \infty$, $(\zeta = 0)$.

2. Substitute a series of the form

$$w = z^m (c_0 + c_1 z + c_2 z^2 + \cdots)$$

into the differential equation $z^2 w'' + z p(z) w' + q(z) w = 0$ and show that two formal solutions can be determined provided m is a root of the indicial equation

$$m(m - 1) + p(0) m + q(0) = 0$$

and the two roots do not differ by an integer.

3. Show that the only second-order linear homogeneous differential equation with but one singular point, a regular singular point at infinity, is $w'' = 0$.

4. Show that the only second-order linear homogeneous differential equation with but two singular points, regular singular points at the origin and infinity, is Euler's equation. Show that if m_1 and m_2 are the roots of the indicial equation at the origin and n_1 and n_2 are the roots of the indicial equation at infinity, then $m_1 + m_2 + n_1 + n_2 = 0$ and $m_1 m_2 = n_1 n_2$.

5. Show that the only second-order linear homogeneous differential equation with but three singular points, regular singular points at the origin, at one, and at infinity, is of the form

$$w'' + \left(\frac{a_0}{z} + \frac{a_1}{z - 1} \right) w' + \left[\frac{b_0}{z^2} + \frac{b_1}{(z - 1)^2} + c_0 \left(\frac{1}{z} - \frac{1}{z - 1} \right) \right] w = 0.$$

If α_1 and α_2, β_1 and β_2, γ_1 and γ_2 are the roots of the indicial equation at $0, 1, \infty$, respectively, show that $a_0 = 1 - \alpha_1 - \alpha_2$, $a_1 = 1 - \beta_1 - \beta_2$, $b_0 = \alpha_1 \alpha_2$, $b_1 = \beta_1 \beta_2$, and $c_0 = \alpha_1 \alpha_2 + \beta_1 \beta_2 - \gamma_1 \gamma_2$, and therefore that $\alpha_1 + \alpha_2 + \beta_1 + \beta_2 + \gamma_1 + \gamma_2 = 1$.

6. Referring to Problem 5, if we let $\alpha_1 = 0$, $\alpha_2 = 1 - \gamma$, $\beta_1 = 0$, $\beta_2 = \gamma - \alpha - \beta$, $\gamma_1 = \alpha$, $\gamma_2 = \beta$, the equation becomes

$$z(1 - z)w'' + [\gamma - (\alpha + \beta + 1)z]w' - \alpha\beta w = 0.$$

If γ is not an integer show that this equation has solutions

$$F(\alpha, \beta, \gamma, z) = \frac{\Gamma(\gamma)}{\Gamma(\alpha)\Gamma(\beta)} \sum_{k=0}^{\infty} \frac{\Gamma(\alpha + k)\Gamma(\beta + k)}{\Gamma(\gamma + k)} \frac{z^k}{k!},$$

and $z^{1-\gamma}F(\alpha - \gamma + 1, \beta - \gamma + 1, 2 - \gamma, z)$ valid for $0 < |z| < 1$. $F(\alpha, \beta, \gamma, z)$ is called the *hypergeometric function*.

7. Show that the Legendre equation

$$(1 - z^2)w'' - 2zw' + n(n + 1)w = 0$$

has regular singular points at $+1$, -1, and ∞ with roots of the indicial equation respectively $0, 0; 0, 0;$ and $n + 1, -n$.

8. Show that the associated Legendre equation

$$(1 - z^2)w'' - 2zw' + \left[n(n + 1) - \frac{m^2}{1 - z^2} \right] w = 0$$

has regular singular points at $+1$, -1, and ∞ with roots of the indicial equation respectively $m/2, -m/2; m/2, -m/2; n + 1, -n$. Make the transformation $\zeta = (z + 1)/2, u = \zeta^{-m/2}(1 - \zeta)^{-m/2}w(\zeta)$ and show that u satisfies the hypergeometric equation

$$\zeta(1 - \zeta)u'' + [m + 1 - (2m + 2)\zeta]u' + (n - m)(n + m + 1)u = 0.$$

Hence, for $m \geq 0$ the associated Legendre equation has a solution

$$w(z) = (z + 1)^{-m/2}(1 - z)^{-m/2}F\left(m + n + 1, m - n, m + 1, \frac{z + 1}{2} \right).$$

9. Show that the Bessel equation $w'' + 1/zw' + (1 - n^2/z^2)w = 0$ has a regular singular point at the origin where the roots of the indicial equation are $\pm n$.

7.5. BESSEL FUNCTIONS

We have already seen (see Section 7.1) that the Bessel differential equation arises in the separation of variables problem for the Helmholtz equation in polar and spherical coordinates. It is one of the most often encountered differential equations in applied mathematics. Therefore, a study of its solutions, the Bessel functions, becomes very important.

The Bessel differential equation is

$$z^2 w'' + zw' + (z^2 - v^2)w = 0,$$

where v is a constant. The equation has a regular singular point at the origin where the indicial equation is $m^2 - v^2$ with roots $m_1 = v$, $m_2 = -v$. Let us assume that $m_1 - m_2 = 2v$ is not an integer, that is that v is not an integer or half integer. We then have case 3 of the previous section, and we know that we can obtain two linearly independent solutions

$$w_1 = z^v \sum_{k=0}^{\infty} c_k z^k,$$

$$w_2 = z^{-v} \sum_{k=0}^{\infty} \gamma_k z^k.$$

Substituting the first of these series in the differential equation, we have

$$\sum_{k=0}^{\infty} k(k + 2v)c_k z^{k+v} + \sum_{k=0}^{\infty} c_k z^{k+v+2} = 0.$$

The coefficient of z^{v+1} is $(1 + 2v)c_1$. If $v \neq -1/2$, then $c_1 = 0$. The remaining coefficients are determined by the recurrence relation

$$c_{k+2} = - \frac{c_k}{(k + 2)(k + 2 + 2v)}.$$

Since $c_1 = 0$, all the coefficients with odd subscripts are also zero. The first coefficient c_0 is arbitrary and

$$c_2 = - \frac{c_0}{2(2 + 2v)} = - \frac{c_0}{2^2(v + 1)},$$

$$c_4 = - \frac{c_2}{2^2 \cdot 2(2 + v)} = \frac{c_0}{2^4 \cdot 2(v + 1)(v + 2)},$$

$$c_6 = - \frac{c_4}{2^2 \cdot 3(3 + v)} = - \frac{c_0}{2^6 \cdot 3 \cdot 2(v + 1)(v + 2)(v + 3)},$$

and so on. In general, for $n = 1, 2, 3, \ldots$

$$c_{2n} = \frac{(-1)^n c_0}{2^{2n} n! (v + 1)(v + 2) \cdots (v + n)}.$$

Let us make a particular choice of c_0, namely

$$c_0 = [2^\nu \Gamma(\nu + 1)]^{-1}.$$

Then

$$c_{2n} = \frac{(-1)^n (\frac{1}{2})^{2n+\nu}}{n! \Gamma(\nu + n + 1)},$$

which proves that

$$J_\nu(z) = \sum_{n=1}^\infty \frac{(-1)^n (\frac{1}{2}z)^{2n+\nu}}{n! \Gamma(\nu + n + 1)},$$

is a solution of the equation. The function is called the *Bessel function of the first kind of order v.* A second linearly independent solution, in case ν is not an integer or a half integer, is

$$J_{-\nu}(z) = \sum_{n=0}^\infty \frac{(-1)^n (\frac{1}{2}z)^{2n-\nu}}{n! \Gamma(-\nu + n + 1)}.$$

This series is obtained for a special choice of γ_0 when the series for w_2 is substituted in the differential equation. If

$$\nu = \frac{2p - 1}{2}, \; p = 1, 2, 3, \ldots,$$

then

$$J_{(2p-1)/2}(z) = \sqrt{\frac{2}{z}} \sum_{n=0}^\infty \frac{(-1)^n (\frac{1}{2}z)^{2n+p}}{n! \Gamma\left(\dfrac{2p + 2n + 1}{2}\right)} = \sqrt{\frac{2}{z}} \, z^p f(z),$$

where $f(z)$ is an entire function. Also,

$$J_{-(2p-1)/2}(z) = \sqrt{\frac{z}{2}} \sum_{n=0}^\infty \frac{(-1)^n (\frac{1}{2}z)^{2n-p}}{n! \Gamma\left(\dfrac{-2p + 2n + 3}{2}\right)} = \sqrt{\frac{z}{2}} \, \frac{1}{z^p} g(z),$$

where $g(z)$ is entire and has no zero at the origin. If $J_{(2p-1)/2}(z)$ and $J_{-(2p-1)/2}(z)$ were linearly dependent, then

$$\sqrt{\frac{2}{z}} \, z^p f(z) = \alpha \sqrt{\frac{z}{2}} \, \frac{1}{z^p} g(z),$$

$$f(z) = \frac{\alpha}{2} \frac{g(z)}{z^{2p-1}}.$$

But this is impossible unless $2p - 1 \leq 0$ or $p \leq 1/2$. Therefore, $J_\nu(z)$ and $J_{-\nu}(z)$ are linearly independent if ν is half an odd integer. By simple differentiation we can show that $J_{-\nu}(z)$ is a solution of the differential equation.

This leaves the case where $\nu = p$, a nonnegative integer. Clearly,

$$J_p(z) = \left(\frac{z}{2}\right)^p \sum_{n=0}^{\infty} \frac{(-1)^n (\frac{1}{2}z)^{2n}}{n!(n+p)!}$$

is an entire function which satisfies the differential equation. On the other hand, since

$$\frac{1}{\Gamma(-p+n+1)} = 0,$$

for $n = 0, 1, 2, \ldots, p-1$,

$$J_{-p}(z) = \sum_{n=p}^{\infty} \frac{(-1)^n (\frac{1}{2}z)^{2n-p}}{n!(n-p)!} = \sum_{m=0}^{\infty} \frac{(-1)^{m+p}(\frac{1}{2}z)^{2m+p}}{(m+p)!m!}$$

$$= (-1)^p J_p(z).$$

Therefore, when $p = 0, 1, 2, \ldots, J_p(z)$ and $J_{-p}(z)$ are not independent. This case requires special treatment. According to case 2 of the last section

$$w_2(z) = J_p(z)\left[\frac{\beta_0 z^{-2p}}{-2p} + \frac{\beta_1 z^{-2p+1}}{-2p+1} + \cdots + \beta_p \log z + A(z)\right],$$

where $A(z)$ is an entire function. In this case, $\beta_p \neq 0$, otherwise $z^{-p}\phi(z)$, where $\phi(z)$ is an entire function, would have to be an independent solution and we have already shown that this assumption leads to $J_{-p}(z) = (-1)^p J_p(z)$.

More specifically there is a solution in the form

$$Y_p(z) = \frac{2}{\pi} J_p(z) \int \left\{ [J_p(z)]^{-2} \exp\left[-\int z^{-1}\,dz\right]\right\} dz$$

$$= \frac{2}{\pi} J_p(z) \int \frac{1}{z[J_p(z)]^2}\,dz,$$

where $J_p(z)$ has a zero of order p at the origin. Therefore, $1/\{z[J_p(z)]^2\}$ has a pole of order $2p+1$ and can be expanded as follows:

$$\frac{1}{z[J_p(z)]^2} = \frac{b_{2p+1}}{z^{2p+1}} + \frac{b_{2p}}{z^{2p}} + \cdots,$$

where

$$b_{2p+1} = \lim_{z\to 0} \frac{z^{2p}}{[J_p(z)]^2} = 2^{2p}(p!)^2.$$

Therefore, $\beta_0 = 2^{2p+1}(p!)^2/\pi$ which shows that $Y_p(z)$ is unbounded at the origin. This is still true if $p = 0$ because of the presence of $\log z$. $Y_p(z)$ is not precisely defined because of the lack of definite limits on the integral. However, to within an additive constant this integral defines the *Bessel function of the second kind of order p.*

Let ν be a complex number, not an integer. Then

$$Y_\nu(z) = \frac{J_\nu(z)\cos\nu\pi - J_{-\nu}(z)}{\sin\nu\pi}$$

is a linear combination of $J_\nu(z)$ and $J_{-\nu}(z)$. Therefore, it is a solution of the Bessel equation of order ν. This solution is called the *Bessel function of the second kind of order ν*. It can be shown that $Y_\nu(z)$ is a continuous function of ν and that

$$Y_p(z) = \lim_{\nu \to p} Y_\nu(z)$$

when p is an integer.*

It is of interest to note that there are solutions of the Bessel differential equation which are defined by the equations

$$H_\nu^{(1)}(z) = J_\nu(z) + i Y_\nu(z),$$

$$H_\nu^{(2)}(z) = J_\nu(z) - i Y_\nu(z).$$

These are known as the *first and second Hankel functions of order ν*.

In the study of the special functions which are solutions of certain ordinary differential equations there are various ideas which reoccur quite frequently, namely recurrence relations, integral representations, and generating functions. We shall illustrate this remark in terms of the Bessel functions. For example, from the series representation we have

$$\frac{d}{dz}[z^\nu J_\nu(z)] = \frac{d}{dz}\sum_{n=0}^\infty \frac{(-1)^n z^{2n+2\nu}}{2^{2n+\nu}n!\,\Gamma(\nu + n + 1)}$$

$$= \sum_{n=0}^\infty \frac{(-1)^n z^{2n+2\nu-1}}{2^{2n+\nu-1}n!\,\Gamma(\nu + n)}$$

$$= z^\nu J_{\nu-1}(z),$$

$$\frac{d}{dz}[z^{-\nu} J_\nu(z)] = \frac{d}{dz}\sum_{n=0}^\infty \frac{(-1)^n z^{2n}}{2^{2n+\nu}n!\,\Gamma(\nu + n + 1)}$$

$$= \sum_{n=1}^\infty \frac{(-1)^n z^{2n-1}}{2^{2n+\nu-1}(n - 1)!\,\Gamma(\nu + n + 1)}$$

$$= \sum_{k=1}^\infty \frac{(-1)^{k+1} z^{2k+1}}{2^{2k+\nu+1}k!\,\Gamma(\nu + k + 2)}$$

$$= -z^{-\nu} J_{\nu+1}(z).$$

---

* See Copson, E. T. *An Introduction to the Theory of Functions of a Complex Variable.* London: Oxford University Press, 1935. Chapter 12.

Expanding these formulas, where the prime means $\dfrac{d}{dz}$, we have

$$\nu z^{\nu-1} J_\nu + z^\nu J_\nu' = z^\nu J_{\nu-1},$$
$$-\nu z^{\nu-1} J_\nu + z^\nu J_\nu' = -z^\nu J_{\nu+1},$$

where in the second we have multiplied by $z^{2\nu}$. Subtracting and dividing by z^ν, we have

$$J_{\nu-1}(z) + J_{\nu+1}(z) = \frac{2\nu}{z} J_\nu(z).$$

Adding and dividing by z^ν, we have

$$J_{\nu-1}(z) - J_{\nu+1}(z) = 2J_\nu'(z).$$

These are just a few of the many interesting relations one can derive connecting the Bessel functions.

To obtain an integral representation of $J_\nu(z)$ we begin with the Hankel contour integral for the Gamma function (see Section 4.10).

$$\frac{1}{\Gamma(\nu + n + 1)} = \frac{1}{2\pi i} \int_C e^\zeta \zeta^{-(\nu+n+1)}\, d\zeta,$$

where the contour comes in from $-\infty$ along the lower side of the negative real ζ axis swings around the origin in the counterclockwise direction and then goes back to $-\infty$ along the top side of the negative real ζ axis. Then

$$(\tfrac{1}{2}z)^{-\nu} J_\nu(z) = \sum_{n=0}^{\infty} (-\tfrac{1}{4}z^2)^n \frac{1}{2\pi i n!} \int_C e^\zeta \zeta^{-(\nu+n+1)}\, d\zeta$$

$$= \frac{1}{2\pi i} \int_C e^\zeta \zeta^{-(\nu+1)} \sum_{n=0}^{\infty} \frac{1}{n!}\left(-\frac{1}{4}\frac{z^2}{\zeta}\right)^n d\zeta$$

$$= \frac{1}{2\pi i} \int_C \exp\left(\zeta - \frac{1}{4}\frac{z^2}{\zeta}\right) \zeta^{-(\nu+1)}\, d\zeta,$$

provided we can show that the interchange of summation and integration is justified. Consider the function

$$f(z) = \frac{1}{2\pi i} \int_C \exp\left(\zeta - \frac{1}{4}\frac{z^2}{\zeta}\right) \zeta^{-(\nu+1)}\, d\zeta.$$

By Theorem 4.9.3, this is an entire function and we can differentiate under the integral sign as many times as we please. Its Maclaurin expansion is therefore

$$\sum_{n=0}^{\infty} (-\tfrac{1}{4}z^2)^n \frac{1}{2\pi i n!} \int_C e^\zeta \zeta^{-(\nu+n+1)}\, d\zeta$$

and

$$J_\nu(z) = \frac{(\tfrac{1}{2}z)^\nu}{2\pi i} \int_C \exp\left(\zeta - \frac{1}{4}\frac{z^2}{\zeta}\right) \zeta^{-(\nu+1)}\, d\zeta.$$

If $\nu = p$, an integer

$$J_p(z) = \frac{(\frac{1}{2}z)^p}{2\pi i} \int_C \exp\left(\zeta - \frac{1}{4}\frac{z^2}{\zeta}\right) \zeta^{-(p+1)} d\zeta,$$

and the integrand is analytic except for an essential singularity at $\zeta = 0$. Therefore, the value of the integral does not change if we replace the contour by a circle with center at the origin with radius $|z|/2$. Having done this, we make the change of variable $\zeta = ze^{i\theta}/2$. Then

$$J_p(z) = \frac{1}{2\pi} \int_{-\pi}^{\pi} e^{-ip\theta + iz\sin\theta} \, d\theta$$

$$= \frac{1}{2\pi} \int_0^{\pi} (e^{-ip\theta + iz\sin\theta} + e^{ip\theta - iz\sin\theta}) \, d\theta$$

$$= \frac{1}{\pi} \int_0^{\pi} \cos(p\theta - z\sin\theta) \, d\theta,$$

which is a fairly concise integral representation.

If in the above integral representation for $J_p(z)$, we replace the contour by a circle of radius $|z|R/2$ and make the change of variables $\zeta = z\omega/2$, where $\omega = Re^{i\theta}$, we have

$$J_p(z) = \frac{1}{2\pi i} \int_{|\omega|=R} e^{z(\omega - \omega^{-1})/2} \frac{d\omega}{\omega^{p+1}}.$$

This means that $J_p(z)$ is the coefficient in the Laurent expansion of $e^{z(\omega - \omega^{-1})/2}$ in powers of ω, that is,

$$e^{z(\omega - \omega^{-1})/2} = \sum_{p=-\infty}^{\infty} \omega^p J_p(z)$$

is a Laurent expansion. We say that the function $e^{z(\omega - \omega^{-1})/2}$ is a *generating function* for the Bessel functions $J_p(z)$, $p = 0, \pm 1, \pm 2, \ldots$.

Exercises 7.5

1. Prove that the Wronskian of $J_\nu(z)$ and $J_{-\nu}(z)$ is $-(2\sin\nu\pi)/\pi z$.

2. Prove that $J_{1/2}(z) = (2/\pi z)^{1/2} \sin z$ and $J_{-1/2}(z) = (2/\pi z)^{1/2} \cos z$ and hence that the general solution of $z^2 w'' + zw' + (z^2 - 1/4)w = 0$ is $(A \sin z + B \cos z)/\sqrt{z}$.

3. Find $J_{3/2}(z)$ and $J_{-3/2}(z)$ in terms of elementary functions.

4. Show that the complete solution of the zero-th order Bessel equation has a logarithmic singularity at the origin.

5. Consider the differential equation $(xy')' - (n^2/x)y' + \lambda xy = 0$. Show that for an integer, n, the only solution finite at the origin is a multiple of $J_n(\sqrt{\lambda}x)$. Let $\sqrt{\lambda_i}b$ be a zero of $J_n(\sqrt{\lambda}b) = 0$. Prove that if $\lambda_i \neq \lambda_j$,

$$\int_0^b x J_n(\sqrt{\lambda_i}\,x) J_n(\sqrt{\lambda_j}\,x)\,dx = 0.$$

6. Prove that $J_n(z) = \dfrac{1}{2\pi i} \displaystyle\int_{-i\pi}^{i\pi} e^{z\,\sinh t - nt}\,dt$.

7. Use the generating function to prove all the recurrence relations of this section for integer orders of Bessel functions.

7.6. LEGENDRE FUNCTIONS

As we have already seen, the Legendre equation and the associated Legendre equation occur in the separation of variables problem for Laplace's equation and Helmholtz's equation in spherical coordinates. The Legendre equation,

$$(1 - z^2)w'' - 2zw' + \lambda w = 0,$$

has an ordinary point at the origin, regular singular points at ± 1, and a regular singular point at infinity. Let us seek a solution analytic at the origin. We assume

$$w = \sum_{k=0}^{\infty} c_k z^k.$$

Substituting into the differential equation, we have

$$(1 - z^2) \sum_{k=2}^{\infty} k(k-1)c_k z^{k-2} - 2z \sum_{k=1}^{\infty} kc_k z^{k-1} + \lambda \sum_{k=0}^{\infty} c_k z^k = 0,$$

$$\sum_{m=0}^{\infty} [(m+2)(m+1)c_{m+2} - m(m+1)c_m + \lambda c_m]z^m = 0.$$

Therefore, the recurrence relation for the coefficients is

$$c_{m+2} = \frac{m(m+1) - \lambda}{(m+2)(m+1)} c_m,$$

$m = 0, 1, 2, \ldots$. If we let $c_1 = 0$, while $c_0 \neq 0$, we obtain a series in even powers of z. If we let $c_0 = 0$ while $c_1 \neq 0$, we obtain a series in odd powers of z. Then

$$w = c_0 w_1(z) + c_1 w_2(z),$$

where w_1 contains only even powers and $w_2(z)$ contains only odd powers. The series converge for $|z| < 1$, the two functions are clearly independent, and therefore this is the complete solution for $|z| < 1$.

Of considerable interest are those solutions which are finite for $z = \pm 1$. Note in the recurrence relation that if $\lambda = n(n + 1)$, where n is a nonnegative integer, then $c_{n+2} = c_{n+4} = \cdots = 0$ and the corresponding series becomes a polynomial. These polynomials are called the *Legendre polynomials*, when they are multiplied by the proper constant so that $P_n(1) = 1$. For example,

$$P_0(z) = 1, P_1(z) = z, P_2(z) = \tfrac{1}{2}(3z^2 - 1), P_3(z) = \tfrac{1}{2}(5z^3 - 3z),$$

and so on. To obtain a general formula consider the solution of

$$(1 - z^2)w'' - 2zw' + n(n + 1)w = 0$$

valid for large values of $|z|$. We assume a solution of the form

$$w = z^{-m} \sum_{k=0}^{\infty} c_k z^{-k} = \sum_{k=0}^{\infty} c_k z^{-k-m}.$$

The indicial equation is $m^2 - m - n(n + 1) = 0$ which has roots $m_1 = -n, m_2 = n + 1$. Let $m = -n$. Then

$$w = \sum_{k=0}^{\infty} c_k z^{n-k}$$

$$w' = \sum_{k=0}^{\infty} (n - k)c_k z^{n-k-1}$$

$$w'' = \sum_{k=0}^{\infty} (n - k)(n - k - 1)c_k z^{n-k-2}.$$

Substituting in the differential equation, we have

$$\sum_{k=0}^{\infty} (n - k)(n - k - 1)c_k z^{n-k-2} - \sum_{k=0}^{\infty} (n - k)(n - k - 1)c_k z^{n-k}$$

$$- \sum_{k=0}^{\infty} 2(n - k)c_k z^{n-k} + \sum_{k=0}^{\infty} n(n + 1)c_k z^{n-k} = 0$$

or

$$2nc_1 z^{n-1} + \sum_{m=0}^{\infty} [(n - m)(n - m - 1)c_m$$

$$- (m + 2)(m - 2n + 1)c_{m+2}]z^{n-m-2} = 0.$$

If $n \neq 0$, then $c_1 = 0$, c_0 is arbitrary and the remaining coefficients are determined by the recurrence relation

$$c_{m+2} = \frac{(n-m)(n-m-1)}{(m+2)(m-2n-1)} c_m$$

$m = 0, 1, 2, \ldots$ Therefore, $c_1 = c_3 = c_5 = \cdots = 0$. Also, we note that $c_{n+2} = c_{n+4} = \cdots = 0$, and therefore, this solution is a polynomial

$$w = c_0 z^n \left[1 - \frac{n(n-1)}{2(2n-1)} z^{-2} + \frac{n(n-1)(n-2)(n-3)}{2 \cdot 4(2n-1)(2n-3)} z^{-4} \right.$$
$$\left. - \cdots + (-1)^j \frac{(n!)^2(2n-2j)!}{(2n)! \, j!(n-2j)!(n-j)!} z^{-2j} + \cdots \right]$$

If n is even the series terminates when $j = n/2$. If n is odd, the series terminates when $j = (n-1)/2$. In particular, if $c_0 = (2n)!/2^n(n!)^2$, we obtain the Legendre polynomials

$$P_n(z) = \sum_{j=0}^{N} \frac{(-1)^j(2n-2j)!}{2^n j!(n-j)!(n-2j)!} z^{n-2j},$$

where $N = n/2$ if n is even and $N = (n-1)/2$ if n is odd. We note that

$$\frac{d^n}{dz^n} z^{2n-2j} = \frac{(2n-2j)!}{(n-2j)!} z^{n-2j}.$$

Therefore,

$$P_n(z) = \sum_{j=0}^{N} \frac{(-1)^j}{2^n j!(n-j)!} \frac{d^n}{dz^n} z^{2n-2j}$$

$$= \frac{1}{2^n n!} \frac{d^n}{dz^n} \sum_{j=0}^{N} \frac{(-1)^j n!}{j!(n-j)!} z^{2n-2j}.$$

In the last expression, the sum is a polynomial of degree $2n$ with the terms of degree $n-1$ or less missing. If we add any terms of degree $n-1$ or less the nth derivative will wipe them out. Therefore,

$$P_n(z) = \frac{1}{2^n n!} \frac{d^n}{dz^n} \sum_{j=0}^{n} \frac{(-1)^j n!}{j!(n-j)!} z^{2n-2j}.$$

But the summation is just the binomial expansion of the function $(z^2 - 1)^n$. Therefore,

$$P_n(z) = \frac{1}{2^n n!} \frac{d^n}{dz^n} (z^2 - 1)^n.$$

This is known as *Rodrigues's formula*. Since $P_n(z)$ is expressible as the nth derivative of an analytic function we have an obvious integral representation using Cauchy's formula for an nth derivative.

$$P_n(z) = \frac{1}{2\pi i} \int_{C+} \frac{(\zeta^2 - 1)^n}{2^n (\zeta - z)^{n+1}} \, d\zeta,$$

where C is any simple closed contour surrounding $\zeta = z$. In particular, for C we can take the circle

$$|\zeta - z| = \sqrt{|z^2 - 1|}.$$

Then

$$\zeta = z + (z^2 - 1)^{1/2} e^{i\theta},$$

$$d\zeta = i(z^2 - 1)^{1/2} e^{i\theta} \, d\theta,$$

$$\zeta^2 - 1 = 2(\zeta - z)[z + (z^2 - 1)^{1/2} \cos \theta].$$

Hence,

$$P_n(z) = \frac{1}{2\pi} \int_{-\pi}^{\pi} [z + (z^2 - 1)^{1/2} \cos \theta]^n \, d\theta$$

$$= \frac{1}{\pi} \int_0^{\pi} [z + (z^2 - 1)^{1/2} \cos \theta]^n \, d\theta.$$

This is known as *Laplace's first integral* for the Legendre polynomials.

To obtain a generating function for $P_n(z)$ we form the sum

$$\sum_{n=0}^{\infty} \omega^n P_n(z) = \frac{1}{\pi} \sum_{n=0}^{\infty} \int_0^{\pi} \omega^n [z + (z^2 - 1)^{1/2} \cos \theta]^n \, d\theta$$

$$= \frac{1}{\pi} \int_0^{\pi} \frac{1}{1 - \omega z - \omega(z^2 - 1)^{1/2} \cos \theta} \, d\theta$$

$$= \frac{1}{(1 - 2\omega z + \omega^2)^{1/2}},$$

provided we can show that the interchange of summation and integration is justified. We have

$$|\omega[z + (z^2 - 1)^{1/2} \cos \theta]| \le |\omega|(|z| + |z^2 - 1|^{1/2}) \le \rho < 1,$$

provided

$$|\omega| \le \frac{\rho}{|z| + |z^2 - 1|^{1/2}}.$$

Therefore, for $|\omega|$ sufficiently small the series

$$\sum_{n=0}^{\infty} \omega^n [z + (z^2 - 1)^{1/2} \cos \theta]^n$$

converges absolutely and uniformly in θ and the interchange is permitted. Now, $(1 - 2\omega z + \omega^2)^{1/2}$ treated as a function of ω, has branch points at $z \pm (z^2 - 1)^{1/2}$ and has a Maclaurin expansion in powers of ω valid for

$$|\omega| < \min\,[|z + (z^2 - 1)^{1/2}|, |z - (z^2 - 1)^{1/2}|]$$

which by uniqueness must be $\sum\limits_{n=0}^{\infty} \omega^n P_n(z)$.

We can find recurrence relations for the Legendre polynomials from the generating function

$$W = (1 - 2z\omega + \omega^2)^{-1/2}.$$

We have

$$\frac{\partial W}{\partial \omega} = (z - \omega)(1 - 2z\omega + \omega^2)^{-3/2},$$

so that

$$(1 - 2z\omega + \omega^2)\frac{\partial W}{\partial \omega} = (z - \omega)W,$$

$$(1 - 2z\omega + \omega^2)\sum_{n=0}^{\infty} n\omega^{n-1}P_n(z) = (z - \omega)\sum_{n=0}^{\infty} \omega^n P_n(z),$$

$$\sum_{m=0}^{\infty} [(m + 1)P_{m+1}(z) - (2m + 1)zP_m(z) + mP_{m-1}(z)]\omega^m = 0.$$

Therefore, since the coefficient of ω^m must vanish for all m

$$(m + 1)P_{m+1}(z) - (2m + 1)zP_m(z) + mP_{m-1}(z) = 0.$$

We also have

$$\frac{\partial W}{\partial z} = \omega(1 - 2z\omega + \omega^2)^{-3/2},$$

so that

$$\omega\,\frac{\partial W}{\partial \omega} = (z - \omega)\frac{\partial W}{\partial z},$$

$$\omega\sum_{n=0}^{\infty} n\omega^{n-1}P_n(z) = (z - \omega)\sum_{n=0}^{\infty} \omega^n P_n'(z),$$

$$\sum_{m=0}^{\infty} [mP_m(z) - zP_m'(z) + P_{m-1}'(z)]\omega^m = 0,$$

if we let $P_{m-1}(z) = 0$. Therefore,

$$mP_m(z) - zP_m'(z) + P_{m-1}'(z) = 0.$$

To find the complete solution of the Legendre equation $(1 - z^2)w'' - 2zw' + n(n + 1)w = 0$, we make the substitution $w = P_n(z)v(z)$ as in Section 7.4 to obtain a second solution

$$w_2(z) = P_n(z) \int_{z_0}^{z} \frac{d\zeta}{(\zeta^2 - 1)P_n^2(\zeta)},$$

where the path of integration is any path not passing through ± 1 or a zero of $P_n(z)$. Now, $P_n(z)$ is not zero at ± 1. Therefore, there is a partial fraction expansion of $(z^2 - 1)^{-1}[P_n(z)]^{-2}$ which contains terms of the form

$$\frac{A}{z - 1} + \frac{B}{z + 1},$$

where

$$A = \lim_{z \to 1} \frac{1}{(z + 1)P_n^2(z)} = \frac{1}{2},$$

$$B = \lim_{z \to -1} \frac{1}{(z - 1)P_n^2(z)} = -\frac{1}{2}.$$

Therefore, the complete solution of the Legendre equation is

$$w = c_1 P_n(z) + c_2 \left[\tfrac{1}{2} P_n(z) \operatorname{Log} \frac{z + 1}{z - 1} + N_{n-1}(z) \right],$$

where $N_{n-1}(z)$ can be shown to be a polynomial of degree $n - 1$. Because of the logarithms, the second solution has a branch cut on the real axis between -1 and 1. The second solution can also be written in the form

$$Q_n(z) = P_n(z) \int_{\infty}^{z} \frac{d\zeta}{(\zeta^2 - 1)P_n^2(\zeta)},$$

where the path of integration does not cross the cut $-1 \leq x \leq 1$. $Q_n(z)$ is called the *Legendre function of the second kind*.

The *associated Legendre equation* is

$$(1 - z^2)w'' - 2zw' + \left[n(n + 1) - \frac{m^2}{1 - z^2} \right] w = 0.$$

We saw that this came out of the separation of variables for Laplace's equation is spherical coordinates. We shall only consider the case where n and m are integers. The equation has regular singular points at ± 1 and ∞. It is easily seen that the indicial equations at ± 1 both have roots $m/2$. Therefore, we make the preliminary transformation $w = (z^2 - 1)^{m/2} v(z)$. Then v satisfies the equation

$$(1 - z^2)v'' - 2(m + 1)zv' + (n - m)(n + m + 1)v = 0.$$

If we now let $v = \dfrac{d^m}{dz^m} u(z)$, we have

$$(1 - z^2)u^{(m+2)} - 2(m + 1)zu^{(m+1)} + (n - m)(n + m + 1)u^{(m)} = 0,$$

and

$$\frac{d^m}{dz^m}(1 - z^2)u'' = (1 - z^2)u^{(m+2)} - 2zmu^{(m+1)} - m(m - 1)u^{(m)},$$

$$\frac{d^m}{dz^m}(-2zu') = -2zu^{(m+1)} - 2mu^{(m)}.$$

Therefore, the equation can be written

$$\frac{d^m}{dz^m}[(1 - z^2)u'' - 2zu' + n(n + 1)u] = 0$$

and, hence if u satisfies the Legendre equation,

$$w = (z^2 - 1)^{m/2}u^{(m)}(z)$$

satisfies the associated Legendre equation. Therefore, we have shown that two linearly independent solutions are

$$P_n^m(z) = (z^2 - 1)^{m/2}\frac{d^m}{dz^m}P_n(z),$$

$$Q_n^m(z) = (z^2 - 1)^{m/2}\frac{d^m}{dz^m}Q_n(z).$$

These functions are called the *associated Legendre functions of the first and second kind.* They are obviously analytic in the plane cut along the real axis between -1 and 1, if we take the appropriate branch for $(z^2 - 1)^{1/2}$. If m is even $P_n^m(z)$ is a polynomial. If m is odd $P_n^m(z)$ has branch points at ± 1. For definiteness, we define for $-1 \leq x \leq 1$

$$P_n^m(x) = (-1)^n(1 - x^2)^{m/2}\frac{d^n}{dx^m}P_n(x),$$

where the positive square root of $1 - x^2$ is taken.

If $m \leq n$, $P_n^m(z)$ is a nontrivial solution of the associated Legendre equation which is finite at ± 1. It can be shown that the only solutions finite at ± 1 are multiples of $P_n^m(z)$. Recall that in separating the variables in Laplace's equation in spherical coordinates we were led to the associated Legendre equation in the variable $x = \cos \phi$, where we required solutions finite for $\phi = 0$ and $\phi = \pi$, that is, $x = \pm 1$. Hence, it is easy to show that solutions of Laplace's equation finite for $\rho = 0$, periodic in θ, and finite for $\phi = 0$ and $\phi = \pi$ are of the form

$$\rho^n[AP_n^m(\cos \phi) \cos n\theta + BP_n^m(\cos \phi) \sin n\theta].$$

The functions

$$AP_n^m(\cos \phi) \cos n\theta + BP_n^m(\cos \phi) \sin n\theta$$

are called *spherical harmonics*. They play a fundamental role in the solution of Laplace's equation in spherical coordinates.

Exercises 7.6

1. Prove that $\int_{-1}^{1} P_m(x)P_n(x) \, dx = 0$ if $m \neq n$.

2. Prove that $\int_{-1}^{1} P_n^2(x) \, dx = \dfrac{2}{2n + 1}$.

3. Use the generating function to show that $P_n(1) = 1$ and $P_n(-1) = (-1)^n$.

4. Prove that the zeros of $P_n(z)$ are all real, distinct, and lie between -1 and 1.

5. Prove that
   (a) $P_n' - zP_{n-1}' = nP_{n-1}$;
   (b) $P_{n+1}' - P_{n-1}' = (2n + 1)P_n$;
   (c) $(z^2 - 1)P_n' = nzP_n - nP_{n-1}$.

6. Prove that

$$\frac{1}{(z^2 - 1)P_n^2(z)} = \frac{1}{2(z - 1)} - \frac{1}{2z + 1} + \sum_{j=1}^{n} \frac{c_j}{(z - x_j)^2},$$

where $x_1, x_2, \ldots, x_n$ are the zeros of $P_n(z)$. Hence, show that

$$Q_n(z) = \tfrac{1}{2}P_n(z) \operatorname{Log} \frac{z + 1}{z - 1} + N_{n-1}(z),$$

where $N_{n-1}(z)$ is a polynomial of degree $n - 1$.

7. Prove that $\int_{-1}^{1} P_n^m(x)P_k^m(x) \, dx = 0$ if $n \neq k$.

8. Prove that $\int_{-1}^{1} [P_n^m(x)]^2 \, dx = \dfrac{2}{2n + 1} \dfrac{(n + m)!}{(n - m)!}$.

9. Prove that $\int_{-1}^{1} P_n^m(x)P_n^k(x) \dfrac{dx}{1 - x^2} = 0$ if $m \neq k$.

10. Prove that $\int_{-1}^{1} \dfrac{[P_n^m(x)]^2}{1 - x^2} \, dx = \dfrac{1}{m} \dfrac{(n + m)!}{(n - m)!}$.

7.7. STURM-LIOUVILLE PROBLEMS

We have already seen several examples of Sturm-Liouville differential equations, such as

$$y'' + n^2 y = 0,$$

$$(xy')' - \frac{\nu^2}{x} + \lambda xy = 0,$$

$$[(1 - x^2)y']' + n(n + 1)y = 0.$$

In general, the Sturm-Liouville equation is written in the form

$$(p(x)y')' - q(x)y + \lambda\rho(x)y = 0,$$

where $p(x)$, some differentiable function, $q(x)$, and $\rho(x)$ are given. This equation is very important in the study of partial differential equations because it results from separation of variables in a large number of cases.

The principal interest in the Sturm-Liouville equation is in connection with the study of boundary value problems. In this section, we shall be concerned only with the regular Sturm-Liouville problem where it is assumed that for $a \leq x \leq b$, a and b finite, and A, B, C, D, real constants:

1. $p'(x)$, $q(x)$, and $\rho(x)$ are continuous;
2. $p(x)$ and $\rho(x)$ are positive;
3. $Ay(a) + By'(a) = 0$ and $Cy(b) + Dy'(b) = 0$.

If a or b are not finite or if one of the three basic assumptions does not hold we may have a singular Sturm-Liouville problem. Many such problems have been studied.* It would be impossible to take up all the various special cases here. Therefore, we shall only study the regular problem.

A variety of boundary conditions are included in the regular case. For example, if $A = C = 1$ and $B = D = 0$, then $y(a) = y(b) = 0$. If $B = D = 1$ and $A = C = 0$, then $y'(a) = y'(b) = 0$. If $B \neq 0$ and $D \neq 0$, then $y'(a) + \mu y(a) = 0$, $y'(b) + \nu y(b) = 0$, where $\mu = A/B$ and $\nu = C/D$. These various boundary conditions have physical significance depending on the particular applications, which are too numerous to go into here.

Before analyzing the regular Sturm-Liouville problem we shall make a preliminary transformation which will simplify the differential equation.

* The reader should see E. C. Titchmarsh, *Eigenfunction Expansions Associated with Second Order Differential Equations.* London: Oxford University Press, 1946.

Let $u = gy$ and $t = \int_a^x (\rho/p)^{1/2}\, dx$, where $g = (\rho p)^{1/4}$. Then if the prime refers to differentiation with respect to x and the dot with respect to t

$$y' = \frac{d}{dt}\left(\frac{u}{g}\right)\frac{dt}{dx} = \left(\frac{\dot{u}}{g} - \frac{u}{g^2}\dot{g}\right)\left(\frac{\rho}{p}\right)^{1/2},$$

$$py' = g\dot{u} - u\dot{g},$$

$$(py')' = (g\ddot{u} - u\ddot{g})\left(\frac{\rho}{p}\right)^{1/2} = \frac{\rho}{g}\ddot{u} - u\ddot{g}\frac{\rho}{g^2},$$

$$(py')' - qy + \lambda\rho y = \frac{\rho}{g}\left[\ddot{u} - \left(\frac{\ddot{g}}{g} + \frac{q}{\rho}\right)u + \lambda u\right] = 0.$$

Since $\rho/g \neq 0$, the differential equation becomes

$$\ddot{u} + [\lambda - Q(t)]u = 0,$$

where*

$$Q(t) = \frac{\ddot{g}}{g} + \frac{q}{\rho}.$$

The boundary conditions become

$$Ay(a) + By'(a) = A\frac{u(0)}{g(0)} + B\left(\frac{\dot{u}(0)}{g(0)} - \frac{u(0)}{g^2(0)}\dot{g}(0)\right)\left[\frac{\rho(a)}{p(a)}\right]^{1/2}$$

$$= A'u(0) + B'\dot{u}(0) = 0,$$

$$Cy(b) + Dy'(b) = C\frac{u(\tau)}{g(\tau)} + B\left(\frac{\dot{u}(\tau)}{g(\tau)} - \frac{u(\tau)}{g^2(\tau)}\dot{g}(\tau)\right)\left[\frac{\rho(b)}{p(b)}\right]^{1/2}$$

$$= C'u(\tau) + D'\dot{u}(\tau) = 0,$$

where $\tau = \int_a^b (\rho/p)^{1/2}\, dx$. Hence, the general form of the boundary conditions does not change.

Having shown that the Sturm-Liouville problem can be simplified in form, we shall henceforth study the following problems:

(a) The homogeneous problem: $y'' + [\lambda - q(x)]y = 0$, $\cos \alpha\, y(a) + \sin \alpha\, y'(a) = 0$, $\cos \beta\, y(b) + \sin \beta\, y'(b)$ where α and β are real and $q(x)$ is a given real-valued continuous function.

(b) The nonhomogeneous problem: $y'' + [\lambda - q(x)]y = f(x)$, $\cos \alpha\, y(a) + \sin \alpha\, y'(a) = 0$, $\cos \beta\, y(b) + \sin \beta\, y'(b) = 0$ where λ is given and $f(x)$ is a given real-valued piecewise continuous function with a piecewise continuous derivative.

---

* The validity of this approach depends on $p(x)$ and $\rho(x)$ being twice continuously differentiable. However, the results we obtain can be made not to depend on this transformation of the differential equation.

Assume that the nonhomogeneous problem has a unique solution $y(x)$. Then the corresponding homogeneous problem cannot have a nontrivial (not identically zero) solution $u(x)$. For if it did $y + cu$ would also be a solution of the nonhomogeneous problem for arbitrary c and the solution y could not possibly be unique.

On the other hand, assume that the homogeneous problem has no nontrivial solution. Then there exist linearly independent solutions $\phi(x, \lambda)$ and $\psi(x, \lambda)$ of $y'' + (\lambda - q)y = 0$ satisfying the conditions

$$\phi(a, \lambda) = \sin \alpha, \qquad \phi'(a, \lambda) = -\cos \alpha,$$
$$\psi(b, \lambda) = \sin \beta, \qquad \psi'(b, \lambda) = -\cos \beta.$$

These functions can be easily constructed from any two linearly independent solutions $u(x, \lambda)$ and $v(x, \lambda)$, for if

$$\phi(x, \lambda) = c_1 u + c_2 v,$$
$$\sin \alpha = c_1 u(a) + c_2 v(a),$$
$$-\cos \alpha = c_1 u'(a) + c_2 v'(a),$$

$$c_1 = \frac{\begin{vmatrix} \sin \alpha & v(a) \\ -\cos \alpha & v'(a) \end{vmatrix}}{W(u, v)} = \frac{\sin \alpha \, v'(a) + \cos \alpha \, v(a)}{W(u, v)},$$

$$c_2 = \frac{\begin{vmatrix} u(a) & \sin \alpha \\ u'(a) & -\cos \alpha \end{vmatrix}}{W(u, v)} = \frac{-\sin \alpha \, u'(a) - \cos \alpha \, u(a)}{W(u, v)},$$

where the Wronskian $W(u, v) = uv' - vu'$ is independent of x and never vanishes. Hence,

$$\phi(x, \lambda) = \frac{[\sin \alpha v'(a) + \cos \alpha v(a)]u - [\sin \alpha u'(a) + \cos \alpha u(a)]v}{W(u, v)}.$$

Interchanging α, β, and a, b, we have

$$\psi(x, \lambda) = \frac{[\sin \beta v'(b) + \cos \beta v(b)]u - [\sin \beta u'(b) + \cos \beta u(b)]v}{W(u, v)}.$$

Now, $\cos \alpha \, \phi(a, \lambda) + \sin \alpha \, \phi'(a, \lambda) = 0$. If the Wronskian $W(\phi, \psi) = \phi\psi' - \psi\phi' = 0$, then $\phi = k\psi$, where k is a nonzero constant. But in this case, $\cos \beta \, \phi(b, \lambda) + \sin \beta \, \phi'(b, \lambda) = 0$ and ϕ is a nontrivial solution of the homogeneous problem, and we have assumed that no such solution exists. Therefore, $\phi(x, \lambda)$ and $\psi(x, \lambda)$ are linearly independent, and by the method of variation of parameters we can show that

$$y(x) = \frac{\psi(x, \lambda)}{W(\phi, \psi)} \int_a^x \phi(t, \lambda)f(t) \, dt + \frac{\phi(x, \lambda)}{W(\phi, \psi)} \int_x^b \psi(t, \lambda)f(t) \, dt$$

is a solution of the nonhomogeneous problem. This solution is unique, because if y_1 and y_2 both solved the nonhomogeneous problem then $y_1 - y_2$ would be a solution of the homogeneous problem, and we have assumed that only the trivial solution $y_1 - y_2 \equiv 0$ exists. Thus we have proved the following theorem:

Theorem 7.7.1. *The nonhomogeneous regular Sturm-Liouville problem (b) has a unique solution if and only if the corresponding homogeneous problem (a) has only the trivial solution.*

It remains only to consider the existence question for the nonhomogeneous problem in the case where the homogeneous problem has a nontrivial solution, that is, there may exist a solution of the nonhomogeneous problem even though it is not unique. If λ_k is a number such that $u_k'' + (\lambda_k - q)u_k = 0$ has a nontrivial solution u_k satisfying the boundary conditions then λ_k is called an *eigenvalue* and u_k is called an *eigenfunction*.

We can show that the eigenvalues are all real. For if $u_k'' + (\lambda_k - q)u_k = 0$, $\cos \alpha\, u_k(a) + \sin \alpha\, u_k'(a) = 0$, $\cos \beta\, u_k(b) + \sin \beta\, u_k'(b) = 0$, then

$$\bar{u}_k'' + (\bar{\lambda} - q)\bar{u}_k = 0,$$
$$\cos \alpha\, \bar{u}_k(a) + \sin \alpha\, \bar{u}_k'(a) = 0,$$
$$\cos \beta\, \bar{u}_k(b) + \sin \beta\, \bar{u}_k(b) = 0,$$

and

$$(\bar{\lambda}_k - \lambda_k) \int_a^b |u_k|^2\, dx = [u_k \bar{u}_k' - \bar{u}_k u_k']_a^b = 0.$$

But $\int_a^b |u_k|^2\, dx$ is positive and therefore $\bar{\lambda}_k = \lambda_k$.

By a similar calculation, if u_j and u_k are eigenfunctions corresponding to distinct eigenvalues $\lambda_j \neq \lambda_k$, then

$$(\lambda_j - \lambda_k) \int_a^b u_j u_k\, dx = [u_j u_k' - u_k u_j']_a^b = 0.$$

Hence, $\int_a^b u_j u_k\, dx = 0$ or u_j and u_k are *orthogonal*.

The functions $\phi(x, \lambda)$ and $\psi(x, \lambda)$ play an important role in the study of the homogeneous problem. In fact, the following theorem establishes a strong connection, between the eigenvalues of the homogeneous problem and the zeros of the Wronskian $W(\phi, \psi)$.

Theorem 7.7.2. *Let $\phi(x, \lambda)$ and $\psi(x, \lambda)$ be solutions of*

$$y'' + (\lambda - q)y = 0$$

satisfying $\phi(a, \lambda) = \sin \alpha$, $\phi'(a, \lambda) = -\cos \alpha$, $\psi(b, \lambda) = \sin \beta$, $\psi'(b, \lambda) = -\cos \beta$. Then the Wronskian $W(\phi, \psi)$ is a function of λ only and λ_i is an eigenvalue of the homogeneous problem (a) if and only if it is a zero of $w(\lambda) = W(\phi, \psi)$.

Proof: To show that $W(\phi, \psi)$ is independent of x we show that

$$\frac{d}{dx} W(\phi, \psi) = \frac{d}{dx} (\phi\psi' - \psi\phi') = \phi\psi'' - \psi\phi'' = \psi(\lambda - q)\phi - \phi(\lambda - q)\psi = 0.$$

If $w(\lambda_i) = 0$, then $\phi(x, \lambda_i)\psi'(x, \lambda_i) - \psi(x, \lambda_i)\phi'(x, \lambda_i) = 0$, which can be used to show that $\phi(x, \lambda_i) = k_i\psi(x, \lambda_i)$, where k_i is a nonzero constant. In this case,

$$\cos \alpha \, \phi(a, \lambda_i) + \sin \alpha \, \phi'(a, \lambda_i) = 0$$
$$\cos \beta \, \phi(b, \lambda_i) + \sin \beta \, \phi'(b, \lambda_i) = 0$$

which shows that $\phi(x, \lambda_i)$ is an eigenfunction and λ_i is an eigenvalue.

Conversely, if $w(\lambda_i) \neq 0$ then $\phi(x, \lambda_i)$ and $\psi(x, \lambda_i)$ are linearly independent and any solution $u(x)$ of the homogeneous equation is a linear combination of ϕ and ψ, that is,

$$u(x) = c_1 \phi(x, \lambda_i) + c_2 \psi(x, \lambda_i).$$

Now if u satisfies the boundary conditions

$$\cos \alpha \, u(a) + \sin \alpha \, u'(a) = c_2[\cos \alpha \, \psi(a, \lambda_i) + \sin \alpha \, \psi'(a, \lambda_i)] = 0,$$
$$\cos \beta \, u(b) + \sin \beta \, u'(b) = c_1[\cos \beta \, \phi(b, \lambda_i) + \sin \beta \, \phi'(b, \lambda_i)] = 0.$$

If either of the square bracketed expressions are zero, then $w(\lambda_i) = 0$ contrary to assumption. Therefore, $c_1 = c_2 = 0$ and $u \equiv 0$ is the trivial solution. This completes the proof.

Theorem 7.7.2. shows that to determine the eigenvalues of the homogeneous problem (a) one can study the zeros of $w(\lambda)$.

Theorem 7.7.3. *Let $\phi(x, \lambda)$ and $\psi(x, \lambda)$ be defined as in theorem 7.7.2. Then $w(\lambda) = W(\phi, \psi)$ is an entire function of λ with only simple zeros.*

Proof: It is easy to show that $\phi(x, \lambda)$ and $\psi(x, \lambda)$ satisfy the following integral equations

$$\phi(x, \lambda) = \sin \alpha - (x - a)\cos \alpha + \int_a^x (q - \lambda)(x - t)\phi(t, \lambda)\, dt,$$

$$\psi(x, \lambda) = \sin \beta - (x - b)\cos \beta + \int_b^x (q - \lambda)(x - t)\psi(t, \lambda)\, dt.$$

We can show that there exist unique solutions of these integral equations by the iteration procedure used in Theorem 7.2.1. We let $\phi_0 = \sin \alpha - (x - a)\cos \alpha$, $\psi_0 = \sin \beta - (x - b)\cos \beta$ and then define

$$\phi_1 = \phi_0 + \int_a^x (q - \lambda)(x - t)\phi_0 \, dt,$$

$$\psi_1 = \psi_0 + \int_b^x (q - \lambda)(x - t)\psi_0 \, dt,$$

or in general

$$\phi_n = \phi_0 + \int_a^x (q - \lambda)(x - t)\phi_{n-1} \, dt,$$

$$\psi_n = \psi_0 + \int_b^x (q - \lambda)(x - t)\psi_{n-1} \, dt.$$

Then

$$\phi(x, \lambda) = \phi_0 + \sum_{n=1}^{\infty} (\phi_n - \phi_{n-1}),$$

$$\psi(x, \lambda) = \psi_0 + \sum_{n=1}^{\infty} (\psi_n - \psi_{n-1}),$$

and we can show that these series converge for each λ uniformly with respect to x for $a \le x \le b$. To show this we have

$$|\phi_1 - \phi_0| = \left| \int_a^x (q - \lambda)(x - t)\phi_0 \, dt \right| \le M(Q + R)(x - a)^2/2,$$

$$|\psi_1 - \psi_0| = \left| \int_b^x (q - \lambda)(x - t)\psi_0 \, dt \right| \le M(Q + R)(b - x)^2/2,$$

where $M = \max [|\phi_0|, |\psi_0|]$, $Q = \max |q|$, and $|\lambda| \le R$. Furthermore,

$$|\phi_2 - \phi_1| = \left| \int_a^x (q - \lambda)(x - t)(\phi_1 - \phi_0) \, dt \right|$$

$$\le \frac{M(Q + R)^2(b - a)}{2} \int_a^x (t - a)^2 \, dt$$

$$\le \frac{M(Q + R)^2(b - a)(x - a)^3}{3!},$$

$$|\psi_2 - \psi_1| = \left| \int_x^b (q - \lambda)(x - t)(\psi_1 - \psi_0) \, dt \right|$$

$$\le \frac{M(Q + R)^2(b - a)}{2} \int_x^b (b - t)^2 \, dt$$

$$\le \frac{M(Q + R)^2(b - a)(b - x)^3}{3!},$$

and by induction

$$|\phi_n - \phi_{n-1}| \le \frac{M(Q + R)^n(b - a)^{n-1}(x - a)^{n+1}}{(n + 1)!},$$

$$|\psi_n - \psi_{n-1}| \le \frac{M(Q + R)^n(b - a)^{n-1}(b - x)^{n+1}}{(n + 1)!}.$$

Hence, the series $\phi_0 + \sum_{n=1}^{\infty} (\phi_n - \phi_{n-1})$ and $\psi_0 + \sum_{n=1}^{\infty} (\psi_n - \psi_{n-1})$ converge uniformly for $a \leq x \leq b$. To show that they satisfy the integral equations one merely substitutes and integrates term by term.

The above inequalities also show that for each x and each R the series converge uniformly in λ for $|\lambda| \leq R$. Since each term in the series is an analytic function of λ the functions $\phi(x, \lambda)$ and $\psi(x, \lambda)$ are entire functions of λ. Therefore, $w(\lambda)$ is also entire.

To show that the zeros of $w(\lambda)$ are simple we should first show that it does not vanish identically. It is easy to show that

$$(\lambda - \lambda') \int_a^b \phi(x, \lambda)\phi(x, \lambda')\, dx = \phi(b, \lambda)\phi'(b, \lambda') - \phi(b, \lambda')\phi'(b, \lambda).$$

Then if $\sin \beta \neq 0$

$$\sin \beta(\lambda - \lambda') \int_a^b \phi(x, \lambda)\phi(x, \lambda')\, dx = w(\lambda')\phi(b, \lambda) - w(\lambda)\phi(b, \lambda')$$

and

$$(\lambda - \lambda') \int_a^b \phi(x, \lambda)\phi(x, \lambda')\, dx = \pm[w(\lambda)\phi'(b, \lambda') - w(\lambda')\phi'(b, \lambda)],$$

if $\sin \beta = 0$, depending on whether $\cos \beta = \pm 1$. In either case if $w(\lambda) \equiv 0$, then for $\lambda \neq \lambda'$, $\int_a^b \phi(x, \lambda)\, \phi(x, \lambda')\, dx = 0$. Hence, if λ and λ' are real

$$\lim_{\lambda' \to \lambda} \int_a^b \phi(x, \lambda)\phi(x, \lambda')\, dx = \int_a^b \phi^2(x, \lambda)\, dx = 0.$$

This implies that $\phi(x, \lambda) \equiv 0$, which is impossible. Therefore, $w(\lambda)$ cannot be identically zero. This result also shows that the zeros of $w(\lambda)$ can have no finite limit point. For if they did $w(\lambda)$ would have to vanish identically.

Let λ_k be a zero of $w(\lambda)$. If this is a zero of order two or higher then $w(\lambda_k) = w'(\lambda_k) = 0$ and

$$w(\lambda_k + i\tau) = -\tau^2 w''(\lambda_k) + \text{higher order terms in } \tau,$$
$$w(\lambda_k - i\tau) = -\tau^2 w''(\lambda_k) + \text{higher order terms in } \tau,$$

where we are assuming that τ is real. Let $\lambda = \lambda_k + i\tau$ and $\lambda' = \lambda_k - i\tau$ in the above and then

$$2i\tau \sin \beta \int_a^b \phi(x, \lambda)\bar{\phi}(x, \lambda)\, dx = -\tau^2 w''(\lambda_k)[\phi(b, \lambda) - \bar{\phi}(b, \lambda)]$$
$$+ \text{ higher order terms in } \tau$$

or

$$2i\tau \int_a^b \phi(x, \lambda)\bar{\phi}(x, \lambda)\, dx = \mp \tau^2 w''(\lambda_k)[\bar{\phi}'(b, \lambda) - \phi'(b, \lambda)]$$

$$+ \text{ higher order terms in } \tau,$$

depending on whether $\sin \beta \neq 0$ or $\sin \beta = 0$. Dividing by τ and taking the limit as τ approaches zero we have in either case $\int_a^b \phi^2(x, \lambda_k)\, dx = 0$, which is a contradiction. This completes the proof of Theorem 7.7.3.

To complete the discussion of the regular Sturm-Liouville problem we shall have to show that an arbitrary piecewise continuous function with a piecewise continuous derivative defined on $a \leq x \leq b$ can be expanded in a series of the eigenfunctions of the homogeneous problem (b). The procedure will be to integrate the solution of the nonhomogeneous problem (a)

$$\frac{1}{2\pi i} \int_{C_{n+}} y(x, \lambda)\, d\lambda,$$

where $\{C_n\}$ is a sequence of closed contours which will be described later. We shall show that for n sufficiently large C_n does not pass through any zeros of $w(\lambda)$. If λ_k is inside C_n then the value of the integral will have a contribution equal to the residue of

$$\frac{\psi(x, \lambda)}{w(\lambda)} \int_a^x \phi(t, \lambda)f(t)\, dt + \frac{\phi(x, \lambda)}{w(\lambda)} \int_x^b \psi(t, \lambda)f(t)\, dt$$

at the simple pole at λ_k. This is easily seen to be

$$\frac{A_k\phi(x, \lambda_k)}{w'(\lambda_k)} \int_a^b \phi(t, \lambda_k)f(t)\, dt,$$

where A_k is the constant in the relation $\psi(x, \lambda_k) = A_k\phi(x, \lambda_k)$. Therefore, provided we can show that the limit exists,

$$\lim_{n \to \infty} \frac{1}{2\pi i} \int_{C_{n+}} y(x, \lambda)\, d\lambda = \sum_{k=1}^{\infty} \frac{A_k\phi(x, \lambda_k)}{w'(\lambda_k)} \int_a^b \phi(t, \lambda_k)f(t)\, dt,$$

where the summation is over all the zeros of $w(\lambda)$. We shall show that the limit exists and is in fact the Fourier series for $f(x)$.* This will show that

$$\frac{f(x^+) + f(x^-)}{2} = \sum_{k=1}^{\infty} \frac{A_k\phi(x, \lambda_k)}{w'(\lambda_k)} \int_a^b \phi(t, \lambda_k)f(t)\, dt,$$

where

$$f(x^+) = \lim_{t \to x^+} f(t), f(x^-) = \lim_{t \to x^-} f(t),$$

$$f(a^-) = \lim_{t \to b^-} f(t), f(b^+) = \lim_{t \to a^+} f(t).$$

* See Section 8.1.

If we define $\Phi_k(x) = \left[\dfrac{A_k}{w'(\lambda_k)} \right]^{1/2} \phi(x, \lambda_k)$, then

$$\frac{f(x^+) + f(x^-)}{2} = \sum_{k=1}^{\infty} c_k \Phi_k(x),$$

where

$$c_k = \int_a^b \Phi_k(x) f(x)\, dx,$$

$$\int_a^b \Phi_j(x)\Phi_k(x)\, dx = 0,\ j \neq k,$$

$$\int_a^b \Phi_k^2(x)\, dx = 1.$$

Theorem 7.7.4. *Let $f(x)$ be an arbitrary piecewise continuous function with a piecewise continuous derivative defined on the interval $a \leq x \leq b$. Then $f(x)$ can be expanded in a series of the eigenfunctions of the homogeneous Sturm-Liouville problem* (a) *as follows:*

$$\frac{f(x^+) + f(x^-)}{2} = \sum_{k=1}^{\infty} c_k \Phi_k(x),$$

where

$$c_k = \int_a^b \Phi_k(x) f(x)\, dx$$

and

$$\int_a^b \Phi_k^2(x)\, dx = 1.$$

Proof: It is easy to show that $\phi(x, \lambda)$ and $\psi(x, \lambda)$ satisfy the following integral equations:

$$\phi(x, \varsigma^2) = \sin \alpha \cos \varsigma(x - a) - \frac{\cos \alpha \sin \varsigma(x - a)}{\varsigma}$$

$$+ \frac{1}{\varsigma} \int_a^x \sin \varsigma(x - t) q(t)\phi(t, \varsigma^2)\, dt,$$

$$\psi(x, \varsigma^2) = \sin \beta \cos \varsigma(x - b) - \frac{\cos \beta \sin \varsigma(x - b)}{\varsigma}$$

$$+ \frac{1}{\varsigma} \int_b^x \sin \varsigma(x - t) q(t)\psi(t, \varsigma^2)\, dt,$$

where we have let $\lambda = \zeta^2$. We let $\zeta = \xi + i\eta$ and

$$\phi(x, \zeta^2) = e^{|\eta|(x-a)}F(x),$$

$$\psi(x, \zeta^2) = e^{|\eta|(b-x)}G(x),$$

$$F(x) = \left[\sin\alpha\cos\zeta(x - a) - \frac{\cos\alpha\sin\zeta(x - a)}{\zeta}\right]e^{-|\eta|(x-a)}$$

$$+ \frac{1}{\zeta}\int_a^x \sin\zeta(x - t)q(t)e^{-|\eta|(x-t)}F(t)\,dt,$$

$$G(x) = \left[\sin\beta\cos\zeta(x - b) - \frac{\cos\beta\sin\zeta(x - b)}{\zeta}\right]e^{-|\eta|(b-x)}$$

$$+ \frac{1}{\zeta}\int_b^x \sin\zeta(x - t)q(t)e^{|\eta|(x-t)}G(t)\,dt.$$

If $M_1 = \max|F(x)|$ and $M_2 = \max|G(x)|$, then

$$M_1 \le |\sin\alpha| + \frac{|\cos\alpha|}{|\zeta|} + \frac{1}{|\zeta|}\int_a^b |q(t)|M_1\,dt,$$

$$M_2 \le |\sin\beta| + \frac{|\cos\beta|}{|\zeta|} + \frac{1}{|\zeta|}\int_a^b |q(t)|M_2\,dt,$$

$$M_1 \le \left(|\sin\alpha| + \frac{|\cos\alpha|}{|\zeta|}\right)\Big/\left(1 - \frac{1}{|\zeta|}\int_a^b |q(t)|\,dt\right),$$

$$M_2 \le \left(|\sin\beta| + \frac{|\cos\beta|}{|\zeta|}\right)\Big/\left(1 - \frac{1}{|\zeta|}\int_a^b |q(t)|\,dt\right),$$

provided $|\zeta|$ is sufficiently large. This shows that*

$$\phi(x, \zeta^2) = \sin\alpha\cos\zeta(x - a) + O(|\zeta|^{-1}e^{|\eta|(x-a)}),$$

if $\sin\alpha \ne 0$ and

$$\phi(x, \zeta^2) = -\frac{\cos\alpha\sin\zeta(x - a)}{\zeta} + O(|\zeta|^{-2}e^{|\eta|(x-a)}),$$

if $\sin\alpha = 0$. Also,

$$\psi(x, \zeta^2) = \sin\beta\cos\zeta(x - b) + O(|\zeta|^{-1}e^{|\eta|(b-x)}),$$

if $\sin\beta \ne 0$ and

$$\psi(x, \zeta^2) = -\frac{\cos\beta\sin\zeta(x - b)}{\zeta} + O(|\zeta|^{-2}e^{|\eta|(b-x)}),$$

---

* The notation $g(z) = O(f(z))$ as $z \to z_0$ means that there is some ϵ-neighborhood of z_0 throughout which $|g(z)| \le K|f(z)|$ for some constant K.

if $\sin \beta = 0$. Similarly, we can show that

$$\phi'(x, \zeta^2) = -\zeta \sin \alpha \sin \zeta(x - a) + O(e^{|\eta|(x-a)}),$$

if $\sin \alpha \neq 0$ and

$$\phi'(x, \zeta^2) = -\cos \alpha \cos \zeta(x - a) + O(|\zeta|^{-1} e^{|\eta|(x-a)}),$$

if $\sin \alpha = 0$. Also,

$$\psi'(x, \zeta^2) = -\zeta \sin \beta \sin \zeta(x - b) + O(e^{|\eta|(b-x)}),$$

if $\sin \beta \neq 0$ and

$$\psi'(x, \zeta^2) = -\cos \beta \cos \zeta(x - b) + O(|\zeta|^{-1} e^{|\eta|(b-x)}).$$

Based on these results, it is easy to show that

$$w(\lambda) = \zeta \sin \alpha \sin \beta \sin \zeta(b - a) + O(e^{|\eta|(b-a)}),$$

if $\sin \alpha \sin \beta \neq 0$, with similar formulas holding if $\sin \alpha$ and/or $\sin \beta$ are zero. To keep the proof short and still give the essentials, we shall continue with the case $\sin \alpha \sin \beta \neq 0$. The reader can supply the details in the other three cases.

To complete the proof we integrate $y(x, \lambda)$, the unique solution of the nonhomogeneous problem (a) with $f(x)$ the right-hand side of the differential equation around a closed contour C_n which we now describe. Consider Figure 7.7.1.

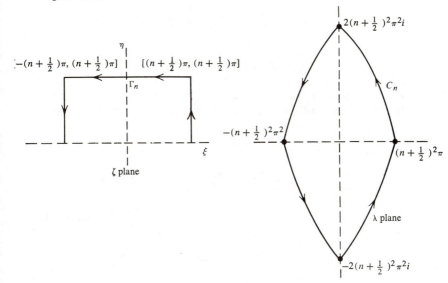

Figure 7.7.1.

The contour in the ζ plane maps into C_n in the λ plane under the transformation $\lambda = \zeta^2$. We have already shown that

$$y(x, \lambda) = \frac{\psi(x, \lambda)}{w(\lambda)} \int_a^x \phi(t, \lambda) f(t)\, dt + \frac{\phi(x, \lambda)}{w(\lambda)} \int_x^b \psi(t, \lambda) f(t)\, dt.$$

On Γ_n, $\eta \geq 0$ and $|\sin \zeta(b - a)| > Ke^{\eta(b-a)}$, where K is a constant. Therefore,

$$\frac{1}{w(\lambda)} = \frac{1}{\zeta \sin \alpha \sin \beta \sin \zeta(b - a)} [1 + O(|\zeta|^{-1})].$$

This shows that for n sufficiently large C_n does not pass through any zeros of $w(\lambda)$. Using our previously determined estimates on $\phi(x, \lambda)$ and $\psi(x, \lambda)$ we have

$$\frac{\psi(x, \lambda)\phi(t, \lambda)}{w(\lambda)} = \frac{\cos \zeta(b - x) \cos \zeta(t - a)}{\zeta \sin \zeta(b - a)} + O(|\zeta|^{-2} e^{\eta(t-x)}),$$

$$\frac{\phi(x, \lambda)\psi(t, \lambda)}{x(\lambda)} = \frac{\cos \zeta(b - t) \cos \zeta(x - a)}{\zeta \sin \zeta(b - a)} + O(|\zeta|^{-2} e^{\eta(x-t)}).$$

Hence,

$$\begin{aligned}
y(x, \lambda) = &\int_a^x \frac{\cos \zeta(b - x) \cos \zeta(t - a)}{\zeta \sin \zeta(b - a)} f(t)\, dt \\
&+ \int_x^b \frac{\cos \zeta(b - t) \cos \zeta(x - a)}{\zeta \sin \zeta(b - a)} f(t)\, dt \\
&+ O\left(|\zeta|^{-2} \int_a^x e^{\eta(t-x)} f(t)\, dt\right) \\
&+ O\left(|\zeta|^{-2} \int_x^b e^{\eta(x-t)} f(t)\, dt\right).
\end{aligned}$$

If we ignore the last two terms for the moment (we shall show that the integral around C_n of these terms goes to zero as $n \to \infty$), we have

$$\begin{aligned}
\lim_{n \to \infty} &\frac{1}{2\pi i} \int_{C_n} y(x, \lambda)\, d\lambda \\
&= \lim_{n \to \infty} \frac{1}{2\pi i} \int_{C_n} \left[\int_a^x \frac{\cos \zeta(b - x) \cos \zeta(t - a)}{\zeta \sin \zeta(b - a)} f(t)\, dt \right. \\
&\qquad\qquad\qquad\qquad \left. + \int_x^b \frac{\cos \zeta(b - t) \cos \zeta(x - a)}{\zeta \sin \zeta(b - a)} f(t)\, dt \right] d\lambda \\
&= \frac{1}{b - a} \int_a^b f(t)\, dt + \frac{2}{b - a} \sum_{n=1}^{\infty} \cos \frac{n\pi}{b - a}(x - a) \\
&\qquad\qquad\qquad\qquad \times \int_a^b \cos \frac{n\pi}{b - a}(t - a) f(t)\, dt,
\end{aligned}$$

where we have computed the sum of the residues at all the poles of the integrand. The last series is just the Fourier cosine series for $f(x)$ (see Section 8.1), which under the given conditions on $f(x)$ is known to converge to $\frac{1}{2}[f(x^+) + f(x^-)]$. It remains to show that the integral around C_n of the other terms approaches zero as $n \to \infty$. Let $0 < \delta < x - a$. Then

$$\int_{C_n} O\left\{|\varsigma|^{-2} \int_a^x e^{\eta(t-x)} f(t)\, dt\right\} d\lambda$$

$$= O\left\{\int e^{-\delta\eta}|\varsigma|^{-1}\, |d\varsigma|\right\} + O\left\{\int_{x-\delta}^x |f(t)|\, dt\right\}.$$

The last integral can be made arbitrarily small by choosing δ small, while the other term behaves as follows

$$O\left\{\frac{1}{n} \int_0^{(2n+1)\pi/2(b-a)} e^{-\delta\eta}\, d\eta\right\}$$

$$+ O\left\{\frac{1}{n} \int_0^{(2n+1)\pi/2(b-a)} \exp\left[-\frac{(2n+1)\pi\delta}{2(b-a)}\right] d\xi\right\}$$

$$= O\left(\frac{1}{n\delta}\right) + O\left[\exp\left(-\frac{(2n+1)\pi\delta}{2(b-a)}\right)\right],$$

which is arbitrarily small for n sufficiently large. This completes the proof of Theorem 7.7.4.

Corollary 7.7.4. Let $f(x)$ be a function satisfying the conditions of Theorem 7.7.4 and $\lambda \neq \lambda_k$, $k = 1, 2, 3, \ldots$, where the λ_k are the eigenvalues of the corresponding homogeneous Sturm-Liouville problem. Then if $y(x, \lambda)$ is the solution of $y'' + [\lambda - q(x)]y = f(x)$ on $a \leq x \leq b$ satisfying appropriate boundary conditions

$$y(x, \lambda) = \sum_{k=1}^{\infty} \frac{c_k \Phi_k(x)}{\lambda - \lambda_k},$$

where $c_k = \int_a^b f(x)\, \Phi_k(x)\, dx$ and the $\Phi_k(x)$ are the normalized eigenfunctions.

Proof: We have already established that

$$y(x, \lambda) = \frac{\psi(x, \lambda)}{w(\lambda)} \int_a^x \phi(t, \lambda) f(t)\, dt + \frac{\phi(x, \lambda)}{w(\lambda)} \int_x^b \psi(t, \lambda) f(t)\, dt$$

from which it can be proved that y is continuously differentiable in $a \leq x \leq b$. This means that $y(x, \lambda)$ has a Sturm-Liouville expansion

$$y(x, \lambda) = \sum_{k=1}^{\infty} b_k \Phi_k(x),$$

where

$$b_k = \int_a^b y(x, \lambda) \Phi_k(x)\, dx.$$

Furthermore,

$$c_k = \int_a^b f(x)\Phi_k(x)\,dx = \int_a^b [y'' + (\lambda - q)y]\Phi_k(x)\,dx$$

$$= [y'\Phi_k - y\Phi_k']_a^b + \int_a^b [\Phi_k'' + (\lambda - q)\Phi_k]y\,dx$$

$$= \int_a^b [\Phi_k'' + (\lambda_k - q)\Phi_k]y\,dx + (\lambda - \lambda_k)\int_a^b y\Phi_k\,dx$$

$$= (\lambda - \lambda_k)b_k.$$

Hence, $b_k = c_k/(\lambda - \lambda_k)$.

If $\lambda = \lambda_k$ for some k, then a necessary condition for $y'' + (\lambda_k - q)y = f(x)$ to have a solution is that $f(x)$ be orthogonal to $\Phi_k(x)$. For suppose the problem has a solution y, then

$$c_k = \int_a^b f(x)\Phi_k(x)\,dx = \int_a^b [y'' + (\lambda_k - q)y]\Phi_k\,dx$$

$$= [y'\Phi_k - y\Phi_k']_a^b + \int_a^b [\Phi_k'' + (\lambda_k - q)\Phi_k]y\,dx$$

$$= 0.$$

Conversely, if $\lambda = \lambda_k$ and $f(x)$ is orthogonal to $\Phi_k(x)$ then, since $c_k = 0 = (\lambda_k - \lambda_k)b_k$ is satisfied for arbitrary b_k, there exists a solution

$$y(x, \lambda_k) = \sum_{\substack{j=1 \\ j\neq k}}^{\infty} \frac{c_j\Phi_j(x)}{\lambda_k - \lambda_j} + b_k\Phi_k(x),$$

where b_k is arbitrary. Therefore, in this case there exists a solution, but it is not unique.

Exercises 7.7

1. Transform the Sturm-Liouville equation from the first form to the second;

$$(e^{-3x}y')' + \lambda e^{-3x}y = f(x), \quad \ddot{u} + [\lambda - q(t)]u = g(t).$$

2. Find the form of the expansion theorem for the problem $y'' + \lambda y = f(x)$, $y(0) = y(\pi) = 0$.

3. Find the form of the expansion theorem for the problem $(xy')' + \lambda xy = f(x)$, $y(1) = y(2) = 0$.

7.8. FREDHOLM INTEGRAL EQUATIONS

Another approach to the study of Sturm-Liouville problems is through integral equations. To illustrate this, let us begin by considering the following nonhomogeneous Sturm-Liouville problem on $a \leq x \leq b$:

$$(py')' - qy + \lambda \rho y = -f(x),$$
$$\cos \alpha \, y(a) + \sin \alpha \, y'(a) = 0,$$
$$\cos \beta \, y(b) + \sin \beta \, y'(b) = 0.$$

We assume that the problem is regular, that is, p is continuously differentiable, q and ρ are continuous, f is piecewise continuous, $p > 0$ and $\rho > 0$, and a and b are finite. To begin with, assume that $\lambda = 0$ is not an eigenvalue of the corresponding homogeneous problem ($f \equiv 0$). Therefore, by Theorem 7.7.1 the equation $(py')' - qy = -f(x)$ has a unique solution subject to the same boundary conditions. We can determine this solution as follows: Let $\phi(x)$ and $\psi(x)$ be solutions of $(py')' - qy = 0$ satisfying

$$\phi(a) = \sin \alpha, \qquad \phi'(a) = -\cos \alpha,$$
$$\psi(b) = \sin \beta, \qquad \psi'(b) = -\cos \beta.$$

By Theorem 7.7.2, $\phi(x)$ and $\psi(x)$ exist, are unique, and are linearly independent. We see that

$$\cos \alpha \, \phi(a) + \sin \alpha \, \phi'(a) = 0,$$
$$\cos \beta \, \psi(b) + \sin \beta \, \psi'(b) = 0.$$

We seek a solution of the nonhomogeneous problem by the method of variation of parameters,

$$y(x) = A(x)\phi(x) + B(x)\psi(x),$$
$$y'(x) = A\phi' + B\psi' + A'\phi + B'\psi.$$

We let $A'\phi + B'\psi = 0$ and then

$$(py')' = A(p\phi')' + B(p\psi')' + A'p\phi' + B'p\psi',$$
$$(py')' - qy = A'p\phi' + B'p\psi' = -f.$$

Solving for A' and B', we have

$$A' = \frac{f(x)\psi(x)}{p(x)(\phi\psi' - \psi\phi')},$$
$$B' = \frac{-f(x)\phi(x)}{p(x)(\phi\psi' - \psi\phi')}.$$

The denominators are constant because

$$\frac{d}{dx}[p(\phi\psi' - \psi\phi')] = \phi(p\psi')' - \psi(p\phi')' = 0.$$

Therefore,

$$A = \frac{1}{k} \int_{c_1}^{x} f(t)\psi(t)\,dt,$$

$$B = -\frac{1}{k} \int_{c_2}^{x} f(t)\phi(t)\,dt,$$

$$y(x) = \frac{\phi(x)}{k} \int_{c_1}^{x} f(t)\psi(t)\,dt - \frac{\psi(x)}{k} \int_{c_2}^{x} f(t)\phi(t)\,dt,$$

$$y'(x) = \frac{\phi'(x)}{k} \int_{c_1}^{x} f(t)\psi(t)\,dt - \frac{\psi'(x)}{k} \int_{c_2}^{x} f(t)\phi(t)\,dt,$$

$$\cos \alpha\, y(a) + \sin \alpha\, y'(a) = [-\cos \alpha\, \psi(a) - \sin \alpha\, \psi'(a)] \int_{c_2}^{a} f(t)\phi(t)\,dt = 0.$$

Since the square bracket does not vanish, $\int_{c_2}^{a} f(t)\phi(t)\,dt = 0$. Also,

$$\cos \beta\, y(b) + \sin \beta\, y'(b) = [\cos \beta\, \phi(b) + \sin \beta\, \phi'(b)] \int_{c_1}^{b} f(t)\psi(t)\,dt = 0,$$

and therefore, $\int_{c_1}^{b} f(t)\psi(t)\,dt = 0$. The result for $y(x)$ is therefore

$$y(x) = -\frac{\psi(x)}{k} \int_{a}^{x} f(t)\phi(t)\,dt - \frac{\phi(x)}{k} \int_{x}^{b} f(t)\psi(t)\,dt.$$

We can write this as follows:

$$y(x) = \int_{a}^{b} G(x,t)f(t)\,dt,$$

where the *Green's function* $G(x,t)$ is defined as

$$G(x,t) = \begin{cases} -\dfrac{\psi(x)\phi(t)}{k}, & a \le t \le x, \\[2mm] -\dfrac{\phi(x)\psi(t)}{k}, & x \le t \le b. \end{cases}$$

If we think of the Green's function as a function of t defined for each fixed value of the parameter x, then it is easy to verify the following properties:

1. $(pG')' - qG = 0$ for $a < t < x$ and $x < t < b$.
2. G is continuous at $t = x$.
3. $\cos \alpha\, G(a) + \sin \alpha\, G'(a) = 0$, $\cos \beta\, G(b) + \sin \beta\, G'(b) = 0$.
4. $G'(x, x^+) - G'(x, x^-) = -1/p(x)$.

Theorem 7.8.1. *If we take Properties 1–4 as defining properties of the Green's function, then $G(x, t)$ is uniquely defined if and only if $\lambda = 0$ is not an eigenvalue of the homogeneous Sturm-Liouville problem defined above.*

Proof: If $\lambda = 0$ is not an eigenvalue then the functions $\phi(x)$ and $\psi(x)$ are uniquely defined and are linearly independent. Then

$$G(x, t) = A\phi(t) + B\psi(t), \qquad a \leq t \leq x,$$
$$G(x, t) = C\phi(t) + D\psi(t), \qquad x \leq t \leq b.$$

Then

$$\cos \alpha \, G(x, a) + \sin \alpha \, G'(x, a) = B\big(\cos \alpha \, \psi(a) + \sin \alpha \, \psi'(a)\big) = 0,$$
$$\cos \beta \, G(x, b) + \sin \beta \, G'(x, b) = C\big(\cos \beta \, \phi(b) + \sin \alpha \, \phi'(b)\big) = 0.$$

Therefore, $B = C = 0$. Also, $G(x, x^{+}) = G(x, x^{-})$ implies that $A\phi(x) = D\psi(x)$, and $G'(x, x^{+}) - G'(x, x^{-}) = -1/p(x)$ implies

$$\frac{A\phi(x)\psi'(x)}{\psi(x)} - A\phi'(x) = -1/p(x),$$

$$A = \frac{-\psi(x)}{p(x)(\phi\psi' - \psi\phi')} = -\frac{\psi(x)}{k},$$

$$D = -\frac{\phi(x)}{k}.$$

Therefore,

$$G(x, t) = \begin{cases} -\dfrac{\psi(x)\phi(t)}{k}, & a \leq t \leq x \\[2ex] -\dfrac{\phi(x)\psi(t)}{k}, & x \leq t \leq b. \end{cases}$$

If $\lambda = 0$ is an eigenvalue of the homogeneous problem, then there is a nontrivial continuously differentiable solution $u(x)$ of $(py')' - qy = 0$ satisfying the boundary conditions. Therefore, $G(x, t) + cu(t)$ satisfies Properties 1–4 for arbitrary c, and hence, G cannot be unique. Actually, no function $G(x, t)$ exists satisfying Properties 1–4 in this case. What happens is that in satisfying Properties 1–3 we produce an eigenfunction which has a continuous derivative and hence will not satisfy Property 4. In this case, we can define a *generalized Green's function* which will satisfy a more general differential equation. This idea will be introduced in the exercises.

We can show that whenever $G(x, t)$ exists, then

$$y(x) = \int_a^b G(x, t)f(t) \, dt$$

solves the nonhomogeneous problem. This follows by direct verification, since

$$y(x) = -\frac{\phi(x)}{k} \int_x^b \psi(t)f(t)\, dt - \frac{\psi(x)}{k} \int_a^x \phi(t)f(t)\, dt$$

$$= A(x)\phi(x) + B(x)\psi(x),$$

where $A'\phi + B'\psi \equiv 0$ and $A'p\phi' + B'p\psi' = -f$. The boundary conditions can also easily be verified.

Now, we return to the case $\lambda \neq 0$ in the equation $(py')' - qy + \lambda\rho y = -f$. We write the equation

$$(py')' - qy = -f - \lambda\rho y$$

and then using the Green's function $G(x, t)$ we have the integral equation

$$y(x) = \int_a^b G(x, t)f(t)\, dt + \lambda \int_a^b \rho(t)G(x, t)y(t)\, dt.$$

Multiplying by $\sqrt{\rho(x)}$, we have

$$\sqrt{\rho(x)}\, y(x) = \int_a^b [\sqrt{\rho(x)\rho(t)}\, G(x, t)f(t)/\sqrt{\rho(t)}]\, dt$$

$$+ \lambda \int_a^b \sqrt{\rho(x)\rho(t)}\, G(x, t)\sqrt{\rho(t)}\, y(t)\, dt.$$

We let $K(x, t) = \sqrt{\rho(x)\rho(t)}\, G(x, t)$, $u(x) = \sqrt{\rho(x)}\, y(x)$, and

$$F(x) = \int_a^b [\sqrt{\rho(x)\rho(t)}\, G(x, t)f(t)/\sqrt{\rho(t)}]\, dt,$$

then

$$u(x) = F(x) + \lambda \int_a^b K(x, t)u(t)\, dt.$$

This is known as *Fredholm's integral equation*.

For certain values of x we can solve this equation by iteration as follows. Let $u_0(x)$ be $F(x)$. Then

$$u_1(x) = F(x) + \lambda \int_a^b K(x, t)u_0(t)\, dt,$$

$$u_2(x) = F(x) + \lambda \int_a^b K(x, t)u_1(t)\, dt,$$

$$\cdots\cdots\cdots\cdots\cdots\cdots\cdots\cdots\cdots\cdots\cdots\cdots$$

$$u_n(x) = F(x) + \lambda \int_a^b K(x, t)u_{n-1}(t)\, dt,$$

can be defined recursively and

$$u_1 - u_0 = \lambda \int_a^b K(x, t) u_0(t)\, dt,$$

$$|u_1 - u_0| \leq |\lambda| M N (b - a),$$

where $M = \max |K(x, t)|$ and $N = \max |F(x)|$. Also,

$$u_2 - u_1 = \lambda \int_a^b K(x, t)[u_1(t) - u_0(t)]\, dt,$$

$$|u_2 - u_1| \leq |\lambda|^2 N M^2 (b - a)^2.$$

By induction we can prove

$$|u_n - u_{n-1}| \leq |\lambda|^n N M^n (b - a)^n.$$

Consider the series $F(x) + \sum_{n=1}^{\infty} [u_n(x) - u_{n-1}(x)]$. This series converges uniformly with respect to x for $|\lambda| < [M(b - a)]^{-1}$. Therefore, we can substitute the series in the integral equation and integrate term by term,

$$F(x) + \lambda \int_a^b K(x, t) \left\{ F(t) + \sum_{n=1}^{\infty} [u_n(t) - u_{n-1}(t)] \right\} dt$$

$$= u_0(x) + u_1(x) + \sum_{n=1}^{\infty} (u_{n+1} - u_n) = u_0(x) + \sum_{n=1}^{\infty} (u_n - u_{n-1}),$$

since clearly both series converge to the same sum.

It is easy to see that this solution is unique. Let $H[f] = \int_a^b K(x, t) f(t)\, dt$. Then $H^2[f] = \int_a^b \int_a^b K(x, t_1) K(t_1, t_2) f(t_2)\, dt_1\, dt_2$, and so on, and our sequence of approximations are

$$u_0(x) = F(x),$$
$$u_1(x) = F(x) + \lambda H[F],$$
$$u_2(x) = F(x) + \lambda H[F] + \lambda^2 H^2[F],$$

$$\cdots\cdots\cdots\cdots\cdots\cdots\cdots\cdots\cdots\cdots\cdots$$

$$u_n(x) = F(x) + \lambda H[F] + \lambda^2 H^2[F] + \cdots + \lambda^n H^n[F].$$

If we started the iteration with some other continuous function say $\tilde{F}$, then

$$u_0(x) = \tilde{F}(x)$$
$$u_1(x) = F(x) + \lambda H[\tilde{F}],$$
$$u_2(x) = F(x) + \lambda H[F] + \lambda^2 H^2[\tilde{F}],$$

$$\cdots\cdots\cdots\cdots\cdots\cdots\cdots\cdots\cdots\cdots\cdots$$

$$u_n(x) = F(x) + \lambda H[F] + \lambda^2 H^2[F] + \cdots + \lambda^{n-1} H^{n-1}[F] + \lambda^n H^n[\tilde{F}].$$

Clearly, both series converge to the same sum. In other words, the continuous function chosen to begin the iteration is completely arbitrary. Let

$$u(x) = F(x) + \sum_{n=1}^{\infty} \lambda^n H^n[F]$$

and let $\bar{u}(x)$ be another solution of the integral equation. Then

$$u_1(x) = F(x) + \lambda \int_a^b K(x, t)\bar{u}(t)\, dt = F(x) + H[\bar{u}] = \bar{u}(x),$$

$$u_2(x) = F(x) + \lambda H[F] + \lambda^2 H^2[\bar{u}] = \bar{u}(x),$$

and so on. Therefore, $u(x) \equiv \bar{u}(x)$, which proves uniqueness.

Remembering that the original Sturm-Liouville problem has a unique solution only if λ is not an eigenvalue of the homogeneous problem, we have proved that there is no eigenvalue with modulus less than $[M(b - a)]^{-1}$, where $M = \max |K(x, t)|$.

We also see that our series solution converges uniformly with respect to λ for $|\lambda| \leq r < [M(b - a)]^{-1}$. Therefore,

$$u(x, \lambda) = \sum_{n=1}^{\infty} [u_n(x, \lambda) - u_{n-1}(x, \lambda)]$$

is an analytic function of λ for $|\lambda| < [M(b - a)]^{-1}$.

Let

$$K^1(x, t) = K(x, t);\ K^2(x, t) = \int_a^b K(x, t_1)K(t_1, t)\, dt_1;$$

$$K^3(x, t) = \int_a^b \int_a^b K(x, t_1)K(t_1, t_2)K(t_2, t)\, dt_1\, dt_2; \ldots;$$

$$K^n(x, t) = \int_a^b \cdots \int_a^b K(x, t_1)K(t_1, t_2) \cdots K(t_{n-1}, t)\, dt_1 \cdots dt_{n-1},$$

and let us define the *reciprocal kernel* as

$$k(x, t, \lambda) = \sum_{n=1}^{\infty} \lambda^{n-1} K^n(x, t).$$

We have already seen that this series converges uniformly with respect to x and t and is an analytic function of λ for $|\lambda| < [M(b - a)]^{-1}$. Also, we can write the solution of the integral equation as follows:

$$u(x) = F(x) + \lambda \int_a^b k(x, t, \lambda)F(t)\, dt.$$

It is easy to show that the reciprocal kernel satisfies the integral equation

$$k(\xi, t, \lambda) = K(\xi, t) + \lambda \int_a^b k(x, t, \lambda)K(\xi, x)\, dx.$$

Conversely, if we can find a reciprocal kernel satisfying this integral equation, then

$$u(x) = F(x) + \lambda \int_a^b k(x, t, \lambda)F(t)\, dt$$

can be shown to satisfy the original Fredholm equation since

$$\lambda \int_a^b K(\xi, x)u(x)\, dx = \lambda \int_a^b F(x)K(\xi, x)\, dx$$

$$+ \lambda^2 \int_a^b \int_a^b k(x, t, \lambda)F(t)K(\xi, x)\, dt\, dx$$

$$= \lambda \int_a^b F(x)K(\xi, x)\, dx$$

$$+ \lambda \int_a^b [k(\xi, t, \lambda) - K(\xi, t)]F(t)\, dt$$

$$= \lambda \int_a^b k(\xi, t, \lambda)F(t)\, dt$$

$$= u(\xi) - F(\xi).$$

The series representation $k(x, t, \lambda) = \sum_{n=1}^{\infty} \lambda^{n-1}K^n(x, t)$ is valid for $|\lambda| < [M(b - a)]^{-1}$. It is possible to obtain an analytic continuation for $k(x, t, \lambda)$ to the rest of the λ plane excluding the eigenvalues of the homogeneous problem. Let $\lambda_1, \lambda_2, \lambda_3, \ldots$ be the eigenvalues corresponding to the eigenfunctions $\phi_1(x), \phi_2(x), \phi_3(x), \ldots$ satisfying

$$\phi_i(x) = \lambda_i \int_a^b K(x, t)\phi_i(t)\, dt,$$

$$\int_a^b \phi_i\phi_j\, dx = \delta_{ij}.$$

It can be shown* that

$$K^n(x, t) = \sum_{i=1}^{\infty} \frac{\phi_i(x)\phi_i(t)}{\lambda_i^n},$$

---

* R. Courant and D. Hilbert, *Methods of Mathematical Physics*, **I**. New York: Interscience Publishers, Inc. pp. 134–138.

for $n = 2, 3, 4, \ldots$ and the convergence is uniform in both x and t. Therefore, if $|\lambda| \le r < [M(b - a)]^{-1}$,

$$
\begin{aligned}
k(x, t, \lambda) &= K(x, t) + \sum_{n=2}^{\infty} \lambda^{n-1} \sum_{i=1}^{\infty} \frac{\phi_i(x)\phi_i(t)}{\lambda_i^n} \\
&= K(x, t) + \sum_{i=1}^{\infty} \left(\sum_{n=2}^{\infty} \frac{\lambda^{n-1}}{\lambda_i^n} \right) \phi_i(x)\phi_i(t) \\
&= K(x, t) + \lambda \sum_{i=1}^{\infty} \frac{\phi_i(x)\phi_i(t)}{\lambda_i(\lambda_i - \lambda)} .
\end{aligned}
$$

This relation has been proved for small $|\lambda|$. However, we now see that the last equation defines $k(x, t, \lambda)$ for all $\lambda \ne \lambda_i$, $i = 1, 2, 3, \ldots$. It is easy to show that it satisfies the integral equation, since

$$
\begin{aligned}
K(\xi, t) + \lambda \int_a^b K(\xi, x) &\left[K(x, t) + \lambda \sum_{i=1}^{\infty} \frac{\phi_i(x)\phi_i(t)}{\lambda_i(\lambda_i - \lambda)} \right] dx \\
&= K(\xi, t) + \lambda K^2(\xi, t) + \lambda^2 \sum_{i=1}^{\infty} \frac{\phi_i(\xi)\phi_i(t)}{\lambda_i^2(\lambda_i - \lambda)} \\
&= K(\xi, t) + \lambda \sum_{i=1}^{\infty} \frac{\phi_i(\xi)\phi_i(t)}{\lambda_i^2} + \lambda^2 \sum_{i=1}^{\infty} \frac{\phi_i(\xi)\phi_i(t)}{\lambda_i^2(\lambda_i - \lambda)} \\
&= K(\xi, t) + \lambda \sum_{i=1}^{\infty} \frac{\phi_i(\xi)\phi_i(t)}{\lambda_i(\lambda_i - \lambda)} = k(\xi, t, \lambda).
\end{aligned}
$$

We notice that $k(x, t, \lambda)$ is a meromorphic function of λ with poles at the eigenvalues $\lambda_1, \lambda_2, \lambda_3, \ldots$. Theorem 5.5.4 tells us that it can be represented as the ratio of two entire functions. These entire functions can be expressed as infinite series in the so-called Fredholm determinants.* We shall not pursue the subject further here.

Exercises 7.8

1. Find the Fredholm integral equation associated with the problem $y'' + \lambda y = -f(x), 0 \le x \le 1, y(0) = y(1) = 0$.

2. Find the Fredholm integral equation associated with the problem $(xy')' + \lambda xy = -f(x), 1 \le x \le e, y(1) = y(e) = 0$.

* J. W. Dettman, *Mathematical Methods in Physics and Engineering.* New York: The McGraw-Hill Book Company, Inc., 1962. pp. 246–258.

3. Show that Properties 1–4 of this section do not define a Green's function for the problem $y'' + \lambda y = -f(x)$, $0 \leq x \leq 1$, $y'(0) = y'(1) = 0$. Find a *generalized Green's function* satisfying

(i) $G'' - 1, 0 \leq t < x, x < t \leq 1$;
(ii) G continuous at $t = x$;
(iii) $G'(0) = G'(1) = 0$;
(iv) $G'(x, x^+) - G'(x, x^-) = -1$;
(v) $G(x, t) = G(t, x)$.

Show that an associated Fredholm equation can be formulated in terms of this generalized Green's function.

4. Consider the Sturm-Liouville problem $((1 - x^2)y')' + \lambda y = -f(x)$, $-1 \leq x \leq 1$, $y(-1)$ and $y(1)$ finite. Show that $\lambda = 0$ is an eigenvalue of the homogeneous equation with eigenfunction $\phi_0(x) = 1/\sqrt{2}$. Show that Properties 1–4 of this section do not define a Green's function for this problem. If v is a solution of $[(1 - x^2)v']' + \lambda v = 0$, $\lambda \neq 0$, $v(-1)$ and $v(1)$ finite, then show that $\int_{-1}^{1} \phi_0 v \, dx = 0$. Because of this orthogonality we can formulate a Fredholm integral equation for v

$$v(x) = \lambda \int_{-1}^{1} G(x, t)v(t) \, dt,$$

where $G(x, t)$ satisfies

(i) $[(1 - t^2)G']' - A\phi_0(t) = 0$, $-1 \leq t < x, x < t \leq 1$;
(ii) G is continuous at $t = x$;
(iii) $G(-1)$ and $G(1)$ finite;
(iv) $G'(x, x^+) - G'(x, x^-) = -1/(1 - x^2)$;
(v) $G(x, t) = G(t, x)$.

Find the constant A and find $G(x, t)$.

5. If $K(x, t)$ is not symmetric; that is, $K(x, t) \neq K(t, x)$ the homogeneous Fredholm equation $y(x) = \lambda \int_a^b K(x, t)y(t) \, dt$ may have no eigenvalues. Show that this is the case when $K(x, t) = \sin x \cos t$, $a = 0$, $b = \pi$.

6. If the kernel $K(x, t)$ is *degenerate*; that is, $K(x, t) = \sum_{n=1}^{N} \alpha_n(x)\beta_n(t)$, then the Fredholm equation $y(x) = F(x) + \lambda \int_a^b K(x, t)y(t) \, dt$ can be solved by algebraic techniques. Do this for $N = 2$ and give criteria for existence and uniqueness of the solution. Express the reciprocal kernel as a rational function of λ.

CHAPTER 8

Fourier Transforms

8.1. FOURIER SERIES

Our main purpose in this chapter will be to study the Fourier transform, especially as it defines an analytic function in the complex plane, and consider some of its applications to the solution of ordinary differential equations, partial differential equations, and integral equations. However, both historically and conceptually, Fourier series come first, and since some readers may not already be familiar with them, we shall review some of the basic theorems about Fourier series in this section.

Let $f(t)$ be a complex-valued function of the real variable t defined for $-\infty < t < \infty$ and periodic with period 2π; that is, for any t,

$$f(t + 2\pi) = f(t).$$

Definition 8.1.1. The function $f(t)$ is *piecewise continuous* in the finite interval* $a \leq t \leq b$, if and only if there exists a finite partition

$$a = t_0 < t_1 < t_2 < \cdots < t_n = b$$

such that $f(t)$ is continuous in each of the open intervals $t_{k-1} < t < t_k$, $k = 1, 2, \ldots, n$, and each of the limits,

$$\lim_{t \to a^+} f(t), \ \lim_{t \to t_1^-} f(t), \ \lim_{t \to t_1^+} f(t), \ldots,$$

and $\lim_{t \to b^-} f(t)$ exist.

We list without proof some of the more obvious facts about piecewise continuous functions.

1. A function continuous on $[a, b]$ is piecewise continous there.

2. The sum, difference, and product of two functions piecewise continuous on $[a, b]$ are piecewise continuous on $[a, b]$.

3. A function piecewise continuous on $[a, b]$ is bounded on $[a, b]$.

---

* From now on we use the abbreviation $[a, b]$ for the closed interval.

4. The definite integral of a piecewise continuous function exists on $[a, b]$ and

$$\int_a^b f(t)\, dt = \sum_{k=1}^n \int_{t_{k-1}}^{t_k} f(t)\, dt.$$

5. The indefinite integral $\int_a^t f(u)\, du$, $a \leq t \leq b$, exists and is continuous on $[a, b]$.

Definition 8.1.2. The function $f(t)$ is piecewise continuously differentiable on $[a, b]$, if and only if its derivative $f'(t)$ exists at all but a finite number of points and is piecewise continuous on $[a, b]$.

If t_k is one of the interior points of $[a, b]$ in the partition used to verify the piecewise continuity of $f'(t)$, then $\lim_{t \to t_k^-} f'(t)$ and $\lim_{t \to t_k^+} f'(t)$ exist. This is not to say that the left-hand and right-hand derivatives of $f(t)$ exist at t_k, for it may be that neither of the following limits exist

$$\lim_{t \to t_k^-} \frac{f(t) - f(t_k)}{t - t_k}, \qquad \lim_{t \to t_k^+} \frac{f(t) - f(t_k)}{t - t_k}.$$

See for example the function whose graph is drawn in Figure 8.1.1. However, if $f(t)$ is piecewise continuous and piecewise differentiable, then the following limits exist

$$\lim_{t \to t_k^-} \frac{f(t) - f(t_k^-)}{t - t_k} = f'(t_k^-),$$

$$\lim_{t \to t_k^+} \frac{f(t) - f(t_k^+)}{t - t_k} = f'(t_k^+),$$

where $f(t_k^-) = \lim_{t \to t_k^-} f(t)$ and $f(t_k^+) = \lim_{t \to t_k^+} f(t)$.

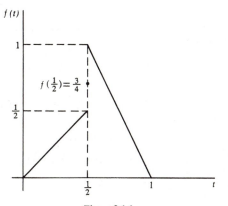

Figure 8.1.1.

Theorem 8.1.1. *Let $f(t)$ be piecewise continuous on $[a, b]$. Then*

$$\lim_{R \to \infty} \int_a^b f(t) \sin Rt \, dt = \lim_{R \to \infty} \int_a^b f(t) \cos Rt \, dt = 0.$$

Proof: We may as well assume that $f(t)$ is continuous on $[a, b]$, since we can always prove the theorem for a finite number of subintervals on which it is continuous. Let $t = \tau + \pi/R$. Then

$$\int_a^b f(t) \sin Rt \, dt = - \int_{a-\pi/R}^{b-\pi/R} f(\tau + \pi/R) \sin R\tau \, d\tau,$$

$$2 \int_a^b f(t) \sin Rt \, dt = \int_a^b f(\tau) \sin R\tau \, d\tau - \int_{a-\pi/R}^{b-\pi/R} f(\tau + \pi/R) \sin R\tau \, d\tau$$

$$= - \int_a^{b-\pi/R} [f(\tau + \pi/R) - f(\tau)] \sin R\tau \, d\tau$$

$$+ \int_{b-\pi/R}^b f(\tau) \sin R\tau \, d\tau$$

$$- \int_{a-\pi/R}^a f(\tau + \pi/R) \sin R\tau \, d\tau.$$

If $f(t)$ is continuous on $[a, b]$, it is uniformly continuous. Therefore, we can pick R large enough that $|f(\tau + \pi/R) - f(\tau)| < \epsilon/(b - a)$ for all τ in $[a, b]$. We can also pick R so large that $\pi/R < \epsilon/2M$, where $|f(t)| \leq M$ in $[a, b]$. Hence,

$$2 \left| \int_a^b f(t) \sin Rt \, dt \right| < \frac{\epsilon}{b - a} (b - a) + \frac{M\epsilon}{2M} + \frac{M\epsilon}{2M} = 2\epsilon.$$

Since ϵ is arbitrary, this completes the proof for the first limit. A similar proof holds for the other limit.

Definition 8.1.3. The Fourier coefficients of $f(t)$ are

$$a_k = \frac{1}{\pi} \int_{-\pi}^{\pi} f(t) \cos kt \, dt \quad \text{and} \quad b_k = \frac{1}{\pi} \int_{-\pi}^{\pi} f(t) \sin kt \, dt,$$

$k = 0, 1, 2, \ldots$, provided these integrals exist.

Corollary 8.1.1. If $f(t)$ is piecewise continuous on $[-\pi, \pi]$, then $\lim_{k \to \infty} a_k = \lim_{k \to \infty} b_k = 0$.

Proof: This follows directly from Theorem 8.1.1.

Theorem 8.1.2. *If $f(t)$ is piecewise continuous on $[-\pi, \pi]$, then*

$$\frac{|a_0|^2}{2} + \sum_{k=1}^{\infty} (|a_k|^2 + |b_k|^2)$$

converges.

Proof:

$$0 \le \int_{-\pi}^{\pi} \left| f(t) - \frac{a_0}{2} - \sum_{k=1}^{n} (a_k \cos kt + b_k \sin kt) \right|^2 dt,$$

$$0 \le \int_{-\pi}^{\pi} |f(t)|^2 \, dt - \pi \left[\frac{|a_0|^2}{2} + \sum_{k=1}^{n} (|a_k|^2 + |b_k|^2) \right].$$

This follows from the well-known orthogonality properties of the trigonometric functions:

$$\int_{-\pi}^{\pi} \cos mt \cos nt \, dt = 0, \qquad m \ne n,$$

$$\int_{-\pi}^{\pi} \sin mt \sin nt \, dt = 0, \qquad m \ne n,$$

$$\int_{-\pi}^{\pi} \cos mt \sin nt \, dt = 0,$$

$$\int_{-\pi}^{\pi} \cos^2 mt \, dt = \int_{-\pi}^{\pi} \sin^2 mt \, dt = \pi, \qquad m = 1, 2, 3, \ldots.$$

We have proved that

$$\frac{|a_0|^2}{2} + \sum_{k=1}^{n} (|a_k|^2 + |b_k|^2) \le \frac{1}{\pi} \int_{-\pi}^{\pi} |f(t)|^2 \, dt < \infty.$$

But this holds independently of n. Therefore, we have *Bessel's inequality*

$$\frac{|a_0|^2}{2} + \sum_{k=1}^{\infty} (|a_k|^2 + |b_k|^2) \le \frac{1}{\pi} \int_{-\pi}^{\pi} |f(t)|^2 \, dt,$$

which proves the convergence of the infinite series.

Theorem 8.1.3. *If $f(t)$ is periodic with period 2π, is continuous and piecewise continuously differentiable in any finite interval, then*

$$\sum_{k=1}^{\infty} (k^2 |a_k|^2 + k^2 |b_k|^2)$$

converges.

Proof: Since $f'(t)$ is piecewise continuous on $[-\pi, \pi]$, we have

$$\frac{|\alpha_0|^2}{2} + \sum_{k=1}^{\infty} (|\alpha_k|^2 + |\beta_k|^2) \leq \frac{1}{\pi} \int_{-\pi}^{\pi} |f'(t)|^2 \, dt,$$

where

$$\alpha_k = \frac{1}{\pi} \int_{-\pi}^{\pi} f'(t) \cos kt \, dt$$

and

$$\beta_k = \frac{1}{\pi} \int_{-\pi}^{\pi} f'(t) \sin kt \, dt.$$

Now, because of the continuity and periodicity of $f(t)$, $f(-\pi) = f(\pi)$, and

$$\alpha_k = \frac{1}{\pi} \int_{-\pi}^{\pi} f'(t) \cos kt \, dt = \left[\frac{1}{\pi} f(t) \cos kt\right]_{-\pi}^{\pi} + \frac{k}{\pi} \int_{-\pi}^{\pi} f(t) \sin kt \, dt = kb_k$$

$$\beta_k = \frac{1}{\pi} \int_{-\pi}^{\pi} f'(t) \sin kt \, dt$$

$$= \left[\frac{1}{\pi} f(t) \sin kt\right]_{-\pi}^{\pi} - \frac{k}{\pi} \int_{-\pi}^{\pi} f(t) \cos kt \, dt = -ka_k.$$

Hence,

$$\sum_{k=1}^{\infty} (k^2|a_k|^2 + k^2|b_k|^2) \leq \frac{1}{\pi} \int_{-\pi}^{\pi} |f'(t)|^2 \, dt,$$

which shows that the series converges.

Theorem 8.1.4. *If $f(t)$ is periodic with period 2π, is continuous and piecewise continuously differentiable in any finite interval, then the Fourier series $\frac{a_0}{2} + \sum_{k=1}^{\infty} (a_k \cos kt + b_k \sin kt)$ converges uniformly to $f(t)$ in any finite interval.*

Proof: Let $S_n = \frac{a_0}{2} + \sum_{k=1}^{\infty} (a_k \cos kt + b_k \sin kt)$. Then

$$|S_{n+p} - S_n|^2 = \left| \sum_{k=n+1}^{n+p} (a_k \cos kt + b_k \sin kt) \right|^2$$

$$= \left| \sum_{k=n+1}^{n+p} \left(ka_k \frac{\cos kt}{k} + kb_k \frac{\sin kt}{k} \right) \right|^2$$

$$\leq \sum_{k=n+1}^{n+p} (k^2|a_k|^2 + k^2|b_k|^2) \sum_{k=n+1}^{n+p} \frac{1}{k^2}$$

$$\leq \frac{1}{\pi} \int_{-\pi}^{\pi} |f'(t)|^2 \, dt \sum_{k=n+1}^{n+p} \frac{1}{k^2}.$$

The inequality follows from the Cauchy-Schwarz inequality (see Exercise 1.3.9). Because the series $\sum_{k=1}^{\infty} \frac{1}{k^2}$ converges $\lim_{n \to \infty} \sum_{k=n+1}^{n+p} \frac{1}{k^2} = 0$, for all positive integers p. This shows that $|S_{n+p} - S_n| \to 0$, as $n \to \infty$, for all positive p. This is the Cauchy criterion for uniform convergence of the Fourier series. This shows that the series converges to a continuous function, but we have yet to show that this function is $f(t)$.

Again consider S_n. We have

$$S_n = \frac{a_0}{2} + \sum_{k=1}^{n} (a_k \cos kt + b_k \sin kt)$$

$$= \frac{1}{\pi} \int_{-\pi}^{\pi} f(t) \left[\frac{1}{2} + \sum_{k=1}^{n} \cos k(t - \tau) \right] d\tau$$

$$= \frac{1}{\pi} \int_{-\pi}^{\pi} f(t) \operatorname{Re} \left[\sum_{k=0}^{n} e^{ik(t-\tau)} - \frac{1}{2} \right] dt$$

$$= \frac{1}{\pi} \int_{-\pi}^{\pi} f(t) \operatorname{Re} \left[\frac{e^{i(n+1)(t-\tau)} - 1}{e^{i(t-\tau)} - 1} - \frac{1}{2} \right] dt$$

$$= \frac{1}{\pi} \int_{-\pi}^{\pi} f(t) \operatorname{Re} \left[\frac{e^{i(n+1/2)(t-\tau)} - e^{-i(t-\tau)/2}}{e^{i(t-\tau)/2} - e^{-i(t-\tau)/2}} - \frac{1}{2} \right] dt$$

$$= \frac{1}{\pi} \int_{-\pi}^{\pi} f(t) \operatorname{Re} \left[\frac{\cos (n + \frac{1}{2})(t - \tau) + i \sin (n + \frac{1}{2})(t - \tau) - \cos (t - \tau)/2 + i \sin (t - \tau)/2}{2i \sin (t - \tau)/2} - \frac{1}{2} \right] dt$$

$$= \frac{1}{2\pi} \int_{-\pi}^{\pi} f(t) \frac{\sin (n + \frac{1}{2})(t - \tau)}{\sin (t - \tau)/2} dt$$

$$= \frac{1}{2\pi} \int_{-\pi-\tau}^{\pi-\tau} f(u + t) \frac{\sin (n + \frac{1}{2})u}{\sin u/2} du$$

$$= \frac{1}{2\pi} \int_{-\pi}^{\pi} f(u + t) \frac{\sin (n + \frac{1}{2})u}{\sin u/2} du,$$

where we have used the periodicity of $f(t)$. We also have

$$\frac{1}{2\pi} \int_{-\pi}^{\pi} f(t) \frac{\sin (n + \frac{1}{2})u}{\sin u/2} du$$

$$= \frac{1}{\pi} \int_{-\pi}^{\pi} f(t)[\tfrac{1}{2} + \cos u + \cos 2u + \cdots + \cos nu] du$$

$$= f(t).$$

Hence,

$$
\begin{aligned}
|S_n - f(t)| &= \left| \frac{1}{2\pi} \int_{-\pi}^{\pi} [f(u + t) - f(t)] \frac{\sin (n + \tfrac{1}{2})u}{\sin u/2} \, du \right| \\
&= \left| \frac{1}{2\pi} \int_{-\pi}^{\pi} \frac{[f(u + t) - f(t)] \cos u/2}{\sin u/2} \sin nu \, du \right. \\
&\qquad \left. + \frac{1}{2\pi} \int_{-\pi}^{\pi} [f(u + t) - f(t)] \cos nu \, du \right| \cdot
\end{aligned}
$$

We let n approach infinity and by Theorem 8.1.1 the two integrals as the right-hand side go to zero. This uses the piecewise continuity of $f'(t)$ to show that

$$
\begin{aligned}
\lim_{u \to 0^-} \frac{[f(u + t) - f(t)] \cos u/2}{\sin u/2} &= \lim_{u \to 0^-} \left[\frac{f(u + t) - f(t)}{u} \right] \frac{u \cos u/2}{\sin u/2} \\
&= 2f'(t^-), \\
\lim_{u \to 0^+} \frac{[f(u + t) - f(t)] \cos u/2}{\sin u/2} &= \lim_{u \to 0^+} \left[\frac{f(u + t) - f(t)}{u} \right] \frac{u \cos u/2}{\sin u/2} \\
&= 2f'(t^+).
\end{aligned}
$$

This completes the proof, for we have shown that $\lim\limits_{n \to \infty} S_n = f(t)$.

Corollary 8.1.2. If $f(t)$ is periodic with period 2π, is continuous and piecewise continuously differentiable in any finite interval, then

$$
\frac{|a_0|^2}{2} + \sum_{k=1}^{\infty} (|a_k|^2 + |b_k|^2) = \frac{1}{\pi} \int_{-\pi}^{\pi} |f(t)|^2 \, dt.
$$

Proof: By the uniform convergence of the Fourier series to $f(t)$ we may multiply by $\dfrac{1}{\pi} f(t)$ and integrate term by term.

Theorem 8.1.5. *If $f(t)$ is periodic with period 2π, is piecewise continuous and piecewise continuously differentiable, then the Fourier series converges to $\tfrac{1}{2}[f(t^+) + f(t^-)]$.*

Proof: We shall prove the theorem in the case where the function has discontinuities at $\pm 2n\pi$, $n = 0, 1, 2, \ldots$ and no others. Simple extensions of this argument will prove the theorem for any finite number of discontinuities in a given period.

Consider the function

$$
F(t) = f(t) - [f(0^+) - f(0^-)]g(t),
$$

where $g(t)$ is periodic with period 2π and defined in $[-\pi, \pi]$ as follows:

$$g(t) = -\frac{1}{2\pi}(\pi + t), \qquad -\pi \leq t < 0,$$

$$= 0, \qquad\qquad\quad t = 0,$$

$$= \frac{1}{2\pi}(\pi - t), \qquad 0 < t \leq \pi.$$

Now, $g(t)$ is continuous except at $t = 0$, where it has a jump of 1. We have constructed $F(t)$ continuous everywhere, since

$$\lim_{t \to 0^-} F(t) = f(0^-) + \tfrac{1}{2}[f(0^+) - f(0^-)] = \tfrac{1}{2}[f(0^+) + f(0^-)],$$

$$\lim_{t \to 0^+} F(t) = f(0^+) - \tfrac{1}{2}[f(0^+) - f(0^-)] = \tfrac{1}{2}[f(0^+) + f(0^-)],$$

and it can have no other discontinuities. Therefore, $F(t)$ has a Fourier series which converges to it everywhere. If we can show that $g(t)$ has a Fourier series which converges to it, we shall have shown that the Fourier series for $f(t)$ converges to $f(t)$ wherever it is continuous and to

$$F(0) = \tfrac{1}{2}[f(0^+) + f(0^-)],$$

where it is discontinuous. In other words, where $f(t)$ is discontinuous its Fourier series converges to the mean value of its right-hand and left-hand limits. It remains to show that $g(t)$ has a Fourier series which converges to it everywhere.

Let

$$c_k = \frac{1}{\pi} \int_{-\pi}^{\pi} g(t) \cos kt\, dt \qquad \text{and} \qquad d_k = \frac{1}{\pi} \int_{-\pi}^{\pi} g(t) \sin kt\, dt.$$

Then

$$c_k = \frac{1}{\pi} \int_{-\pi}^{0} g(t) \cos kt\, dt + \frac{1}{\pi} \int_{0}^{\pi} g(t) \cos kt\, dt$$

$$= \frac{1}{\pi} \int_{0}^{\pi} g(-t) \cos kt\, dt + \frac{1}{\pi} \int_{0}^{\pi} g(t) \cos kt\, dt = 0,\, k = 0, 1, 2, \ldots,$$

$$d_k = \frac{1}{\pi} \int_{-\pi}^{0} g(t) \sin kt\, dt + \frac{1}{\pi} \int_{0}^{\pi} g(t) \sin kt\, dt$$

$$= \frac{-1}{\pi} \int_{0}^{\pi} g(-t) \sin kt\, dt + \frac{1}{\pi} \int_{0}^{\pi} g(t) \sin kt\, dt$$

$$= \frac{2}{\pi} \int_{0}^{\pi} \frac{1}{2\pi}(\pi - t) \sin kt\, dt = \frac{1}{\pi k},\, k = 1, 2, 3, \ldots.$$

The Fourier series for $g(t)$ is then $\dfrac{1}{\pi} \displaystyle\sum_{k=1}^{\infty} \dfrac{\sin kt}{k}$, which by simple tests can be shown to converge. We still have to show that it converges to $g(t)$.

Consider $G(t) = g(t)(1 - \cos t)$ which is everywhere continuous. The Fourier series for $G(t)$, which converges uniformly to it in every closed interval, has coefficients

$$\gamma_k = \frac{1}{\pi} \int_{-\pi}^{\pi} G(t) \cos kt \, dt = 0, \, k = 0, 1, 2, \ldots,$$

$$\delta_k = \frac{1}{\pi} \int_{-\pi}^{\pi} g(t)(1 - \cos t) \sin kt \, dt$$

$$= \frac{1}{\pi} \int_{-\pi}^{\pi} g(t) \sin kt \, dt - \frac{1}{2\pi} \int_{-\pi}^{\pi} g(t)[\sin (k + 1)t + \sin (k - 1)t] \, dt$$

$$= d_k - \tfrac{1}{2}(d_{k+1} + d_{k-1}), \, k = 1, 2, 3, 4, \ldots,$$

if we define $d_0 = 0$.

Let $S_n = \displaystyle\sum_{k=1}^{n} d_k \sin kt$. Then

$$(1 - \cos t)S_n$$

$$= \sum_{k=1}^{n} (1 - \cos t) \, d_k \sin kt = \sum_{k=1}^{n} [d_k - \tfrac{1}{2}(d_{k+1} + d_{k-1})] \sin kt$$

$$= \sigma_n + \frac{1}{(n + 1)\pi} \sin nt - \frac{1}{n\pi} \sin (n + 1)t,$$

where $\sigma_n = \displaystyle\sum_{k=1}^{n} \delta_k \sin kt$. Therefore,

$$|S_n - g(t)| \le \frac{|\sigma_n - G(t)| + 2/n\pi}{|1 - \cos t|}$$

and if $t \ne \pm 2m\pi$, $m = 0, 1, 2, \ldots,$ $\lim\limits_{n \to \infty} S_n = g(t)$, since $\sigma_n \to G(t)$, as $n \to \infty$. When $t = \pm 2m\pi$, $m = 0, 1, 2, \ldots,$ $\sin kt = 0$, for all k. Hence, $S_n (\pm 2m\pi) = 0$ and $0 = \lim\limits_{n \to \infty} S_n(\pm 2m\pi) = g(\pm 2m\pi)$. This completes the proof.

Theorem 8.1.6. *Let $f(t)$ be piecewise continuous on $[-\pi, \pi]$. Then*

$$a_k = \frac{1}{\pi} \int_{-\pi}^{\pi} f(t) \cos kt \, dt$$

and

$$b_k = \frac{1}{\pi} \int_{-\pi}^{\pi} f(t) \sin kt \, dt$$

exist and the formal Fourier series $\frac{a_0}{2} + \sum_{k=1}^{\infty} (a_k \cos kt + b_k \sin kt)$ *can be integrated term by term, that is, the integrated series converges to*

$$\int_{-\pi}^{t} f(\tau) \, d\tau, \quad -\pi \leq t \leq \pi.$$

Proof: If we formally integrate the series, we obtain the series

$$\frac{a_0}{2}(t + \pi) + \sum_{k=1}^{\infty} \left(\frac{a_k}{k} \sin kt - \frac{b_k}{k} \cos kt + \frac{(-1)^k b_k}{k} \right).$$

Now, consider the function $G(t) = \int_{-\pi}^{t} f(\tau) \, d\tau - (a_0/2)(t + \pi)$. Clearly, $G(t)$ is continuous on $[-\pi, \pi]$ and piecewise continuously differentiable, since $G'(t) = f(t) - a_0/2$. Furthermore, $G(-\pi) = 0$ and

$$G(\pi) = \int_{-\pi}^{\pi} f(\tau) \, d\tau - \pi a_0 = 0.$$

Therefore,

$$G(t) = \frac{\alpha_0}{2} + \sum_{k=0}^{\infty} (\alpha_k \cos kt + \beta_k \sin kt),$$

where for $k = 1, 2, 3, \ldots,$

$$\alpha_k = \frac{1}{\pi} \int_{-\pi}^{\pi} G(t) \cos kt \, dt$$

$$= \frac{1}{\pi} \frac{G(t) \sin kt}{k} \Big|_{-\pi}^{\pi} - \frac{1}{\pi k} \int_{-\pi}^{\pi} [f(t) - a_0/2] \sin kt \, dt$$

$$= -\frac{b_k}{k}.$$

$$\beta_k = \frac{1}{\pi} \int_{-\pi}^{\pi} G(t) \sin kt \, dt$$

$$= \frac{-1}{\pi} \frac{G(t) \cos kt}{k} \Big|_{-\pi}^{\pi} + \frac{1}{\pi k} \int_{-\pi}^{\pi} [f(t) - a_0/2] \cos kt \, dt$$

$$= \frac{a_k}{k}.$$

Hence,

$$G(t) = \frac{\alpha_0}{2} + \sum_{k=1}^{\infty} \left(\frac{a_k}{k} \sin kt - \frac{b_k}{k} \cos kt \right),$$

$$G(\pi) = \frac{\alpha_0}{2} - \sum_{k=1}^{\infty} \frac{(-1)^k}{k} b_k = 0,$$

$$\int_{-\pi}^{t} f(\tau)\, d\tau - \frac{a_0}{2} (t + \pi) = \sum_{k=1}^{\infty} \frac{(-1)^k b_k}{k} + \sum_{k=1}^{\infty} \left(\frac{a_k}{k} \sin kt - \frac{b_k}{k} \cos kt \right),$$

$$\int_{-\pi}^{t} f(\tau)\, d\tau = \frac{a_0}{2} (t + \pi) + \sum_{k=1}^{\infty} \left(\frac{a_k}{k} \sin kt - \frac{b_k}{k} \cos kt + \frac{(-1)^k b_k}{k} \right),$$

which shows that the formally integrated series converges to $\int_{-\pi}^{t} f(\tau)\, d\tau$. Note the result is independent of whether the formal series converges or not.

Exercises 8.1

1. Let $f(t)$ be periodic with period $2L$, be piecewise continuous on $[-L, L]$, and have a piecewise continuous derivative. Prove that

$$\frac{f(t^+) + f(t^-)}{2} = \frac{a_0}{2} + \sum_{k=1}^{\infty} \left(a_k \cos \frac{k\pi t}{L} + b_k \sin \frac{k\pi t}{L} \right),$$

where

$$a_k = \frac{1}{L} \int_{-L}^{L} f(t) \cos \frac{k\pi t}{L}\, dt, \quad b_k = \frac{1}{L} \int_{-L}^{L} f(t) \sin \frac{k\pi t}{L}\, dt.$$

2. Let $f(t)$ be periodic with period $2L$, be piecewise continuous on $[-L, L]$, and have a piecewise continuous derivative. Prove that

$$\frac{f(t^+) + f(t^-)}{2} = \sum_{k=-\infty}^{\infty} c_k e^{ik\pi t/L},$$

where

$$c_k = \frac{1}{2L} \int_{-L}^{L} f(t) e^{-ik\pi t/L}\, dt.$$

3. Let $f(t)$ be periodic with period $2L$, be piecewise continuous on $[-L, L]$, and have a piecewise continuous derivative. Let $f(t)$ be an odd function; that is, $f(-t) = -f(t)$. Prove that

$$\frac{f(t^+) + f(t^-)}{2} = \sum_{k=1}^{\infty} b_k \sin \frac{k\pi t}{L},$$

where

$$b_k = \frac{2}{L} \int_0^L f(t) \sin \frac{k\pi t}{L} \, dt.$$

4. Let $f(t)$ be periodic with period $2L$, be piecewise continuous on $[-L, L]$, and have a piecewise continuous derivative. Let $f(t)$ be an even function, that is, $f(t) = f(-t)$. Prove that

$$\frac{f(t^+) + f(t^-)}{2} = \frac{a_0}{2} + \sum_{k=1}^{\infty} a_k \cos \frac{k\pi t}{L},$$

where

$$a_k = \frac{2}{L} \int_0^L f(t) \cos \frac{k\pi t}{L} \, dt.$$

5. Let $f(t) = t$, $0 \le t \le \pi$. Find Fourier series which converge to $f(t)$ on the interval $0 < t < \pi$ with the following properties:

(a) contains terms in $\cos kt$, $k = 0, 1, 2, \ldots$ only;
(b) contains terms in $\sin kt$, $k = 1, 2, 3, \ldots$ only;
(c) contains terms in both $\cos 2kt$ and $\sin 2kt$, $k = 0, 1, 2, \ldots$;
(d) contains terms in both $\cos kt$ and $\sin kt$, $k = 0, 1, 2, \ldots$.

What do these series converge to at $t = 0$ and $t = \pi$?

6. Prove that, if $f(t)$ is periodic and continuous and has piecewise continuous first and second derivatives, then except at points of discontinuity of $f'(t)$ its derivative can be computed by termwise differentiation of its Fourier series.

7. Prove that, if $f(t)$ is periodic and has continuous derivatives up to the $(n-1)$st and a piecewise continuous nth derivative, then its Fourier coefficients have the properties

$$\lim_{k \to \infty} k^n a_k = 0,$$

$$\lim_{k \to \infty} k^n b_k = 0.$$

8.2. THE FOURIER INTEGRAL THEOREM

We saw in Exercises 8.1.1 and 8.1.2 that if $f(t)$ is periodic with period $2L$, is piecewise continuous on $[-L, L]$, and has a piecewise continuous derivative, then

$$\frac{f(t^+) + f(t^-)}{2} = \sum_{k=-\infty}^{\infty} c_k e^{ik\pi t/L},$$

where

$$c_k = \frac{1}{2L} \int_{-L}^{L} f(\tau)e^{-ik\pi\tau/L} \, d\tau.$$

Substituting c_k into the above expression, we have

$$\frac{f(t^+) + f(t^-)}{2} = \frac{1}{2L} \sum_{k=-\infty}^{\infty} \int_{-L}^{L} f(\tau)e^{ik\pi(t-\tau)/L} \, d\tau.$$

If $f(t)$ is not periodic, then we could not hope to represent it by a Fourier series for all values of t. However, as a purely formal procedure we might consider the exponential form as the half period L approaches infinity. Let

$$x_k = k\pi/L \quad \text{and} \quad G(x_k) = \frac{1}{\sqrt{2\pi}} \int_{-L}^{L} f(\tau)e^{-ix_k\tau} \, d\tau.$$

Then

$$\frac{f(t^+) + f(t^-)}{2} = \frac{1}{2\pi} \sum_{k=-\infty}^{\infty} \left(\int_{-L}^{L} f(\tau)e^{-ix_k\tau} \, d\tau \right) e^{ix_k t}(x_{k+1} - x_k)$$

$$= \frac{1}{\sqrt{2\pi}} \sum_{k=-\infty}^{\infty} G(x_k)e^{ix_k t} \, \Delta x_k.$$

As $L \to \infty$, $\Delta x_k \to 0$, and ignoring all difficulties we arrive at

$$\frac{f(t^+) + f(t^-)}{2} = \frac{1}{\sqrt{2\pi}} \int_{-\infty}^{\infty} G(x)e^{ixt} \, dx$$

$$= \frac{1}{2\pi} \int_{-\infty}^{\infty} \int_{-\infty}^{\infty} f(\tau)e^{-ix(\tau-t)} \, d\tau \, dx.$$

This is the formal statement of the *Fourier integral theorem* which we now prove rigorously for a large class of functions.

We begin by proving an important lemma.

Lemma 8.2.1. If $f(t)$ is piecewise continuous and has a piecewise continuous derivative in any finite interval and $\int_{-\infty}^{\infty} |f(t)| \, dt$ exists, then

$$\lim_{R \to \infty} \int_{-T}^{T} f(t + \tau)\frac{\sin R\tau}{\tau} \, d\tau = \frac{\pi}{2}[f(t^+) + f(t^-)],$$

where T may be finite or infinite.

Proof: We first prove it for T finite. Breaking up the interval into four, we have

$$\int_{-T}^{T} f(t + \tau) \frac{\sin R\tau}{\tau} d\tau$$

$$= \int_{-T}^{-\delta} \frac{f(t + \tau)}{\tau} \sin R\tau \, d\tau + \int_{-\delta}^{0} \frac{f(t + \tau)}{\tau} \sin R\tau \, d\tau$$

$$+ \int_{0}^{\delta} \frac{f(t + \tau)}{\tau} \sin R\tau \, d\tau + \int_{\delta}^{T} \frac{f(t + \tau)}{\tau} \sin R\tau \, d\tau$$

$$= \int_{\delta}^{T} \frac{f(t - \tau)}{\tau} \sin R\tau \, d\tau + \int_{0}^{\delta} \frac{f(t - \tau)}{\tau} \sin R\tau \, d\tau$$

$$+ \int_{0}^{\delta} \frac{f(t + \tau)}{\tau} \sin R\tau \, d\tau + \int_{\delta}^{T} \frac{f(t + \tau)}{\tau} \sin R\tau \, d\tau.$$

We also have the following:

$$\lim_{R \to \infty} \int_{0}^{\delta} \frac{\sin Rt}{t} dt = \lim_{R \to \infty} \int_{0}^{R\delta} \frac{\sin \tau}{\tau} d\tau = \int_{0}^{\infty} \frac{\sin \tau}{\tau} d\tau = \frac{\pi}{2}.$$

Hence,

$$\lim_{R \to \infty} \int_{-T}^{T} f(t + \tau) \frac{\sin R\tau}{\tau} d\tau - \frac{\pi}{2} [f(t^{+}) + f(t^{-})]$$

$$= \lim_{R \to \infty} \int_{\delta}^{T} \frac{f(t - \tau)}{\tau} \sin R\tau \, d\tau + \lim_{R \to \infty} \int_{\delta}^{T} \frac{f(t + \tau)}{\tau} \sin R\tau \, d\tau$$

$$+ \lim_{R \to \infty} \int_{0}^{\delta} \frac{f(t - \tau) - f(t^{-})}{\tau} \sin R\tau \, d\tau$$

$$+ \lim_{R \to \infty} \int_{0}^{\delta} \frac{f(t + \tau) - f(t^{+})}{\tau} \sin R\tau \, d\tau.$$

Each of these limits is zero by Theorem 8.1.1, where in the first two we use the piecewise continuity of the function, and in the last two we use the piecewise continuity of the derivative.

To take care of the case $T = \infty$, we merely observe that for S sufficiently large

$$\int_{S}^{\infty} \left| \frac{f(t) \sin Rt}{t} \right| dt \leq \int_{S}^{\infty} |f(t)| \, dt < \epsilon,$$

$$\int_{-\infty}^{-S} \left| \frac{f(t) \sin Rt}{t} \right| dt \leq \int_{-\infty}^{-S} |f(t)| \, dt < \epsilon,$$

for arbitrary $\epsilon > 0$.

Theorem 8.2.1. *Fourier Integral Theorem. Let $f(t)$ be piecewise continuous and have a piecewise continuous first derivative in any finite interval. Let $f(t)$ be absolutely integrable, that is, $\int_{-\infty}^{\infty} |f(t)|\, dt$ exists. Then*

$$\frac{f(t^+) + f(t^-)}{2} = \frac{1}{2\pi} \int_{-\infty}^{\infty} \int_{-\infty}^{\infty} f(\tau) e^{ix(t-\tau)}\, d\tau\, dx,$$

where the integration with respect to x is in the Cauchy principal value sense.

Proof: We note first that $\int_{-\infty}^{\infty} |f(\tau)e^{-ix\tau}|\, d\tau = \int_{-\infty}^{\infty} |f(\tau)|\, d\tau < \infty$. Hence, the improper integral $\int_{-\infty}^{\infty} f(\tau)e^{-ix\tau}\, d\tau$ converges absolutely and uniformly for x in $[-R, R]$. Then

$$\lim_{R \to \infty} \frac{1}{2\pi} \int_{-R}^{R} \int_{-\infty}^{\infty} f(\tau) e^{ix(t-\tau)}\, d\tau\, dx = \lim_{R \to \infty} \frac{1}{2\pi} \int_{-\infty}^{\infty} f(\tau) \int_{-R}^{R} e^{ix(t-\tau)}\, dx\, d\tau$$

$$= \lim_{R \to \infty} \frac{1}{\pi} \int_{-\infty}^{\infty} f(\tau) \frac{\sin R(t - \tau)}{(t - \tau)}\, d\tau$$

$$= \lim_{R \to \infty} \frac{1}{\pi} \int_{-\infty}^{\infty} f(t + u) \frac{\sin Ru}{u}\, du$$

$$= \tfrac{1}{2}[f(t^+) + f(t^-)],$$

by Lemma 8.2.1.

The Fourier integral theorem can be put in the following form. Let

$$G(x) = \frac{1}{\sqrt{2\pi}} \int_{-\infty}^{\infty} f(\tau) e^{-ix\tau}\, d\tau$$

be the *Fourier transform* of $f(t)$. Then if $f(t)$ satisfies the hypotheses of Theorem 8.2.1,

$$\frac{f(t^+) + f(t^-)}{2} = \frac{1}{\sqrt{2\pi}} \int_{-\infty}^{\infty} G(x) e^{ixt}\, dx.$$

If $f(t)$ is continuous and we denote the Fourier transform by $F[f]$ then

$$f(t) = F^{-1}\{F[f]\},$$

where F^{-1} denotes the inverse transform $\dfrac{1}{\sqrt{2\pi}} \int_{-\infty}^{\infty} G(x) e^{ixt}\, dx.$

EXAMPLE 8.2.1.　Find the Fourier transform of $f(t) = e^{-|t|}$ and verify the Fourier integral theorem in this case. The function is

$$f(t) = \begin{cases} e^t, & t < 0, \\ e^{-t}, & t \geq 0. \end{cases}$$

Then

$$G(x) = \frac{1}{\sqrt{2\pi}} \int_{-\infty}^{0} e^{t(1-ix)} \, dt + \frac{1}{\sqrt{2\pi}} \int_{0}^{\infty} e^{-t(1+ix)} \, dt$$

$$= \frac{1}{\sqrt{2\pi}} \left(\frac{1}{1-ix} + \frac{1}{1+ix} \right) = \frac{1}{\sqrt{2\pi}} \frac{2}{1+x^2} \cdot$$

We can invert the transform in this case using contour integration. First consider $t > 0$. Then

$$\frac{1}{2\pi} \int_{-\infty}^{\infty} \frac{2}{1+x^2} e^{ixt} \, dx = \lim_{R \to \infty} \frac{1}{\pi} \int_{C} \frac{e^{izt}}{1+z^2} \, dz,$$

where C is the contour shown in Figure 8.2.1. There is a simple pole at $z = i$, and by the residue theorem

$$\frac{1}{\pi} \int_{-\infty}^{\infty} \frac{e^{ixt}}{1+x^2} \, dx = 2\pi i \left(\frac{e^{-t}}{2\pi i} \right) = e^{-t},$$

provided that we can show that the contribution to the contour integral of the semicircular arc goes to zero as $R \to \infty$. Letting $z = Re^{i\theta}$, we have

$$\left| \frac{1}{\pi} \int_{0}^{\pi} \frac{1}{1 + R^2 e^{2i\theta}} e^{it(R\cos\theta + iR\sin\theta)} \, iRe^{i\theta} \, d\theta \right| \leq \frac{R}{|R^2 - 1|} \to 0,$$

as $R \to \infty$. For $t < 0$, we close the contour with a semicircular arc lying below the x axis. Then the simple pole at $z = -i$ contributes a residue of $-e^t/2\pi i$. But in this case counterclockwise integration gives the negative of the inverse transform. Hence, the desired result is $2\pi i e^t/2\pi i = e^t$. If $t = 0$, we can integrate directly, that is,

$$\frac{1}{\pi} \int_{-\infty}^{\infty} \frac{1}{1+x^2} \, dx = \frac{1}{\pi} \tan^{-1} x \Big|_{-\infty}^{\infty} = 1.$$

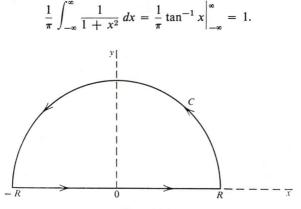

Figure 8.2.1.

Theorem 8.2.2. *If $f(t)$ and $g(t)$ are both complex-valued functions satisfying the hypotheses of theorem 8.2.1, and which have the same Fourier transform, then they may differ at only a finite number of points in any finite interval.*

Proof: By hypothesis

$$G(x) = \frac{1}{\sqrt{2\pi}} \int_{-\infty}^{\infty} f(t)e^{-ixt}\, dt = \frac{1}{\sqrt{2\pi}} \int_{-\infty}^{\infty} g(t)e^{-ixt}\, dt,$$

and by Theorem 8.2.1

$$\frac{f(t^+) + f(t^-)}{2} = \frac{g(t^+) + g(t^-)}{2} = \frac{1}{\sqrt{2\pi}} \int_{-\infty}^{\infty} G(x)e^{ixt}\, dx.$$

At points where $f(t)$ and $g(t)$ are continuous, $f(t) = g(t)$. In any finite interval there may be a finite number of points where $f(t)$ and/or $g(t)$ are discontinuous. At these points, and only at these points, $f(t)$ and $g(t)$ may differ.

Exercises 8.2

1. Find the Fourier transform of $f(t) = e^{-at}$, $t \geq 0$, $f(t) = 0$, $t < 0$, with $a > 0$. Verify the inverse transformation by direct integration. Note that the transform is not absolutely integrable, although it is square integrable.

2. Show that $F[e^{-t^2/2}] = e^{-x^2/2}$.

3. Show that, if $f(t)$ is real and satisfies the conditions of Theorem 8.2.1, then

$$\frac{f(t^+) + f(t^-)}{2} = \frac{1}{\pi} \int_0^{\infty} \int_{-\infty}^{\infty} f(x) \cos \tau(t - x)\, dx\, d\tau.$$

If $f(t)$ is even, show that

$$\frac{f(t^+) + f(t^-)}{2} = \frac{2}{\pi} \int_0^{\infty} \cos t\tau \int_0^{\infty} f(x) \cos x\tau\, dx\, d\tau.$$

If $f(t)$ is odd, show that

$$\frac{f(t^+) + f(t^-)}{2} = \frac{2}{\pi} \int_0^{\infty} \sin t\tau \int_0^{\infty} f(x) \sin x\tau\, dx\, d\tau.$$

Note: $\sqrt{\dfrac{2}{\pi}} \displaystyle\int_0^{\infty} f(t) \cos xt\, dt$ is called the *Fourier cosine transform*, and $\sqrt{\dfrac{2}{\pi}} \displaystyle\int_0^{\infty} f(t) \sin xt\, dt$ is called the *Fourier sine transform*.

8.3. THE COMPLEX FOURIER TRANSFORM

In the Fourier integral theorem we are dealing with functions which are absolutely integrable, that is, $\int_{-\infty}^{\infty} |f(t)|\, dt < \infty$. This is a rather severe restriction. It means, for example, that functions like 1, $\sin t$, $\cos t$, e^t, and so on, do not have Fourier transforms. In an attempt to remove this restriction we are led to a generalization of the Fourier transform as a function of the complex variable z.

Consider a complex-valued function $f(t)$ which is continuous and has a piecewise continuous derivative in any finite interval. Suppose $g(t) = e^{yt}f(t)$ is absolutely integrable for some y. Then

$$g(t) = \frac{1}{2\pi} \int_{-\infty}^{\infty} e^{itx} \int_{-\infty}^{\infty} g(\tau)e^{-ix\tau}\, d\tau\, dx,$$

$$e^{yt}f(t) = \frac{1}{2\pi} \int_{-\infty}^{\infty} e^{itx} \int_{-\infty}^{\infty} f(\tau)e^{-i\tau(x+iy)}\, d\tau\, dx,$$

$$f(t) = \frac{1}{2\pi} \int_{-\infty}^{\infty} e^{it(x+iy)} \int_{-\infty}^{\infty} f(\tau)e^{-i\tau(x+iy)}\, d\tau\, dx$$

$$= \frac{1}{2\pi} \int_{-\infty+iy}^{\infty+iy} e^{itz} \int_{-\infty}^{\infty} f(\tau)e^{-i\tau z}\, d\tau\, dz,$$

where the integration in the z plane is along a straight line $x + iy$, y fixed, and $-\infty < x < \infty$, such that $f(t)e^{yt}$ is absolutely integrable. More generally, we have the following theorem:

Theorem 8.3.1. *Let $f(t)$ be a complex-valued function which is piecewise continuous and has a piecewise continuous derivative in any finite interval. Let $f(t)e^{\gamma t}$ be absolutely integrable for some real γ. Then*

$$\frac{f(t^+) + f(t^-)}{2} = \frac{1}{2\pi} \int_{-\infty+i\gamma}^{\infty+i\gamma} e^{itz} \int_{-\infty}^{\infty} f(\tau)e^{-i\tau z}\, d\tau\, dz,$$

where the integration in the z plane is along the line $x + i\gamma$, $-\infty < x < \infty$.

We define the *complex Fourier transform* of $f(t)$ as follows:

$$G(z) = \frac{1}{\sqrt{2\pi}} \int_{-\infty}^{\infty} f(\tau)e^{-i\tau z}\, d\tau.$$

Then if $f(t)$ satisfies the hypotheses of Theorem 8.3.1, the inverse transform is

$$\frac{f(t^+) + f(t^-)}{2} = \frac{1}{\sqrt{2\pi}} \int_{-\infty+i\gamma}^{\infty+i\gamma} e^{itz}G(z)\, dz$$

for some real γ.

Theorem 8.3.2. *Let* $f(t)$ *be a piecewise continuous function such that* $|f(t)| \leq Ke^{-bt}$, $0 \leq t < \infty$, *and* $|f(t)| \leq Me^{-at}$, $-\infty < t \leq 0$, $a < b$. *Then the Fourier transform of* $f(t)$ *exists and is an analytic function of* z *for* $a < \text{Im}(z) < b$. *Also,* $G'(z) = (1/\sqrt{2\pi}) \int_{-\infty}^{\infty} [-i\tau f(\tau)]e^{-i\tau z} d\tau$.

Proof: This is the result of Example 4.9.1. By Theorem 4.9.4

$$G'(z) = \frac{1}{\sqrt{2\pi}} \int_{-\infty}^{\infty} \frac{\partial}{\partial z}[e^{-i\tau z}f(\tau)] d\tau = \frac{1}{\sqrt{2\pi}} \int_{-\infty}^{\infty} [-i\tau f(\tau)]e^{-i\tau z} d\tau.$$

EXAMPLE 8.3.1.   Find the complex Fourier transform of $f(t) = \sin \omega t$, $0 \leq t < \infty$, $\omega > 0$; $f(t) = 0$, $-\infty < t \leq 0$, and verify the inverse transform. Since $|f(t)| \leq 1$ for $0 \leq t < \infty$, and $|f(t)| = 0$ for $-\infty < t \leq 0$, we may take $b = 0$ and $a = -\infty$. Therefore, the transform is analytic for $-\infty < \text{Im}(z) < 0$.

$$G(z) = \frac{1}{\sqrt{2\pi}} \int_0^{\infty} \left(\frac{e^{i\omega t} - e^{-i\omega t}}{2i}\right) e^{-izt} dt$$

$$= \frac{1}{2\sqrt{2\pi}} \left(\frac{1}{\omega - z} - \frac{1}{\omega + z}\right) = \frac{1}{\sqrt{2\pi}} \frac{\omega}{\omega^2 - z^2}.$$

Actually, $\omega/\sqrt{2\pi}(\omega^2 - z^2)$ is analytic in the extended z plane except at $z = \pm\omega$. Therefore, this function is the analytic continuation of the Fourier transform to the rest of the plane. Let $\gamma < 0$. If $t > 0$ we evaluate by contour integration

$$\frac{1}{2\pi} \int_{-\infty+i\gamma}^{\infty+i\gamma} \frac{\omega e^{izt}}{\omega^2 - z^2} dz = \lim_{R \to \infty} \frac{1}{2\pi} \int_{C+} \frac{\omega e^{izt}}{\omega^2 - z^2} dz,$$

where C is the contour shown in Figure 8.3.1.

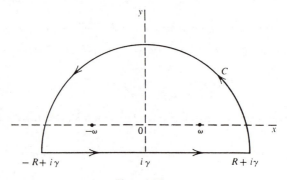

Figure 8.3.1.

There are two poles inside the contour. By the residue theorem

$$\frac{1}{2\pi} \int_{-\infty+i\gamma}^{\infty+i\gamma} \frac{\omega e^{izt}}{\omega^2 - z^2} \, dz = i\left(\frac{\omega e^{i\omega t}}{2\omega} - \frac{\omega e^{-i\omega t}}{2\omega}\right) = \sin \omega t,$$

provided we can show that the contribution of the semicircular arc goes to zero as R approaches infinity. On the semicircular contour $z = i\gamma + Re^{i\theta}$, $0 \leq \theta \leq \pi$, and

$$|e^{izt}| = |e^{-\gamma t}e^{itR(\cos\theta + i\sin\theta)}| = e^{-\gamma t}e^{-tR\sin\theta} \leq e^{-\gamma t}.$$

Hence, on this part of the contour

$$\left|\frac{1}{2\pi} \int \frac{\omega e^{izt}}{\omega^2 - z^2} \, dz\right| \leq \frac{\omega e^{-\gamma t}}{2\pi} \frac{\pi R}{(R - \gamma)^2 - \omega^2} \to 0,$$

as $R \to \infty$. If $t \leq 0$ we close the contour with a semicircle below the x axis and, since there are no poles inside the contour the result is zero. Also, on the semicircular arc $z = i\gamma + Re^{i\theta}$, $\pi \leq \theta \leq 2\pi$, and

$$\left|\frac{1}{2\pi} \int \frac{\omega e^{izt}}{\omega^2 - z^2} \, dz\right| \leq \frac{\omega e^{-\gamma t}}{2\pi} \frac{\pi R}{(R - \gamma)^2 - \omega^2} \to 0,$$

as $R \to \infty$.

In Example 8.3.1, it is easily seen that the inversion can be carried out for any $\gamma < 0$. In other words, the inverse transform gives back the original function regardless of what line in the lower half-plane parallel to the real axis is used for the integration. This is not surprising since $e^{\gamma t}f(t)$ is absolutely integrable for any negative γ, and therefore by Theorem 8.3.1 the inversion should not depend on γ, provided it is negative. In fact, we can state an obvious corollary to Theorem 8.3.1.

Corollary 8.3.1. Let $f(t)$ be piecewise continuous and have a piecewise continuous derivative in any finite interval. Let $|f(t)| \leq Ke^{-bt}, 0 \leq t < \infty$, and $|f(t)| \leq Me^{-at}$, $-\infty < t \leq 0$, $a < b$. Then $e^{\gamma t}f(t)$ is absolutely integrable for $a < \gamma < b$, and the inversion integral can be computed along any line $x + i\gamma$, $-\infty < x < \infty$, in the strip of analyticity of the Fourier transform of $f(t)$.

We can now raise the following question: If we define any function $G(z)$ analytic in a strip $a < \text{Im}(z) < b$, how do we know if it is a Fourier transform? Obviously, we need more than the condition of analyticity in a strip to guarantee that the function is a Fourier transform. We can easily obtain necessary conditions for a function to be a Fourier transform.

Theorem 8.3.3. *Let $f(t)$ be piecewise continuous in any finite interval. Let $|f(t)| \leq Ke^{-bt}$, $0 \leq t < \infty$, $|f(t)| \leq Me^{-at}$, $-\infty < t < 0$, $a < b$. Then $G(z) = (1/\sqrt{2\pi})\int_{-\infty}^{\infty} f(t)e^{-izt} \, dt$ is bounded in every strip $a < \rho_1 \leq \text{Im}(z) \leq \rho_2 < b$.*

Proof:

$$|G(z)| \leq \frac{K}{\sqrt{2\pi}} \int_0^\infty e^{(y-b)t}\, dt + \frac{M}{\sqrt{2\pi}} \int_{-\infty}^0 e^{(y-a)t}\, dt$$

$$= \frac{1}{\sqrt{2\pi}} \left(\frac{K}{b-y} + \frac{M}{y-a} \right) \leq \frac{1}{\sqrt{2\pi}} \left(\frac{K}{b-\rho_2} + \frac{M}{\rho_1-a} \right).$$

Theorem 8.3.4. *Let $f(t)$ be continuous and have a piecewise continuous derivative in any finite interval. Let $|f(t)| \leq Ke^{-bt}$, $|f'(t)| \leq K'e^{-b't}$, $0 \leq t < \infty$; $|f(t)| \leq Me^{-at}$, $|f'(t)| \leq M'e^{-a't}$, $-\infty < t < \infty$, $a' < a < b < b'$. Then in the strip $a < \mathrm{Im}(z) < b$, there is a constant N such that $|G(z)| \leq N/|z|$.*

Proof: Under the hypotheses, the Fourier transform of $f'(t)$ exists and is analytic in the strip $a' < \mathrm{Im}(z) < b'$. In this strip we may integrate by parts as follows:

$$H(z) = \frac{1}{\sqrt{2\pi}} \int_{-\infty}^\infty f'(t)e^{-izt}\, dt$$

$$= \frac{1}{\sqrt{2\pi}} \int_{-\infty}^\infty f'(t)e^{yt} (\cos xt - i \sin xt)\, dt$$

$$= \frac{-y+ix}{\sqrt{2\pi}} \int_{-\infty}^\infty f(t)e^{yt} (\cos xt - i \sin xt)\, dt$$

$$= \frac{iz}{\sqrt{2\pi}} \int_{-\infty}^\infty f(t)e^{-izt}\, dt = izG(z).$$

In the strip $a < \mathrm{Im}(z) < b$, $H(z)$ is bounded. Therefore, there is an N such that $|H(z)| \leq N$, and $|G(z)| \leq N/|z|$.

If we can evaluate the inversion integral for some γ such that $a < \gamma < b$, then we have a function

$$f(t) = \frac{1}{\sqrt{2\pi}} \int_{-\infty+i\gamma}^{\infty+i\gamma} G(z)e^{izt}\, dz.$$

Then we can investigate the properties of $f(t)$ and evaluate, if possible, its Fourier transform

$$G^*(z) = \frac{1}{\sqrt{2\pi}} \int_{-\infty}^\infty f(t)e^{-izt}\, dt.$$

If $G^*(z) \equiv G(z)$, then $G(z)$ must be the Fourier transform of a function which may differ from $f(t)$ at only a finite number of points in any finite interval.

On the other hand, if we build up a table of Fourier transforms of known functions, then we can search the table for our given function. If found in the table, we know that we have the transform of a known function. In the next section we shall investigate more of the properties of Fourier transforms and start to build up a catalog of known transforms.

Exercises 8.3

1. Find the complex Fourier transform of $e^{-|t|}$ and show that it is analytic for $-1 < \text{Im}(z) < 1$. Invert the transform by contour integration.

2. Find the complex Fourier transform of $u(t) \cos \omega t$, where $u(t) = 1$ for $t \geq 0$ and $u(t) = 0$ for $t < 0$. Show that the transform is analytic for $\text{Im}(z) < 0$. Invert the transform by contour integration.

3. Prove that the complex Fourier transform of $e^{-t^2/2}$ is $e^{-z^2/2}$.

4. Let $G(z) = -i/\sqrt{2\pi}\, z$. Evaluate $(1/\sqrt{2\pi}) \int_{-\infty+i\gamma}^{\infty+i\gamma} G(z)e^{izt}\, dz$, for $\gamma > 0$ and for $\gamma < 0$. Interpret the result.

5. Let $G(z)$ be analytic for $a < \text{Im}(z) < b$. Suppose for any ρ_1 and ρ_2 such that $a < \rho_1 \leq \text{Im}(z) \leq \rho_2 < b$ there exists an N such that $|G(z)| \leq N/|z|$. Prove that if $(1/\sqrt{2\pi}) \int_{-\infty+i\gamma}^{\infty+i\gamma} G(z)e^{izt}\, dz$ exists for some γ satisfying $a < \gamma < b$, then the value of the integral is independent of γ, provided it satisfies the same inequality.

6. Suppose we restrict our attention to functions $f(t)$ such that $f(t) = 0$, for $-\infty < t < 0$. Then

$$G(z) = \frac{1}{\sqrt{2\pi}} \int_0^\infty f(t)e^{-izt}\, dt$$

in the *one-sided Fourier transform*. If $f(t)$ is piecewise continuous and satisfies $|f(t)| \leq Ke^{-bt}$, $0 \leq t < \infty$, prove that $G(z)$ is analytic in a half-plane $\text{Im}(z) < b$.

8.4. PROPERTIES OF THE FOURIER TRANSFORM

In this section we shall investigate some of the more important properties of the Fourier transform and build up a short table of transforms of known functions.

Theorem 8.4.1. *The Fourier transform is linear; that is, if $f_1(t)$ and $f_2(t)$ are both functions with Fourier transforms analytic in the strip $a < \text{Im}(z) < b$, and c_1 and c_2 are any complex numbers, then*

$$F[c_1f_1 + c_2f_2] = c_1F[f_1] + c_2F[f_2].$$

Proof:

$$F[c_1f_1 + c_2f_2] = \frac{1}{\sqrt{2\pi}} \int_{-\infty}^{\infty} [c_1f_1(t) + c_2f_2(t)]e^{-izt}\, dt$$

$$= \frac{c_1}{\sqrt{2\pi}} \int_{-\infty}^{\infty} f_1(t)e^{-izt}\, dt + \frac{c_2}{\sqrt{2\pi}} \int_{-\infty}^{\infty} f_2(t)e^{-izt}\, dt$$

$$= c_1F[f_1] + c_2F[f_2].$$

Theorem 8.4.2. *If $f(t)$ is a function with a Fourier transform $F[f(t)]$, then $F[f(t - \tau)] = e^{-i\tau z}F[f(t)]$.*

Proof:

$$F[f(t - \tau)] = \frac{1}{\sqrt{2\pi}} \int_{-\infty}^{\infty} f(t - \tau)e^{-izt}\, dt$$

$$= \frac{1}{\sqrt{2\pi}} \int_{-\infty}^{\infty} f(u)e^{-iz(u+\tau)}\, du$$

$$= e^{-i\tau z} \frac{1}{\sqrt{2\pi}} \int_{-\infty}^{\infty} f(u)e^{-izu}\, du$$

$$= e^{-i\tau z}F[f(t)].$$

Theorem 8.4.3. *Let $f(t)$ be a function with a Fourier transform $G(z)$, analytic in the strip $a < \text{Im}(z) < b$. Let k be a real constant $\neq 0$. Then $F[f(kt)] = G(z/k)/|k|$ analytic in the strip $ka < \text{Im}(z) < kb$ if $k > 0$, or the strip $kb < \text{Im}(z) < ka$, if $k < 0$.*

Proof:

$$F[f(kt)] = \frac{1}{\sqrt{2\pi}} \int_{-\infty}^{\infty} f(kt)e^{-izt}\, dt$$

$$= \frac{1}{|k|\sqrt{2\pi}} \int_{-\infty}^{\infty} f(u)e^{-i(z/k)u}\, du$$

$$= \frac{1}{|k|} G\left(\frac{z}{k}\right).$$

If $G(z)$ is analytic for $a < \text{Im}(z) < b$, then $G(z/k)$ is analytic for $a < \text{Im}(z/k) < b$, or $ka < \text{Im}(z) < kb$ for $k > 0$, and $kb < \text{Im}(z) < ka$, for $k < 0$.

Theorem 8.4.4. *Let $f(t)$ be continuous and have a piecewise continuous derivative in any finite interval. Let $|f(t)| \leq Ke^{-bt}$, $|f'(t)| \leq Ke^{-bt}$, $0 \leq t < \infty$, $|f(t)| \leq Me^{-at}$, $|f'(t)| \leq Me^{-at}$, $-\infty < t \leq 0$, $a < b$. Then $F[f]$ and $F[f']$ are analytic in the strip $a < \mathrm{Im}(z) < b$, where $F[f'] = izF[f]$.*

Proof: In the strip $a < \mathrm{Im}(z) < b$, the transform $F[f']$ exists and we can integrate by parts as follows:

$$F[f'] = \frac{1}{\sqrt{2\pi}} \int_{-\infty}^{\infty} f'(t)e^{-izt} \, dt$$

$$= \frac{1}{\sqrt{2\pi}} \int_{-\infty}^{\infty} f'(t)e^{yt} (\cos xt - i \sin xt) \, dt$$

$$= \frac{-y + ix}{\sqrt{2\pi}} \int_{-\infty}^{\infty} f(t)e^{yt} (\cos xt - i \sin xt) \, dt$$

$$= \frac{iz}{\sqrt{2\pi}} \int_{-\infty}^{\infty} f(t)e^{-izt} \, dt = izF[f].$$

Theorem 8.4.5. *Let $f^{(k)}(t)$, $k = 0, 1, 2, \ldots, n - 1$, be continuous and $f^{(n)}(t)$ be piecewise continuous. Let $|f^{(k)}(t)| \leq Ke^{-bt}$, $0 \leq t < \infty$, $|f^{(k)}(t)| \leq Me^{-at}$, $-\infty < t \leq 0$, $k = 0, 1, 2, \ldots, n$. Then $F[f^{(n)}] = (iz)^n F[f]$.*

Proof: By Theorem 8.4.4,

$$F[f^{(n)}] = izF[f^{(n-1)}] = (iz)^2 F[f^{(n-2)}] = \cdots = (iz)^n F[f].$$

Theorem 8.4.6. *Let $f(t)$ be piecewise continuous and satisfy $|f(t)| \leq Ke^{-bt}$, $0 \leq t < \infty$, $|f(t)| \leq Me^{-at}$, $-\infty < t \leq 0$, $a < b$. Then in the strip $a < \mathrm{Im}(z) < b$*

$$\frac{d^n}{dz^n} F[f] = (-i)^n F[t^n f(t)],$$

$n = 1, 2, 3, \ldots$.

Proof: $F[f]$ is analytic in the strip and we may differentiate under the integral sign as many times as we please. Hence,

$$\frac{d}{dz} F[f] = \frac{1}{\sqrt{2\pi}} \int_{-\infty}^{\infty} (-it)f(t)e^{-izt} \, dt = (-i)F[tf(t)],$$

$$\frac{d^2}{dz^2} F[f] = \frac{1}{\sqrt{2\pi}} \int_{-\infty}^{\infty} (-it)^2 f(t)e^{-izt} \, dt = (-i)^2 F[t^2 f(t)],$$

and so on.

Theorem 8.4.7. *Let $f(t)$ be piecewise continuous in any finite interval. Let $h(t) = \int_{t_0}^{t} f(\tau)\, d\tau$. Let $h(t)$ and $h'(t) = f(t)$ satisfy the hypotheses of Theorem 8.4.4. Then $F[h] = F[f]/iz$*

Proof: By Theorem 8.4.4, $F[f] = izF[f'] = izF[h]$.

In Theorems 8.4.4 and 8.4.7 we see the result on the transform of differentiating and integrating the function. Differentiation has the effect of multiplying the transform by iz, while integration has the effect of dividing the transform by iz. In terms of the transforms, we thus replace these calculus operations by algebraic operations. This becomes of fundamental importance in solving differential and integral equations, as we shall see later in the applications.

A certain integral plays a special role in the theory of the Fourier transform. Let $f(t)$ and $g(t)$ be piecewise continuous and satisfy the inequalities $|f(t)| \leq Ke^{-bt}$, $|g(t)| \leq Ke^{-bt}$, $0 \leq t < \infty$; $|f(t)| \leq Me^{-at}$, $|g(t)| \leq Me^{-at}$, $-\infty < t \leq 0$, $a < b$. Then there is a γ, satisfying $a < \gamma < b$, such that $e^{\gamma t}f(t)$ and $e^{\gamma t}g(t)$ are absolutely integrable. Let

$$h(t) = \frac{1}{\sqrt{2\pi}} \int_{-\infty}^{\infty} f(t - \tau)g(\tau)\, d\tau$$

$$= \frac{1}{\sqrt{2\pi}} \int_{-\infty}^{\infty} f(\tau)g(t - \tau)\, d\tau.$$

We call $h(t)$ the *convolution of f and g*, which we denote by $h = f * g$. It is easy to show that $e^{\gamma t}h(t)$ is piecewise continuous and absolutely integrable. Let $t_1 \leq t \leq t_2$ be an interval in which $f(t)$ is continuous if we redefine $f(t_1) = f(t_1^+)$ and $f(t_2) = f(t_2^-)$ if necessary. Now, for t in this interval $e^{\gamma(t-\tau)}|f(t - \tau)|$ is bounded uniformly in t. Let $e^{\gamma(t-\tau)}|f(t - \tau)| \leq N$. Then since

$$e^{\gamma t}h(t) = \frac{1}{\sqrt{2\pi}} \int_{-\infty}^{\infty} e^{\gamma(t-\tau)}f(t - \tau)e^{\gamma\tau}g(\tau)\, d\tau,$$

and

$$\int_{-\infty}^{\infty} e^{\gamma(t-\tau)}|f(t - \tau)|e^{\gamma\tau}|g(\tau)|\, d\tau \leq N \int_{-\infty}^{\infty} e^{\gamma t}|g(\tau)|\, d\tau,$$

the integral is uniformly convergent. Therefore, $e^{\gamma t}h(t)$ is continuous for $t_1 \leq t \leq t_2$. But the argument can be repeated for any finite interval in which $f(t)$ is continuous.

To show that $e^{\gamma t}h(t)$ is absolutely integrable we do the following:

$$\int_{-R}^{R} e^{\gamma t}|h(t)|\,dt \leq \frac{1}{\sqrt{2\pi}} \int_{-\infty}^{\infty} e^{\gamma \tau}|g(\tau)| \int_{-R}^{R} e^{\gamma(t-\tau)}|f(t-\tau)|\,dt\,d\tau$$

$$\leq \frac{1}{\sqrt{2\pi}} \int_{-\infty}^{\infty} e^{\gamma \tau}|g(\tau)| \int_{-R-\tau}^{R-\tau} e^{\gamma u}|f(u)|\,du\,d\tau$$

$$\leq \frac{1}{\sqrt{2\pi}} \int_{-\infty}^{\infty} e^{\gamma \tau}|g(\tau)|\,d\tau \int_{-\infty}^{\infty} e^{\gamma u}|f(u)|\,du < \infty.$$

Taking transforms we have

$$F[e^{\gamma t}h(t)] = \lim_{R\to\infty} \frac{1}{2\pi} \int_{-R}^{R} \int_{-\infty}^{\infty} e^{\gamma(t-\tau)}f(t-\tau)e^{\gamma\tau}g(\tau)e^{-izt}\,d\tau\,dt$$

$$= \lim_{R\to\infty} \frac{1}{2\pi} \int_{-\infty}^{\infty} e^{\gamma\tau}g(\tau) \int_{-R}^{R} e^{\gamma(t-\tau)}f(t-\tau)e^{-izt}\,dt\,d\tau$$

$$= \lim_{R\to\infty} \frac{1}{2\pi} \int_{-\infty}^{\infty} e^{\gamma\tau}g(\tau)e^{-iz} \int_{-R-\tau}^{R-\tau} e^{\gamma u}f(u)e^{-izu}\,du\,d\tau$$

$$= \frac{1}{\sqrt{2\pi}} \int_{-\infty}^{\infty} e^{\gamma\tau}g(\tau)e^{-iz}\,d\tau \frac{1}{\sqrt{2\pi}} \int_{-\infty}^{\infty} e^{\gamma u}f(u)e^{-izu}\,du$$

$$= F[e^{\gamma t}g(t)]\,F[e^{\gamma t}f(t)],$$

where the equation holds for $a - \gamma < \text{Im}(z) < b - \gamma$. Now, if $G(z) = F[f]$, is analytic for $a < \text{Im}(z) < b$, then

$$F[e^{\gamma t}f(t)] = \frac{1}{\sqrt{2\pi}} \int_{-\infty}^{\infty} f(t)e^{-it(z+i\gamma)}\,dt = G(z + i\gamma)$$

is analytic for $a < \text{Im}(z + i\gamma) < b$, that is, $a - \gamma < \text{Im}(z) < b - \gamma$. Therefore, by making the translation $z + i\gamma \to z$, we have the equation $F[h] = F[f]\,F[g]$ valid for $a < \text{Im}(z) < b$. This also proves that $F[h]$ is analytic in the same strip.

We conclude this section with several examples to show how these properties can be used to determine transforms of elementary functions.

EXAMPLE 8.4.1. Let $u(t) = 1, 0 \leq t < \infty$ and $u(t) = 0, -\infty < t < 0$. Then

$$F[u] = \frac{1}{\sqrt{2\pi}} \int_{0}^{\infty} e^{-izt}\,dt = \frac{-i}{\sqrt{2\pi}\,z},$$

provided $\text{Im}(z) < 0$. The function $-i/\sqrt{2\pi}\, z$ is analytic except at $z = 0$. It is therefore the analytic continuation of $F[u]$ to the rest of the plane. To invert the transform using the inversion integral, we must take $\gamma < 0$. The function $u(t)$ is the *unit step* function.

EXAMPLE 8.4.2. Let $f(t) = u(t)t^n$, n a positive integer. Then

$$f^{(k)}(t) = u(t)n(n-1)\cdots(n-k+1)t^{n-k}, k = 1, 2, 3, \ldots, n.$$

$f^{(k)}(t)$ is continuous if $k < n$ and $f^{(n)}(t) = n!u(t)$ is piecewise continuous. Therefore, by Theorem 8.4.5

$$F[f^{(n)}(t)] = F[n!\, u(t)] = \frac{n!}{\sqrt{2\pi}\, iz} = (iz)^n F[f]$$

$$F[f(t)] = F[u(t)t^n] = \frac{n!}{\sqrt{2\pi}\,(iz)^{n+1}}.$$

EXAMPLE 8.4.3. Let $f(t) = u(t)e^{kt}$. Then

$$F[f] = \frac{1}{\sqrt{2\pi}} \int_0^\infty e^{(k-iz)t}\, dt = \frac{1}{\sqrt{2\pi}} \frac{1}{iz - k},$$

provided $\text{Im}(z) < -k$.

EXAMPLE 8.4.4. Let $f(t) = u(t - \tau)e^{kt}$. Then

$$f(t) = u(t - \tau)e^{k(t-\tau)}e^{k\tau} = e^{k\tau}g(t - \tau),$$

where $g(t) = u(t)e^{kt}$. Then by Theorem 8.4.2 and Example 8.4.3

$$F[f] = e^{k\tau}F[g(t - \tau)] = \frac{e^{k\tau}e^{-i\tau z}}{\sqrt{2\pi}\,(iz - k)}.$$

EXAMPLE 8.4.5. Let $f(t) = u(t)\cos \omega t$. We have

$$u(t)\sin \omega t = \omega \int_0^t u(t)\cos \omega \tau\, d\tau$$

and from Theorem 8.4.7 and Example 8.3.1,

$$F[u(t)\sin \omega t] = \frac{1}{\sqrt{2\pi}} \frac{\omega}{\omega^2 - z^2} = \frac{\omega}{iz} F[u(t)\cos \omega t].$$

Therefore,

$$F[u(t)\cos \omega t] = \frac{1}{\sqrt{2\pi}} \frac{iz}{\omega^2 - z^2}.$$

EXAMPLE 8.4.6. Let $f(t) = u(t)e^{-kt} \cos \omega t$. Then

$$F[f] = \frac{1}{\sqrt{2\pi}} \int_0^\infty \cos \omega t \, e^{-t(k+iz)} \, dt$$

$$= F[u(t) \cos \omega t]_{iz \to k+iz}$$

$$= \frac{1}{\sqrt{2\pi}} \frac{(k + iz)}{\omega^2 + (k + iz)^2} \cdot$$

EXAMPLE 8.4.7. Let $f(t) = tu(t) \sin \omega t$. Then if $g(t) = u(t) \sin \omega t$,
then $F[g] = \dfrac{1}{\sqrt{2\pi}} \dfrac{\omega}{\omega^2 - z^2}$ and by Theorem 8.3.2

$$\frac{d}{dz} F[g] = \frac{\omega}{\sqrt{2\pi}} \frac{2z}{(\omega^2 - z^2)^2} = -iF[f],$$

$$F[f] = \frac{2iz\omega}{\sqrt{2\pi} (\omega^2 - z^2)^2} \cdot$$

EXAMPLE 8.4.8. Let $F[f(t)] = \dfrac{1}{z(z^2 - 1)}$, where $f(t) = 0$ for $t < 0$.
Find $f(t)$. We know that $F^{-1}[1/z] = \sqrt{2\pi} \, iu(t)$ and

$$F^{-1}\left(\frac{1}{z^2 - 1}\right) = -\sqrt{2\pi} \, u(t) \sin t.$$

Therefore, using the convolution integral we have

$$F^{-1}\left[\frac{1}{z(z^2 - 1)}\right] = -i\sqrt{2\pi} \, u(t) \int_0^t \sin (t - \tau) \, d\tau$$

$$= -i\sqrt{2\pi} \, u(t) \, [\cos (t - \tau)]_0^t$$

$$= -i\sqrt{2\pi} \, u(t)[1 - \cos t].$$

Another way to proceed is to expand $\dfrac{1}{z(z^2 - 1)}$ in a partial fraction expansion

$$\frac{1}{z(z^2 - 1)} = \frac{A}{z} + \frac{B}{z - 1} + \frac{C}{z + 1} \cdot$$

Then $A = \lim\limits_{z \to 0} \dfrac{1}{z^2 - 1} = -1$, $B = \lim\limits_{z \to 1} \dfrac{1}{z(z + 1)} = \dfrac{1}{2}$, and

$$C = \lim_{z \to -1} \frac{1}{z(z - 1)} = \frac{1}{2} \cdot$$

By the linearity of the transform

$$F^{-1}\left[\frac{-1}{z} + \frac{1}{2}\frac{1}{z-1} + \frac{1}{2}\frac{1}{z+1}\right] = -i\sqrt{2\pi}\,u(t)\left[1 - \frac{1}{2}(e^{it} + e^{-it})\right]$$
$$= -i\sqrt{2\pi}\,u(t)[1 - \cos t].$$

Exercises 8.4

1. Prove that $F[u(t)t^{\alpha}] = \dfrac{\Gamma(\alpha + 1)}{(iz)^{\alpha+1}}$, where $\alpha > 0$ and $(iz)^{\alpha+1}$ is defined as a branch which is analytic for $\text{Im}(z) < 0$.

2. Find $F[u(t)\cosh \omega t]$ and $F[u(t)\sinh \omega t]$.

3. Find $F[u(t)t \sinh \omega t]$ and $F[u(t)t \cosh \omega t]$.

4. Find $F[u(t - \tau)te^{at}]$, where $\tau > 0$.

5. Find $F[u(t - \tau)te^{at} \sin \omega t]$, where $\tau > 0$.

6. Let $f(t) = 0$ for $t < 0$. Find $f(t)$ if $F[f] = \dfrac{1}{z(z^2 - 1)(z^2 + 1)}$.

8.5. THE SOLUTION OF ORDINARY DIFFERENTIAL EQUATIONS

In this section we shall consider the application of the Fourier transform technique to the solution of ordinary differential equations. Consider the linear nth order differential equation with constant coefficients

$$a_n \frac{d^n y}{dt^n} + a_{n-1} \frac{d^{n-1}y}{dt^{n-1}} + \cdots + a_1 \frac{dy}{dt} + a_0 y = f(t).$$

Assume $f(t)$ has a Fourier transform $F[f]$. Then taking the transform of both sides of the equation we have

$$[a_n(iz)^n + a_{n-1}(iz)^{n-1} + \cdots + a_1(iz) + a_0]\,F[y] = F[f],$$

where $F[y]$ is the Fourier transform of y, if it exists. Hence

$$F[y] = F[f]/P(iz),$$

where $P(D) = a_n D^n + a_{n-1}D^{n-1} + \cdots + a_1 D + a_0$ is the operator on the left-hand side of the differential equation. Let us assume that $1/P(iz)$ has an inverse transform

$$g(t) = \frac{1}{\sqrt{2\pi}} \int_{-\infty+i\gamma}^{\infty+i\gamma} \frac{e^{izt}}{P(iz)}\,dz.$$

Using the convolution, we have

$$y(t) = \frac{1}{\sqrt{2\pi}} \int_{-\infty}^{\infty} f(\tau)g(t - \tau) \, d\tau.$$

If we can determine a function $y(t)$ in this way then we can easily test it in the differential equation to see if it is a solution. Alternatively, we can invoke the following theorem:

Theorem 8.5.1. *Let $f(t)$ be a continuous function for which the Fourier integral theorem applies and let $G(z) = F[f]$ be analytic in some strip $a < \text{Im}(z) < b$. Let $P(iz)$ have no zeros in this strip. Let*

$$f(t) = \frac{1}{\sqrt{2\pi}} \int_{-\infty+i\gamma}^{\infty+i\gamma} G(z)e^{izt} \, dz$$

converge uniformly in t for some γ satisfying $a < \gamma < b$. Then

$$y(t) = \frac{1}{\sqrt{2\pi}} \int_{-\infty+i\gamma}^{\infty+i\gamma} \frac{G(z)e^{izt}}{P(iz)} \, dz$$

is a solution of

$$P(D)y = f(t).$$

Proof: Under the given hypotheses

$$\frac{1}{\sqrt{2\pi}} \int_{-\infty+i\gamma}^{\infty+i\gamma} \frac{G(z)e^{izt}}{P(iz)} \, dz$$

converges uniformly in t. Therefore, we can differentiate under the integral sign with respect to t, and

$$a_1 Dy + a_0 = \frac{1}{\sqrt{2\pi}} \int_{-\infty+i\gamma}^{\infty+i\gamma} \frac{[a_1(iz) + a_0] G(z)e^{izt}}{P(iz)} - dt.$$

Likewise, the higher derivatives exist and we have

$$P(D)y = \frac{1}{\sqrt{2\pi}} \int_{-\infty+i\gamma}^{\infty+i\gamma} \frac{P(iz)G(z)e^{izt}}{P(iz)} \, dz$$

$$= \frac{1}{\sqrt{2\pi}} \int_{-\infty+i\gamma}^{\infty+i\gamma} G(z)e^{izt} \, dz = f(t).$$

EXAMPLE 8.5.1. Find a solution of $L \, dI/dt + RI = E_0 e^{-|t|}$, where L, R, and E_0 are positive constants. This is the equation satisfied by the current I in a series inductance-resistance circuit with input voltage $E_0 e^{-|t|}$. Here,

$f(t) = E_0 e^{-|t|}$ and $G(z) = F[f] = 2E_0/\sqrt{2\pi}(1 + z^2)$ is analytic for $-1 < \text{Im}(z) < 1$. $P(iz) = izL + R$ has one zero at iR/L. By Theorem 8.5.1

$$I(t) = \frac{E_0}{\pi} \int_{-\infty}^{\infty} \frac{e^{izt}}{(1 + z^2)(izL + R)} \, dz,$$

which can be evaluated by contour integration. Let $t > 0$. Then summing residues at the singularities i and iR/L in the upper half-plane, we have

$$I(t) = 2E_0 i \left[\frac{e^{-t}}{2i(R - L)} + \frac{e^{-Rt/L}}{iL(1 - R^2/L^2)} \right]$$

$$= E_0 \left[\frac{e^{-t}}{R - L} + \frac{2Le^{-Rt/L}}{L^2 - R^2} \right].$$

If $t < 0$, we must close the contour in the lower half-plane and hence the residue at $-i$ is the only one that enters. Then

$$I(t) = 2E_0 i \frac{e^t}{2i(R + L)} = \frac{E_0 e^t}{R + L}.$$

At $t = 0$, the current $I(t)$ is continuous and hence

$$I(0) = \lim_{t \to 0} I(t) = \frac{E_0}{R + L}.$$

Notice that in this example the solution is not unique because any multiple of a solution of the homogeneous equation

$$L \frac{dI}{dt} + RI = 0$$

could have been added to the solution found. The function $e^{-Rt/L}$ is the only independent solution of this equation. By Theorem 7.2.1 we have a unique solution of the following problem

$$L \frac{dI}{dt} + RI = E_0 e^{-t},$$

for $0 < t$ and $I_0 = \lim_{t \to 0} I(t)$. In fact, we could find this solution by taking the above solution for $t > 0$, adding a constant C times $e^{-Rt/L}$, and then determining C to satisfy the initial condition; that is,

$$I(t) = Ce^{-Rt/L} + E_0 \left[\frac{e^{-t}}{R - L} + \frac{2Le^{-Rt/L}}{L^2 - R^2} \right]$$

$$I_0 = C + E_0/(R + L)$$

Hence,

$$C = I_0 - E_0/(R + L)$$

and

$$I(t) = I_0 e^{-Rt/L} + \frac{E_0}{R - L}[e^{-t} - e^{-Rt/L}].$$

A direct way to get this last solution is to use the one-sided Fourier transform, that is, if $I(t) = 0$ for $t < 0$ and $\lim_{t \to 0^+} I(t) = I_0$. Then if $F_+[I]$ is the one-sided transform

$$F_+[I] = \frac{1}{\sqrt{2\pi}} \int_0^\infty I(t)e^{-izt}\, dt,$$

$$F_+[I'] = \frac{1}{\sqrt{2\pi}} \int_0^\infty I'(t)e^{-izt}\, dt$$

$$= \frac{I(t)e^{-izt}}{\sqrt{2\pi}}\bigg|_0^\infty + \frac{iz}{\sqrt{2\pi}} \int_0^\infty I(t)e^{-izt}\, dt$$

$$= -\frac{I_0}{\sqrt{2\pi}} + izF_+[I].$$

Transforming the differential equation, we have

$$-\frac{I_0 L}{\sqrt{2\pi}} + [(iz)L + R]\,F_+[I] = F_+[f] = G_+(z),$$

where $G_+(z)$ is the one-sided transform of $E_0 e^{-t}$, that is,

$$G_+(z) = \frac{1}{\sqrt{2\pi}} \int_0^\infty E_0 e^{-t} e^{-izt}\, dt$$

$$= \frac{E_0}{\sqrt{2\pi}} \frac{1}{1 + iz}.$$

Then

$$F_+[I] = \frac{E_0}{\sqrt{2\pi}} \frac{1}{(izL + R)(1 + iz)} + \frac{I_0 L}{\sqrt{2\pi}\,(izL + R)}.$$

Inverting the transform, we have for $t > 0$

$$I(t) = \frac{E_0}{R - L}[e^{-t} - e^{-Rt/L}] + I_0 e^{-Rt/L},$$

and $I_0 = \lim_{t \to 0^+} I(t)$.

We see that the one-sided Fourier transform has some advantage over the two-sided Fourier transform in that it can be used effectively to introduce initial values and hence give the complete solution of the differential equation. Furthermore, since the right-hand side of the differential equation can be assumed to be zero for $t < 0$, we find that many simple functions like $u(t) \sin \omega t$, $u(t) \cos \omega t$, $u(t)e^{at}$, $u(t)t^n$, and so on can be taken as inputs to the differential equation, whereas $\sin \omega t$, $\cos \omega t$, e^{at}, t^n, and so on, do not have two-sided Fourier transforms. For these reasons the one-sided transform is more effective in solving initial value problems for ordinary differential equations. However, traditionally the Laplace transform has been used for this job. The Laplace transform, which we shall study in the next chapter, is essentially the one-sided Fourier transform with iz replaced by z, that is, with the complex plane rotated through $\pi/2$. With this change, the transform will be analytic in a right half-plane and will exist on most of the positive real axis. We shall therefore postpone further discussion of the solution of ordinary differential equations until we have introduced the Laplace transform.

Exercises 8.5

1. Find a solution of $\dfrac{d^2y}{dt^2} + 3\dfrac{dy}{dt} + 2y = e^{-|t|}$ using the Fourier transform.

2. Find a solution of $\dfrac{d^2y}{dt^2} + 3\dfrac{dy}{dt} + 2y = u(t) \sin \omega t$, for $t > 0$ satisfying $\lim\limits_{t \to 0^+} y(t) = 0$ and $\lim\limits_{t \to 0^+} y(t) = 1$.

3. Solve the integrodifferential equation

$$\frac{dy}{dt} + \int_0^t y(\tau)\, d\tau = e^{-t},$$

for $t > 0$ subject to $\lim\limits_{t \to 0^+} y(t) = y_0$.

4. The differential equation satisfied by the current I in a series inductance, resistance, capacitance circuit is $L\dfrac{d^2I}{dt^2} + R\dfrac{dI}{dt} + \dfrac{1}{C}I = \dfrac{dE}{dt}$, where L, R, and C are positive constants and $E(t)$ is the input voltage. Let $A(t)$ be the solution with a unit step input voltage. Show that

$$I(t) = \int_{-\infty}^{\infty} \dot{E}(\tau)A(t - \tau)\, d\tau.$$

$A(t)$ is called the *indicial admittance*.

8.6. THE SOLUTION OF PARTIAL DIFFERENTIAL EQUATIONS

In this section we shall consider the application of the Fourier transform technique to the solution of certain partial differential equations.

EXAMPLE 8.6.1. Find the displacement of an infinitely long elastic string which has an initial displacement $u(x, 0) = f(x)$, which is zero outside of a finite interval, and an initial velocity $u_t(x, 0) = g(x)$, which is also zero outside of a finite interval. Under the assumption of small displacements the problem can be formulated as follows:

$$\frac{\partial^2 u}{\partial x^2} = \frac{1}{a^2} \frac{\partial^2 u}{\partial t^2}, \qquad -\infty < x < \infty, \quad t > 0,$$

$$u(x, 0) = f(x),$$

$$u_t(x, 0) = g(x),$$

$$\lim_{|x| \to \infty} u(x, t) = 0, \qquad t > 0.$$

Assuming that $u(x, t), f(x)$, and $g(x)$ have Fourier transforms in the variable x, we have if $U(z, t) = (1/\sqrt{2\pi}) \int_{-\infty}^{\infty} u(x, t) e^{-izx} \, dx$,

$$(iz)^2 U(z, t) = \frac{1}{a^2} \frac{\partial^2 U}{\partial t^2},$$

$$U(z, 0) = \frac{1}{\sqrt{2\pi}} \int_{-\infty}^{\infty} f(x) e^{-izx} \, dx = F(z),$$

$$U_t(z, 0) = \frac{1}{\sqrt{2\pi}} \int_{-\infty}^{\infty} g(x) e^{-izx} \, dx = G(z),$$

or

$$U_{tt} + a^2 z^2 U = 0,$$

$$U(z, t) = A \sin azt + B \cos azt,$$

$$U(z, 0) = B = F(z),$$

$$U_t(z, 0) = azA = G(z).$$

Therefore,

$$U(z, t) = \frac{G(z)}{az} \sin azt + F(z) \cos azt$$

$$= \frac{G(z)}{2iaz} (e^{iazt} - e^{-iazt}) + \frac{F(z)}{2} (e^{iazt} + e^{-iazt}).$$

Applying the inversion integral

$$u(x, t) = \frac{1}{2\sqrt{2\pi}} \int_{-\infty+i\gamma}^{\infty+i\gamma} [F(z)e^{iz(x+at)} + F(z)e^{iz(x-at)}] \, dz$$

$$+ \frac{1}{2a\sqrt{2\pi}} \int_{-\infty+i\gamma}^{\infty+i\gamma} \frac{G(z)}{iz} [e^{iz(x+at)} - e^{iz(x-at)}] \, dz$$

$$= \tfrac{1}{2}[f(x + at) + f(x - at)] + \frac{1}{2a} \int_{x-at}^{x+at} g(\xi) \, d\xi.$$

In obtaining this result, we have glossed over many details. The point is that once the result is obtained one can test it directly in the differential equation. As a matter of fact, if f is twice differentiable and g is once differentiable it is easy to verify that this is a solution. The reader should carry out the details.

EXAMPLE 8.6.2. Find the temperature in an infinite uniform rod if the initial temperature is $u(x, 0) = f(x)$, which vanishes outside a finite interval. This problem can be formulated as follows:

$$\frac{\partial^2 u}{\partial x^2} = \frac{1}{a^2} \frac{\partial u}{\partial t}, \quad -\infty < x < \infty, \quad t > 0,$$

$$u(x, 0) = f(x),$$

$$\lim_{|x|\to\infty} u(x, t) = 0.$$

Assuming that $u(x, t)$ and $f(x)$ have Fourier transforms in the variable x, we have

$$U(z, t) = \frac{1}{\sqrt{2\pi}} \int_{-\infty}^{\infty} u(x, t)e^{-izx} \, dx,$$

$$(iz)^2 U(z, t) = \frac{1}{a^2} \frac{\partial U}{\partial t},$$

$$U(z, 0) = \frac{1}{\sqrt{2\pi}} \int_{-\infty}^{\infty} f(x)e^{-izx} \, dx = G(z).$$

Then

$$\frac{\partial U}{\partial t} + a^2 z^2 U = 0,$$

$$U(z, t) = U_0 e^{-z^2 a^2 t}$$

$$U_0 = U(z, 0) = G(z)$$

$$U(z, t) = G(z)e^{-z^2 a^2 t}.$$

Recalling from Exercise 8.3.3 that

$$\frac{1}{\sqrt{2\pi}} \int_{-\infty}^{\infty} e^{-x^2/2} e^{-i\zeta x} \, dx = e^{-\zeta^2/2},$$

we let $\zeta^2/2 = z^2 a^2 t$. Then

$$e^{-z^2 a^2 t} = \frac{1}{\sqrt{2\pi}} \int_{-\infty}^{\infty} e^{-x^2/2} e^{-iz\sqrt{2t} \, ax} \, dx$$

$$= \frac{1}{2a\sqrt{\pi t}} \int_{-\infty}^{\infty} e^{-u^2/4a^2 t} e^{-izu} \, du$$

$$= F\left[\frac{1}{a\sqrt{2t}} e^{-u^2/4a^2 t} \right].$$

Therefore, using the convolution theorem

$$u(x, t) = \frac{1}{2a\sqrt{\pi t}} \int_{-\infty}^{\infty} f(\tau) e^{-(x-\tau)^2/4a^2 t} \, d\tau.$$

It can be verified directly that this is a solution to the problem.

EXAMPLE 8.6.3. Next, we consider the corresponding problem for a semi-infinite heat conducting rod. Let the temperature at $x = 0$ be maintained at zero and the initial temperature distribution be $f(x)$ with $f(0) = 0$. Then find the temperature for $x > 0$ and $t > 0$. The problem can be formulated as follows:

$$\frac{\partial^2 u}{\partial x^2} = \frac{1}{a^2} \frac{\partial u}{\partial t}, \qquad 0 < x < \infty, \qquad t > 0,$$

$$u(x, 0) = f(x), \qquad 0 < x < \infty,$$

$$u(0, t) = 0, \qquad\qquad\qquad t \geq 0,$$

$$\lim_{x \to \infty} u(x, t) = 0.$$

In this case, we cannot transform with respect to x because, using the one-sided Fourier transform, we need to know $u(0, t)$ and $u_x(0, t)$. On the other hand, if we transform with respect to t, we can determine the transformed equation because only the first partial derivative with respect to t is involved.

However, in this particular case, we can get a solution directly from the solution to the last problem. Note that if $f(x)$ is odd

$$u(x, t) = \frac{1}{2a\sqrt{\pi t}} \int_{-\infty}^{\infty} f(\tau)e^{-(x-\tau)^2/4a^2 t}\, d\tau$$

is odd; that is, if $\sigma = -\tau$,

$$u(x, t) = \frac{1}{2a\sqrt{\pi t}} \int_{-\infty}^{\infty} f(-\sigma)e^{-(x+\sigma)^2/4a^2 t}\, d\sigma$$

$$= -\frac{1}{2a\sqrt{\pi t}} \int_{-\infty}^{\infty} f(\sigma)e^{-(-x-\sigma)^2/4a^2 t}\, d\sigma$$

$$= -u(-x, t).$$

Also,

$$u(0, t) = \frac{1}{2a\sqrt{\pi t}} \int_{-\infty}^{\infty} f(\tau)e^{-\tau^2/4a^2 t}\, d\tau$$

$$= -\frac{1}{2a\sqrt{\pi t}} \int_{-\infty}^{\infty} f(\sigma)e^{-\sigma^2/4a^2 t}\, d\sigma = -u(0, t).$$

Therefore, $u(0, t) = 0$, and in the present example if we continue $f(x)$ as an odd function we obtain our solution for $x > 0$.

EXAMPLE 8.6.4. Find the temperature in a semiinfinite heat conducting rod if the temperature at $x = 0$ is maintained at $g(t)$ and the initial temperature is $u(x, 0) = f(x)$. The problem can be formulated as follows:

$$\frac{\partial^2 u}{\partial x^2} = \frac{1}{a^2}\frac{\partial u}{\partial t}, \qquad x > 0, \qquad t > 0,$$

$$u(x, 0) = f(x),$$

$$u(0, t) = g(t),$$

$$\lim_{x \to \infty} u(x, t) = 0.$$

By the linearity of the differential equation we can split the solution as follows:

$$u(x, t) = u_1(x, t) + u_2(x, t),$$

where u_1 and u_2 both satisfy the differential equation and $u_1(x, 0) = f(x)$, $u_1(0, t) = 0$, $u_2(x, 0) = 0$, $u_2(0, t) = g(t)$, $\lim_{x \to \infty} u_1 = \lim_{x \to \infty} u_2 = 0$. Then

$$u_1(x, t) = \frac{1}{2a\sqrt{\pi t}} \int_{-\infty}^{\infty} f(\tau)e^{-(x-\tau)^2/4a^2 t}\, d\tau,$$

where $f(-x) = -f(x)$, $x > 0$. Finally, we have to solve for $u_2(x, t)$. In this case, it is convenient to use the Fourier sine transform so we again seek a solution which is an odd function in x; that is, $u_2(-x, t) = -u_2(x, t)$. Then

$$F[u_2(x, t)] = \frac{1}{\sqrt{2\pi}} \int_{-\infty}^{\infty} u_2(x, t)e^{-izx} \, dx$$

$$= \frac{1}{\sqrt{2\pi}} \int_{0}^{\infty} u_2(x, t)e^{-izx} \, dx - \frac{1}{\sqrt{2\pi}} \int_{0}^{\infty} u_2(x, t)e^{izx} \, dx$$

$$= -\sqrt{\frac{2}{\pi}} \, i \int_{0}^{\infty} u_2(x, t) \sin zx \, dx,$$

$$F[u_{2xx}] = -\sqrt{\frac{2}{\pi}} \, i \int_{0}^{\infty} u_{2xx} \sin zx \, dx,$$

$$= -\sqrt{\frac{2}{\pi}} \, i[u_{2x} \sin zx]_0^{\infty} + \sqrt{\frac{2}{\pi}} \, iz \int_{0}^{\infty} u_{2x} \cos zx \, dx$$

$$= \sqrt{\frac{2}{\pi}} \, iz[u_2 \cos zx]_0^{\infty} + \sqrt{\frac{2}{\pi}} \, iz^2 \int_{0}^{\infty} u_2 \sin zx \, dx$$

$$= -\sqrt{\frac{2}{\pi}} \, izg(t) - z^2 F[u_2].$$

If $U_2(z, t) = F[u_2]$, then

$$\frac{\partial U_2}{\partial t} + a^2 z^2 U_2 = -\sqrt{\frac{2}{\pi}} \, iza^2 g(t)$$

subject to $U_2(z, 0) = 0$. Solving this equation, we have

$$U_2(z, t) = -\sqrt{\frac{2}{\pi}} \, iza^2 \int_{0}^{t} g(\tau)e^{-a^2 z^2(t-\tau)} \, d\tau.$$

We know from Example 8.6.2 that

$$e^{-a^2 z^2(t-\tau)} = F\left[\frac{1}{a\sqrt{2(t-\tau)}} e^{-x^2/4a^2(t-\tau)}\right].$$

Therefore,

$$ize^{-a^2 z^2(t-\tau)} = F\left[\frac{\partial}{\partial x} \frac{1}{a\sqrt{2(t-\tau)}} e^{-x^2/4a^2(t-\tau)}\right]$$

$$= F\left[\frac{-x}{[2a^2(t-\tau)]^{3/2}} e^{-x^2/4a^2(t-\tau)}\right].$$

Therefore,

$$u_2(x, t) = \frac{x}{2a\sqrt{\pi}} \int_0^t \frac{g(\tau)e^{-x^2/4a^2(t-\tau)}}{(t-\tau)^{3/2}} \, d\tau$$

and

$$u(x, t) = \frac{1}{2a\sqrt{\pi t}} \int_{-\infty}^{\infty} f(\tau)e^{-(x-\tau)^2/4a^2 t} \, d\tau + \frac{x}{2a\sqrt{\pi}} \int_0^t \frac{g(\tau)e^{-x^2/4a^2(t-\tau)}}{(t-\tau)^{3/2}} \, d\tau.$$

It is possible to verify directly that this is a solution to the problem.

We conclude this section by an example involving the wave equation again, this time over a finite interval in x.

EXAMPLE 8.6.5. Find the displacement of an elastic string of length π, fixed at $x = 0$ and $x = \pi$, subject to a force proportional to $\sin x$ but constant in t, assuming that the initial velocity and displacement are zero. The problem can be formulated as follows:

$$\frac{\partial^2 u}{\partial x^2} = \frac{1}{a^2} \frac{\partial^2 u}{\partial t^2} - b \sin x, \qquad 0 < x < \pi, \quad t > 0,$$

$$u(x, 0) = u_t(x, 0) = 0,$$

$$u(0, t) = u(\pi, t) = 0.$$

Here, we take the one-sided Fourier transform with respect to t. Then if $U(x, z) = F[u]$, we have

$$U_{xx} + \frac{z^2}{a^2} U = -\frac{b \sin x}{\sqrt{2\pi} \, iz}.$$

The solution of this equation is

$$U(x, z) = A \sin \frac{zx}{a} + B \cos \frac{zx}{a} - \frac{a^2 b \sin x}{\sqrt{2\pi} \, iz(z^2 - a^2)}.$$

Using the boundary conditions, we have $A = B = 0$, and

$$U(x, z) = \frac{-a^2 b \sin x}{\sqrt{2\pi} \, iz(z^2 - a^2)}.$$

The inversion integral gives

$$u(x, t) = \frac{-a^2 b \sin x}{2\pi i} \int_{-\infty+i\gamma}^{\infty+i\gamma} \frac{e^{izt}}{z(z^2 - a^2)} \, dz,$$

where $\gamma < -a$. This can be evaluated by contour integration. If $t > 0$ we close the contour by a semicircular path in the upper half-plane and we have residues at the origin and at $z = \pm a$ to compute. If $t < 0$ the integral is zero since, in this case, we close the contour in the lower half-plane where there are no poles. The final result is

$$u(x, t) = b \sin x(1 - \cos at), \qquad t > 0,$$
$$= 0, \qquad\qquad\qquad t < 0.$$

These examples illustrate the utility of the Fourier transform in solving certain problems in partial differential equations. This does not exhaust the possible types to which the technique will apply. The general procedure is to apply the transform to the differential equation with respect to one of the variables. This usually reduces the problem to an ordinary differential equation in the other variable, or a partial differential equation in the remaining variables. The new problem is solved, with the transform variable carried along as a parameter, using the boundary conditions or initial conditions imposed on the transform. In this way the transform of the solution is found and inverted to find the solution. In general, the solution found by this technique is checked directly against the stated conditions of the original problem rather than trying to justify the transform technique at every step of the way. If the original problem is correctly formulated, there is usually a uniqueness theorem to assure that the solution found by the transform technique is the only solution to the problem.

Exercises 8.6

1. Solve for the displacement of an elastic string fixed at $x = 0$ and $x = \pi$, if the initial displacement and velocity are zero, when a force $bx(\pi - x)$ is applied along the string.

2. Find the displacement of a semiinfinite string with zero initial displacement and velocity, if the displacement at $x = 0$ is given by $u(0, t) = g(t)$, $t \geq 0$.

3. Find the displacement of a semiinfinite string with initial displacement $u(x, 0) = f(x)$, $x \geq 0$, and zero initial velocity, if the string is fixed at $x = 0$. Hint: show that a solution can be obtained from the result of Example 8.6.1 by continuing $f(x)$ as an odd function. How can the following problem be solved: $a^2 u_{xx} = u_{tt}$, $x > 0$, $t > 0$, $u(x, 0) = f(x)$, $u_t(x, 0) = 0$, $u(0, t) = g(t)$?

4. Find the steady state temperature in a semiinfinite heat conducting plate ($x \geq 0, 0 \leq y \leq 1$) if $u(x, 0) = u(x, 1) = 0$ and $u(0, y) = f(y)$.

8.7. THE SOLUTION OF INTEGRAL EQUATIONS

Quite often in the applications an integral equation will arise in which the unknown occurs in a convolution integral. Suppose

$$f(t) = \int_{-\infty}^{\infty} k(t - \tau)u(\tau) \, d\tau$$

is an integral equation to be solved for the unknown u where f and k are given functions, each of which has a Fourier transform. Let

$$F(z) = \frac{1}{2\pi} \int_{-\infty}^{\infty} f(t)e^{-izt} \, dt \quad \text{and} \quad K(z) = \frac{1}{\sqrt{2\pi}} \int_{-\infty}^{\infty} k(t)e^{-izt} \, dt.$$

Then

$$F(z) = K(z)U(z),$$

where $U(z)$ is the transform of the unknown. Therefore,

$$U(z) = F(z)/K(z),$$

$$u(t) = \frac{1}{\sqrt{2\pi}} \int_{-\infty+i\gamma}^{\infty+i\gamma} [F(z)/K(z)]e^{izt} \, dt.$$

EXAMPLE 8.7.1. Solve the integral equation

$$\sqrt{2\pi} \, e^{-t^2/2} = \int_{-\infty}^{\infty} e^{-|t-\tau|}u(\tau) \, d\tau.$$

The transformed equation is

$$e^{-z^2/2} = \sqrt{\frac{2}{\pi}} \frac{U(z)}{1 + z^2}.$$

Therefore,

$$U(z) = \sqrt{\frac{\pi}{2}} (1 + z^2)e^{-z^2/2}$$

$$= \sqrt{\frac{\pi}{2}} [e^{-z^2/2} - (iz)^2 e^{-z^2/2}].$$

Inverting, we have

$$u(t) = \sqrt{\frac{\pi}{2}} \left[e^{-t^2/2} - \frac{d^2}{dt^2} e^{-t^2/2} \right]$$

$$= \sqrt{2\pi} \left[e^{-t^2/2} - \frac{t^2}{2} e^{-t^2/2} \right].$$

The integral equation we have been discussing is usually called the *Wiener-Hopf* equation. Unfortunately, in many applications, the left-hand side of the equation is not completely known. A typical situation is the following:

$$f(t) = h(t) = \int_{-\infty}^{\infty} k(t - \tau)u(\tau) \, d\tau, \qquad t > 0,$$

$$f(t) = g(t) = \int_{-\infty}^{\infty} k(t - \tau)u(\tau) \, d\tau, \qquad t < 0,$$

where $h(t)$ is known but $g(t)$ is not. In this case, a more elaborate procedure, known as the *Wiener-Hopf technique*, is needed to solve the integral equation. We shall outline this method and then show how it can be applied to a particular problem.

Suppose

$$H_-(z) = \frac{1}{2\pi} \int_0^{\infty} h(t)e^{-izt} \, dt$$

is analytic in a lower half-plane $\text{Im}(z) < b$ and that

$$G_+(z) = \frac{1}{2\pi} \int_{-\infty}^0 g(t)e^{-izt} \, dt$$

is analytic in an upper half-plane $a < \text{Im}(z)$. Then

$$H_-(z) + G_+(z) = K(z)U(z),$$

where $K(z)$ and $U(z)$ are the transforms of $k(t)$ and $u(t)$, respectively. Now, suppose $K(z)$ can be written as follows

$$K(z) = \frac{K_-(z)}{K_+(z)},$$

where $K_-(z)$ is analytic for $\text{Im}(z) < b$ and $K_+(z)$ is analytic for $a < \text{Im}(z)$. Then

$$K_+(z)H_-(z) + K_+(z)G_+(z) = K_-(z)U(z).$$

Suppose further that $K_+(z)H_-(z)$ can be written as

$$K_+(z)H_-(z) = P_+(z) + Q_-(z),$$

where $P_+(z)$ is analytic for $a < \text{Im}(z)$ and $Q_-(z)$ is analytic for $\text{Im}(z) < b$. Then

$$E(z) = K_+(z)G_+(z) + P_+(z) = K_-(z)U(z) - Q_-(z).$$

We also assume that $U(z)$ is analytic for $\text{Im}(z) < b$. We have now defined a function $E(z)$ which is analytic in the strip $a < \text{Im}(z) < b$. Furthermore, the last equation gives the analytic continuation of $E(z)$ to the whole plane. Hence, $E(z)$ is an entire function. If it can be shown that $E(z)$ is bounded, then by Liouville's theorem $E(z)$ is a constant. If $|E(z)| \to 0$ as $|z| \to \infty$ in any part of the plane, then $E(z) \equiv 0$. Hence,

$$K_-(z)U(z) - Q_-(z) \equiv 0,$$

$$U(z) = \frac{Q_-(z)}{K_-(z)}.$$

By this device we have determined the transform of $u(t)$ without specific knowledge of $G_+(z)$, the transform of $g(t)$. However, the procedure requires the decomposition of $K(z)$ and of $K_+(z)H_-(z)$, which is not always easy. For a more general discussion of the Wiener-Hopf technique see Benjamin Noble, *The Wiener-Hopf Technique*. New York: Pergamon Press, Inc., 1958.

We shall now apply the Wiener-Hopf technique to a specific problem in diffraction theory. Consider the diffraction of an acoustic plane wave by a semiinfinite flat plate occupying half the xz plane ($x < 0$, $-\infty < z < \infty$). We shall assume that the plane wave strikes the edge of the plate perpendicular to the edge ($x = 0$) so that the solution to the problem does not depend on the z coordinate, and therefore is two-dimensional. The acoustic velocity potential of the plane wave has the form

$$u_p(x, y, t) = e^{i[k(x\cos\alpha + y\sin\alpha) - \omega t]}.$$

It can easily be shown that u_p satisfies the wave equation

$$\nabla^2 u_p = \frac{k^2}{\omega^2} \frac{\partial^2 u_p}{\partial t^2}.$$

Now, the entire acoustic field, the plane wave plus the scattered field, must satisfy the wave equation

$$\nabla^2 u = \frac{1}{a^2} \frac{\partial^2 u}{\partial t^2}.$$

Let $u(x, y, t) = u_p(x, y, t) + u_s(x, y, t)$, where u_s stands for the scattered field and $u_s = \phi_s(x, y)e^{-i\omega t}$. Then

$$\nabla^2 \phi_s + k^2 \phi_s = 0,$$

where $k^2 = \omega^2/a^2$. We can therefore write

$$u(x, y, t) = \phi(x, y)e^{-i\omega t} = (\phi_p + \phi_s)e^{-i\omega t},$$

where $\phi_p = e^{ik(x\cos\alpha + y\sin\alpha)}$.

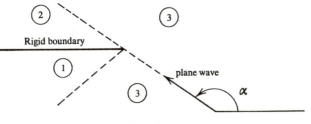

Figure 8.7.1.

Since the plate is a rigid barrier to sound there can be no velocity component at right angles to it. Therefore, ϕ must satisfy the boundary condition $d\phi/dn = 0$, for $y = 0$ and $x < 0$. For this reason ϕ_s must satisfy

$$\frac{d\phi_s}{dn} = -\frac{d\phi_p}{dn}$$

on the plate. The problem then is to determine ϕ_s satisfying

$$\nabla^2\phi_s + k^2\phi_s = 0$$

away from the plate with $d\phi_s/dn$ given on the plate.

We shall make a further splitting of ϕ_s as follows. Geometrical optics predicts that in Region 1 of Figure 8.7.1 there should be a reflected wave

$$u_r(x, y, t) = e^{i[k(x\cos\alpha - y\sin\alpha) - \omega t]}$$

and a "shadow" in Region 2. The solution predicted by geometrical optics is the following:

Region 1: $u = u_p + u_r$

Region 2: $u \equiv 0$

Region 3: $u = u_p$.

This solution is unsatisfactory on physical grounds, because experiment shows diffraction effects not contained in it, and on mathematical grounds, because it is not continuous everywhere away from the plate. We shall say that the diffraction effects are contained in that part of the solution not predicted by geometrical optics. Therefore, we have

Region 1: $u = u_p + u_r + u_d$

Region 2: $u = u_d$

Region 3: $u = u_p + u_d$

and since u_d consists of outward traveling cylindrical waves it must satisfy a radiation condition

$$\lim_{r\to\infty} \sqrt{r}\left[\frac{\partial u_d}{\partial r} - iku_d\right] = 0,$$

where $r = \sqrt{x^2 + y^2}$.

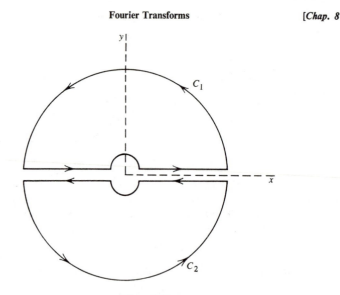

Figure 8.7.2.

To develop the appropriate Wiener-Hopf integral equation for this problem we use the following Green's function:

$$G(\xi, \eta, x, y) = \frac{i}{4}[H_0^{(1)}(k\sqrt{(x - \xi)^2 + (y - \eta)^2})$$

$$+ H_0^{(1)}(k\sqrt{(x - \xi)^2 + (y + \eta)^2})],$$

which satisfies $\nabla^2 G + k^2 G = 0$ except at (ξ, η), $G_y(\xi, \eta, x, 0) = 0$, the radiation condition as $r \to \infty$, and behaves like

$$-(1/2\pi) \ln \sqrt{(x - \xi)^2 + (y - \eta)^2},$$

as $(x, y) \to (\xi, \eta)$. Consider the two contours in Figure 8.7.2. In the upper half-plane we apply the Green's identity in the usual way to the whole field, $\phi^{(1)}(x, y)$, and we obtain using the properties of the Green's function*

$$\phi^{(1)}(\xi, \eta) = \int_{C_1} \left(G\frac{d\phi^{(1)}}{dn} - \phi^{(1)}\frac{dG}{dn} \right) ds$$

$$= -\int_{-\infty}^{\infty} G\left(\frac{\partial \phi^{(1)}}{\partial y}\right)_{y=0^+} dx.$$

---

* One must show that the integrals on the small and large semicircles go to zero as their radii go to zero and infinity, respectively. For the details see J. W. Dettman, *Mathematical Methods in Physics and Engineering*. New York: McGraw-Hill Book Company, Inc., 1962.

In the lower half-plane we apply the Green's identity to the whole field $\phi^{(2)}$ minus the plane wave and the reflected wave. We obtain

$$\phi^{(2)}(\xi, \eta) - e^{ik(\xi \cos \alpha + \eta \sin \alpha)} - e^{ik(\xi \cos \alpha - \eta \sin \alpha)}$$

$$= \int_{-\infty}^{\infty} G \left(\frac{\partial \phi^{(2)}}{\partial y} \right)_{y=0^-} dx.$$

Now, for $\xi > 0$

$$\phi^{(2)}(\xi, \eta^-) - \phi^{(1)}(\xi, \eta^+) = 0,$$

$$\left(\frac{\partial \phi^{(2)}}{\partial y} \right)_{y=0^+} = \left(\frac{\partial \phi^{(1)}}{\partial y} \right)_{y=0^-}$$

and, therefore

$$-2e^{ik\xi \cos \alpha} = 2 \int_{-\infty}^{\infty} G \left(\frac{\partial \phi}{\partial y} \right)_{y=0} dx$$

$$= i \int_{-\infty}^{\infty} H_0^{(1)}(k|x - \xi|)I(x) \, dx,$$

where $I(x) = (\partial \phi / \partial y)_{y=0}$. For $\xi < 0$

$$\phi^{(2)}(\xi, 0^-) - \phi^{(1)}(\xi, 0^+) - 2e^{ik\xi \cos \alpha} = i \int_{-\infty}^{\infty} H_0^{(1)}(k|x - \xi|)I(x) \, dx.$$

Hence, we have the Wiener-Hopf integral equation

$$f(\xi) = i \int_{-\infty}^{\infty} H_0^{(1)}(k|x - \xi|)I(x) \, dx$$

to be solved for $I(x)$, where

$$f(\xi) = -2e^{ik\xi \cos \alpha}, \qquad \xi > 0,$$

$$f(\xi) = \phi(\xi, 0^-) - \phi(\xi, 0^+) - 2e^{ik\xi \cos \alpha}, \qquad \xi < 0,$$

and $I(x) = 0$, for $x < 0$.

This is a particular example of the general situation described above. We do not know $\phi(\xi, 0^-) - \phi(\xi, 0^+)$ and therefore, we shall have to use the

Wiener-Hopf technique. To identify the present case with the general discussion we write

$$U(z) = \frac{1}{\sqrt{2\pi}} \int_0^\infty I(x)e^{-izz} \, dx,$$

$$H_-(z) = -\frac{1}{\pi} \int_0^\infty e^{ikx\cos\alpha} e^{-izx} \, dx$$

$$= \frac{i}{\pi} \frac{1}{z - k\cos\alpha},$$

$$G_+(z) = \frac{i}{2\pi} \int_{-\infty}^\infty e^{-izx} \int_{-\infty}^\infty H_0^{(1)}(k|x - \xi|)I(\xi) \, d\xi \, dx$$

$$K(z) = \frac{i}{\sqrt{2\pi}} \int_{-\infty}^\infty H_0^{(1)}(k|x|)e^{-izx} \, dx$$

$$= \sqrt{\frac{2}{\pi}} \frac{1}{\sqrt{k^2 - z^2}}.$$

In the present case, $k = k_1 + ik_2$, where $k_2 > 0$. $K(z)$ has branch points at $\pm k$, but is analytic in the strip $-k_2 < \text{Im}(z) < k_2$ provided the branch cuts are not in this strip. We assume that $I(x)$ behaves like $e^{\delta x}$ as $x \to \infty$, and therefore $U(z)$ is analytic for $\text{Im}(z) < -\delta$. It can be shown that $H_0^{(1)}(k|x - \xi|)$ behaves like $e^{ik|x - \xi|}/\sqrt{k|x - \xi|}$, as $|x - \xi| \to \infty$. Therefore, for x large and negative

$$i \int_0^\infty H_0^{(1)}(k|x - \xi|)I(\xi) \, d\xi$$

behaves like e^{-ikx} provided $\int_0^\infty \xi^{-1/2}I(\xi)e^{ik\xi} \, d\xi$ exists. This last integral exists provided $-k_2 < -\delta$. From this we conclude that $G(z)$ is analytic for $-k_2 < \text{Im}(z)$ and

$$H_-(z) + G_+(z) = K(z)U(z)$$

holds in the strip $-k_2 < \text{Im}(z) < \min\left[-\delta, k_2 \cos\alpha\right]$. Therefore,

$$\frac{i}{\pi(z - k\cos\alpha)} + G_+(z) = \sqrt{\frac{2}{\pi}} \frac{U(z)}{\sqrt{k^2 - z^2}}.$$

Since

$$K(z) = \sqrt{\frac{2}{\pi}} \frac{1}{\sqrt{k^2 - z^2}} = \sqrt{\frac{2}{\pi}} \frac{1}{\sqrt{k - z}\sqrt{k + z}},$$

we have

$$\sqrt{\frac{2}{\pi}} \frac{U(z)}{\sqrt{k - z}} = \frac{i\sqrt{k + z}}{\pi(z - k\cos\alpha)} + \sqrt{k + z} \, G_+(z).$$

Next, we write

$$\frac{i\sqrt{k+z}}{\pi(z-\cos\alpha)} = \frac{i\sqrt{k+k\cos\alpha}}{\pi(z-k\cos\alpha)} + \frac{i\sqrt{k+z}-i\sqrt{k+k\cos\alpha}}{\pi(z-k\cos\alpha)}$$

$$= Q_-(z) + P_+(z),$$

where we define $P_+(z)$ at $k\cos\alpha$ by $i/2\pi\sqrt{k+k\cos\alpha}$. It is easy to show that $P_+(z)$ is analytic at $k\cos\alpha$. It has a branch point at $z = -k$ but is analytic in the upper half-plane $-k_2 < \text{Im}(z)$. As a result

$$\sqrt{\frac{2}{\pi}}\frac{U(z)}{\sqrt{k-z}} - \frac{i}{\pi}\frac{\sqrt{k+k\cos\alpha}}{z-k\cos\alpha} = P_+(z) + \sqrt{k+z}\,G_+(z),$$

where the left-hand side is analytic for $\text{Im}(z) < \min[-\delta, k_2\cos\alpha]$ and the right-hand side is analytic for $-k_2 < \text{Im}(z)$. Therefore, as in the general case this equation defines by analytic continuation an entire function. Clearly, the functions involved approach zero as $|z| \to \infty$, except possibly $\sqrt{k+z}\,G_+(z)$. The latter can grow, no faster than $\sqrt{z}$, as $z \to \infty$. But a nonconstant entire function must grow at least as fast as z. Therefore, this function is everywhere zero and from this we derive the transform

$$U(z) = \frac{i}{\sqrt{2\pi}}\frac{\sqrt{k-z}\sqrt{k+k\cos\alpha}}{(z-k\cos\alpha)}.$$

By the inversion integral

$$I(x) = \frac{i}{2\pi}\sqrt{k+k\cos\alpha}\int_{-\infty+i\gamma}^{\infty+i\gamma}\frac{e^{izx}\sqrt{k-z}}{z-k\cos\alpha}\,dz,$$

where $-k_2 < \gamma < k_2\cos\alpha$. If $x < 0$, we can integrate along a line $x + i\gamma$, $-R < x < R$ and close the contour with a semicircle of radius R in the lower half-plane. It is easy to show that the contribution of the semicircle approaches zero as $R \to \infty$. Hence, for $x < 0$, $I(x) = 0$ because there are no singularities of the integral inside the contour. For $x > 0$ consider Figure 8.7.3. In this case,

$$\int_{C+}\frac{e^{izx}\sqrt{k-z}}{z-k\cos\alpha}\,dz = 0,$$

because there are no singularities inside the contour. Also, as $R \to \infty$ the contribution of the semicircular arc approaches zero. Hence,

$$\int_{C+}\frac{e^{izx}\sqrt{k-z}}{z-k\cos\alpha}\,dz + \int_{\Gamma+}\frac{e^{izx}\sqrt{k-z}}{z-k\cos\alpha}\,dz = 2\pi i e^{ikx\cos\alpha}\sqrt{k-k\cos\alpha},$$

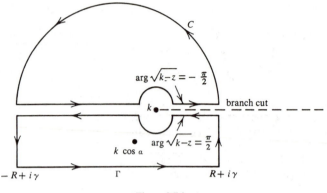

Figure 8.7.3.

because of the simple pole at $k \cos \alpha$ inside Γ. We let $R \to \infty$ to obtain

$$I(x) = -\sqrt{k^2 - k^2 \cos \alpha} \, e^{ikx \cos \alpha} - \frac{i}{2\pi} \sqrt{k + k \cos \alpha} \int_L \frac{e^{ixz}\sqrt{k - z}}{z - k \cos \alpha} \, dz,$$

where L is a hook type contour coming in from infinity along the lower side
of the branch cut, circling the point k on a small circle and returning to
infinity along the top of the cut. This is because the contributions of C and
Γ on the lines to the left of k cancel and the other parts drop out as $R \to \infty$.
The contribution of the small circle of radius ρ centered on k goes to zero as
$\rho \to 0$. On the lower side of the cut we let $k - z = u^2 e^{i\pi}$. Then $dz =
2u \, du$ and $\sqrt{k - z} = iu$. On the upper side of the cut we let $k - z =
u^2 e^{-i\pi}$. Then $dz = 2u \, du$ and $\sqrt{k - z} = -iu$. Therefore,

$$\int_L \frac{e^{ixz}\sqrt{k - z}}{z - k \cos \alpha} \, dz = -4i \int_0^\infty \frac{e^{ix(k+u^2)}u^2 \, du}{k + u^2 - k \cos \alpha}$$

$$= -4ie^{ixk} \int_0^\infty \frac{e^{ixu^2}u^2 \, du}{k + u^2 - k \cos \alpha},$$

$$I(x) = -\sqrt{k^2 - k^2 \cos \alpha} \, e^{ikx \cos \alpha} - \frac{2}{\pi} \sqrt{k + k \cos \alpha} \, e^{ikx} g(x),$$

where

$$g(x) = \int_0^\infty \frac{u^2 e^{ixu^2}}{k + u^2 - k \cos \alpha} \, du$$

$$= \int_0^\infty e^{ixu^2} \, du - (k - k \cos \alpha) \int_0^\infty \frac{e^{ixu^2}}{u^2 + k - k \cos \alpha} \, du.$$

In the first integral, we let $v = xu^2$. Then $du = dv/2\sqrt{x}\sqrt{v}$ and

$$\int_0^\infty e^{ixu^2} \, du = \frac{1}{2\sqrt{x}} \int_0^\infty \frac{e^{iv}}{\sqrt{v}} \, dv$$

and the latter is a convergent integral. It remains to determine

$$h(x) = \int_0^\infty \frac{e^{ixu^2}}{u^2 + k - k \cos \alpha} \, du.$$

It can be shown that $h'(x) + i\beta h(x) = iA/2\sqrt{x}$, where $\beta = k - k \cos \alpha$ and $A = \int_0^\infty (e^{iv}/\sqrt{v}) \, dv$. Also, $h(0) = \int_0^\infty (u^2 + \beta)^{-1} \, du = \pi/2\sqrt{\beta}$. Solving for $h(x)$, we have

$$h(x) = e^{-i\beta x} \left[\frac{\pi}{2\sqrt{\beta}} + \frac{iA}{2} \int_0^x \frac{e^{i\beta t}}{\sqrt{t}} \, dt \right]$$

$$= e^{-i\beta x} \left[\frac{\pi}{2\sqrt{\beta}} + \frac{iA}{2} \int_0^\infty \frac{e^{i\beta t}}{\sqrt{t}} \, dt - \int_x^\infty \frac{e^{i\beta t}}{\sqrt{t}} \, dt \right]$$

$$= e^{-i\beta x} \left[\frac{\pi}{2\sqrt{\beta}} + \frac{iA}{2} \int_0^\infty \frac{e^{i\beta t}}{\sqrt{t}} \, dt + \frac{e^{i\beta x}}{i\beta\sqrt{x}} - \frac{1}{2i\beta} \int_x^\infty \frac{e^{i\beta t}}{t^{3/2}} \, dt \right].$$

The last integral behaves as $e^{i\beta x}/x^{3/2}$. Therefore, for $x > 0$

$$I(x) = -\sqrt{k^2 - k^2 \cos \alpha} \, e^{ikx \cos \alpha}$$
$$- \frac{2}{\pi} \sqrt{k + k \cos \alpha} \, e^{ikx} \left[\frac{A}{2\sqrt{x}} - (k - k \cos \alpha)h(x) \right],$$

which shows that as $x \to \infty$, $I(x)$ behaves as $e^{ikx \cos \alpha}$. Hence,

$$\delta = k_2 \cos \alpha < k_2,$$

justifying the assumption made in the derivation of the integral equation.

To determine the acoustic field, we must substitute $I(x)$ in the expressions

$$\phi^{(1)}(\xi, \eta) = -\frac{i}{2} \int_{-\infty}^\infty H_0^{(1)}(k\sqrt{(x - \xi)^2 + \eta^2}) \, I(x) \, dx$$

$$\phi^{(2)}(\xi, \eta) - e^{ik(\xi \cos \alpha + \eta \sin \alpha)} - e^{ik(\xi \cos \alpha - \eta \sin \alpha)}$$

$$= \frac{i}{2} \int_{-\infty}^\infty H_0^{(1)}(k\sqrt{(x - \xi)^2 + \eta^2}) \, I(x) \, dx.$$

By the convolution theorem the transform of

$$\frac{i}{2} \int_{-\infty}^{\infty} H_0^{(1)}(k\sqrt{(x - \xi)^2 + \eta^2})\, I(x)\, dx$$

is

$$\frac{i}{\sqrt{k^2 - z^2}}\, e^{i\sqrt{k^2-z^2}|\eta|}\, U(z) = \frac{-1}{\sqrt{2\pi}}\, \frac{\sqrt{k + k \cos \alpha}\; e^{i\sqrt{k^2-z^2}|\eta|}}{\sqrt{k + z}(z - k \cos \alpha)}$$

and

$$\frac{i}{2} \int_{-\infty}^{\infty} H_0^{(1)}(k\sqrt{(x - \xi)^2 + \eta^2})\, I(x)\, dx$$

$$= \frac{-1}{2\pi} \int_{-\infty+i\gamma}^{\infty+i\gamma} \frac{e^{iz\xi}\sqrt{k + k \cos \alpha}\; e^{i\sqrt{k^2-z^2}|\eta|}}{\sqrt{k + z}\,(z - k \cos \alpha)}\, dz.$$

One can obtain the properties of the acoustic field by investigating this contour integral. We shall not carry out the details.

Exercises 8.7

1. Show that $H_0^{(1)}(k|x|)$ satisfies the equation $xy'' + y' + k^2xy = 0$ for all $x \neq 0$. Find the Fourier transform of $H_0^{(1)}(k|x|)$ to within a constant of integration.

2. Derive the appropriate Wiener-Hopf integral equation for the diffraction of a plane wave by a half-plane if the boundary condition on the plane is $u(x, 0) = 0$.

CHAPTER 9

Laplace Transforms

9.1. FROM FOURIER TO LAPLACE TRANSFORM

Let $f(t)$ be a complex-valued function of the real variable t. Let $f(t)$ be piecewise continuous and have a piecewise continuous derivative in any finite interval. Let $|f(t)| \leq Ke^{-bt}$, $0 \leq t < \infty$, and $|f(t)| \leq Me^{-at}$, $-\infty < t \leq 0$, $a < b$. Then the complex Fourier transform

$$G(z) = \frac{1}{\sqrt{2\pi}} \int_{-\infty}^{\infty} f(t)e^{-itz}\, dt$$

is an analytic function of z for $a < \operatorname{Im}(z) < b$. Furthermore, the inverse transform exists and

$$\frac{f(t^+) + f(t^-)}{2} = \frac{1}{\sqrt{2\pi}} \int_{-\infty+i\gamma}^{\infty+i\gamma} G(z)e^{itz}\, dz,$$

for $a < \gamma < b$. Suppose we perform a simple rotation of the complex z plane through 90°, that is, $\zeta = iz$. Then

$$H(\zeta) = \sqrt{2\pi}\, G(-i\zeta) = \int_{-\infty}^{\infty} f(t)e^{-t\zeta}\, dt$$

is analytic for $-b < \operatorname{Re}(\zeta) < -a$. The inverse is

$$\frac{f(t^+) + f(t^-)}{2} = \frac{1}{2\pi i} \int_{-\gamma-i\infty}^{-\gamma+i\infty} H(\zeta)e^{t\zeta}\, d\zeta.$$

$H(\zeta)$ is usually called the *two-sided Laplace transform*. There is a complete theory for it which parallels very closely that for the Fourier transform. There is considerably more interest in the one-sided Laplace transform or simply Laplace transform.

Definition 9.1.1. Let $f(t)$ be a complex-valued function of the real variable t which is zero for t negative. Let $f(t)$ be piecewise continuous in any finite interval and $|f(t)| \leq Ke^{bt}$, $0 \leq t < \infty$. Then the Laplace transform of $f(t)$ is

$$H(z) = L[f] = \int_0^\infty f(t)e^{-zt}\, dt.$$

Theorem 9.1.1. *Let $f(t)$ be a complex-valued function of the real variable t which is zero for t negative. Let $f(t)$ be piecewise continuous and have a piecewise continuous derivative in any finite interval. Let $|f(t)| \leq Ke^{bt}$, $0 \leq t < \infty$. Then the Laplace transform of $f(t)$ is analytic for $b < \text{Re}(z)$ and*

$$\frac{f(t^+) + f(t^-)}{2} = \frac{1}{2\pi i}\int_{\gamma - i\infty}^{\gamma + i\infty} H(z)e^{zt}\, dz,$$

where the integration is carried out along a line $\gamma + iy$, $-\infty < y < \infty$, with $b < \gamma$.

Proof: This obviously follows from our work on Fourier transforms upon performing the rotation mentioned above.

EXAMPLE 9.1.1. Find the Laplace transform of $f(t) = u(t)\sin \omega t$, where $u(t)$ is the unit step function, and verify the inversion.

$$L[u(t)\sin \omega t] = \frac{1}{2i}\int_0^\infty (e^{i\omega t} - e^{-i\omega t})e^{-zt}\, dt$$

$$= \frac{1}{2i}\left[\frac{1}{z - i\omega} - \frac{1}{z + i\omega}\right]$$

$$= \frac{\omega}{z^2 + \omega^2},$$

provided $0 < \text{Re}(z)$. The function $\omega/(z^2 + \omega^2)$, analytic except at $\pm i\omega$, is the analytic continuation of the Laplace transform to the rest of the plane. The inversion integral can be evaluated by contour integration

$$\frac{\omega}{2\pi i}\int_{\gamma - i\infty}^{\gamma + i\infty} \frac{e^{zt}}{z^2 + \omega^2}\, dz = \frac{\omega}{2\pi i}\int_C \frac{e^{zt}}{z^2 + \omega^2}\, dz,$$

where C is the contour shown in Figure 9.1.1. There are two simple poles inside the contour at $i\omega$ and $-i\omega$. Therefore, by the residue theorem

$$\frac{\omega}{2\pi i}\int_C \frac{e^{zt}}{z^2 + \omega^2}\, dz = \omega\left[\frac{e^{i\omega t} - e^{-i\omega t}}{2i\omega}\right] = \sin \omega t.$$

We still have to show that the integral over the semicircular contour goes to zero as R approaches infinity. Let $z = \gamma + Re^{i\theta}$, $\pi/2 \leq \theta \leq 3\pi/2$.

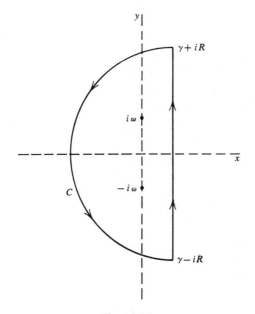

Figure 9.1.1.

Then

$$\left| \int_{\pi/2}^{3\pi/2} \frac{e^{\gamma t} e^{tR(\cos\theta + i\sin\theta)} Re^{i\theta}}{(\gamma + Re^{i\theta})^2 + \omega^2} \, d\theta \right| \leq \frac{\pi e^{\gamma t} R}{R^2 - \gamma^2 - \omega^2} \to 0,$$

as $R \to \infty$.

It may be possible to relax the hypotheses of Theorem 9.1.1 in some special cases.

EXAMPLE 9.1.2.   Find the Laplace transform of $u(t)t^c$, $c > -1$. If $-1 < c < 0$, then the given function approaches infinity as t approaches zero from the right. Nevertheless, the Laplace transform exists and is analytic for $\mathrm{Re}(z) > 0$, since

$$L[u(t)t^c] = \int_0^\infty t^c e^{-zt} \, dt$$

and the integral is analytic for $\mathrm{Re}(z) > 0$. To evaluate the integral let $z = x$ and let $s = xt$. Then

$$\int_0^\infty t^c e^{-zt} \, dt = x^{-c-1} \int_0^\infty s^c e^{-s} \, ds = \frac{\Gamma(c+1)}{x^{c+1}}.$$

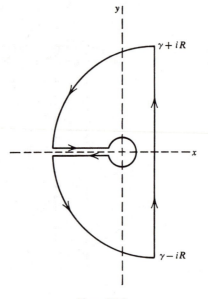

Figure 9.1.2.

Now the function $\Gamma(c + 1)/z^{c+1}$ is analytic for Re$(z) > 0$ if a branch is chosen with a branch cut along the negative real axis. Therefore,

$$L[u(t)t^c] = \int_0^\infty t^c e^{-zt}\, dt = \frac{\Gamma(c + 1)}{z^{c+1}},$$

since both sides of the equation are analytic in the same half-plane and agree along the positive real axis. To verify the inversion integral we have to integrate around a contour which does not cross the branch cut. (See Figure 9.1.2.) Since there are no singularities inside C

$$\frac{\Gamma(c + 1)}{2\pi i} \int_C \frac{e^{zt}}{z^{c+1}}\, dz = 0.$$

It can be shown that the contributions of the circular arcs of radius R go to zero as R approaches infinity. Therefore

$$\frac{\Gamma(c + 1)}{2\pi i} \int_{\gamma-i\infty}^{\gamma+i\infty} \frac{e^{zt}}{z^{c+1}}\, dz = \frac{\Gamma(c + 1)}{2\pi i} \int_\Gamma \frac{e^{zt}}{z^{c+1}}\, dz,$$

where Γ is the same hook type contour used in the Hankel integral of Section 4.10. Since $t > 0$ we can make the change of variable $\zeta = zt$.

Then

$$\frac{\Gamma(c+1)}{2\pi i} \int_{\gamma-i\infty}^{\gamma+i\infty} \frac{e^{zt}}{z^{c+1}}\,dz = t^c\,\frac{\Gamma(c+1)}{2\pi i} \int_\Gamma e^\zeta \zeta^{-(c+1)}\,d\zeta = t^c,$$

since $\dfrac{1}{2\pi i}\displaystyle\int_\Gamma e^\zeta \zeta^{-(c+1)}\,d\zeta$ is the Hankel integral for $\dfrac{1}{\Gamma(c+1)}$.

Exercises 9.1

1. Find the Laplace transform of the unit step function; $u(t) = 1, t \geq 0$, $u(t) = 0, t < 0$; and verify the inversion integral for both $t > 0$ and $t < 0$. What is the value of the inversion integral when $t = 0$?

2. Find the Laplace transform of $f(t) = u(t) \cos \omega t$ and verify the inversion integral.

3. Under the hypotheses of Theorem 9.1.1, the Laplace transform of $f(t)$ is analytic for $\text{Re}(z) > b$, and therefore can be differentiated under the integral sign. Show that

$$-\frac{d}{dz} L[f(t)] = L[tf(t)].$$

Use this result to obtain the Laplace transform of $tu(t) \sin \omega t$.

9.2. PROPERTIES OF THE LAPLACE TRANSFORM

Before taking up some of the more important applications of the Laplace transform, we must study its properties and some of the techniques for inverting the transform.

Theorem 9.2.1. *The Laplace transform is a linear transformation; that is, if $f(t)$ and $g(t)$ are both functions with Laplace transforms and a and b are any constants, then*

$$L[af + bg] = aL[f] + bL[g].$$

Proof:

$$L[af + bg] = \int_0^\infty [af(t) + bg(t)]e^{-zt}\,dt$$

$$= a \int_0^\infty f(t)e^{-zt}\,dt + b \int_0^\infty g(t)e^{-zt}\,dt$$

$$= aL[f] + bL[g].$$

Theorem 9.2.2. *If* $f(t)$ *is continuous for* $t > 0$, $\lim\limits_{t \to 0+} f(t) = f(0^+)$ *exists,* $f'(t)$ *is piecewise continuous in any finite interval, and* $|f'(t)| \leq Ke^{bt}$, $0 \leq t < \infty$, *then*

$$L[f'] = zL[f] - f(0^+)$$

holds in $\text{Re}(z) > a$, *where* $a = \max[0, b]$.

Proof: For $t > 0$

$$f(t) = f(0^+) + \int_0^t f'(\tau)\, d\tau,$$

$$|f(t)| \leq |f(0^+)| + \int_0^t |f'(\tau)|\, d\tau$$

$$\leq |f(0^+)| + \int_0^t Ke^{b\tau}\, d\tau$$

$$\leq |f(0^+)| + K\frac{e^{bt} - 1}{b} \leq Me^{at},$$

where $a = \max[0, b]$. Therefore, $L[f]$ and $L[f']$ exist and are analytic for $\text{Re}(z) > a$. Let $z = x > a$. Then

$$\int_0^\infty f'(t)e^{-xt}\, dt = -\lim_{\epsilon \to 0+} f(\epsilon)e^{-x\epsilon} + \lim_{T \to \infty} f(T)e^{-xT} + x\int_0^\infty f(t)e^{-xt}\, dt$$

$$= -f(0^+) + x\int_0^\infty f(t)e^{-xt}\, dt.$$

Hence, $L[f'] = zL[f] - f(0^+)$, since both sides are analytic for $\text{Re}(z) > a$ and are equal on the real axis.

Corollary 9.2.1. Let $f(t)$ satisfy the hypotheses of Theorem 9.2.2. Then there is a constant M and a half-plane $\text{Re}(z) \geq \rho$ such that $|L[f]| \leq M/|z|$.

Proof: Referring to Theorem 9.2.2, $L[f']$ and $L[f]$ exist for $\text{Re}(z) > a$. Let $\rho > a$. Then

$$|L[f']| \leq \int_0^\infty Ke^{bt}e^{-xt}\, dt = \frac{K}{x - b} \leq \frac{K}{\rho - b}.$$

Let $M = 2\max\left[\dfrac{K}{\rho - b}, |f(0^+)|\right]$. Then

$$|L[f]| \leq \frac{|f(0^+)| + |L[f']|}{|z|}$$

$$\leq \frac{|f(0^+)| + K/(\rho - b)}{|z|} \leq \frac{M}{|z|}.$$

Theorem 9.2.3. *Let $f(t)$ and its first $n - 1$ derivatives be continuous for $t > 0$. Let $\lim_{t \to 0^+} f(t) = f(0^+)$, $\lim_{t \to 0^+} f'(t) = f'(0^+)$, ..., $\lim_{t \to 0^+} f^{(n-1)}(t) = f^{(n-1)}(0^+)$, all exist. Let $f^{(n)}(t)$ be piecewise continuous in any finite interval and $|f^{(n)}(t)| \le Ke^{bt}$, for $0 \le t < \infty$. Then in the half-plane $\mathrm{Re}(z) > a$, where $a = \max[0, b]$, the following holds:*

$$L[f^{(n)}] = z^n L[f] - z^{n-1}f(0^+) - z^{n-2}f'(0^+)$$
$$- \cdots - zf^{(n-2)}(0^+) - f^{(n-1)}(0^+).$$

Proof: For $t > 0$

$$f^{(n-1)}(t) = f^{(n-1)}(0^+) + \int_0^t f^{(n)}(\tau)\, d\tau,$$

$$|f^{(n-1)}(t)| \le |f^{(n-1)}(0^+)| + \int_0^t Ke^b\, r\, d\tau$$

$$\le |f^{(n-1)}(0^+)| + K\frac{e^{bt} - 1}{b} \le Me^{at},$$

where $a = \max[0, b]$. By Theorem 9.2.2,

$$L[f^{(n)}] = zL[f^{(n-1)}] - f^{(n-1)}(0^+)$$

in the half-plane $\mathrm{Re}(z) > a$. Similarly,

$$L[f^{(n-1)}] = zL[f^{(n-2)}] - f^{(n-2)}(0^+)$$
$$L[f^{(n)}] = z\{zL[f^{(n-2)}] - f^{(n-2)}(0^+)\} - f^{(n-1)}(0^+)$$
$$= z^2 L[f^{(n-2)}] - zf^{(n-2)}(0^+) - f^{(n-1)}(0^+).$$
$$L[f^{(n)}] = z^n L[f] - z^{n-1}f(0^+) - z^{n-2}f'(0^+)$$
$$- \cdots - z^{(n-2)}(0^+) - f^{(n-1)}(0^+).$$

Theorem 9.2.4. *Let $f(t)$ be piecewise continuous in any finite interval, and $|f(t)| \le Ke^{bt}$, $0 \le t < \infty$. Then $F(t) = \int_0^t f(\tau)\, d\tau$ is continuous for $t > 0$, $F(0) = 0$, and*

$$L[F] = \frac{1}{z} L[f]$$

in the half-plane $\mathrm{Re}(z) > a$, where $a = \max[0, b]$.

Proof: For $t > 0$

$$|F(t)| \le \int_0^t |f(\tau)|\, d\tau \le \int_0^t Ke^{b\tau}\, d\tau = K\frac{e^{bt} - 1}{b}$$

$$\le Me^{at},$$

where $a = \max [0, b]$. By Theorem 9.2.2,

$$L[f] = L[F'] = zL[F] - F(0) = zL[F].$$

Then for $\operatorname{Re}(z) > a$, $L[F] = \dfrac{1}{z} L[f]$.

Theorem 9.2.5. *Let $f(t)$ be piecewise continuous in any finite interval and $|f(t)| \leq Ke^{bt}$ for $0 \leq t < \infty$. Then if $H(z)$ is the Laplace transform of $f(t)$,*

$$L[e^{ct}f(t)] = H(z - c)$$

for $\operatorname{Re}(z) > b + c$.

Proof: The Laplace transform of $f(t)$ exists and is analytic for $\operatorname{Re}(z) > b$. Also, $|e^{ct}f(t)| \leq Ke^{(b+c)t}$. Hence, $L[e^{ct}f(t)]$ is analytic for $\operatorname{Re}(z) > b + c$. Let $z = x > b$. Then

$$\int_0^\infty e^{ct}f(t)e^{-xt}\, dt = \int_0^\infty f(t)e^{-(x-c)t}\, dt = H(x - c).$$

Therefore, for $\operatorname{Re}(z) > b + c$, $L[e^{ct}f(t)] = H(z - c)$, since both sides of the equation are analytic in the same half-plane and agree on the real axis.

Theorem 9.2.6. *Let $f(t)$ be a function with a Laplace transform for $\operatorname{Re}(z) > b$. Let $u(t)$ be the unit step function, $u(t) = 1$ for $t \geq 0$, $u(t) = 0$, for $t < 0$. Then if $a > 0$*

$$L[u(t - a)f(t - a)] = e^{-az}L[f(t)].$$

Proof:

$$L[u(t - a)f(t - a)] = \int_a^\infty f(t - a)e^{-zt}\, dt = \int_0^\infty f(\tau)e^{-z(\tau+a)}\, d\tau$$

$$= e^{-az}\int_0^\infty f(\tau)e^{-z\tau}\, d\tau = e^{-az}L[f(t)].$$

Theorem 9.2.7. *Let $f(t)$ and $g(t)$ be zero for $t < 0$, continuous for $t > 0$, $|f(t)| \leq Ke^{bt}$, and $|g(t)| \leq Ke^{bt}$, $0 \leq t < \infty$. Then the convolution*

$$f * g = h(t) = \int_0^t f(\tau)g(t - \tau)\, d\tau = \int_0^t g(\tau)f(t - \tau)\, d\tau$$

is zero for $t < 0$, continuous, and satisfies $|h(t)| \leq Me^{at}$, for $0 \leq t < \infty$, where M is a constant and $a > b$. Moreover,

$$L[h] = L[f]L[g].$$

Proof: That $h(t) = 0$ for $t < 0$ is obvious from the fact that $f(t) = 0$ for $t < 0$. Also,

$$\lim_{t \to 0^+} h(t) = \lim_{t \to 0^+} \int_0^t f(\tau)g(t - \tau)\, d\tau = 0,$$

since $f(\tau)$ and $g(t - \tau)$ are bounded for sufficiently small τ. For $t > 0$ the integral is a continuous function of the upper limit and of the parameter. Now, consider

$$\left| e^{-at}h(t) \right| = e^{-at} \left| \int_0^t f(\tau)g(t - \tau)\, d\tau \right|$$

$$\leq e^{-at} \int_0^t K^2 e^{b\tau} e^{b(t-\tau)}\, d\tau$$

$$= K^2 e^{(b-a)t} \int_0^t d\tau = K^2 t e^{(b-a)t} \to 0,$$

as $t \to \infty$ provided $a > b$. Therefore, there exists an M such that

$$|h(t)| \leq M e^{at}.$$

Let $\operatorname{Re}(z) > b$. Then there exists an $a > b$ such that $\operatorname{Re}(z) > a > b$. Hence, $L[h]$ is analytic at z. This proves that $L[h]$ is analytic for $\operatorname{Re}(z) > b$. Also, $L[f]$ and $L[g]$ are analytic for $\operatorname{Re}(z) > b$. Therefore, to prove that $L[h] = L[f]L[g]$ for $\operatorname{Re}(z) > b$, we have to show that the equality holds on the real axis $z = x > b$. Using the fact that $g(t - \tau) = 0$, for $\tau > t$, we have

$$\int_0^\infty h(t)e^{-xt}\, dt = \lim_{T \to \infty} \int_0^T \int_0^t f(\tau)e^{-xt}g(t - \tau)\, d\tau\, dt$$

$$= \lim_{T \to \infty} \int_0^\infty f(\tau) \int_0^T e^{-xt}g(t - \tau)\, dt\, d\tau$$

$$= \lim_{T \to \infty} \int_0^\infty f(\tau)e^{-x\tau} \int_0^{T-\tau} g(u)e^{-xu}\, du\, d\tau$$

$$= \int_0^\infty f(\tau)e^{-x\tau}\, d\tau \int_0^\infty g(u)e^{-xu}\, du.$$

This completes the proof.

We conclude this section with a number of examples showing how these properties can be used to find the Laplace transforms of some elementary functions.

EXAMPLE 9.2.1. Let $u(t)$ be the unit step function. Then $L[u(t)] = 1/z$. Since $|u(t)| \leq 1$, we have $b = 0$ and $L[u(t)]$ is analytic for $\text{Re}(z) > 0$. The function $1/z$ is analytic for $\text{Re}(z) > 0$, and

$$\int_0^\infty u(t)e^{-zt}\, dt = \int_0^\infty e^{-zt}\, dt = \frac{1}{x},$$

for $x > 0$.

EXAMPLE 9.2.2. Let $f(t) = u(t)t^n$, $n = 1, 2, 3, \ldots$. Prove that

$$L[f(t)] = n!\, z^{-(n+1)}.$$

We have $f'(t) = nu(t)t^{n-1}$, $f''(t) = n(n - 1)u(t)$, $t^{n-2}, \ldots, f^{(n-1)}(t) = n!\, u(t)t$, $f^{(n)}(t) = n!\, u(t)$. Therefore, $f, f', f'', \ldots, f^{(n-1)}$ are continuous and $f^{(n)}$ is piecewise continuous. By Theorem 9.2.3,

$$z^n L[f] = L[f^{(n)}] = L[n!\, u(t)] = n!\, z^{-1},$$
$$L[f] = n!\, z^{-(n+1)}.$$

EXAMPLE 9.2.3. Let $f(t) = u(t)\cos \omega t$. Prove that $L[f(t)] = z/(z^2 + \omega^2)$. We have for $t > 0$, $(1/\omega)\sin \omega t = \int_0^t u(\tau)\cos \omega\tau\, d\tau$. Therefore, by Theorem 9.2.4 and Example 9.1.1,

$$L\left[\frac{u(t)\sin \omega t}{\omega}\right] = \frac{1}{z} L[f(t)],$$

$$\frac{z}{z^2 + \omega^2} = L[f(t)].$$

EXAMPLE 9.2.4. Let $f(t) = u(t)e^{at}$. Prove that $L[f(t)] = (z - a)^{-1}$. We have that $|f(t)| \leq e^{at}$. Therefore, $L[f(t)]$ is analytic for $\text{Re}(z) > a$, as is $(z - a)^{-1}$. Also,

$$\int_0^\infty e^{at}e^{-zt}\, dt = \frac{1}{x - a}.$$

Therefore, the equality holds in the half-plane $\text{Re}(z) > a$.

EXAMPLE 9.2.5. Find the Laplace transform of $u(t)e^{at}\sin \omega t$. By Theorem 9.2.5 and Example 9.1.1,

$$L[u(t)e^{at}\sin \omega t] = \frac{\omega}{(z - a)^2 + \omega^2}.$$

EXAMPLE 9.2.6. Find the Laplace transform of $u(t - a)\cos \omega(t - a)$, where $a > 0$. By Theorem 9.2.6 and Example 9.2.3,

$$L[u(t - a)\cos \omega(t - a)] = \frac{ze^{-az}}{z^2 + \omega^2}.$$

EXAMPLE 9.2.7. Find the Laplace transform of $u(t)te^{at}$. We have that $L[u(t)e^{at}] = (z - a)^{-1}$, analytic for $\mathrm{Re}(z) > a$. The transform can be differentiated under the integral sign. Hence,

$$-\frac{d}{dz}\left(\frac{1}{z-a}\right) = \frac{1}{(z-a)^2} = -\frac{d}{dz}\int_0^\infty e^{at}e^{-zt}\,dt$$

$$= -\int_0^\infty e^{at}\frac{d}{dz}e^{-zt}\,dt$$

$$= \int_0^\infty te^{at}e^{-zt}\,dt = L[u(t)te^{at}].$$

EXAMPLE 9.2.8. Find the Laplace transform of $u(t)J_0(kt)$, where $J_0(kt)$ is the Bessel function of order zero. Now, $J_0(kt)$ is the unique solution of

$$tJ_0''(kt) + J_0'(kt) + k^2tJ_0(kt) = 0,$$

$J_0(0) = 1$, $J_0'(0) = 0$. Taking the transform term by term

$$-\frac{d}{dz}[z^2G(z) - z] + zG(z) - 1 - k^2\frac{d}{dz}G(z) = 0,$$

where $G(z) = L[u(t)J_0(kt)]$. Hence, G satisfies the differential equation

$$G' + \frac{z}{k^2 + z^2}G = 0,$$

which can be solved to give

$$G(z) = \frac{C}{\sqrt{k^2 + z^2}},$$

where C is a constant of integration and the square root is a branch which is analytic in the right half-plane $\mathrm{Re}(z) > 0$ and is positive on the real axis. To evaluate C, we recall that $J_0'(kt) = -kJ_1(kt)$ and that $J_1(kt)$ is bounded for all t. Therefore, by Corollary 9.2.1

$$\lim_{x\to\infty} L[-ku(t)J_1(kt)] = \lim_{x\to\infty}[xG(x) - J_0(0)]$$

$$= \lim_{x\to\infty}\left[\frac{Cx}{\sqrt{k^2 + x^2}} - 1\right] = 0.$$

Therefore, $C = 1$ and

$$L[u(t)J_0(kt)] = \frac{1}{\sqrt{k^2 + z^2}}.$$

Incidentally, we have also proved that

$$L[u(t)J_1(kt)] = \frac{1}{k}\left(1 - \frac{z}{\sqrt{k^2 + z^2}}\right).$$

Exercises 9.2

1. Using the general properties of the Laplace transform find the transforms of the following functions:

(a) $u(t)t^n e^{at}$, $n = 1, 2, 3, \ldots$,   Ans: $n!/(z - a)^{n+1}$;

(b) $u(t)\cosh \omega t$,   Ans: $z/(z^2 - \omega^2)$;

(c) $tu(t - a)e^{bt}$, $a > 0$,   Ans: $[e^{ab}e^{-az}/(z - b)^2] + [(ae^{ab}e^{-az}/(z - b)]$;

(d) $\int_0^t e^{-\tau} \cos \omega(t - \tau)\, d\tau$,   Ans: $z/(z - 1)(z^2 + \omega^2)$.

2. Let $G(z) = L[f(t)]$, $\mathrm{Re}(z) > b$.   Then $H(x) = \int_x^\infty G(t)\, dt$, $x > b$, where the integration is along the real axis, is a function of x with all derivatives on the x axis. Let $H(z)$ be the analytic continuation of $H(x)$ into the half-plane $\mathrm{Re}(z) > b$. Prove that $H(z) = L[f(t)/t]$ provided that $\lim_{t \to 0^+} f(t)/t$ exists.

3. Use the result of Problem 2 to find the Laplace transform of

(a) $u(t) \sin \omega t/t$,   Ans: $\tan^{-1}(\omega/z)$;

(b) $u(t) \sinh \omega t/t$,   Ans: $\frac{1}{2}\log\left(\dfrac{z + 1}{z - 1}\right)$;

(c) $u(t)J_1(kt)/t$,   Ans: $\frac{1}{k}(\sqrt{k^2 + z^2} - z)$.

4. Let $f(t)$ satisfy the hypotheses of Theorem 9.1.1 and for some $\tau > 0$, $f(t + \tau) = f(t)$, for $t > 0$. Show that

$$L[f] = \frac{1}{1 - e^{-\tau z}} \int_0^\tau e^{-zt} f(t)\, dt.$$

Find the Laplace transform of $f(t) = u(t)[t]$, where $[t]$ is the greatest integer in t.

9.3. INVERSION OF LAPLACE TRANSFORMS

In a sense the inversion problem is solved by the inversion integral which was stated in Theorem 9.1.1. However, in practice the art of inverting the Laplace transform is usually the ability to recognize certain general types as the transforms of known functions. As an aid in this art we have the following uniqueness theorem.

Theorem 9.3.1. *Let $f(t)$ and $g(t)$ both satisfy the hypotheses of Theorem 9.1.1 and both have the same Laplace transform $G(z)$. Then $f(t)$ and $g(t)$ may differ at only a finite number of points in any finite interval. If $f(t)$ and $g(t)$ are continuous then $f(t) \equiv g(t)$.*

Proof: Since f and g both have the same transform

$$\frac{f(t^+) - f(t^-)}{2} = \frac{g(t^+) - g(t^-)}{2}.$$

But in any finite interval, f and g have but a finite number of discontinuities. At a point where f and g are continuous $f(t) = g(t)$.

In many applications the transform is a rational function, that is,

$$L[f] = \frac{P(z)}{Q(z)}.$$

If f is a function which satisfies the hypothesis of Theorem 9.2.2, then by Corollary 9.2.1 the degree of $Q(z)$ must be greater than the degree of $P(z)$. We can also assume that P and Q have no common zeros since the transform must be analytic in a right half-plane. We shall prove a set of theorems showing how to invert transforms of this type based on the partial fraction expansions of Section 4.6.

Theorem 9.3.2. *Let $f(t)$ satisfy the hypotheses of Theorem 9.2.2,*

$$f(0) = \lim_{t \to 0^+} f(t),$$

and $L[f] = P(z)/Q(z)$, where $Q(z)$ has simple zeros $r_1, r_2, \ldots, r_n$. Then

$$f(t) = u(t) \sum_{k=1}^{n} A_k e^{r_k t},$$

where $u(t)$ is the unit step function, and

$$A_k = \lim_{z \to r_k} (z - r_k)P(z)/Q(z).$$

Proof: We can write

$$L[f] = P(z)/Q(z) = \sum_{k=1}^{n} A_k/(z - r_k)$$

by Theorem 4.6.6. Since $A_k/(z - r_k) = L[u(t)A_k e^{r_k t}]$ the result follows from Theorem 9.3.1.

Theorem 9.3.3. *Let $f(t)$ satisfy the hypotheses of Theorem 9.2.2,*

$$f(0) = \lim_{t \to 0+} f(t),$$

and $L[f] = P(z)/Q(z)$, *where* $Q(z)$ *has roots* $r_1, r_2, \ldots, r_m$ *with multiplicities* $k_1, k_2, \ldots, k_m$, *respectively, such that* $\sum_{j=1}^{m} k_j = n$. *Then*

$$f(t) = u(t) \sum_{j=1}^{m} f_j(t),$$

$$f_j(t) = \sum_{i=1}^{k_j} A_{ij} t^{i-1} e^{r_j t},$$

$$A_{ij} = \lim_{z \to r_j} \frac{1}{(k_j - i)!} \frac{d^{k_j - i}}{dz^{k_j - i}} [(z - r_j)^{k_j} P(z)/Q(z)].$$

Proof: We can write

$$L[f] = P(z)/Q(z) = \sum_{j=1}^{m} \sum_{i=1}^{k_j} A_{ij}/(z - r_j)^i,$$

by Theorem 4.6.6. Since $A_{ij}/(z - r_j)^i = L[u(t) A_{ij} t^{i-1} e^{r_j t}]$, the result follows from Theorem 9.3.1.

EXAMPLE 9.3.1. Find the function satisfying the hypotheses of Theorem 9.3.3 which has the Laplace transform

$$L[f] = \frac{z + 1}{(z - 1)^2 (z^2 + 2z + 2)}.$$

Here, $P(z) = z + 1$ and $Q(z) = (z - 1)^2 (z^2 + 2z + 2) = (z - 1)^2 \times (z + 1 - i)(z + 1 + i)$. Therefore, $r_1 = 1, r_2 = -1 + i, r_3 = -1 - i$, with $k_1 = 2, k_2 = 1, k_3 = 1$. Then

$$A_{11} = \lim_{z \to 1} \frac{d}{dz} \left(\frac{z + 1}{z^2 + 2z + 2} \right) = -\frac{3}{25},$$

$$A_{21} = \lim_{z \to 1} \frac{z + 1}{z^2 + 2z + 2} = \frac{2}{5},$$

$$A_{12} = \lim_{z \to -1+i} \frac{z + 1}{(z - 1)^2 (z + 1 + i)} = \frac{3 + 4i}{50},$$

$$A_{13} = \lim_{z \to -1-i} \frac{z + 1}{(z - 1)^2 (z + 1 - i)} = \frac{3 - 4i}{50},$$

and

$$f(t) = u(t)\left[-\frac{3}{25}e^t + \frac{2}{5}te^t + \frac{3+4i}{50}e^{-t+it} + \frac{3-4i}{50}e^{-t-it}\right]$$

$$= u(t)\left[-\frac{3}{25}e^t + \frac{2}{5}te^t + \frac{3}{25}e^{-t}\cos t - \frac{4}{25}e^{-t}\sin t\right].$$

In this example we could have anticipated the terms $e^{-t}\cos t$ and $e^{-t}\sin t$ for the following reason:

$$\frac{az+b}{z^2+2z+2} = \frac{az+b}{z^2+2z+1+1} = \frac{a(z+1)}{(z+1)^2+1} + \frac{b-a}{(z+1)^2+1}$$

$$= aL[u(t)e^{-t}\cos t] + (b-a)L[u(t)e^{-t}\sin t].$$

As a matter of fact, Corollary 4.6.2 tells us that in the partial fraction expansion of a rational function $P(z)/Q(z)$, where P and Q have real coefficients, there is a term of the form

$$\frac{Az+B}{(z-a)^2+b^2},$$

corresponding to the nonrepeated roots $a + ib$ of $Q(z)$, and

$$A = \text{Im}\left(\frac{1}{b}\lim_{z\to a+ib}\{[(z-a)^2+b^2]P(z)/Q(z)\}\right),$$

$$B = \text{Re}\left(\lim_{z\to a+ib}\{[(z-a)^2+b^2]P(z)/Q(z)\}\right) - aA.$$

Clearly, A and B are real, and therefore

$$\frac{Az+B}{(z-a)^2+b^2} = L\left[\frac{A(z-a)+B+aA}{(z-a)^2+b^2}\right]$$

$$= L[u(t)Ae^{at}\cos bt] + L\left[u(t)\frac{B+aA}{b}e^{at}\sin bt\right].$$

Theorem 9.3.4. *Let $f(t)$ satisfy the hypotheses of Theorem 9.2.2,*

$$f(0) = \lim_{t\to 0+}f(t),$$

and $L[f] = P(z)/Q(z)$, where P and Q have real coefficients and Q has nonrepeated roots $a \pm ib$. Then $f(t)$ contains terms

$$u(t)Ae^{at}\cos bt + u(t)A'e^{at}\sin bt$$

corresponding to the terms $C/(z - a - ib)$ *and* $D/(z - a + ib)$ *in the partial fraction expansion of* $P(z)/Q(z)$, *where*

$$A = \text{Im}\left(\frac{1}{b} \lim_{z \to a+ib} \{[(z - a)^2 + b^2]P(z)/Q(z)\}\right),$$

$$A' = \text{Re}\left(\frac{1}{b} \lim_{z \to a+ib} \{[(z - a)^2 + b^2]P(z)/Q(z)\}\right).$$

Proof: This result follows from the above discussion plus the observation that $(B + aA)/b = A'$.

EXAMPLE 9.3.2. Find the function satisfying the hypotheses of Theorem 9.3.4 which has the Laplace transform

$$L[f] = \frac{z - 1}{[(z - 2)^2 + 1][(z + 2)^2 + 4]}.$$

Let

$$L_1 = \lim_{z \to 2+i} \frac{z - 1}{(z + 2)^2 + 4} = \frac{27 + 11i}{525},$$

$$L_2 = \lim_{z \to -2+2i} \frac{z - 1}{(z - 2)^2 + 1} = \frac{-71 - 22i}{425}.$$

Let $A_1 = (1/b_1)\,\text{Im}(L_1)$, $A_1' = (1/b_1)\,\text{Re}(L_1)$, $A_2 = (1/b_2)\,\text{Im}(L_2)$, and $A_2' = (1/b_2)\,\text{Re}(L_2)$. Then

$$f(t) = u(t)\left[\frac{27}{525} e^{2t} \cos t + \frac{11}{525} e^{2t} \sin t - \frac{71}{850} e^{-2t} \cos 2t - \frac{11}{425} e^{-2t} \sin 2t\right].$$

There is a theorem corresponding to Theorem 9.3.4 for repeated quadratic factors in $Q(z)$ but we shall not go into that here.

Many transforms can be recognized as transforms of known functions using properties derived in the last section.

EXAMPLE 9.3.3. Find a function with Laplace transform $\log(z - a) - \log(z - b)$, where a and b are real and branches are chosen which are analytic in a right half-plane. Referring to Exercise 9.2.2, we have

$$\frac{d}{dx}[\log(x - a) - \log(x - b)] = \frac{1}{x - a} - \frac{1}{x - b}.$$

Now,

$$\frac{1}{z - b} - \frac{1}{z - a} = L[u(t)(e^{bt} - e^{at})] = G(z).$$

Hence,

$$H(x) = \int_x^\infty G(t)\,dt = \log(x - a) - \log(x - b).$$

Therefore, using the result of Exercise 9.2.2, we have

$$\log (z - a) - \log (z - b) = L\left[u(t)\frac{e^{bt} - e^{at}}{t}\right].$$

EXAMPLE 9.3.4. Find a function with Laplace transform $1/z\sqrt{z^2 + k^2}$, where $k > 0$ and a branch of the square root is chosen which is analytic in the half-plane $\text{Re}(z) > 0$ and takes on the value $\sqrt{x^2 + k^2}$ on the real axis. We can write

$$\frac{1}{z\sqrt{z^2 + k^2}} = L[u(t)]L[u(t)J_0(kt)]$$

$$= L\left[u(t)\int_0^t J_0(k\tau)\, d\tau\right].$$

Hence, the desired function is $u(t)\int_0^t J_0(k\tau)\, d\tau$.

We conclude this section with a brief table of Laplace transforms

$G(z) = L[f(t)]$		$f(t)$
1. $1/z^c$,	$c < 0$	$u(t)t^{c-1}/\Gamma(c)$
2. $1/(z - a)$		$u(t)e^{at}$
3. $1/(z - a)^k$,	$k > 0$	$u(t)t^{k-1}e^{at}/\Gamma(k)$
4. $1/(z^2 + \omega^2)$,	$\omega > 0$	$u(t)\sin \omega t/\omega$
5. $z/(z^2 + \omega^2)$,	$\omega > 0$	$u(t)\cos \omega t$
6. $1/(z^2 - \omega^2)$,	$\omega > 0$	$u(t)\sinh \omega t/\omega$
7. $z/(z^2 - \omega^2)$,	$\omega > 0$	$u(t)\cosh \omega t$
8. $1/[(z - a)^2 + \omega^2]$,	$\omega > 0$	$u(t)e^{at}\sin \omega t/\omega$
9. $(z - a)/[(z - a)^2 + \omega^2]$,	$\omega > 0$	$u(t)e^{at}\cos \omega t$
10. $1/\sqrt{z^2 + k^2}$		$u(t)J_0(kt)$
11. $\tan^{-1}(\omega/z)$	$\omega > 0$	$u(t)\sin \omega t/t$
12. $\log (z - a) - \log (z - b)$		$u(t)[(e^{bt} - e^{at})/t]$
13. $\log (z^2 + \omega^2) - \log z^2$,	$\omega > 0$	$2u(t)[1 - \cos \omega t]/t$
14. $\log (z^2 - \omega^2) - \log z^2$,	$\omega > 0$	$2u(t)[1 - \cosh \omega t]/t$

Exercises 9.3

1. Find a function with Laplace transform

$$\frac{z}{(z + 1)(z - 2)^2(z^2 + 2z + 5)}.$$

2. Find a function with Laplace transform

$$\frac{z^2 + 2z + 5}{(z^2 + 2z + 2)^2}.$$

Hint: $L[tf(t)] = -\dfrac{d}{dz} L[f(t)]$.

3. Verify entries 13 and 14 in the table of Laplace transforms.

4. Let $L[f] = P(z)/[(z - a)^2 + b^2]^2$, where $P(z)$ has real coefficients and is of degree 3, and a and b are real. Show that

$$\frac{P(z)}{Q(z)} = \frac{A_1 z + b_1}{(z - a)^2 + b^2} + \frac{A_2 z^2 + b_2 z + C_2}{[(z - a)^2 + b^2]^2},$$

where A_1, B_1, A_2, B_2, and C_2 are real. Find the coefficients and $f(t)$ such that $P(z)/Q(z) = L[f(t)]$.

9.4. THE SOLUTION OF ORDINARY DIFFERENTIAL EQUATIONS

One of the most important applications of the Laplace transform is in the solution of linear ordinary differential equations. We begin this study by considering the general nth order linear equation with constant coefficients

$$a_n \frac{d^n y}{dt^n} + a_{n-1} \frac{d^{n-1} y}{dt^{n-1}} + \cdots + a_1 \frac{dy}{dt} + a_0 y = f(t),$$

where $f(t)$ is defined for $t \geq 0$ and has a Laplace transform. We seek a solution $y(t)$ which is n times continuously differentiable for $t > 0$ and for which $\lim_{t \to 0+} y^{(k)}(t) = b_k$, $k = 0, 1, 2, \ldots, n - 1$. If we take the Laplace transform of the jth derivative, we have

$$L[y^{(j)}(t)] = z^j Y(z) - z^{j-1} b_0 - z^{j-2} b_1 - \cdots - z b_{j-2} - b_{j-1}$$

$$= z^j Y(z) - \sum_{i=0}^{j-1} b_i z^{j-1-i},$$

where $Y(z) = L[y(t)]$. Transforming the entire equation, we have

$$Y(z) \sum_{k=0}^{n} a_k z^k = F(z) + \sum_{k=0}^{n} a_k \sum_{i=0}^{k-1} b_i z^{k-1-i}$$

$$= F(z) + P(z),$$

where $F(z) = L[f(t)]$ and $P(z)$ is a polynomial of degree $n - 1$. Then

$$Y(z) = F(z)/Q(z) + P(z)/Q(z),$$

where $Q(z) = \sum\limits_{k=0}^{n} a_k z^k$. Let $1/Q(z) = L[q(t)]$, where $q(t)$, by Theorem 9.3.2, contains terms of the form $t^m e^{\alpha t}$, where m is a nonnegative integer. Then by the convolution theorem

$$y(t) = \int_0^t f(\tau)q(t - \tau)\, d\tau + L^{-1}[P(z)/Q(z)].$$

If we let $y_p(t) = \int_0^t f(\tau)q(t - \tau)\, d\tau$, we find that for $f(t)$ continuous for $t \geq 0$, $y_p(t)$ is a *particular solution* of the original differential equation satisfying $\lim\limits_{t \to 0^+} y_p^{(k)}(t) = 0$, $k = 0, 1, 2, \ldots, n - 1$. On the other hand $y_c(t) = L^{-1}[P(z)/Q(z)]$ is a solution of the homogeneous equation

$$a_n \frac{d^n y}{dt^n} + a_{n-1} \frac{d^{n-1} y}{dt^{n-1}} + \cdots + a_1 \frac{dy}{dt} + a_0 y = 0$$

satisfying $y_c^{(k)}(0) = b_k$, $k = 0, 1, 2, \ldots, n - 1$. By Theorem 9.3.2 $y_c(t)$ has terms of the form $t^m e^{\alpha t}$, where m is a nonnegative integer. We call $y_c(t)$ the *complementary solution*.

EXAMPLE 9.4.1. Consider the electrical circuit shown in Figure 9.4.1.

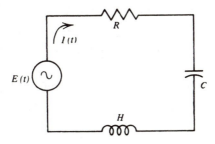

Figure 9.4.1.

The differential equation satisfied by the current $I(t)$ is

$$H \frac{d^2 I}{dt^2} + R \frac{dI}{dt} + \frac{I}{C} = \frac{d}{dt} E(t),$$

where the inductance H, the resistance R, and the capacitance C are constant and $E(t)$ is the input voltage. Let $E(t) = E_0 \sin \omega t$, where E_0 and ω are constants. Solve for $I(t)$ if $I(0) = I_0$ and $I'(0) = I_0'$. Transforming the equation, we have

$$H[z^2 G(z) - zI_0 - I_0'] + R[zG(z) - I_0] + \frac{1}{C} G(z) = \frac{\omega E_0 z}{z^2 + \omega^2},$$

where $G(z)$ is the Laplace transform of $I(t)$, Then

$$G(z) = \frac{\omega E_0 z}{(z^2 + \omega^2)\left(Hz^2 + Rz + \dfrac{1}{C}\right)} + \frac{HI_0 z + HI_0' + RI_0}{\left(Hz^2 + Rz + \dfrac{1}{C}\right)}.$$

The complementary solution is the part with Laplace transform

$$\frac{HI_0 z + HI_0' + RI_0}{Hz^2 + Rz + \dfrac{1}{C}} = \frac{I_0 z + I_0' + RI_0/H}{[z + (R/2H)]^2 + [(1/HC) - R^2/4H^2]}$$

$$= \frac{I_0 z + I_0' + RI_0/H}{(z + \alpha)^2 + \omega_0^2}.$$

We have to distinguish three cases:

Case I. *Underdamping.* $\omega_0^2 = 1/HC - R^2/4H^2 > 0.$

In this case,

$$\frac{I_0 z + I_0' + RI_0/H}{(z + \alpha)^2 + \omega_0^2} = \frac{I_0(z + \alpha)}{(z + \alpha)^2 + \omega_0^2} + \frac{[(I_0' - I_0\alpha)/\omega_0 + RI_0/\omega_0 H]\omega_0}{(z + \alpha)^2 + \omega_0^2}$$

$$= L[u(t)I_0 e^{-\alpha t} \cos \omega_0 t] + L[u(t)A e^{-\alpha t} \sin \omega_0 t],$$

where $A = (I_0' - I_0\alpha)/\omega_0 + RI_0/\omega_0 H$. If $R > 0$, then $\alpha = R/2H > 0$ and this part of the solution damps out exponentially. If α is large, it damps out very rapidly.

Case II. *Critical damping.* $\omega_0^2 = 0$; that is, $1/HC = R^2/4H$.

In this case,

$$\frac{I_0 z + I_0' + RI_0/H}{(z + \alpha)^2} = \frac{I_0(z + \alpha) + I_0' - \alpha I_0 + RI_0/H}{(z + \alpha)^2}$$

$$= L[u(t)I_0 e^{-\alpha t}] + L[u(t)B t e^{-\alpha t}],$$

where $B = I_0' - \alpha I_0 + RI_0/H$. Since $\alpha = R/2H > 0$ this part of the solution damps out exponentially as in Case I, but does not have the same oscillatory nature.

Case III. *Overdamping.* $\omega_0^2 = \dfrac{1}{HC} - R^2/4H^2 = -\beta^2 < 0.$

In this case,

$$\frac{I_0 z + I_1' + RI_0/H}{(z - \alpha)^2 - \beta^2} = \frac{I_0(z + \alpha) + I_0' - \alpha I_0 + RI_0/H}{(z + \alpha)^2 - \beta^2}$$

$$= L[u(t)I_0 e^{-\alpha t} \cosh \beta t] + L[u(t)D e^{-\alpha t} \sinh \beta t],$$

where $D = (I_0' - I_0\alpha)/\beta + RI_0/\beta H$. This part of the solution damps out exponentially because

$$e^{-\alpha t} \cosh \beta t = \tfrac{1}{2}[e^{-(\alpha-\beta)t} + e^{-(\alpha+\beta)t}],$$
$$e^{-\alpha t} \sinh \beta t = \tfrac{1}{2}[e^{-(\alpha-\beta)t} - e^{-(\alpha+\beta)t}],$$

and $\alpha + \beta = R/2H + \sqrt{R^2/4H^2 - 1/HC} > 0$ and $\alpha - \beta = R/2H - \sqrt{R^2/4H^2 - 1/HC} > 0$. Clearly, the larger R the faster this part of the solution damps out.

The part of the solution corresponding to

$$\omega E_0 z/(Hz^2 + Rz + 1/C)(z^2 + \omega^2)$$

can be determined using Theorem 9.3.4. There will be terms of the form $\omega/(z^2 + \omega^2)$ and $z/(z^2 + \omega^2)$ in the transform corresponding to $\sin \omega t$ and $\cos \omega t$, respectively. To determine the coefficients of these terms in the solution we have to compute

$$L_1 = \lim_{z \to i\omega} \frac{\omega E_0 z}{Hz^2 + Rz + 1/C} = \frac{i\omega^2 E_0}{-\omega^2 H + R\omega i + 1/C}$$

$$= \frac{i\omega E_0}{\left(\dfrac{1}{\omega C} - \omega H\right) + Ri}$$

$$= \omega E_0 \left[\frac{R}{R^2 + \left(\omega H - \dfrac{1}{\omega C}\right)^2} - \frac{\left(\omega H - \dfrac{1}{\omega C}\right) i}{R^2 + \left(\omega H - \dfrac{1}{\omega C}\right)^2} \right].$$

Let $X = \omega H - \dfrac{1}{\omega C}$ (we call this the *reactance*) and $Z = R + iX$ (we call this the *complex impedance*). Then

$$L_1 = \frac{\omega E_0}{|Z|} \left[\frac{R}{|Z|} - \frac{iX}{|Z|} \right],$$

and

$$\frac{\omega E_0 z}{(Hz^2 + Rz + 1/C)(z^2 + \omega^2)} = L\left[u(t) \frac{E_0}{|Z|} \left(\frac{R}{|Z|} \sin \omega t - \frac{X}{|Z|} \cos \omega t \right) \right]$$
$$+ \text{ other terms}$$
$$= L\left[u(t) \frac{E_0}{|Z|} \sin (\omega t - \phi) \right]$$
$$+ \text{ other terms},$$

where $\cos \phi = R/|Z|$ and $\sin \phi = X/|Z|$, that is, $\phi = \arg Z$, and "other terms" refers to the same types of terms encountered in Cases I, II, and III.

For our purposes we shall lump these "other terms" with the complementary solution and write

$$I(t) = \frac{E_0}{|Z|} \sin (\omega t - \phi) + \text{transient solution,}$$

where the *transient solution* is of the form:

Case I. $Me^{-\alpha t} \cos \omega_0 t + Ne^{-\alpha t} \sin \omega_0 t$

Case II. $Me^{-\alpha t} + Nte^{-\alpha t}$

Case III. $Me^{-(\alpha+\beta)t} + Ne^{(-\alpha+\beta)t}$

depending on whether we have the underdamped, critically damped, or over-damped case. The part of the solution $(E_0/|Z|) \sin (\omega t - \phi)$ is called the *steady state* solution because it is the part which remains after a long period of time when the transient has damped out.

The Laplace transform can be used equally well to solve systems of ordinary differential equations. Consider the electrical circuit shown in Figure 9.4.2.

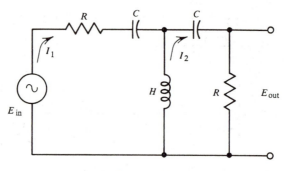

Figure 9.4.2.

EXAMPLE 9.4.2. Let us assume that initially the currents are zero and the capacitors are uncharged. Hence, the equations which must be satisfied are

$$E_{\text{in}} = RI_1 + (1/C) \int_0^t I_1 \, d\tau + H\frac{dI_1}{dt} - H\frac{dI_2}{dt},$$

$$0 = RI_2 + H\frac{dI_2}{dt} - H\frac{dI_1}{dt} + (1/C) \int_0^t I_2 \, d\tau.$$

Taking Laplace transforms in both equations, we have

$$E_{\text{in}} = RI_1 + \frac{1}{Cz}I_1 + HzI_1 - HzI_2,$$

$$0 = RI_2 + \frac{1}{Cz}I_2 + HzI_2 - HzI_1,$$

where, for simplicity we have let E_{in}, I_1, and I_2 stand for the Laplace transforms of the corresponding functions. Solving for I_1 and I_2, we have

$$I_1 = \frac{E_{in}\left[Hz + R + \dfrac{1}{Cz}\right]}{\left(Hz + R + \dfrac{1}{Cz}\right)^2 - H^2z^2},$$

$$I_2 = \frac{E_{in}zH}{\left(Hz + R + \dfrac{1}{Cz}\right)^2 - H^2z^2},$$

$$E_{out} = RI_2 = \frac{zHR}{\left(Hz + R + \dfrac{1}{Cz}\right)^2 - H^2z^2}E_{in}$$

$$= T(z)E_{in},$$

where

$$T(z) = \frac{z^3HR}{2HRz^3 + \left(\dfrac{2H}{C} + R^2\right)z^2 + \dfrac{2R}{C}z + \dfrac{1}{C^2}}$$

is called the *transfer function of the network*. The formulation of the problem in the form $E_{out} = T(z)E_{in}$ has certain advantages in that given the Laplace transform of the input voltage we can easily find the Laplace transform of the output voltage, which in turn will give the output by inversion. The important thing to note here is that the use of the Laplace transform has reduced the problem of solving a system of ordinary differential equations to the problem of solving a system of linear algebraic equations. For the analysis of more complicated networks a knowledge of linear algebra and matrix theory is essential. Therefore, we shall not pursue the subject further.

The transfer function has the following interpretation. Suppose $E_{in} = u(t)$. Then the Laplace transform of E_{in} is $1/z$ and the Laplace transform of E_{out} is $T(z)/z$. Therefore,

$$\frac{T(z)}{z} = L[A(t)],$$

where $A(t)$ is the response of the network when the input is a unit step function. If we differentiate the original equations, we have

$$\frac{dE_{in}}{dt} = R\frac{dI_1}{dt} + \frac{1}{C}I_1 + H\frac{d^2I_1}{dt^2} - H\frac{d^2I_2}{dt^2},$$

$$0 = R\frac{dI_2}{dt} + H\frac{d^2I_2}{dt^2} - H\frac{d^2I_1}{dt^2} + \frac{1}{C}I_2,$$

and taking transforms

$$E'_{in} = zRI_1 + \frac{1}{C}I_1 + z^2HI_1 - z^2HI_2,$$

$$0 = zRI_2 + \frac{1}{C}I_2 + z^2HI_2 - z^2HI_1,$$

$$E_{out} = RI_2 = E'_{in}\frac{T(z)}{z}.$$

Applying the convolution theorem, we have

$$E_{out}(t) = \int_0^t E'_{in}(\tau)A(t - \tau)\,d\tau.$$

$A(t)$ is called the *admittance* of the network.

One of the important problems of network analysis is the so-called "black box" problem. Suppose we have a "black box" which contains a network which is known to have only passive elements, that is, resistors, capacitors, and inductors. Suppose there are two input terminals and two output terminals. Under quite general conditions one can obtain the admittance of the network by determining the response of the network to a unit step input. Then, in general, the output is given by convolution

$$E_{out}(t) = \int_0^t E'_{in}(\tau)A(t - \tau)\,d\tau.$$

The problem of determining the transfer function of networks is usually called *network analysis*, whereas the problem of realizing networks with a given transfer function is called *network synthesis*.

Exercises 9.4

1. Solve the differential equation $y'' + 3y' + 2y = e^{-t}$, $0 < t$, subject to $y(0^+) = y_0$ and $y'(0^+) = y'_0$, using the Laplace transform.

2. Solve the differential $y'' + \omega_0^2 y = \sin \omega t$, $0 < t$, $\omega \neq \omega_0$, subject to $y(0^+) = y'(0^+) = 0$, using the Laplace transform. Also, consider the case $\omega = \omega_0$.

3. Solve the integrodifferential equation $y' + \int_0^t y(\tau)\,d\tau = e^{-t}$, $0 < t$, subject to $y(0^+) = y_0$, using the Laplace transform.

4. Solve the system of differential equations: $x' + y' - x + 3y = e^{-t}$, $x' + y' + 2x + y = e^{-2t}$ for $t > 0$ subject to $x(0^+) = x_0$, $y(0^+) = y_0$.

9.5. STABILITY

If the output of a given linear system is represented by the solution of a linear ordinary differential equation with constant coefficients, then its Laplace transform is of the form

$$Y(z) = F(z)/Q(z) + P(z)/Q(z),$$

where $Q(z) = a_n z^n + a_{n-1} z^{n-1} + \cdots + a_1 z + a_0$, the a's are the constant coefficients in the differential equation which we assume to be real, $F(z)$ is the transform of the forcing term (input) in the differential equation, and $P(z)$ is a polynomial of degree $n - 1$ which contains the initial conditions. We have called the inverse transform of $P(z)/Q(z)$ the complementary solution and we have seen that corresponding to a real multiple root r of $Q(z)$ of order m there is a term in it of the form

$$(A_0 + A_1 t + A_2 t^2 + \cdots + A_{m-1} t^{m-1}) e^{rt},$$

where the A's depend on the initial conditions. If $r < 0$ then whatever the initial conditions this term will approach zero as $t \to \infty$. The larger $|r|$ the faster this term will damp out. If $r + iq$ is a multiple complex root of order m then there are terms of the form

$$[A_0 \cos qt + B_0 \sin qt + t(A_1 \cos qt + B_1 \sin qt) + \cdots \\ + t^{m-1}(A_{m-1} \cos qt + B_{m-1} \sin qt)] e^{rt}$$

in the complementary solution. Again, if $r < 0$ these terms will damp in time. As we have seen in Example 9.4.1, there are also terms of these forms in

$$y_p(t) = L^{-1}[F(z)/Q(z)],$$

although in general there are also terms which do not damp out, even if the real parts of the zeros of $Q(z)$ are all negative. These terms which do not damp out we have called the *steady state solution*. The terms which contain $e^{r_j t}$, where $r_j = \text{Re}(z_j)$, z_j a zero of $Q(z)$, made up what we have called the *transient solution*. The transient solution will really be transient if $r_j < 0$ for all j, and in this case the steady state solution becomes the dominant part of the output after a reasonable period of time.

Most systems are subject to random fluctuations which are not part of the input. These take the form of a new set of initial conditions and tend to introduce transients. If these transients damp out leaving the system in the steady state then we say that the system is *stable*. From what we have already said we have the following stability criterion:

If the output of a linear system is the solution of the ordinary differential equation with constant coefficients

$$a_n \frac{d^n y}{dt^n} + a_{n-1} \frac{d^{n-1}y}{dt^{n-1}} + \cdots + a_1 \frac{dy}{dt} + a_0 y = f(t),$$

then the system is stable if and only if all of the roots of

$$Q(z) = a_n z^n + a_{n-1} z^{n-1} + \cdots + a_1 z + a_0$$

have negative real parts.

EXAMPLE 9.5.1. Show that the system with output $y(t)$, where

$$ay'' + by' + cy = f(t)$$

is stable if and only if a, b, and c are all of the same sign. We can assume that $a > 0$. If not we multiply by -1. If $az^2 + bz + c = 0$ has real roots r_1 and r_2, then

$$z^2 + \left(\frac{b}{a}\right) z + \left(\frac{c}{a}\right) = (z - r_1)(z - r_2) = z^2 - (r_1 + r_2) z + r_1 r_2$$

and $(b/a) = -(r_1 + r_2)$, $(c/a) = r_1 r_2$. If r_1 and r_2 are both negative, then $b > 0$ and $c > 0$. Conversely, if $a > 0$, $b > 0$, $c > 0$, then r_1 and r_2 must be negative. If $az^2 + bz + c = 0$ has complex roots $r \pm iq$, then

$$z^2 + \left(\frac{b}{a}\right) z + \left(\frac{c}{a}\right) = (z - r - iq)(z - r + iq) = z^2 - 2rz + r^2 + q^2.$$

Clearly, $r < 0$ if and only if $b > 0$ and $c > 0$.

In Example 9.4.2, we are dealing with a more complicated system. Nevertheless, we found that we could formulate the solution as follows

$$E_{\text{out}} = T(z)E_{\text{in}}, \qquad T(z) = \frac{R(z)}{Q(z)},$$

where $T(z)$ is a rational function of z with real coefficients. In this example, we put all the initial conditions equal to zero. However, if we had started with nonzero initial values we would only have introduced a term of the form $P(z)/Q(z)$ into the solution, where $P(z)$ and $Q(z)$ are polynomials with real coefficients and $P(z)$ has lower degree than $Q(z)$. Therefore, the same considerations arise here as in the above discussion. There will be a steady state solution coming from the term $T(z)E_{\text{in}}(z)$ and a transient solution coming from both $T(z)E_{\text{in}}(z)$ and $P(z)/Q(z)$ and the stability criterion will be, as before, that the real parts of the roots of the polynomial $Q(z)$ be negative.

EXAMPLE 9.5.2. Show that the system of Example 9.4.2 is stable. In this case,

$$Q(z) = 2HRz^3 + \left(\frac{2H}{C} + R^2\right)z^2 + \frac{2R}{C}z + \frac{1}{C^2}.$$

If $Q(z) = az^3 + bz^2 + cz + d$, it can be shown (Exercise 9.5.2) that $Q(z)$ has roots with negative real parts if and only if $a > 0$, $c > 0$, $d > 0$, and $bc - ad > 0$. Here, $a = 2HR > 0$, $c = 2R/C > 0$, $d = \frac{1}{C^2} > 0$, and

$$bc - ad = \left(\frac{2H}{C} + R^2\right)\frac{2R}{C} - \frac{2HR}{C^2} = \frac{2R^3}{C} + \frac{2HR}{C^2} > 0.$$

We shall give two more equivalent stability criteria, one which is algebraic in nature and the other which is function theoretic. We saw in each case above that the stability of a system depended on the location of the zeros of a polynomial $Q(z)$. We shall call the equation $Q(z) = 0$ the *characteristic equation* of the system.

Hurwitz Stability Criterion: Let

$$Q(z) = a_n z^n + a_{n-1} z^{n-1} + \cdots + a_1 z + a_0 = 0$$

be the characteristic equation of a linear system where the a's are real and $a_0 > 0$. Then the system is stable if and only if the following determinants are positive:

$$\begin{vmatrix} a_1 & a_0 \\ a_3 & a_2 \end{vmatrix}, \quad \begin{vmatrix} a_1 & a_0 & 0 \\ a_3 & a_2 & a_1 \\ a_5 & a_4 & a_3 \end{vmatrix}, \quad \begin{vmatrix} a_1 & a_0 & 0 & 0 \\ a_3 & a_2 & a_1 & a_0 \\ a_5 & a_4 & a_3 & a_2 \\ a_7 & a_6 & a_5 & a_4 \end{vmatrix},$$

and so on.

The proof of this consists of showing that it is a necessary and sufficient condition for the real parts of all the roots of $Q(z)$ to be negative. It is algebraic in nature and we shall not give it here.* The case $n = 2$ is in Example 9.5.1, and the case $n = 3$ is in Exercise 9.5.2.

Nyquist Stability Criterion:† Let

$$Q(z) = a_n z^n + z_{n-1} z^{n-1} + \cdots + a_1 z + a_0 = 0$$

be the characteristic equation of a linear system where the a's are real, and $Q(z)$ has no zeros on the imaginary axis. Then the system is stable if and

---

* See J. V. Uspensky, *Theory of Equations*. New York: McGraw-Hill Book Company, Inc., 1948, pp. 304–309.

† This is not the exact form in which the Nyquist criterion is usually given, but since it uses the same basic idea we have taken the liberty to extend the meaning and hence the credit.

only if the change in the argument of $Q(iy)$ when y changes from $-\infty$ to ∞ is $n\pi$.

The proof is based on the principle of the argument, Theorem 5.3.1. If the system is stable then the zeros of $Q(z)$ are in the left half-plane $\text{Re}(z) < 0$. We apply the argument principle along the imaginary axis and around a large semicircle of radius R in the right half-plane. Then $\Delta \arg Q(z) = 0$. For large R the dominant term is $a_n R^n e^{in\theta}$ and $\Delta \arg a_n R^n e^{in\theta} = n\pi$. Therefore, $\Delta \arg Q(iy) = n\pi$, when y changes from $-\infty$ to ∞. Clearly, the converse is also true.

EXAMPLE 9.5.3.   Show that the system governed by the characteristic equation $Q(z) = z^3 + 2z^2 + z + 1 = 0$ is stable. Since $Q(iy) = -iy^3 - 2y^2 + iy + 1$, $\arg Q(iy) = \tan^{-1}(y - y^3)/(1 - 2y^2)$. To start with, we take $\arg Q(iy) = -\pi/2$ when $y = -\infty$ and then consider the changes in $\arg Q(iy)$ as y increases to ∞, making sure that we take values of $\tan^{-1}$ resulting in continuous changes in $\arg Q(iy)$. The result is that

$$\Delta \arg Q(iy) = 3\pi$$

which means that the system is stable.

For the control engineer the important stability problems usually arise when there is feedback in the system. Let us indicate a system by the following block diagram: (see Figure 9.5.1).

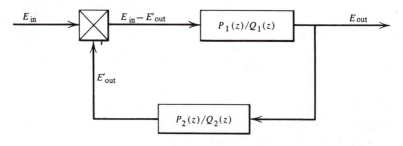

Figure 9.5.1.

E_{in} is the input which is fed into a component which subtracts the *feedback* E'_{out} from the input. The result is put through a linear device which has a transfer function $P_1(z)/Q_1(z)$. The output E_{out} is fed back through a device with transfer function $P_2(z)/Q_2(z)$ to produce the feedback E'_{out}. The system equations are

$$E_{\text{out}} = \frac{P_1(z)}{Q_1(z)} (E_{\text{in}} - E'_{\text{out}}),$$

$$E'_{\text{out}} = \frac{P_2(z)}{Q_2(z)} E_{\text{out}},$$

where the E's now stand for transforms. We are assuming that the initial conditions are all zero. Solving for E_{out} we have,

$$E_{\text{out}} = \frac{P_1 Q_2}{P_1 P_2 + Q_1 Q_2} E_{\text{in}}.$$

Therefore, the characteristic equation of the system is

$$P_1(z)P_2(z) + Q_1(z)Q_2(z) = 0$$

and we can determine stability by any one of the above criteria.

Suppose that the subtracting element is eliminated so that we have the following system: (see Figure 9.5.2)

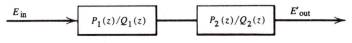

$E_{\text{in}} \longrightarrow \boxed{P_1(z)/Q_1(z)} \longrightarrow \boxed{P_2(z)/Q_2(z)} \longrightarrow E'_{\text{out}}$

Figure 9.5.2.

Hence,

$$E'_{\text{out}} = \frac{P_1(z)P_2(z)}{Q_1(z)Q_2(z)} E_{\text{in}}.$$

This is called the open loop system. Therefore, the characteristic equation of the open loop system is

$$Q_1(z)Q_2(z) = 0.$$

Let us assume that the open loop is stable. Therefore, there are no zeros of $Q_1(z)Q_2(z)$ for $\operatorname{Re}(z) \geq 0$. Consider

$$R(z) = \frac{P_1(z)P_2(z) + Q_1(z)Q_2(z)}{Q_1(z)Q_2(z)} = \frac{P_1(z)P_2(z)}{Q_1(z)Q_2(z)} + 1.$$

If we apply the principle of the argument to $R(z)$ around a closed contour consisting of the imaginary axis plus a large semicircle in the right half-plane, we have

$$\underset{-\infty < y < \infty}{\Delta \arg R(iy)} + \underset{\text{semicircle}}{\Delta \arg R(z)} = 2\pi(N - M),$$

where N is the number of zeros of $Q_1 Q_2$ in the right half-plane and M is the number of zeros of $P_1 P_2 + Q_1 Q_2$ in the right half-plane. Since $N = 0$, for stability of the closed loop we should obtain zero for M. In practice, the degree of P_1 is less than or equal to the degree of Q_1 and the degree of P_2 is less than or equal to the degree of Q_2. Therefore, the degree of P_1P_2 is less than the degree of Q_1Q_2, and

$$\lim_{z \to \infty} R(z) = 1 + \lim_{z \to \infty} \frac{P_1(z)P_2(z)}{Q_1(z)Q_2(z)},$$

which is finite. For this reason the image of the imaginary axis under $w = R(z)$ is a closed curve and the feedback system is stable if the open loop is stable and $\Delta \arg_{-\infty < y < \infty} R(iy) = 0$. This will be true if the closed curve $w = R(iy)$ does not enclose the origin in the w plane. It is customary to plot the curve $\tilde{w} = \tilde{R}(iy)$, where

$$\tilde{R}(z) = \frac{P_1(z)P_2(z)}{Q_1(z)Q_2(z)}.$$

But since $w = \tilde{w} + 1$, we have that *the feedback system is stable if the open loop is stable and the curve* $\tilde{w} = \tilde{R}(iy)$ *does not enclose* -1. This is the Nyquist criterion for feedback systems.

Exercises 9.5

1. Prove that a polynomial $Q(z)$ has a zero $r - iq$ if it has a zero $r + iq$ and has real coefficients.

2. Prove that the zeros of $az^3 + bz^2 + cz + d$ are in the left half-plane if and only if a, c, d, and $bc - ad$ are of the same sign.

3. Use the Nyquist criterion to determine the stability of a system with characteristic equation $z^4 + z^3 + 4z^2 + 2z + 3 = 0$.

4. Is it possible that a feedback system could be stable when the open loop is unstable? Hint: consider the discussion for the case where $N \neq 0$. Generalize the Nyquist criterion to handle this situation.

9.6. THE SOLUTION OF PARTIAL DIFFERENTIAL EQUATIONS

In Section 8.6 we illustrated how the Fourier transform could be used in solving certain problems in partial differential equations. Since the Laplace transform is just the one-sided Fourier transform with a rotation in the complex plane, it is not surprising that the Laplace transform is also useful in solving partial differential equations. As a matter of fact, the method is essentially the same as that employed earlier. However, we give two more examples to illustrate the method using the Laplace transform.

EXAMPLE 9.6.1. Find the displacement of a semiinfinite elastic string which is initially at rest in the equilibrium position and then displaced at $x = 0$ starting at $t = 0$ as follows: $u(0, t) = f(t)$. The problem can be formulated as follows:

$$u_{xx} = \frac{1}{a^2} u_{tt}, \qquad x > 0, \qquad t > 0,$$

$$u(x, 0) = 0, \qquad\qquad x \geq 0,$$

$$u_t(x, 0) = 0, \qquad\qquad x \geq 0,$$

$$u(0, t) = f(t), \qquad\qquad t \geq 0,$$

$$\lim_{x \to \infty} u(x, t) = 0, \qquad\qquad t \geq 0.$$

We transform with respect to the time variable. Let

$$U(x, z) = \int_0^\infty u(x, t)e^{-zt}\, dt.$$

Then

$$\frac{\partial^2 U}{\partial x^2} - \frac{z^2}{a^2} U = 0.$$

Solving for $U(x, z)$, we have

$$U(x, z) = Ae^{(z/a)x} + Be^{-(z/a)x}.$$

The transform should exist in a right half-plane. Therefore, for

$$\mathrm{Re}(z) > c > 0, \lim_{x \to \infty} U(x, z) = 0,$$

and this implies that $A = 0$. Now

$$U(0, z) = \int_0^\infty f(t)e^{-zt}\, dt = F(z) = B,$$

and

$$U(x, z) = F(z)e^{-(z/a)x},$$

$$u(x, t) = f\left(t - \frac{x}{a}\right)u\left(t - \frac{x}{a}\right),$$

using Theorem 9.2.6. The interpretation is that the displacement propagates down the string at the velocity a; that is, there is no displacement until $x = at$ and then the displacement at that point follows the displacement produced at the end. It can be shown that this is a solution to the problem provided $f(t)$ is twice differentiable.

EXAMPLE 9.6.2. Find the displacement of a semiinfinite elastic string which is fixed at the ends, has initial displacement $u(x, 0) = g(x)$, and no initial velocity. The problem can be formulated as follows:

$$u_{xx} = \frac{1}{a^2} u_{tt}, \qquad x > 0, \qquad t > 0,$$

$$u(x, 0) = g(x), \qquad\qquad\qquad x \geq 0,$$

$$u_t(x, 0) = 0, \qquad\qquad\qquad x \geq 0,$$

$$u(0, t) = 0, \qquad\qquad\qquad t \geq 0,$$

$$\lim_{x \to \infty} u(x, t) = 0, \qquad\qquad t \geq 0.$$

We transform with respect to the time variable,

$$U(x, z) = \int_0^\infty u(x, t)e^{-zt}\, dt,$$

$$\frac{\partial^2 U}{\partial x^2} - \frac{z^2}{a^2} U = -\frac{z}{a^2} g(x).$$

This equation can be solved by variation of parameters, using the boundary conditions $U(0, z) = 0$ and $\lim_{x \to \infty} U(x, z) = 0$, to obtain

$$U(x, z) = \frac{1}{2a} \int_0^x g(\xi)[e^{-(z/a)(x-\xi)} - e^{-(z/a)(x+\xi)}]\, d\xi$$

$$+ \frac{1}{2a} \int_x^\infty g(\xi)[e^{-(z/a)(\xi-x)} - e^{-(z/a)(x+\xi)}]\, d\xi$$

$$= \frac{1}{2} \int_0^\infty g(x + a\eta)e^{-z\eta}\, d\eta + \frac{1}{2} \int_0^{(x/a)} g(x - a\eta)e^{-z\eta}\, d\eta$$

$$- \frac{1}{2} \int_{(x/a)}^\infty g(a\eta - x)e^{-z\eta}\, d\eta.$$

At this point it is convenient to continue $g(x)$ as an odd function; that is, we define $g(x) = -g(-x)$ for $x < 0$. Then

$$U(x, z) = \frac{1}{2} \int_0^\infty g(x + a\eta)e^{-z\eta}\, d\eta + \frac{1}{2} \int_0^\infty g(x - a\eta)e^{-z\eta}\, d\eta.$$

We immediately recognize this as the transform of

$$u(x, t) = \tfrac{1}{2}g(x + at) + \tfrac{1}{2}g(x - at),$$

where $g(x)$ must be understood in the second term to be an odd function. It can be shown that this is a solution to the problem provided $g(x)$ is twice differentiable.

Exercises 9.6

1. Find the solution of $U_{xx} - (z^2/a^2)U = -(z/a^2)g(x)$ for $U(x, z)$ subject to $U(0, z) = 0$ and $\lim_{x \to 0} U(x, z) = 0$ for $\text{Re}(z) > c \geq 0$.

2. Solve Example 9.6.2 with initial velocity $u_t(x, 0) = h(x)$ and no initial displacement.

3. Using Examples 9.6.1, 9.6.2 and Exercise 9.6.2 show how the following problem can be solved:

$$u_{xx} = (1/a^2)u_{tt}, \qquad x > 0, \qquad t > 0;$$

$$u(x, 0) = g(x), u_t(x, 0) = h(x), u(0, t) = f(t), \lim_{x \to \infty} u(x, t) = 0.$$

4. Using the Laplace transform find the displacement of an elastic string fixed at $x = 0$ and $x = 1$, with no initial velocity, and with initial displacement $u(x, t) = A \sin \pi x$.

9.7. THE SOLUTION OF INTEGRAL EQUATIONS

Because of the different form that the convolution theorem (see Theorem 9.2.7) takes for the Laplace transform compared with that for the Fourier transform (see Section 8.3), the type of integral equation which can easily be solved using the Laplace transform is again different. Let us consider the following Volterra integral equation

$$u(t) = f(t) + \int_0^t k(t - \tau)u(\tau) \, d\tau$$

to be solved for $t > 0$. Let $f(t)$ and $k(t)$ have Laplace transforms $F(z)$ and $K(z)$ valid for $\text{Re}(z) > a$. Then, if $U(z)$ is the Laplace transform of $u(t)$,

$$U(z) = F(z) + K(z)U(z),$$

$$U(z) = \frac{F(z)}{1 - K(z)}.$$

EXAMPLE 9.7.1. Solve the integral equation

$$u(t) = 1 - \int_0^t (t - \tau)u(\tau) \, d\tau,$$

for $t > 0$. Here, $f(t) = 1$ and $K(t) = t$. Therefore,

$$U(z) = \frac{1}{z} - \frac{1}{z^2} U(z),$$

$$U(z) = \frac{z}{z^2 + 1},$$

$$u(t) = \cos t.$$

The Laplace transform can also be used effectively to solve certain integral equations of the first kind, that is, where the unknown appears only under the integral sign. Consider

$$f(t) = \int_0^t k(t - \tau)u(\tau)\, d\tau$$

to be solved for $u(t)$ for $t > 0$. Assume $f(t)$ and $k(t)$ have Laplace transforms for $\text{Re}(z) > a$. Then if $F(z)$ and $K(z)$ are their transforms

$$F(z) = K(z)U(z),$$

where $U(z)$ is the transform of $u(t)$. Then

$$U(z) = \frac{F(z)}{K(z)}.$$

EXAMPLE 9.7.2. Solve *Abel's integral equation*

$$f(t) = \int_0^t \frac{u(\tau)}{\sqrt{t - \tau}}\, d\tau.$$

Here, $f(t)$ is given and is assumed to have a Laplace transform $F(z)$. Also $k(t) = t^{-1/2}$ has a transform $\Gamma(1/2)z^{-1/2}$. Hence,

$$F(z) = \frac{\sqrt{\pi}\, U(z)}{\sqrt{z}},$$

$$U(z) = \frac{\sqrt{z}}{\sqrt{\pi}} F(z) = \frac{z}{\pi}\left(\frac{\sqrt{\pi}}{\sqrt{z}} F(z)\right) = \frac{z}{\pi} K(z)F(z).$$

Therefore, by Theorems 9.2.2 and 9.2.7

$$u(t) = \frac{1}{\pi} \frac{d}{dt} \int_0^t \frac{f(\tau)}{\sqrt{t - \tau}}\, d\tau.$$

One cannot explicitly carry out the differentiation until $f(t)$ is known. Suppose $f(t) = t$, for example. Then

$$\int_0^t \frac{\tau}{\sqrt{t - \tau}}\, d\tau = \frac{4}{3} t^{3/2},$$

$$u(t) = \frac{2}{\pi} t^{1/2}.$$

Exercises 9.7

1. Solve the integral equation $u(t) = 1 + \int_0^t (t - \tau)^2 u(\tau)\, d\tau,\ t \geq 0$.

2. Solve the generalized Abel integral equation

$$f(t) = \int_0^t \frac{u(\tau)}{(t - \tau)^\alpha}\, d\tau, \qquad 0 < \alpha < 1.$$

3. Describe a method for solving the system of integral equations:

$$u_1(t) = f(t) + \int_0^t k_1(t - \tau) u_1(\tau)\, d\tau + \int_0^t k_2(t - \tau) u_2(\tau)\, d\tau,$$

$$u_2(t) = g(t) + \int_0^t k_3(t - \tau) u_1(\tau)\, d\tau + \int_0^t k_4(t - \tau) u_2(\tau)\, d\tau.$$

CHAPTER 10

Asymptotic Expansions

10.1. INTRODUCTION AND DEFINITIONS

To introduce the idea of an asymptotic expansion, let us consider the function $\int_0^\infty e^{-zt}(1 + t^2)^{-1}\,dt$. This is the Laplace transform of $(1 + t^2)^{-1}$ and therefore is analytic for $\mathrm{Re}(z) > 0$. A naive analyst might be tempted to expand $(1 + t^2)^{-1} = 1 - t^2 + t^4 - \cdots$ and integrate term by term. The nth term would be

$$\int_0^\infty e^{-zt}(-1)^{n-1}t^{2n-2}\,dt = \frac{(-1)^{n-1}(2n - 2)!}{z^{2n-1}}.$$

The result would be the series $\displaystyle\sum_{n=1}^\infty \frac{(-1)^{n-1}(2n - 2)!}{z^{2n-1}}$. This is nonsense because the expansion $(1 + t^2)^{-1} = 1 - t^2 + t^4 - \cdots$ converges only for $|t| < 1$ and, as a matter of fact, the resulting series diverges for all z.

On the other hand, suppose we write

$$\frac{1}{1 + t^2} = 1 - t^2 + t^4 - \cdots + (-1)^{n-1}t^{2n-2} + \frac{(-1)^n t^{2n}}{1 + t^2}.$$

Since this is a finite sum we may integrate term by term. Hence,

$$\int_0^\infty \frac{e^{-zt}}{1 + t^2}\,dt = \frac{1}{z} - \frac{2!}{z^3} + \frac{4!}{z^5} - \cdots + \frac{(-1)^{n-1}(2n - 2)!}{z^{2n-1}} + R_n(z),$$

where

$$R_n(z) = (-1)^n \int_0^\infty \frac{e^{-zt}t^{2n}}{1 + t^2}\,dt.$$

Now,

$$|R_n(z)| = \left| \int_0^\infty \frac{e^{-zt}t^{2n}}{1 + t^2}\,dt \right| \le \int_0^\infty e^{-xt}t^{2n}\,dt = \frac{(2n)!}{x^{2n+1}}.$$

Let $z \neq 0$ and $-(\pi/2) + \alpha \leq \arg z \leq (\pi/2) - \alpha$, $0 < \alpha < \pi/2$. Then $x \geq |z| \sin \alpha$. Hence,

$$|R_n(z)| \leq \frac{(2n)!}{(\sin \alpha)^{2n+1}} \frac{1}{|z|^{2n+1}}.$$

For a fixed n, $\lim_{|z| \to \infty} |R_n(z)| = 0$. Therefore, even though the series diverges for all z, nevertheless, for $|z|$ sufficiently large a finite number of terms of the series is still a good approximation to the function. In this case, we say that the series $\dfrac{1}{z} - \dfrac{2!}{z^3} + \dfrac{4!}{z^5} - \cdots$ is an *asymptotic expansion* of the given function and write

$$\int_0^\infty \frac{e^{-zt}}{1 + t^2} dt \sim \frac{1}{z} - \frac{2!}{z^3} + \frac{4!}{z^5} - \cdots,$$

for $z \neq 0$ and $-(\pi/2) + \alpha \leq \arg z \leq (\pi/2) - \alpha$.

Another way to obtain this expansion is by integration by parts. For $\text{Re}(z) > 0$ we have, if $f(t) = (1 + t^2)^{-1}$

$$\int_0^\infty e^{-zt} f(t) \, dt = \frac{e^{-zt}}{-z} \Big|_0^\infty + \frac{1}{z} \int_0^\infty e^{-zt} f'(t) \, dt$$

$$= \frac{1}{z} + \frac{1}{z^2} f'(0) + \frac{1}{z^2} \int_0^\infty e^{-zt} f''(t) \, dt$$

$$= \frac{1}{z} + \frac{1}{z^3} f''(0) + \frac{1}{z^3} \int_0^\infty e^{-zt} f'''(t) \, dt$$

$$= \frac{1}{z} - \frac{2!}{z^3} + \frac{1}{z^4} f'''(0) + \frac{1}{z^4} \int_0^\infty e^{-zt} f^{(4)}(t) \, dt$$

$$= \frac{1}{z} - \frac{2!}{z^3} + \frac{4!}{z^5} - \cdots + \frac{1}{z^{2n-1}} f^{(2n-2)}(0)$$

$$+ \frac{1}{z^{2n}} \int_0^\infty e^{-zt} f^{(2n)}(t) \, dt.$$

It is easy to show that $|f^{(2n)}(t)| \leq (2n)!$ Hence,

$$\left| \frac{1}{z^{2n}} \int_0^\infty e^{-zt} f^{(2n)}(t) \, dt \right| \leq \frac{(2n)!}{|z|^{2n}} \int_0^\infty e^{-xt} \, dt \leq \frac{(2n)!}{(|z| \sin \alpha)^{2n+1}},$$

for $z \neq 0$, $-\pi/2 + \alpha \leq \arg z \leq \pi/2 - \alpha$.

We now give a careful definition of what we mean by an asymptotic expansion. Let R be a region in the extended complex plane. Let z_0 be a limit point of R.

Definition 10.1.1. A finite or infinite sequence of functions $\{w_n(z)\}$ is an asymptotic sequence for $z \to z_0$ in R if each $w_n(z)$ is defined in R and $w_{n+1}(z) = o(w_n(z))$, that is, $\lim_{z \to z_0} |w_{n+1}(z)/w_n(z)| = 0$.

Definition 10.1.2. Let $f(z)$ be defined in R and let $\{w_n(z)\}$ be an asymptotic sequence for $z \to z_0$ in R. Then the series $\sum\limits_{n=1}^{N} a_n w_n(z)$ is an asymptotic expansion to N terms of $f(z)$ for $z \to z_0$ in R if

$$f(z) = \sum_{n=1}^{N} a_n w_n(z) + o(w_N), \quad \text{as} \quad z \to z_0.$$

We shall write

$$f(z) \sim \sum_{n=1}^{N} a_n w_n(z),$$

if the series is an asymptotic representation to N terms. If $N = 1$, we shall have

$$f(z) \sim a_1 w_1(z) = g(z)$$

and say that $g(z)$ is an *asymptotic formula* for $f(z)$ as $z \to z_0$.

Definition 10.1.3. If $\{w_n(z)\}$ is an infinite asymptotic sequence for $z \to z_0$ in R, and

$$f(z) \sim \sum_{n=1}^{N} a_n w_n(z),$$

for every N, then $\sum\limits_{n=1}^{\infty} a_n w_n(z)$ is an asymptotic series or expansion for $z \to z_0$, and we write

$$f(z) \sim \sum_{n=1}^{\infty} a_n w_n(z).$$

Theorem 10.1.1. *The coefficients in an asymptotic expansion* $\sum\limits_{n=1}^{\infty} a_n w_n(z)$ *for $f(z)$, as $z \to z_0$, can be determined as follows:*

$$a_N = \lim_{z \to z_0} \left\{ \frac{f(z) - \sum\limits_{n=1}^{N-1} a_n w_n(z)}{w_N(z)} \right\}.$$

Proof: Beginning with $N = 1$ we have

$$f(z) = a_1 w_1(z) + g_1(z),$$

where $g_1(z) = o(w_1)$ as $z \to z_0$. Dividing by w_1, we have

$$a_1 = \frac{f(z)}{w_1(z)} - \frac{g_1(z)}{w_1(z)}$$

$$= \lim_{z \to z_0} \frac{f(z)}{w_1(z)}.$$

Having determined $a_1, a_2, \ldots, a_{N-1}$, we have

$$f(z) = a_N w_N(z) + \sum_{n=1}^{N-1} a_n w_n(z) + g_N(z),$$

where $g_N(z) = o(w_N)$ as $z \to z_0$. Then

$$a_N = \frac{f(z) - \sum_{n=1}^{N-1} a_n w_n(z)}{w_N(z)} - \frac{g_N(z)}{w_N(z)}$$

$$= \lim_{z \to z_0} \left\{ \frac{f(z) - \sum_{n=1}^{N-1} a_n w_n(z)}{w_N(z)} \right\}.$$

Theorem 10.1.2. *Let $f(z)$, $w_1(z)$, $w_2(z)$, $\ldots$, $w_N(z)$ be defined in R. Let*

$$a_k = \lim_{z \to z_0} \left\{ \frac{f(z) - \sum_{n=1}^{k-1} a_n w_n(z)}{w_k(z)} \right\}$$

be defined and not zero for $k = 1, 2, \ldots, N$. Then $\sum_{n=1}^{N} a_n w_n(z)$ is an asymptotic expansion to N terms for $f(z)$ as $z \to z_0$.

Proof: We must first show that $\{w_n(z)\}$ is an asymptotic sequence. It follows from the definition of a_k that

$$f(z) - \sum_{n=1}^{k} a_n w_n(z) = g_k(z) = o(w_k),$$

$$f(z) - \sum_{n=1}^{k} a_n w_n(z) = a_{k+1} w_{k+1}(z) + h_k(z),$$

where $h_k = o(w_{k+1})$. Therefore,

$$(a_{k+1} + h_k/w_{k+1}) w_{k+1} = g_k = o(w_k).$$

But $\lim_{z \to z_0} |h_k/w_{k+1}| = 0$ and $a_{k+1} \neq 0$. Therefore, there exists an ϵ-neighborhood of z_0 such that when z is in R and in this ϵ-neighborhood

$$a_{k+1} + h_k/w_{k+1} \neq 0.$$

Hence, $w_{k+1} = o(w_k)$. Finally, the a_k's are defined in such a way that

$$f(z) = \sum_{n=1}^{k} a_n w_n(z) + o(w_k),$$

$k = 1, 2, 3, \ldots, N$.

Theorem 10.1.3. *Let $\{w_n(z)\}$ be a given asymptotic sequence in R for $z \to z_0$. Let $\sum_{n=1}^{N} a_n w_n(z)$ be an asymptotic expansion to N terms for $f(z)$ as $z \to z_0$. Then the expansion is unique.*

Proof: If $\sum_{n=1}^{N} b_n w_n(z)$ were another asymptotic expansion then

$$b_1 = \lim_{z \to z_0} \frac{f(z)}{w_1(z)} = a_1,$$

$$b_2 = \lim_{z \to z_0} \frac{f(z) - a_1 w_1(z)}{w_2(z)} = a_2,$$

and so on.

This theorem asserts the uniqueness of the expansion for $f(z)$ in terms of a given asymptotic sequence. However, there may be many asymptotic expansions for $f(z)$ involving different asymptotic sequences. For example,

$$\frac{z+1}{1-z} \sim 1 + 2z + 2z^2 + \cdots$$

$$\frac{z+1}{1-z} = \frac{(z+1)^2}{1-z^2} \sim (z+1)^2 + (z+1)^2 z^2 + (z+1)^2 z^4 + \cdots$$

are both asymptotic expansions for $z \to 0$. These happen to be convergent for $|z| < 1$, but in general asymptotic expansions can be either convergent or divergent.

A given asymptotic expansion can be the expansion for more than one function. For example, let $w_n(z) = z^{-n}$. Then $\{w_n(z)\}$ is an asymptotic sequence for $z \to \infty$ since $\lim_{z \to \infty} |z^n/z^{n+1}| = 0$. Let $R = \{z \mid \mathrm{Re}(z) > 0\}$. Then $g(z) = e^{-1/z}$ has the asymptotic expansion zero in R as $z \to \infty$, since $\lim_{z \to \infty} z^n e^{-1/z} = 0$, $n = 1, 2, 3, \ldots$. Therefore, if any other function $f(z)$ has an asymptotic expansion in R as $z \to \infty$ in terms of $\{w_n(z)\}$, then $f(z) + g(z)$ has the same asymptotic expansion. In other words, an asymptotic expansion does not determine the function. It only determines approximately the values of the function near z_0 in R.

Definition 10.1.4. Let $w_n(z) = (z - z_0)^n$, $n = 0, 1, 2, \ldots$. Then $\sum_{n=0}^{\infty} a_n(z - z_0)^n$ is an asymptotic power series expansion for $f(z)$ as $z \to z_0$ in R if

$$f(z) = \sum_{n=0}^{N} a_n(z - z_0)^n + o[(z - z_0)^N],$$

for $N = 0, 1, 2, \ldots$, as $z \to z_0$ in R. We write

$$f(z) \sim \sum_{n=0}^{\infty} a_n(z - z_0)^n.$$

It is easy to verify that $\{w_n(z)\} = \{(z - z_0)^n\}$ is an asymptotic sequence as $z \to z_0$. Therefore, this is just a special case of the general definition of an asymptotic expansion. An important special case is when $z_0 = \infty$, and we write

$$f(z) \sim \sum_{n=0}^{\infty} \frac{a_n}{z^n},$$

as $z \to \infty$ in R if

$$f(z) = \sum_{n=0}^{N} \frac{a_n}{z^n} + o(z^{-N}),$$

as $z \to \infty$ in R.

Exercises 10.1

1. Prove that if $\sum_{n=1}^{N} a_n w_n(z)$ is an asymptotic expansion to N terms for $f(z)$ as $z \to z_0$, then it is an asymptotic expansion to $1, 2, \ldots, N - 1$ terms.

2. Prove that $\dfrac{1}{z} - \dfrac{2!}{z^3} + \dfrac{4!}{z^5} - \cdots$ is an asymptotic expansion for

$$\int_0^{\infty} e^{-zt}(1 + t^2)^{-1}\, dt,$$

as $z \to \infty$ in $R = \{z \mid z \neq 0, -\pi/2 + \alpha \leq \arg z \leq \pi/2 - \alpha\}, 0 < \alpha < \pi/2$.

3. Let $f(z)$ be analytic at z_0. Prove that the Taylor series expansion of $f(z)$ is an asymptotic expansion for $f(z)$ as $z \to z_0$.

4. Let $f(z) \sim \sum_{n=1}^{\infty} a_n w_n(z)$, as $z \to z_0$ in R. Prove that

$$f(z) = \sum_{n=0}^{N} a_n w_n(z) + O(w_{N+1}),$$

as $z \to z_0$ in R. In other words the error one makes in stopping with the Nth term is of the order of the first term omitted.

5. Prove that

$$\int_0^\infty \frac{e^{-t}dt}{z+t} \sim \sum_{n=0}^\infty \frac{(-1)^{n+1}(n-1)!}{z^n},$$

as $z \to \infty$ in $R = \{z \mid z \neq 0, -\pi/2 + \alpha \leq \arg z \leq \pi/2 - \alpha, 0 < \alpha < \pi/2\}$.

6. Prove that $\int_z^\infty e^{-t}t^{-1} dt \sim e^{-z}/z$, as $z \to \infty$ in

$$R = \{z \mid z \neq 0, -\pi/2 + \alpha \leq \arg z \leq \pi/2 - \alpha, 0 < \alpha < \pi/2\}.$$

10.2. OPERATIONS ON ASYMPTOTIC EXPANSIONS

In this section we shall discuss some of the common operations on asymptotic expansions such as addition, multiplication, division, differentiation, and integration. We shall be concerned mainly with asymptotic power series.

Theorem 10.2.1. *Let* $f(z) \sim \sum_{n=1}^N a_n w_n(z)$ *and* $g(z) \sim \sum_{n=1}^N b_n w_n(z)$*, both to* N *terms in* R*, as* $z \to z_0$*. Then if* α *and* β *are any complex constants*

$$\alpha f(z) + \beta g(z) \sim \sum_{n=1}^N (\alpha a_n + \beta b_n)w_n(z)$$

to N *terms in* R *as* $z \to z_0$*.*

Proof: This result follows immediately from

$$f(z) = \sum_{n=1}^N a_n w_n(z) + o(w_N),$$

$$g(z) = \sum_{n=1}^N b_n w_n(z) + o(w_N),$$

$$\alpha f(z) + \beta g(z) = \sum_{n=1}^N (\alpha a_n + \beta b_n)w_n(z) + \alpha o(w_N) + \beta o(w_N),$$

since $\alpha o(w_N) + \beta o(w_N) = o(w_N)$.

Corollary 10.2.1. *If* $f(z) \sim \sum_{n=1}^\infty a_n w_n(z)$ *and* $g(z) \sim \sum_{n=1}^\infty b_n w_n(z)$ *in* R *as* $z \to z_0$*, then* $\alpha f(z) + \beta g(z) \sim \sum_{n=1}^\infty (\alpha a_n + \beta b_n)w_n(z)$*.*

Theorem 10.2.2. *Let* $f(z) \sim \sum\limits_{n=0}^{N} a_n(z - z_0)^n$ *and* $g(z) \sim \sum\limits_{n=0}^{N} b_n(z - z_0)^n$ *in R as* $z \to z_0$. *Then in R as* $z \to z_0$,

$$f(z)g(z) \sim \sum_{n=0}^{N} c_n(z - z_0)^n,$$

where $c_n = \sum\limits_{k=0}^{n} a_k b_{n-k}$.

Proof: Without loss of generality we may take $z_0 = 0$. Then

$$f(z) = \sum_{n=0}^{N} a_n z^n + f_N(z),$$

$$g(z) = \sum_{n=0}^{N} b_n z^n + g_N(z),$$

where $f_N(z)$ and $g_N(z)$ are $o(z^N)$

$$
\begin{aligned}
f(z)g(z) = a_0 b_0 &+ (a_0 b_1 + b_0 a_1)z + \cdots \\
&+ (a_0 b_N + a_1 b_{N-1} + \cdots + a_{N-1} b_1 + a_N b_0)z^N \\
&+ g_N \sum_{n=0}^{N} a_n z^n + f_N \sum_{n=0}^{N} b_n z^n + f_N(z)g_N(z) \\
&+ (a_1 b_N + a_2 b_{N-1} + \cdots + a_{N-1} b_2 + a_N b_1)z^{N+1} \\
&+ \cdots
\end{aligned}
$$

All the terms after $(a_0 b_N + a_1 b_{N-1} + \cdots + a_{N-1} b_1 + a_N b_0)z^N$ are clearly $o(z^N)$. This proves the theorem.

Corollary 10.2.2. *Let* $f(z) \sim \sum\limits_{n=0}^{\infty} a_n(z - z_0)^n$ *and* $g(z) \sim \sum\limits_{n=0}^{\infty} b_n(z - z_0)^n$ *in R as* $z \to z_0$. *Then if* $c_n = \sum\limits_{k=0}^{n} a_k b_{n-k}$,

$$f(z)g(z) \sim \sum_{n=2}^{\infty} c_n(z - z_0)^n$$

in R as $z \to z_0$.

Theorem 10.2.3. *Let* $f(z) \sim \sum\limits_{n=0}^{N} a_n(z - z_0)^n$ *and* $g(z) \sim \sum\limits_{n=0}^{N} b_n(z - z_0)^n$ *in R as* $z \to z_0$, *where* $b_0 \neq 0$. *Then the asymptotic series for* $f(z)$ *can be divided formally by the asymptotic power series for* $g(z)$, *retaining terms up to degree N. The resulting series is an asymptotic power series to N terms for* $f(z)/g(z)$.

Proof: Without loss of generality we can assume that $z_0 = 0$. In R we have

$$f(z) = \sum_{n=0}^{N} a_n z^n + f_N(z),$$

$$g(z) = \sum_{n=0}^{N} b_n z^n + g_N(z)$$

$$= \left(\sum_{n=0}^{N} b_n z^n \right) \left[1 + G_N(z) \right],$$

where f_N and g_N are $o(z^N)$ as $z \to 0$, and

$$G_N(z) = g_N(z) \bigg/ \sum_{n=0}^{N} b_n z^n$$

is $o(z^N)$, since $b_0 \neq 0$. Therefore,

$$\frac{f(z)}{g(z)} = \frac{a_0 + a_1 z + \cdots + a_N z^N}{b_0 + b_1 z + \cdots + b_N z^N} [1 + o(z^N)] + \frac{f_N(z)}{\displaystyle\sum_{n=0}^{N} b_n z^n} [1 + o(z^N)].$$

Therefore, up to terms $o(z^N)$,

$$f(z)/g(z) \sim \frac{a_0 + a_1 z + \cdots + a_N z^N}{b_0 + b_1 z + \cdots + b_N z^N}.$$

Dividing and retaining terms up to $o(z^N)$ we have the result.

Corollary 10.2.3.   Let

$$f(z) \sim \sum_{n=0}^{\infty} a_n (z - z_0)^n \quad \text{and} \quad g(z) \sim \sum_{n=0}^{\infty} b_n (z - z_0)^n,$$

$b_0 \neq 0$, in R as $z \to z_0$. Then

$$\frac{f(z)}{g(z)} \sim \frac{a_0}{b_0} + \frac{a_1 b_0 - a_0 b_1}{b_0} z + \cdots$$

in R as $z \to z_0$.

Theorem 10.2.4.   *Let* $f(z) \sim \sum_{n=0}^{N+1} a_n (z - z_0)^n$ *in R as $z \to z_0$. If there exists a straight line path from z_0 to z in R when z is close to z_0, then*

$$\int_{z_0}^{z} f(\zeta) \, d\zeta \sim \sum_{n=1}^{N+1} \frac{a_{n-1}}{n} (z - z_0)^n,$$

as $z \to z_0$ in R, where the integration is along a straight line path.

Proof:

$$f(\zeta) = \sum_{n=0}^{N} a_n(z - z_0)^n + g(z),$$

where $g(z) = O[(z - z_0)^{N+1}]$, as $z \to z_0$. Then

$$\int_{z_0}^{z} f(\zeta)\,d\zeta = \sum_{n=1}^{N+1} \frac{a_{n-1}}{n}(z - z_0)^n + \int_{z_0}^{z} g(\zeta)\,d\zeta.$$

For z sufficiently close to z_0, $|g(\zeta)| \leq K|z - z_0|^{N+1}$ and

$$\left| \int_{z_0}^{z} g(\zeta)\,d\zeta \right| \leq K \int_{0}^{|z-z_0|} r^{N+1}\,dr = K \frac{|z - z_0|^{N+2}}{N + 2}.$$

Therefore,

$$\int_{z_0}^{z} f(\zeta)\,d\zeta = \sum_{n=1}^{N+1} \frac{a_{n-1}}{n}(z - z_0)^n + o[(z - z_0)^{N+1}],$$

which proves the theorem.

Corollary 10.2.4. Let $f(z) \sim \sum_{n=0}^{\infty} a_n(z - z_0)^n$ in R as $z \to z_0$. If there exists a straight line path from z_0 to z in R when z is close to z_0, then

$$\int_{z_0}^{z} f(\zeta)\,d\zeta \sim \sum_{n=1}^{\infty} \frac{a_{n-1}}{n}(z - z_0)^n$$

in R as $z \to z_0$.

An important special case, occurs when $z_0 = \infty$.

Theorem 10.2.5. Let $f(z) \sim \sum_{n=0}^{N+1} a_n z^{-n}$ as $z \to \infty$ for $\alpha \leq \arg z \leq \beta$. *Then*

$$\int_{z}^{\infty} \left[f(\zeta) - a_0 - \frac{a_1}{\zeta} \right] d\zeta \sim \sum_{n=1}^{N-1} \frac{a_{n+1}}{n} z^{-n},$$

where the integration is along a line with fixed argument between α and β.

Proof:

$$f(\zeta) - a_0 - a_1 \zeta^{-1} = a_2 \zeta^{-2} + a_3 \zeta^{-3} + \cdots + a_N \zeta^{-N} + g(\zeta),$$

where $g(\zeta) = O[\zeta^{-(N+1)}]$, as $\zeta \to \infty$. Then

$$\int_{z}^{\infty} [f(\zeta) - a_0 - a_1 \zeta^{-1}]\,d\zeta = \frac{a_2}{z} + \frac{a_3}{2z^2} + \cdots + \frac{a_N}{(N-1)z^{N-1}}$$
$$+ \int_{z}^{\infty} g(\zeta)\,d\zeta.$$

For $|\zeta|$ sufficiently large $|g(\zeta)| \leq K|\zeta|^{-(N+1)}$ and

$$\left| \int_z^\infty g(\zeta)\, d\zeta \right| \leq K \int_{|z|}^\infty r^{-(N+1)}\, dr = \frac{K}{N} |z|^{-N}.$$

Therefore,

$$\int_z^\infty [f(\zeta) - a_0 - a_1\zeta^{-1}]\, d\zeta = \sum_{n=1}^{N-1} \frac{a_{n+1}}{n} z^{-n} + o(z^{-(N-1)}),$$

which proves the theorem.

Corollary 10.2.5. Let $f(z) \sim \sum_{n=0}^\infty a_n z^{-n}$ as $z \to \infty$ for $\alpha \leq \arg z \leq \beta$. Then

$$\int_z^\infty [f(\zeta) - a_0 - a_1\zeta^{-1}]\, d\zeta \sim \sum_{n=1}^\infty \frac{a_{n+1}}{n} z^{-n},$$

where the integration is along a line with fixed argument between α and β.

Theorem 10.2.6. Let $f(z) \sim \sum_{n=0}^N a_n(z - z_0)^n$ in R as $z \to z_0$. Let $f'(z)$ exist in R and $f'(z) \sim \sum_{n=0}^{N-1} b_n(z - z_0)^n$ as $z \to z_0$. Let R satisfy the condition of Theorem 10.2.4 *for term by term integration of an asymptotic power series. Then* $b_0 = a_1, b_1 = 2a_2, b_2 = 3a_3, \ldots, b_{N-1} = Na_N$.

Proof: Integrating the series for $f'(z)$, we have

$$f(z) - \lim_{z \to z_0} f(z) = \int_{z_0}^z f'(\zeta)\, d\zeta \sim \sum_{n=1}^N \frac{b_{n-1}}{n} (z - z_0)^n$$

and by the uniqueness of the asymptotic power series for $f(z)$, we have

$$a_0 = \lim_{z \to z_0} f(z), b_0 = a_1, b_1 = 2a_2, b_2 = 3a_3, \ldots, b_{N-1} = Na_N.$$

Corollary 10.2.6. Let $f(z) \sim \sum_{n=0}^\infty a_n(z - z_0)^n$ in R, as $z \to z_0$. Let $f'(z)$ exist in R and $f'(z) \sim \sum_{n=0}^\infty b_n(z - z_0)^n$. Let R satisfy the condition of theorem 10.2.4 for term by term integration of an asymptotic power series. Then $b_0 = a_1, b_1 = 2a_2, b_2 = 3a_3, \ldots$.

Corollary 10.2.7. Let $f(z) \sim \sum_{n=0}^\infty a_n z^{-n}$ as $z \to \infty$ for $\alpha \leq \arg z \leq \beta$. Let $f'(z)$ exist in the same region and $f'(z) \sim - \sum_{n=1}^\infty nb_n z^{-(n+1)}$ as $z \to \infty$. Then $a_n = b_n, n = 1, 2, 3, \ldots$.

For analytic functions we can make a more definitive statement than the last corollary.

Theorem 10.2.7. *Let $f(z)$ be analytic in $R = \{z \mid |z| > r, \alpha \le \arg z \le \beta\}$. Let $f(z) \sim \sum_{n=0}^{\infty} a_n z^{-n}$ in R as $z \to \infty$. Then $f'(z) \sim -\sum_{n=1}^{\infty} n a_n z^{-(n+1)}$ as $z \to \infty$ in $R' = \{z \mid |z| > r, \alpha < \alpha_1 \le \arg z \le \beta_1 < \beta\}$.*

Proof: Consider $g(z) = f(z) - a_0 - a_1 z^{-1} - \cdots - a_N z^{-N}$ for a fixed N. This function is analytic in R and $O[z^{-(N+1)}]$ as $z \to \infty$. Therefore there exists an $r^* \ge r$ such that $|g(z)| \le M|z|^{-(N+1)}$ for $|z| > r^*$. Now, consider Figure 10.2.1.

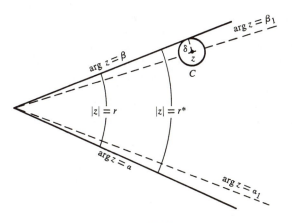

Figure 10.2.1.

We pick a z in R' with sufficiently large modulus that its minimum distance to the boundary of R is the perpendicular distance to one of the lines $\arg z = \alpha$ or $\arg z = \beta$. Let this distance be δ. We note that δ is proportional to $|z|$ and $\delta \to \infty$ as $|z| \to \infty$. Let $\delta = k|z|$. With center at z we draw a circle C of radius δ. Now,

$$g'(z) = \frac{1}{2\pi i} \int_C \frac{g(\zeta)}{(z - \zeta)^2} \, d\zeta,$$

$$|g'(z)| \le \frac{M}{\delta(|z| - \delta)^{N+1}} = \frac{M}{k|z|^{N+2}(1 - k)^{N+1}},$$

$$\lim_{|z| \to \infty} |z|^{N+1} |g'(z)| = 0.$$

This proves that $g'(z) = o[z^{-(N+1)}]$ which proves the theorem since

$$g'(z) = f'(z) + a_1 z^{-2} + 2a_2 z^{-3} + \cdots + N a_N z^{-(N+1)}.$$

Exercises 10.2

Let $f(z) \sim g(z)$, as $z \to z_0$ in R. Does this imply that $[f(z)]^n \sim [g(z)]^n$, $e^{f(z)} \sim e^{g(z)}$, and $\log f(z) \sim \log g(z)$? Explain.

10.3 ASYMPTOTIC EXPANSION OF INTEGRALS

As we have already indicated in Chapter 7, there are integral representations of certain special functions. If the function is the solution of a differential equation we may be able to obtain an integral representation of it in terms of its Laplace transform (Chapter 9) or its Fourier transform (Chapter 8). Therefore, the problem of finding asymptotic expansions of integrals plays an important role in analytic function theory.

In the example of Section 10.1, we obtained an asymptotic expansion for an integral by two procedures; that is, by expanding a part of the integrand in a series and formally integrating term by term, and by repeatedly integrating by parts. In this section, we shall generalize these methods to fit a much larger class of integral.

Suppose we have an integral of the form $\int_0^\infty F(t)e^{-zt}\, dt$, a Laplace transform. Suppose further that $F(t) = f(t^a)t^b$, $a > 0$, $b > -1$. If $f(x)$ has a Maclaurin expansion for $|x| < \delta$, then

$$f(x) = \sum_{n=0}^\infty \frac{f^{(n)}(0)}{n!} x^n = \sum_{n=0}^\infty a_n x^n$$

and

$$F(t) = \sum_{n=0}^\infty a_n t^{na+b}.$$

If we formally substitute this series into the integral and integrate term by term we obtain the series

$$\sum_{n=0}^\infty \frac{a_n \Gamma(na + b + 1)}{z^{na+b+1}}.$$

This procedure is not, in general, valid, because the resulting series usually diverges. However, we shall show that under quite general circumstances the series obtained in an asymptotic expansion for the integral. The following is known as Watson's lemma.

Theorem 10.3.1. *Let* $F(t) = f(t^a)t^b$, $a > 0$, $b > -1$. *Let* $f(x)$ *have a Maclaurin expansion for* $|x| < \delta$. *Let* $|F(t)| \leq Me^{ct}$ *for some constants* M *and* c *as* $t \to \infty$. *Let* $f(x)$ *be continuous for all values of* x. *Then*

$$\int_0^\infty F(t)e^{-zt}\, dt \sim \sum_{n=0}^\infty \frac{a_n \Gamma(na + b + 1)}{z^{na+b+1}},$$

as $z \to \infty$ *for* $|\arg z| \leq \pi/2 - \alpha$, $0 < \alpha < \pi/2$, *where* $a_n = \dfrac{f^{(n)}(0)}{n!}$.

Proof: The given conditions guarantee that $\int_0^\infty F(t)e^{-zt}\,dt$ exists for $\mathrm{Re}(z) > c$. Now, for a fixed N

$$\left| F(t) - \sum_{n=0}^N a_n t^{na+b} \right| \le K t^{(N+1)a+b} e^{ct}$$

for some K. Therefore, if $G(z) = \int_0^\infty F(t)e^{-zt}\,dt$,

$$G(z) = \sum_{n=0}^N \int_0^\infty a_n t^{na+b} e^{-zt}\,dt + R_N$$

$$= \sum_{n=0}^N a_n \Gamma(na + b + 1) z^{-(na+b+1)} + R_N,$$

and

$$|R_N| \le \int_0^\infty K t^{(N+1)a+b} e^{(c-x)t}\,dt = \frac{K\Gamma[(N + 1)a + b + 1]}{(x - c)^{(N+1)a+b+1}},$$

provided $\mathrm{Re}(z) > c$. If $|\arg z| \le \pi/2 - \alpha$, then $x > |z| \sin \alpha$, and $x - c > 0$ if $|z| > c \cos \alpha$. Hence,

$$|z^{Na+b+1} R_N| = \frac{K\Gamma[(N + 1)a + b + 1]z^{Na+b+1}}{(|z| \sin \alpha - c)^{(N+1)a+b+1}} \to 0,$$

as $|z| \to \infty$. This proves the theorem since the result holds for $N = 0, 1, 2, \ldots$.

EXAMPLE 10.3.1. Find an asymptotic expansion for the gamma function $\Gamma(z) = \int_0^\infty e^{-t} t^{z-1}\,dt$ for $z \to \infty$, $|\arg z| \le \pi/2 - \alpha$, $0 < \alpha < \pi/2$. Let $z = x > 0$ and let $t = sx$. Then

$$\Gamma(x) = \int_0^\infty e^{-t} t^{x-1}\,dt = \frac{1}{x}\Gamma(x + 1) = x^x e^{-x} \int_0^\infty (se^{1-s})^x\,ds.$$

The formula

$$\Gamma(z) = z^z e^{-z} \int_0^\infty (se^{1-s})^z\,ds$$

holds for $\mathrm{Re}(z) > 0$ since both sides are analytic in this domain and they agree on the positive real axis. We shall obtain our asymptotic expansion by applying Watson's lemma to the integral in the last expression. First we note that the function

$$\eta(s) = se^{1-s}$$

has a maximum of one at $s = 1$ and is monotonic in the intervals $0 \le s \le 1$ and $1 \le s < \infty$. See Figure 10.3.1.

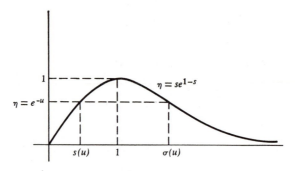

Figure 10.3.1.

For a given value of $\eta = e^{-u}$, there are two solutions of

$$e^{-u} = se^{1-s},$$
$$u = s - 1 - \ln s.$$

Let these solutions be $s(u)$ and $\sigma(u)$, where $0 \le s(u) \le 1$ and $1 \le \sigma(u) < \infty$. Therefore,

$$
\begin{aligned}
\int_0^\infty (se^{1-s})^z \, ds &= \int_0^1 (se^{1-s})^z \, ds + \int_1^\infty (\sigma e^{1-\sigma})^z \, d\sigma \\
&= \int_\infty^0 e^{-uz} \frac{ds}{du} \, du + \int_0^\infty e^{-uz} \frac{d\sigma}{du} \, du \\
&= \int_0^\infty e^{-uz} \left(\frac{d\sigma}{du} - \frac{ds}{du} \right) du.
\end{aligned}
$$

A simple calculation shows that

$$\frac{d\sigma}{du} - \frac{ds}{du} = \frac{\sigma}{\sigma - 1} + \frac{s}{1 - s} = \frac{1}{\sigma - 1} + \frac{1}{1 - s}$$

and since $\sigma \ne 1$ and $s \ne 1$ when $u \ne 0$, it follows that $\dfrac{d\sigma}{du} - \dfrac{ds}{du}$ is bounded for $u \ge \epsilon > 0$. It remains to show that

$$\frac{d\sigma}{du} - \frac{ds}{du} = f(u^a)u^b,$$

where $f(v)$ has a Maclaurin expansion for small $|v|$. ' Consider the implicit relation $(1/2)\zeta^2 = \omega - \log(1 + \omega)$. The reason for considering this is that $u = s - 1 - \ln s$ and if we put $2u = \zeta^2$, $s - 1 = \omega$, we obtain the relation between ζ and ω. Now, $\omega - \log(1 + \omega)$ is analytic at $\omega = 0$; in fact for $|\omega| < 1$,

$$\tfrac{1}{2}\zeta^2 = \omega - \log(1 + \omega) = \frac{\omega^2}{2} - \frac{\omega^3}{3} + \frac{\omega^4}{4} - \cdots,$$

$$\zeta = \pm\omega\left(1 - \frac{2}{3}\omega + \frac{2}{4}\omega^2 - \cdots\right)^{1/2},$$

and we see that ζ has two branches in the neighborhood of $\omega = 0$. Therefore, it follows from Theorem 5.3.3 that the equation

$$\zeta = \omega\left(1 - \frac{2}{3}\omega + \frac{2}{4}\omega^2 - \cdots\right)^{1/2}$$

has a unique solution

$$\omega(\zeta) = \zeta + b_2\zeta^2 + b_3\zeta^3 + \cdots,$$

where kb_k is the residue of ζ^{-k} at $\omega = 0$. It is an easy calculation to show that $b_2 = 1/3$, $b_3 = 1/36$, $b_4 = -1/270$, and so on. The other branch is $\omega_2(\zeta) = \omega_1(-\zeta)$. Putting $\omega_1 = \sigma - 1$, $\omega_2 = s - 1$, $\zeta^2 = 2u$, we have for $|u|$ small enough

$$\sigma = 1 + (2u)^{1/2} + \frac{1}{3}(2u) + \frac{1}{36}(2u)^{3/2} - \frac{1}{270}(2u)^2 + \cdots,$$

$$s = 1 - (2u)^{1/2} + \frac{1}{3}(2u) - \frac{1}{36}(2u)^{3/2} - \frac{1}{270}(2u)^2 + \cdots,$$

$$\frac{d\sigma}{du} - \frac{ds}{du} = \sqrt{2}\,u^{-1/2} + \frac{\sqrt{2}}{6}u^{1/2} + \cdots$$

$$= u^{-1/2}\left(\sqrt{2} + \frac{\sqrt{2}}{6}u + \cdots\right).$$

Putting the last expression into the integrand and formally integrating term by term, we have by Watson's lemma

$$\Gamma(z) \sim \sqrt{2\pi}\,e^{-z}z^{z-1/2}\left(1 + \frac{1}{12z} + \cdots\right),$$

as $z \to \infty$, $|\arg z| \leq \pi/2 - \alpha$, $0 < \alpha < \pi/2$. It can be shown that this expression is actually valid for $|\arg z| \leq \pi - \alpha$, $0 < \alpha < \pi$. (See Exercise 10.3.1.)

Next, we consider the development of asymptotic expansions by integration by parts. Consider the integral

$$\int_0^\infty g(t)e^{zh(t)}\,dt.$$

Proceeding formally, we have

$$\int_0^\infty g(t)e^{zh(t)}\,dt = \int_0^\infty zh'(t)e^{zh(t)}\left[\frac{g(t)}{zh'(t)}\right]dt$$

$$= \frac{g(t)}{zh'(t)}e^{zh(t)}\bigg|_0^\infty - \frac{1}{z}\int_0^\infty \frac{d}{dt}\frac{g(t)}{h'(t)}e^{zh(t)}\,dt$$

$$= -\frac{g(0)}{zh'(0)}e^{zh(0)} - \frac{1}{z}\int_0^\infty f(t)e^{zh(t)}\,dt,$$

provided $\mathrm{Re}(z) > 0$ and $h(t) \to -\infty$ as $t \to \infty$. We see that the resulting integral is of the first type considered, and therefore there is a chance that the process can be repeated. We begin by proving that under certain hypotheses on $h(t)$ and $g(t)$ the first term in our formula is an asymptotic formula for the integral as $z \to \infty$.

Theorem 10.3.2. *Let $g(t)$ be bounded and continuous for $0 \le t < \infty$ and $g(0) \ne 0$. Let $h(t)$ be real and continuous for $0 \le t < \infty$. Let $h'(0)$ exist and $h'(0) < 0$. Let $h(t) < h(0)$ for all positive t and $h(t) \to -\infty$, as $t \to \infty$. Let $\int_0^\infty e^{zh(t)}\,dt$ exist for $\mathrm{Re}(z) > 0$. Then*

$$\int_0^\infty g(t)e^{zh(t)}\,dt \sim -\frac{g(0)}{zh'(0)}e^{zh(0)}$$

as $z \to \infty$, $|\arg z| \le \pi/2 - \alpha$, $0 < \alpha < \pi/2$.

Proof: We first show that we may as well treat only the case $g(t) \equiv 1$, since if

$$\int_0^\infty e^{zh(t)}\,dt \sim -\frac{e^{zh(0)}}{zh'(0)},$$

then we may write

$$\int_0^\infty g(t)e^{zh(t)}\,dt = \int_0^\infty g(0)e^{zh(t)}\,dt + \int_0^\infty (g(t) - g(0))e^{zh(t)}\,dt.$$

Since $g(t)$ is continuous, given $\epsilon > 0$, there exists $\delta > 0$ such that

$$|g(t) - g(0)| < \epsilon \quad \text{for} \quad t < \delta,$$

and

$$\left|\int_0^\infty [g(t) - g(0)]e^{zh(t)}\,dt\right| \le \int_0^\infty |g(t) - g(0)|e^{zh(t)}\,dt$$

$$\le \epsilon \int_0^\delta e^{zh(t)}\,dt + 2M\int_0^\infty e^{zh(t)}\,dt,$$

where $|g(t) - g(0)| \le |g(t)| + |g(0)| \le 2M, 0 \le t < \infty$. Now,

$$\int_\delta^\infty e^{zh(t)}\, dt = \int_\delta^\infty e^{(x-1)h(t)}e^{h(t)}\, dt \le e^{-\eta(x-1)} \int_0^\infty e^{h(t)}\, dt,$$

since for some $\eta > 0$, $h < -\eta$ for $t \ge \delta$. Therefore, since ϵ is arbitrary and $e^{-\eta(x-1)} \to 0$, as $x \to \infty$, $\int_0^\infty g(t)e^{zh(t)}\, dt$ and $\int_0^\infty g(0)e^{zh(t)}\, dt$ have the same asymptotic formulas. We also have shown that $\int_0^\infty e^{zh(t)}\, dt$ and $\int_0^\delta e^{zh(t)}\, dt$ have the same asymptotic formula if $\delta > 0$. Now

$$\lim_{t\to 0^+} \frac{h(t) - h(0)}{t} = h'(0).$$

Hence, $|h(t) - h(0) - th'(0)| < \epsilon t$ for arbitrary $\epsilon > 0$ and some $\delta(\epsilon)$ such that $0 \le t \le \delta$. Therefore,

$$\int_0^\delta e^{z[h(0)+th'(0)-\epsilon t]}\, dt < \int_0^\delta e^{zh(t)}\, dt < \int_0^\delta e^{z[h(0)+th'(0)+\epsilon t]}\, dt.$$

All three of these integrals differ from the corresponding integral from 0 to ∞ by terms $O(e^{-\beta z})$, $\beta > 0$. Furthermore,

$$\int_0^\infty e^{z[h(0)+th'(0)-\epsilon t]}\, dt = \frac{-e^{zh(0)}}{z[h'(0) - \epsilon]},$$

$$\int_0^\infty e^{z[h(0)+th'(0)+\epsilon t]}\, dt = \frac{-e^{zh(0)}}{z[h'(0) + \epsilon]},$$

for $\text{Re}(z) > 0$. It follows, since ϵ is arbitrary, that

$$\int_0^\infty e^{zh(t)}\, dt \sim \frac{-e^{zh(0)}}{zh'(0)},$$

as $z \to \infty$ $|\arg z| \le \pi/2 - \alpha, 0 < \alpha < \pi/2$. This completes the proof.

EXAMPLE 10.3.2. Find an asymptotic expansion for the error function $E(z) = \int_z^\infty e^{-t^2}\, dt$, as $z \to \infty$, $|\arg z| \le \pi/2 - \alpha, 0 < \alpha < \pi/2$. Let $z = x > 0$. Then, letting $u = t - x$,

$$E(x) = \int_x^\infty e^{-t^2}\, dt = e^{-x^2} \int_0^\infty e^{-u^2}e^{-2ux}\, du.$$

This implies that $E(z) = e^{-z^2} \int_0^\infty e^{-u^2} e^{-2uz}\, du$ for $\text{Re}(z) > 0$, since in this domain both sides of the equation are analytic and agree on the real axis.

Now, the integral $\int_0^\infty e^{-u^2} e^{-2uz} \, du$ fits the hypotheses of Theorem 10.3.2 and by repeated application of this theorem, we have

$$E(z) \sim e^{-z^2} \left(\frac{1}{2z} - \frac{1}{2^2 z^3} + \frac{1\cdot3}{2^3 z^5} - \cdots \right),$$

for $z \to \infty$ $|\arg z| \leq \pi/2 - \alpha, 0 < \alpha < \pi/2$.

If, for the kind of integral covered by Theorem 10.3.2, $h'(0) = 0$, then the procedure must be modified. Therefore, we shall generalize to the case $h'(0) = 0$.

Theorem 10.3.3. *Let $g(t)$ be bounded and continuous for $0 \leq t < \infty$ and $g(0) \neq 0$. Let $h(t)$ be real and continuous for $0 \leq t < \infty$, and $h(t) < h(0)$ for all $t > 0$. Let $h'(t)$ exist for some δ such that $0 \leq t \leq \delta$, and $h'(0) = 0$. Let $h''(0) < 0$. Let $\int_0^\infty e^{zh(t)} \, dt$ exist for $\mathrm{Re}(z) > 0$. Then*

$$\int_0^\infty g(t)e^{zh(t)} \, dt \sim g(0) \sqrt{\frac{\pi}{-2zh''(0)}} \, e^{zh(0)},$$

as $z \to \infty$, $|\arg z| \leq \pi/2 - \alpha, 0 < \alpha < \pi/2$.

Proof: As in the proof of Theorem 10.3.2 we need only consider the asymptotic formula for $\int_0^\infty e^{zh(t)} \, dt$. By the extended mean value theorem for functions of the real variable t, we have for $\epsilon > 0$

$$|h(t) - h(0) - \tfrac{1}{2}h''(0)t^2| \leq \epsilon t^2,$$

for some $\delta > 0$ and $t \leq \delta$. As in the previous theorem, we need only consider the interval $0 \leq t \leq \delta$, where

$$\int_0^\delta e^{z[h(0)+(1/2)t^2h''(0)-\epsilon t^2]} \, dt < \int_0^\delta e^{zh(t)} \, dt < \int_0^\delta e^{z[h(0)+(1/2)t^2h''(0)+\epsilon t^2]} \, dt.$$

Now all three integrals have the same asymptotic formula as their counterparts integrated from 0 to ∞. Also it is well known that

$$\int_0^\infty e^{-u^2/2} \, du = \sqrt{\frac{\pi}{2}}.$$

Hence,

$$\int_0^\infty e^{z[h(0)+(1/2)t^2h''(0)-\epsilon t^2]} \, dt = e^{zh(0)} \sqrt{\frac{\pi}{-2z[h''(0) + \epsilon]}},$$

$$\int_0^\infty e^{z[h(0)+(1/2)t^2h''(0)+\epsilon t^2]} \, dt = e^{zh(0)} \sqrt{\frac{\pi}{-2z[h''(0) - \epsilon]}},$$

provided $z \neq 0$, $\mathrm{Re}(z) > 0$, and the branch cut of $\sqrt{z}$ is in the left half-plane. It follows since ϵ is arbitrary, that

$$\int_0^\infty e^{zh(t)}\,dt \sim \sqrt{\frac{\pi}{-2zh''(0)}}\,e^{zh(0)},$$

as $z \to \infty$, $|\arg z| \le \pi/2 - \alpha$, $0 < \alpha < \pi/2$. This completes the proof.

This last theorem is the basis for a method of asymptotic expansion for certain types of integrals known as the method of *steepest descent*. Suppose we have an integral of the form

$$\int_C g(\zeta)e^{zh(\zeta)}\,d\zeta,$$

where the integration is along some contour in the ζ plane on which $g(\zeta)$ and $h(\zeta)$ are analytic. If the integral is known to exist for $\mathrm{Re}(z) > 0$ and that its value does not change if the contour C is deformed, within certain limitation, then the method consists of changing the path to a new path to which Theorem 10.3.3 may be applied. Consider the equation

$$\sigma = F(\xi, \eta) = \mathrm{Re}[h(\zeta)].$$

This represents some sort of surface in the (ξ, η, σ) space. Now, σ cannot have a true maximum or minimum in the domain where $h(\zeta)$ is analytic since it is harmonic there. However, $\sigma(\xi, \eta)$ can have a saddle point, where $\sigma_\xi = \sigma_\eta = 0$. At such a point $h'(\zeta) = \sigma_x - i\sigma_n = 0$. At a saddle point the curves of steepest descent and steepest ascent are perpendicular to the curves of constant elevation, $\sigma = c$. Therefore, on curves of steepest descent and steepest ascent $\mathrm{Im}[h(\zeta)]$ is constant. On curves of steepest ascent $\mathrm{Re}[h(\zeta)]$ is increasing and hence positive.* But we have assumed that the integral exists for $\mathrm{Re}(z) > 0$; therefore, the path cannot follow curves of steepest ascent. In summary, we look for a path which passes through a zero ζ_0 of $h'(\zeta)$ and on which $\mathrm{Im}[h(\zeta)]$ is constant. Let us say that ζ_0 is at the end of the path (if not, we can express the result as the sum of integrals with this property). Then near the end point $h(\zeta) = h(\zeta_0) - t$, where t is real, positive, and increases as we move away from the point ζ_0. We are thus led to the consideration of integrals of the type

$$\int_0^\infty e^{-zt}g[\zeta(t)]\frac{d\zeta}{dt}\,dt.$$

We illustrate the method with the following example.

---

* We can assume, without loss of generality, that $h(\zeta_0) = 0$, where ζ_0 is the saddle point.

EXAMPLE 10.3.3.   Find an asymptotic expansion of the Airy function, $A(z)$ which is a particular solution of the Airy differential equation, $w'' - zw = 0$. We look for solutions of the form

$$w = \frac{1}{2\pi i} \int_C e^{z\zeta} W(\zeta)\, d\zeta,$$

where C is a contour in the ζ-plane to be determined. The motivation for this is that this is the general form for the inverse Laplace transform of the solution. Now

$$w'' = \frac{1}{2\pi i} \int_C \zeta^2 e^{z\zeta} W(\zeta)\, d\zeta,$$

$$zw = \frac{1}{2\pi i} \int_C z e^{z\zeta} W(\zeta)\, d\zeta = \frac{e^{z\zeta} W(\zeta)}{2\pi i}\bigg|_a^b - \frac{1}{2\pi i} \int_C e^{z\zeta} \frac{dW}{d\zeta}\, d\zeta,$$

where a and b are the end points of the contour. We shall assume that the contour is chosen so that

$$\frac{e^{z\zeta} W(\zeta)}{2\pi i}\bigg|_a^b = 0.$$

Hence, W satisfies

$$\frac{dW}{d\zeta} + \zeta^2 W = 0,$$

which has a solution $W(\zeta) = e^{-\zeta^3/3}$. Therefore,

$$w(z) = \frac{1}{2\pi i} \int_C e^{z\zeta} e^{-\zeta^3/3}\, d\zeta,$$

where the contour has to be chosen so that $e^{z\zeta} e^{-\zeta^3/3} \to 0$ at the ends of C. For this reason we pick paths which approach asymptotically the lines $\arg \zeta^3 = 0, 2\pi, 4\pi$, or $\arg \zeta = 0, 2\pi/3, 4\pi/3$. A particular path is shown in Figure 10.3.2. We shall define $A(z)$ as that solution of $w'' - zw = 0$

$$A(z) = \frac{1}{2\pi i} \int_C e^{z\zeta} e^{-\zeta^3/3}\, d\zeta,$$

where we shall pick the path of steepest descent in order to obtain the asymptotic expansion. First we make the change of variables $\zeta = z^{1/2} u$. Then

$$A(z) = \frac{\sqrt{z}}{2\pi i} \int_\Gamma e^{z^{3/2}(u - u^3/3)}\, du$$

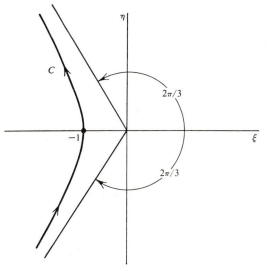

Figure 10.3.2.

and Γ is the path of steepest descent in the u plane. Let $h(u) = u - u^3/3$. Then $h'(u) = 1 - u^2 = 0$ at $u = \pm 1$. We choose -1 as our saddle point. At $u = -1$, $h(-1) = -1 + (1/3)$ is real. Therefore, we seek a path through -1 with $\text{Im}[h(u)] = 0$. If $u = \xi + i\eta$,

$$h(u) = \xi - \frac{\xi^3}{3} + \xi\eta^2 + i\left(\eta - \xi^2\eta + \frac{\eta^3}{3}\right),$$

$$\text{Im}[h(u)] = \frac{\eta}{3}(3 - 3\xi^2 + \eta^2) = 0.$$

The curve $\eta = 0$ is the real axis which does not approach the asymptotes $\arg u = \pm 2\pi/3$. The curve $\eta^2 - 3\xi^2 - 3 = 0$ does approach the asymptotes so we use this curve as the path of integration. We can then introduce the variable

$$t = h(-1) - \text{Re}[h(u)],$$

when u is on Γ. Then

$$A(z) = \frac{\sqrt{z}}{2\pi i}e^{-(2/3)z^{3/2}}\left[\int_0^{\infty} e^{-tz^{3/2}}\frac{du_1}{dt}\,dt - \int_0^{\infty} e^{-tz^{3/2}}\frac{du_2}{dt}\,dt\right],$$

where u_1 and u_2 correspond to the upper and lower halves of Γ, respectively. Actually, in this case, we can reduce the problem to an application of

Watson's lemma and get the whole expansion. Let $v = h(-1) - h(u) = -(2/3) - u + u^3/3 = (u + 1)^2[-1 + (u + 1)/3]$. Then

$$\pm v^{1/2} = (u + 1)[-1 + (u + 1)/3]^{1/2}.$$

By Theorem 5.3.3, $u + 1$ has an expansion in powers of $\pm v^{1/2}$, that is,

$$u + 1 = \sum_{n=1}^{\infty} b_n(\pm v^{1/2})^n,$$

where nb_n is the residue of $(u + 1)^{-n}[-1 + (u + 1)/3]^{-n/2}$, at $u = -1$. Therefore,

$$b_n = \frac{1}{n!}\left\{\frac{d^{n-1}}{dz^{n-1}}[-1 + (u + 1)/3]^{-n/2}\right\}_{u=-1}.$$

From this we obtain $b_1 = -i, b_2 = -1/6, b_3 = 5i/72$, and so on.

$$u_1 + 1 = it^{1/2} - (1/6)t - (5i/72)t^{3/2} + \cdots,$$

$$u_2 + 1 = -it^{1/2} - (1/6)t + (5i/72)t^{3/2} + \cdots,$$

$$\frac{du_1}{dt} - \frac{du_2}{dt} = it^{-1/2} - \frac{5i}{48}t^{1/2} + \cdots,$$

$$A(z) \sim \frac{\sqrt{z}}{2\pi}e^{-(2/3)z^{3/2}}\left(\frac{\Gamma(\frac{1}{2})}{z^{3/4}} - \frac{5}{24}\frac{\Gamma(\frac{3}{2})}{z^{9/4}} + \cdots\right)$$

$$\sim \frac{e^{-(2/3)z^{3/2}}}{2\sqrt{\pi}z^{1/4}}\left(1 - \frac{5}{48}\frac{1}{z^{3/2}} + \cdots\right).$$

The expansion is valid in the sector $|\arg z^{3/2}| \leq \pi/2 - \alpha, 0 < \alpha < \pi/2$, or $|\arg z| \leq \pi/3 - \beta, 0 < \beta < \pi/3$.

So far we have used the integration by parts technique only on integrals where the upper limit is ∞. Now, consider the following where a and b are finite and $a < b$:

$$\int_a^b g(t)e^{zh(t)}\,dt = \frac{g(t)}{zh'(t)}e^{zh(t)}\Big|_a^b - \frac{1}{z}\int_a^b e^{zh(t)}\frac{d}{dt}\left[\frac{g(t)}{h'(t)}\right]dt$$

$$= \frac{g(b)}{zh'(b)}e^{zh(b)} - \frac{g(a)}{zh'(a)}e^{zh(a)} - \frac{1}{z}\int_a^b e^{zh(t)}\frac{d}{dt}\left[\frac{g(t)}{h'(t)}\right]dt.$$

Suppose $h(b) < h(a)$, $h'(b) \neq 0$, $h'(a) \neq 0$, and that

$$\int_a^b g(t)e^{zh(t)}\,dt \sim \frac{g(b)}{zh'(b)}e^{zh(b)} - \frac{g(a)e^{zh(a)}}{zh'(a)},$$

as $z \to \infty$, $|\arg z| \leq \pi/2 - \alpha, 0 < \alpha < \pi/2$.

Then

$$\int_a^b g(t)e^{zh(t)}\, dt \sim -\frac{g(a)e^{zh(a)}}{zh'(a)},$$

because

$$\frac{e^{zh(b)}}{e^{zh(a)}} = e^{z[h(b)-h(a)]} \to 0,$$

as $z \to \infty$, $\mathrm{Re}(z) > 0$.

A similar situation arises when one wishes to obtain an asymptotic formula for $\int_a^b g(t)e^{ixh(t)}\, dt$, as $x \to \infty$, x real. We shall assume that $g(t)$ is continuous in the interval $a \leq t \leq b$, that $h(t)$ is real and twice continuously differentiable. We shall also assume that $h'(a) = 0$ but that $h'(t) \neq 0$ elsewhere in the interval. The method used to obtain the required formula is known as *the method of stationary phase*. It consists in showing that the principal contribution for large x is from the immediate neighborhood of $t = a$. The rational for this is that where $h(t)$ is changing and x is large $e^{ixh(t)}$ oscillates rapidly and the positive and negative swings in value tend to cancel out in the integration. On the other hand, where $h(t)$ is stationary, that is, $h'(t) = 0$, $e^{ixh(t)}$ does not oscillate and here the largest effect is felt. Let us be a little more precise.

Given $\epsilon > 0$ there is a $\delta > 0$ such that

$$|h(t) - h(a) - \tfrac{1}{2}h''(a)(t - a)^2| \leq \epsilon(t - a)^2,$$

when $t - a \leq \delta$. Then

$$\int_a^b g(t)e^{ixh(t)}\, dt = \int_a^{a+\delta} g(t)e^{ixh(t)}\, dt + \int_{a+\delta}^b g(t)e^{ixh(t)}\, dt.$$

In the interval $a + \delta \leq t \leq b$, $h'(t) \neq 0$. Therefore,

$$\int_{a+\delta}^b g(t)e^{ixh(t)}\, dt = \int_{h(a+\delta)}^{h(b)} g(t)e^{ixh}\frac{dt}{dh}\, dh$$

$$= \int_\alpha^\beta F(h)\,(\cos xh + i \sin xh)\, dh = O(x^{-1})$$

from the result of Theorem 8.8.1. On the other hand, letting $u = t - a$ in the other interval

$$\int_a^{a+\delta} g(t)e^{ix[h(a)+(1/2)h''(a)(t-a)^2]}\, dt = \int_0^\delta g(u + a)e^{ix[h(a)+(1/2)h''(a)u^2]}\, du.$$

Now, this last integral has the same asymptotic formula as*

$$\int_0^\infty g(a)e^{ixh(a)}e^{(ix/2)h''(a)u^2}\,du = \frac{g(a)e^{ixh(a)}}{\sqrt{\frac{1}{2}xh''(a)}}\int_0^\infty e^{iv^2}\,dv$$

$$= \frac{\sqrt{\pi}\,g(a)e^{ixh(a)}}{\sqrt{2xh''(a)}}e^{i\pi/4}.$$

Our asymptotic formula is then

$$\int_a^b g(t)e^{ixh(t)}\,dt \sim \frac{\sqrt{\frac{\pi}{2}}\,g(a)e^{ixh(a)}}{\sqrt{xh'(a)}}e^{i\pi/4}.$$

EXAMPLE 10.3.4.   Find an asymptotic formula for the Bessel function of the first kind of order n, n an integer. In Section 7.5, we proved the following formula

$$J_n(x) = \frac{1}{\pi}\int_0^\pi \cos(n\theta - x\sin\theta)\,d\theta$$

$$= \frac{1}{\pi}\,\mathrm{Re}\left[\int_0^\pi e^{in\theta}e^{-ix\sin\theta}\,d\theta\right].$$

Here, $h(\theta) = -\sin\theta$, $h'(\theta) = -\cos\theta$, $h''(\theta) = \sin\theta$, and the point of stationary phase is $\theta = \pi/2$, where $h(\pi/2) = -1$ and $h''(\pi/2) = 1$. Now,

$$\int_0^\pi e^{in\theta}e^{-ix\sin\theta}\,d\theta = \int_0^{\pi/2} e^{in\theta}e^{-ix\sin\theta}\,d\theta$$

$$+ \int_{\pi/2}^\pi e^{in\theta}e^{-ix\sin\theta}\,d\theta \sim \frac{\sqrt{2\pi}\,e^{in\pi/2}e^{-ix}}{\sqrt{x}}e^{i\pi/4}.$$

Therefore,

$$J_n(x) \sim \sqrt{\frac{2}{\pi x}}\cos\left(x - \frac{n\pi}{2} - \frac{\pi}{4}\right),$$

as $x \to \infty$.

Exercises 10.3

1. Prove that the asymptotic expansion for $\Gamma(z)$ of Example 10.3.1 is valid for $|\arg z| \le \pi - \alpha$, $0 < \alpha < \pi$. Hint: Show that

$$\Gamma(z) = z^z e^{-z}\int_0^\infty e^{-zt}F(t)\,dt$$

$$= z^z e^{-z}\int_0^\infty e^{-zte^{\beta i}}F(te^{\beta i})e^{\beta i}\,dt,$$

* We can assume, without loss of generality, that $h''(a) > 0$. If not we can consider $\int_a^b g(t)e^{-ixh(t)}\,dt$ and replace i by $-i$ in the formula.

if $|\beta| < \pi/2$ and $\mathrm{Re}(ze^{\beta i}) > 0$, and find an asymptotic expansion for the last integral.

2. Obtain *Stirling's formula* $n! \sim n^n\sqrt{2\pi n}\, e^{-n}$.

3. Show that $\Gamma(z) = z^z e^{-z} \int_0^\infty e^{z(1-t-\ln t)}\, dt$ and hence obtain the asymptotic formula $\Gamma(z) \sim z^z e^{-z} \sqrt{2\pi}/\sqrt{z}$, as $z \to \infty$, $|\arg z| \leq \pi/2 - \alpha$, $\alpha > 0$.

4. Obtain an asymptotic expansion of $\int_0^\infty t^z e^{-zt}\, dt$.

5. The Fresnel integrals are $C(x) = \int_0^x \cos t^2\, dt$ and $S(x) = \int_0^x \sin t^2\, dt$. Then $C(x) + iS(x) = \int_0^x e^{it^2}\, dt$. If

$$C(x) = \tfrac{1}{2}\sqrt{\pi/2} - P(x)\cos x^2 + Q(x)\sin x^2,$$
$$S(x) = \tfrac{1}{2}\sqrt{\pi/2} - P(x)\sin x^2 - Q(x)\cos x^2,$$

show that

$$P(x) \sim \frac{1}{2}\left(\frac{1}{2x^3} - \frac{1 \cdot 3 \cdot 5}{2^3 x^7} + \cdots\right),$$
$$Q(x) \sim \frac{1}{2}\left(\frac{1}{x} - \frac{1 \cdot 3}{2^2 x^5} + \frac{1 \cdot 3 \cdot 5 \cdot 7}{2^4 x^9} - \cdots\right).$$

6. Let n be an integer and y be real. Prove that

$$J_n(iy) \sim \frac{i^n e^y}{\sqrt{2\pi y}},$$

as $y \to \infty$.

7. Show that $\int_z^\infty e^{-t} t^{-\nu}\, dt \sim e^{-z} \sum_{k=0}^\infty (-1)^k a_k z^{-\nu-k}$, as $z \to \infty$, $|\arg z| \leq \pi/2 - \alpha$, $0 < \alpha < \pi/2$, where $a_0 = 0$ and $a_k = \nu(\nu + 1)\cdots(\nu + k - 1)$.

10.4. ASYMPTOTIC SOLUTIONS OF ORDINARY DIFFERENTIAL EQUATIONS

As we have seen, many of the important functions of mathematical physics come to us as the solutions of ordinary differential equations. In many applications, it is important to have asymptotic expansions of these functions. In the last section, we showed that in certain cases, for example, the Airy equation and the Bessel equation, we were able to obtain asymptotic expansions from the integral representation of the solutions. In this section, we shall show how we can obtain asymptotic solutions directly from the differential equation.

We shall consider the linear second order differential equation

$$w'' + p(z)w' + q(z)w = 0.$$

Most of the time we shall be interested in asymptotic expansions as $z \to \infty$. Therefore, we shall consider solutions of the differential equation near $z = \infty$. The result of Exercise 7.3.1 shows that, if $2z - z^2p(z)$ and $z^4q(z)$ are analytic at ∞, then ∞ is an ordinary point of the differential equation and there are two linearly independent solutions in the form of power series in $1/z$ which converge in a neighborhood of ∞. These series are asymptotic series which give us a means of evaluating the solutions for large $|z|$. The result of Exercise 7.4.1 shows that, if $zp(z)$ and $z^2q(z)$ are analytic at $z = \infty$, then ∞ is a regular singular point of the differential equation, and the Frobenius method in general gives two linearly independent solutions of the form

$$w_1 = z^{-\alpha_1}(c_0 + c_1z^{-1} + c_2z^{-2} + \cdots),$$
$$w_2 = z^{-\alpha_2}(\gamma_0 + \gamma_1z^{-1} + \gamma_2z^{-2} + \cdots),$$

if α_1 and α_2, the roots of the indicial equation, do not differ by an integer. If α_1 and α_2 differ by an integer the general solution may contain a term in $\log z$ and possibly positive powers of z. In either case, the solutions are asymptotic expansions as $z \to \infty$. It remains to consider the irregular singularity at ∞.

The result of Exercise 10.4.1 will show that we may consider the simpler differential equation

$$w'' + q(z)w = 0.$$

If $q(z) = O(z^{-2})$, as $z \to \infty$, we have a regular singular point at ∞. If $q(z) = O(z^{-1})$ or $q(z) = O(1)$, as $z \to \infty$ we have an irregular singular point at ∞. We shall only consider the case where $q(z)$ is analytic at ∞, and therefore we shall assume that

$$q(z) = q_0 + q_1z^{-1} + q_2z^{-2} + \cdots,$$

where q_0 and q_1 are not both zero.

The Bessel equation of order ν, for example, is

$$w'' + \frac{1}{z} w' + \frac{z^2 - \nu^2}{z^2} w = 0.$$

By the transformation $u = \sqrt{z}\, w$ it is changed to

$$u'' + \left(1 - \frac{4\nu^2 - 1}{4z^2}\right) u = 0.$$

Therefore, there is an irregular singularity at ∞, even though

$$q = 1 - \frac{4\nu^2 - 1}{4z^2}$$

is analytic at ∞. We have already seen some asymptotic formulas for Bessel functions. These have contained a term of the form $e^{\alpha z} z^{-\beta}$. Therefore, we first try for a solution of the form*

$$w(z) = e^{\alpha z} \sum_{k=0}^{\infty} c_k z^{-\beta-k}.$$

Proceeding formally, we have

$$w'(z) = \alpha e^{\alpha z} \sum_{k=0}^{\infty} c_k z^{-\beta-k} - e^{\alpha z} \sum_{k=0}^{\infty} (\beta + k)c_k z^{-\beta-k-1},$$

$$w''(z) = \alpha^2 e^{\alpha z} \sum_{k=0}^{\infty} c_k z^{-\beta-k} - 2\alpha e^{\alpha z} \sum_{k=0}^{\infty} (\beta + k)c_k z^{-\beta-k-1}$$

$$+ e^{\alpha z} \sum_{k=0}^{\infty} (\beta + k)(\beta + k + 1)c_k z^{-\beta-k-2}.$$

Substituting in the equation $w'' + q(z)w = 0$, we have

$$\alpha^2 \sum_{k=0}^{\infty} c_k z^{-\beta-k} - 2\alpha \sum_{k=0}^{\infty} (\beta + k)c_k z^{-\beta-k-1}$$

$$+ \sum_{k=0}^{\infty} (\beta + k)(\beta + k + 1)c_k z^{-\beta-k-2}$$

$$+ \left(\sum_{k=0}^{\infty} q_k z^{-k} \right)\left(\sum_{k=0}^{\infty} c_k z^{-\beta-k} \right) = 0.$$

Comparing coefficients, we have

$$\alpha^2 c_k - 2\alpha(\beta + k - 1)c_{k-1} + (\beta + k - 1)(\beta + k - 2)c_{k-2}$$
$$+ \sum_{j=0}^{k} q_j c_{k-j} = 0,$$

for $k = 0, 1, 2, \ldots$, if we define $c_{-1} = c_{-2} = 0$. Since $c_0 \neq 0$

$$\alpha^2 + q_0 = 0, \quad -2\alpha\beta + q_1 = 0,$$

and for $k = 1, 2, 3, \ldots$,

$$2\alpha k c_k = (\beta + k)(\beta + k - 1)c_{k-1} + \sum_{i=2}^{k+1} q_i c_{k+1-i}.$$

---

* These are called *normal solutions*.

If $q_0 \neq 0$, then $\alpha = \pm\sqrt{-q_0}$ and $\beta = \mp q_1/2\sqrt{-q_0}$, and the coefficients $c_1, c_2, c_3, \ldots$ are determined recursively by the other equation. In this case, two normal solutions are formally determined. If $q_0 = q_1 = 0$, then we have the case of the regular singular point, $\alpha = 0$ and β must satisfy the indicial equation $\beta(\beta + 1) + q_2 = 0$. In this case, one proceeds as in the Frobenius method. If $q_0 = 0$ and $q_1 \neq 0$, then no normal solutions exist. However, in this case *subnormal solutions** can be determined by making the change of variables.

$$\zeta = \sqrt{z}, \qquad u(\zeta) = w(z)/\sqrt{\zeta}.$$

In the cases where normal solutions exist, they exist only in the formal sense, that is, the necessary constants and coefficients can be determined. In general, the series diverge. However, we can show that they are asymptotic expansions of certain solutions of the differential equation.

The first step is to transform the differential equation by the change of variable $w(z) = e^{\alpha z} z^{-\beta} u(z)$, where α and β are the constants determined above. It turns out that the differential equation satisfied by $u(z)$ is

$$u'' + 2\left(\alpha - \frac{\beta}{z}\right)u' + \left[\alpha^2 - \frac{2\alpha\beta}{z} + \frac{\beta(\beta + 1)}{z^2} + q(z)\right]u = 0.$$

Since $\alpha^2 = -q_0$ and $2\alpha\beta = q_1$, we can write

$$u'' + 2\left(\alpha - \frac{\beta}{z}\right)u' + \frac{1}{z^2}Q(z)u = 0,$$

where $Q(z) = z^2[q(z) - q_0 - q_1 z^{-1}] + \beta(\beta + 1)$ is bounded as $z \to \infty$. Multiplying by $e^{2\alpha z} z^{-2\beta}$, we can write

$$\frac{d}{dz}\left(e^{2\alpha z} z^{-2\beta}\frac{du}{dz}\right) + e^{2\alpha z} z^{-2\beta-2} Q(z)u = 0.$$

Integrating, we have

$$\frac{du}{dz} + e^{-2\alpha z} z^{2\beta}\int_{z_1}^{z} e^{2\alpha\zeta} \zeta^{-2\beta-2} Q(\zeta)u(\zeta)\, d\zeta = \gamma_1 e^{-2\alpha z} z^{2\beta}$$

and integrating again, we have

$$u(z) = \gamma_2 + \gamma_1\int_{z_1}^{z} e^{-2\alpha\zeta} \zeta^{2\beta}\, d\zeta - \int_{z_1}^{z} K(z, \eta)Q(\eta)u(\eta)\eta^{-2}\, d\eta,$$

where $K(z, \eta) = -\int_{z_1}^{\eta} e^{2\alpha\eta} e^{-2\alpha\zeta}(\zeta/\eta)^{2\beta}\, d\zeta$. We have expressed the general solution of the differential equation in terms of the solution of a Volterra

* See Exercise 10.4.2.

integral equation.* It can be shown that a solution of the integral equation is a solution of the differential equation. Particular solutions of the differential equation can be obtained by the proper choices of γ_1, γ_2, and z_1. In this section, we are interested in solutions near ∞. Therefore, we put $z_1 = \infty$ and consider the equation

$$u(z) = \gamma_2 - \gamma_1 \int_z^\infty e^{-2\alpha\zeta} \zeta^{2\beta}\, d\zeta + \int_z^\infty K(z, \eta)Q(\eta)u(\eta)\eta^{-2}\, d\eta.$$

Now, γ_2 is a constant and we can obtain an asymptotic expansion for $\int_z^\infty e^{-2\alpha\zeta} \zeta^{2\beta}\, d\zeta$ (Exercise 10.3.7). Therefore, it remains to find an asymptotic expansion for the solution of an equation of the form

$$w(z) = 1 + \int_z^\infty K(z, \eta)Q(\eta)w(\eta)\eta^{-2}\, d\eta.$$

This is done by a method of successive approximations. We shall not carry out the general case,† but instead will illustrate the method using the Bessel equation.

EXAMPLE 10.4.1. Find asymptotic expansions of solutions of the Bessel equation

$$w'' + \frac{1}{z} w' + \left(1 - \frac{\nu^2}{z^2}\right) w = 0.$$

We first make the change of variable $u = \sqrt{z}\, w$. Then u satisfies

$$u'' + \left(1 - \frac{\nu^2 - \frac{1}{4}}{z^2}\right) u = 0.$$

In this case, $q = 1 - (\nu^2 - 1/4)/z^2$, so that $q_0 = 1$ and $q_1 = 0$. Therefore $\alpha = \pm i$ and $\beta = 0$. The recurrence relation for the formal solutions is

$$\pm 2ikc_k = k(k - 1)c_{k-1} - (\nu^2 - \tfrac{1}{4})c_{k-1},$$

$k = 1, 2, 3, \ldots$.

Turning to the integral equation, we have, if we make the further change $u(z) = e^{\pm iz} v(z)$, that v satisfies

$$\frac{d}{dz}\left(e^{\pm 2iz} \frac{dv}{dz}\right) = \frac{\nu^2 - \frac{1}{4}}{z^2} e^{\pm 2iz}v.$$

---

* The ideas behind this formulation of the solution in terms of a Volterra integral equation have been developed in Exercises 7.2. This is also similar to what was done in developing the Fredholm formulation of boundary value problems in Section 7.8.

† See A. Erdelyi, *Asymptotic Expansions.* New York: Dover Publications, Inc., 1956, pp. 64–72.

and the integral equation is

$$v(z) = \gamma_2 + \gamma_1 e^{\mp 2iz} + \frac{1}{2i}\int_z^\infty [1 - e^{\pm 2i(\eta - z)}]\frac{(\nu^2 - \frac{1}{4})}{\eta^2}v(\eta)\,d\eta.$$

In particular, let us take $\gamma_1 = 0$, $\gamma_2 = 1$, and $\alpha = i$, and then

$$v(z) = 1 + \int_z^\infty K(z, \eta)\frac{\nu^2 - \frac{1}{4}}{\eta^2}v(\eta)\,d\eta,$$

and we shall investigate the asymptotic behavior of the solution of this equation, where

$$K(z, \eta) = \frac{1}{2i}[1 - e^{2i(\eta - z)}].$$

Let us specify the path of integration in the integral equation to be along the line $\arg \eta = \arg z$. Then $\eta - z = (|\eta| - |z|)e^{i\theta}$, where θ is constant and

$$|(\nu^2 - \tfrac{1}{4})K(z, \eta)| \le \frac{|\nu^2 - \frac{1}{4}|}{2}[1 + e^{-2(|\eta| - |z|)\sin\theta}] \le M,$$

where M is a constant and provided $0 < \alpha \le \arg z \le \pi - \alpha$. We begin the iteration procedure by letting $v_0 = 1$ and

$$v_1(z) = 1 + \int_z^\infty K(z, \eta)\frac{\nu^2 - \frac{1}{4}}{\eta^2}\,d\eta.$$

Then

$$|v_1(z) - v_0| = \left|\int_z^\infty K(z, \eta)\frac{\nu^2 - \frac{1}{4}}{\eta^2}\,d\eta\right| \le \frac{M}{|z|}.$$

Next, we let

$$v_2(z) = 1 + \int_z^\infty K(z, \eta)\frac{\nu^2 - \frac{1}{4}}{\eta^2}v_1(\eta)\,d\eta.$$

Then

$$|v_2(z) - v_1(z)| = \left|\int_z^\infty K(z, \eta)\frac{\nu^2 - \frac{1}{4}}{\eta^2}(v_1 - v_0)\,d\eta\right| \le \frac{M^2}{2!}\frac{1}{|z|^2}.$$

In general, we let

$$v_n(z) = 1 + \int_z^\infty K(z, \eta)\frac{\nu^2 - \frac{1}{4}}{\eta^2}v_{n-1}(\eta)\,d\eta$$

and by an obvious induction

$$|v_n(z) - v_{n-1}(z)| \le \frac{M^n}{n!}\frac{1}{|z|^n}.$$

Now, consider the series $v_0 + \sum_{n=1}^{\infty} [v_n(z) - v_{n-1}(z)]$. Let $|z| > R$. Then by comparison with the series of constants

$$\sum_{n=1}^{\infty} \frac{M^n}{n! \, R^n},$$

we have uniform convergence of the series. Clearly, each term in the series is an analytic function. Therefore, if

$$v(z) = v_0 + \sum_{n=1}^{\infty} [v_n(z) - v_{n-1}(z)],$$

then $v(z)$ is analytic in the sector $0 < \alpha \leq \arg z \leq \pi - \alpha$. We can show that $v(z)$ is a solution of the integral equation by substitution, using the uniform convergence to integrate term by term.

$$1 + \int_z^{\infty} K(z, \eta) \frac{v^2 - \frac{1}{4}}{2} \left\{ 1 + \sum_{n=1}^{\infty} [v_n(\eta) - v_{n-1}(\eta)] \right\} d\eta$$

$$= 1 + v_1 + \sum_{n=1}^{\infty} [v_{n+1}(z) - v_n] = v(z).$$

The uniqueness of the solution will be covered in the exercises.

Now, consider

$$|v(z) - v_N(z)| = \left| \sum_{n=N+1}^{\infty} [v_n(z) - v_{n-1}(z)] \right| \leq \sum_{n=N+1}^{\infty} \frac{M^n}{n! \, |z|^n} = O\left(\frac{1}{|z|^{N+1}}\right).$$

This proves that $v(z) = v_N(z) + O(z^{-N-1})$. If we can find an asymptotic expansion of $v_N(z)$ to N terms, then this will give us an asymptotic expansion to N terms of $v(z)$. Consider, for example

$$v_1(z) = 1 + \int_z^{\infty} K(z, \eta) \frac{v^2 - \frac{1}{4}}{\eta^2} d\eta$$

$$= 1 + \frac{v^2 - \frac{1}{4}}{2i} \int_z^{\infty} \frac{1 - e^{2i(\eta - z)}}{\eta^2} d\eta$$

$$= 1 + \frac{v^2 - \frac{1}{4}}{2i} \frac{1}{\eta^2} \left[\eta - \frac{e^{2i(\eta - z)}}{2i} \right]_z^{\infty} + \frac{v^2 - \frac{1}{4}}{i} \int_z^{\infty} \frac{1}{\eta^3} \left[\eta - \frac{e^{2i(\eta - z)}}{2i} \right] d\eta$$

$$= 1 - \frac{v^2 - \frac{1}{4}}{2iz} + O\left(\frac{1}{|z|^2}\right).$$

Therefore, $v(z) \sim 1 - (v^2 - 1/4)/2iz$, as $|z| \to \infty$. Notice that in the formal solutions with $\alpha = i, c_0 = 1, c_1 = -(v^2 - 1/4)/2i$. Hence, the formal solution gives us the asymptotic expansion to this many terms. It can be shown that the formal solutions are asymptotic expansions of solutions of the differential equation. If we take the other case $\alpha = -i$, we shall obtain an asymptotic solution in the sector $-\pi + \alpha \leq \arg z \leq -\alpha < 0$. This completes our discussion of the example.

There are times when we have a differential equation in the real variable x which contains a parameter λ, and we wish an asymptotic expansion of the solution as $\lambda \to \infty$. Consider, for example, the differential equation

$$y'' + [\lambda^2 p(x) + q(z)]y = 0$$

in the interval $a \leq x \leq b$, where $q(x)$ is continuous, and $p(x)$ is positive and has two continuous derivatives. We can make the change of variables

$$\xi = \int_a^x \sqrt{p(t)}\, dt, \qquad \eta = [p(x)]^{1/4}y.$$

The new interval becomes $0 \leq \xi \leq \beta = \int_a^b \sqrt{p(t)}\, dt$ and the differential equation becomes

$$\frac{d^2\eta}{d\xi^2} + \lambda^2\eta = r(\xi)\eta,$$

where $r(\xi)$ is the continuous function

$$r(\xi) = \frac{1}{p^3}\left[\frac{1}{4}\,pp'' - \frac{5}{16}\,(p')^2 - qp^2\right].$$

By the method outlined in Exercises 7.2, we can show that η satisfies the integral equation

$$\eta(\xi) = c_1 \sin \lambda\xi + c_2 \cos \lambda\xi + \frac{1}{\lambda}\int_0^\xi \sin \lambda(\xi - t)r(t)\eta(t)\, dt,$$

where $c_1 = \eta(0)$ and $c_2 = \eta'(0)/\lambda$. We solve the integral equation by successive approximations as follows:

$$\eta_0(\xi) = c_1 \sin \lambda\xi + c_2 \cos \lambda\xi,$$

$$\eta_n(\xi) = c_1 \sin \lambda\xi + c_2 \cos \lambda\xi + \frac{1}{\lambda}\int_0^\xi \sin \lambda(\xi - t)r(t)\eta_{n-1}(t)\, dt,$$

$n = 1, 2, 3, \ldots$. Then if $|r(\xi)| \leq M$,

$$|\eta_n(\xi) - \eta_{n-1}(\xi)| \leq \frac{M^n(|c_1| + |c_2|)\xi^n}{n!\,\lambda^n} \leq \frac{M^n(|c_1| + |c_2|)\beta^n}{n!\,\lambda^n}.$$

We consider the series

$$\eta(\xi) = \eta_0 + \sum_{n=1}^\infty [\eta_n(\xi) - \eta_{n-1}(\xi)].$$

By comparison with

$$\sum_{n=0}^{\infty} \frac{M^n(|c_1| + |c_2|)\beta^n}{n!\,\lambda^n},$$

we have uniform convergence of the series with respect to x, and we can prove that this is a solution by direct substitution. Next, we consider

$$|\eta(\xi) - \eta_N(\xi)| = \left| \sum_{n=N+1}^{\infty} [\eta_n(\xi) - \eta_{n-1}] \right| \le \sum_{n=N+1}^{\infty} \frac{M^n(|c_1| + |c_2|\beta^n}{n!\,\lambda^n}$$

$$= O\left(\frac{1}{\lambda^{N+1}}\right).$$

This shows that to N terms $\eta(\xi)$ and $\eta_N(\xi)$ have the same asymptotic expansion, considered as a function of λ.

Exercises 10.4

1. Show that under the transformation $w = ue^{-(1/2)\int p\,dz}$ the differential equation $w'' + pw' + qw = 0$ takes the form

$$u'' + (q - \tfrac{1}{2}p' - \tfrac{1}{4}p^2)\,u = 0$$

and that an ordinary point at ∞ becomes a regular singular point at ∞ while a regular singular point at ∞ becomes a regular singular point at ∞.

2. Consider the differential equation $w'' + q(z)w = 0$, where $q(z) = q_1 z^{-1} + q_2 z^{-2} + \cdots, q_1 \ne 0$. Make the change of variables $u(\zeta) = \sqrt{z}\,w(z)$, $\zeta = \sqrt{z}$ and show that subnormal solutions of the form

$$w = e^{\alpha\sqrt{z}} \sum_{k=0}^{\infty} c_k z^{(1/4)-(1/2)k}$$

exist.

3. Let $f(z)$ be any analytic function bounded in the sector

$$0 < \alpha \le \arg z \le \pi - \alpha.$$

Show that, in Example 10.4.1, if $v_0 = f(z)$ the iteration will lead to the same solution $v(z)$. How does this prove uniqueness of the solution of the integral equation?

4. In Example 10.4.1, obtain an asymptotic expansion of $v_2(z)$ to two terms past the constant term and show that it has the same coefficients as the formal solution up to the term $c_2 z^{-2}$.

5. Using the power series expansion of the Bessel function $J_\nu(x)$, show that for fixed x, $J_\nu(x) \sim \dfrac{1}{\Gamma(\nu + 1)}\left(\dfrac{x}{2}\right)^\nu$, as $\nu \to \infty$.

References

Analytic Function Theory

1. L. V. Ahlfors, *Complex Analysis*. New York: McGraw-Hill Book Company, Inc., 1953.
2. C. Caratheodory, *Theory of Functions*, I and II. New York: Chelsea Publishing Company, 1954.
3. R. V. Churchill, *Introduction to Complex Variables and Applications*, 2nd ed. New York: McGraw-Hill Book Company, Inc., 1960.
4. E. T. Copson, *An Introduction to the Theory of Functions of a Complex Variable*. London: Oxford University Press, 1935.
5. P. Franklin, *Functions of Complex Variables*. Englewood Cliffs, N.J.: Prentice-Hall, Inc., 1958.
6. B. A. Fuchs and B. V. Shabat, *Functions of a Complex Variable and some of their Applications*, I. Reading, Mass.: Addison-Wesley Publishing Company, Inc., 1964.
7. E. Hille, *Analytic Function Theory*, I and II. Boston: Ginn and Company, 1959.
8. R. Knopp, *Theory of Functions*, I and II plus 2 vols. of problems. New York: Dover Publications, Inc., 1945.
9. W. R. LePage, *Complete Variables and the Laplace Transform for Engineers*. New York: McGraw-Hill Book Company, Inc., 1961.
10. T. M. MacRobert, *Functions of a Complex Variable*. London: Macmillan and Company, Ltd., 1938.
11. R. S. Miller, *Advanced Complex Calculus*. New York: Harper and Brothers, 1960.
12. Z. Nehari, *Introduction to Complex Analysis*. Boston: Allyn and Bacon, Inc., 1961.
13. L. L. Pennisi, *Elements of Complex Variables*. New York: Holt, Rinehart, and Winston, 1963.

14. E. G. Phillips, *Functions of a Complex Variable with Applications*, 6th ed. Edinburgh: Oliver and Boyd, Ltd., 1949.

15. W. J. Thron, *Introduction to the Theory of Functions of a Complex Variable*. New York: John Wiley and Sons, Inc., 1953.

16. E. C. Titchmarsh, *The Theory of Functions*, 2nd ed. London: Oxford University Press, 1939.

17. E. T. Whittaker and G. N. Watson, *A Course of Modern Analysis*, 4th ed. London: Cambridge University Press, 1952.

Potential Theory and Conformal Mapping

1. L. Bieberbach, *Conformal Mapping*. New York: Chelsea Publishing Company, 1953.

2. C. Caratheodory, *Conformal Representations*. London: Cambridge University Press, 1941.

3. O. D. Kellogg, *Foundations of Potential Theory*. New York: Dover Publications, Inc., 1953.

4. Kober, *Dictionary of Conformal Representations*. London: Admiralty Computing Service, British Admiralty, 1945.

5. S. G. Mikhlin, *Integral Equations*. New York: Pergamon Press, 1957.

6. N. I. Muskhelishvili, *Singular Integral Equations*. Groningen, Holland: P. Noordhoff N. V., 1953.

7. Z. Nehari, *Conformal Mapping*. New York: McGraw-Hill Book Company, Inc., 1952.

Differential Equations

1. G. Birkhoff and G. C. Rota, *Ordinary Differential Equations*. Boston: Ginn and Company, 1962.

2. E. A. Coddington and N. Levinson, *Theory of Ordinary Differential Equations*. New York: McGraw-Hill Book Company, Inc., 1955.

3. J. W. Dettman, *Mathematical Methods in Physics and Engineering*. New York: McGraw-Hill Book Company, Inc., 1962.

4. B. A. Fuchs and V. I. Levin, *Functions of a Complex Variable and some of their Applications*, II. Reading, Mass.: Addison-Wesley Publishing Company, Inc., 1961.

5. E. L. Ince, *Ordinary Differential Equations*, New York: Dover Publications, Inc., 1956.

6. E. C. Titchmarsh, *Eigenfunctions Expansions Associated with Second Order Differential Equations*, I and II. London: Oxford University Press, 1946.

7. G. N. Watson, *A Treatise on the Theory of Bessel Functions*, 2nd ed. New York: The Macmillan Company, 1944.

Fourier Transforms

1. G. A. Campbell and R. M. Foster, *Fourier Integrals for Practical Applications.* New York: D. Van Nostrand Company, 1948.
2. H. S. Carslaw, *Introduction to the Theory of Fourier's Series and Integrals,* 3rd ed. New York: Dover Publications, Inc., 1930.
3. J. W. Dettman, *Mathematical Methods in Physics and Engineering.* New York: McGraw-Hill Book Company, Inc., 1962.
4. B. Noble, *Methods Based on the Wiener-Hopf Technique for the Solution of Partial Differential Equations.* New York: Pergamon Press, 1958.
5. I. N. Sneddon, *Fourier Transforms.* New York: McGraw-Hill Book Company, Inc., 1951.
6. E. C. Titchmarsh, *Introduction to the Theory of Fourier Integrals,* London: Oxford University Press, 1937.

Laplace Transforms

1. H. S. Carslaw and J. C. Jeager, *Operational Methods in Applied Mathematics.* New York: Dover Publications, Inc., 1963.
2. R. V. Churchill, *Modern Operational Mathematics in Engineering,* 2nd ed. New York: McGraw-Hill Book Company, Inc., 1958.
3. D. L. Holl, C. G. Maple, and B. Vinograde, *Introduction to the Laplace Transform,* New York: Appleton-Century-Crofts, Inc., 1959.
4. W. R. LePage, *Complex Variables and the Laplace Transform for Engineers.* New York: McGraw-Hill Book Company, Inc., 1961.
5. N. W. McLachlan, *Complex Variable Theory and Operational Calculus with Technical Applications,* 2nd ed. London: Cambridge University Press, 1953.
6. N. W. McLachlan, *Modern Operational Calculus with Applications in Technical Mathematics.* New York: Dover Publications, Inc., 1962.
7. E. J. Scott, *Transform Calculus with an Introduction to Complex Variables,* New York: Harper and Brothers, 1955.

Asymptotic Expansions

1. L. Cesari, *Asymptotic Behavior and Stability Problems in Ordinary Differential Equations.* New York: Academic Press, 1963.
2. N. G. DeBruijn, *Asymptotic Methods in Analysis.* Amsterdam: North-Holland Publishing Company, 1958.
3. A. Erdelyi, *Asymptotic Expansions.* New York: Dover Publications, Inc., 1956.
4. T. Heading, *An Introduction to Phase—Integral Methods,* London: Methuen and Company Ltd., 1962.
5. H. Jeffreys, *Asymptotic Approximations.* London: Oxford University Press, 1962.

Index

A CATALOG OF SELECTED
DOVER BOOKS
IN SCIENCE AND MATHEMATICS

A CATALOG OF SELECTED

DOVER BOOKS

IN SCIENCE AND MATHEMATICS

Astronomy

BURNHAM'S CELESTIAL HANDBOOK, Robert Burnham, Jr. Thorough guide to the stars beyond our solar system. Exhaustive treatment. Alphabetical by constellation: Andromeda to Cetus in Vol. 1; Chamaeleon to Orion in Vol. 2; and Pavo to Vulpecula in Vol. 3. Hundreds of illustrations. Index in Vol. 3. 2,000pp. 6⅛ x 9¼.
23567-X, 23568-8, 23673-0 Three-vol. set

THE EXTRATERRESTRIAL LIFE DEBATE, 1750–1900, Michael J. Crowe. First detailed, scholarly study in English of the many ideas that developed from 1750 to 1900 regarding the existence of intelligent extraterrestrial life. Examines ideas of Kant, Herschel, Voltaire, Percival Lowell, many other scientists and thinkers. 16 illustrations. 704pp. 5⅜ x 8½. 40675-X

A HISTORY OF ASTRONOMY, A. Pannekoek. Well-balanced, carefully reasoned study covers such topics as Ptolemaic theory, work of Copernicus, Kepler, Newton, Eddington's work on stars, much more. Illustrated. References. 521pp. 5⅜ x 8½.
65994-1

AMATEUR ASTRONOMER'S HANDBOOK, J. B. Sidgwick. Timeless, comprehensive coverage of telescopes, mirrors, lenses, mountings, telescope drives, micrometers, spectroscopes, more. 189 illustrations. 576pp. 5⅝ x 8¼. (Available in U.S. only.)
24034-7

STARS AND RELATIVITY, Ya. B. Zel'dovich and I. D. Novikov. Vol. 1 of *Relativistic Astrophysics* by famed Russian scientists. General relativity, properties of matter under astrophysical conditions, stars, and stellar systems. Deep physical insights, clear presentation. 1971 edition. References. 544pp. 5⅝ x 8¼. 69424-0

Chemistry

CHEMICAL MAGIC, Leonard A. Ford. Second Edition, Revised by E. Winston Grundmeier. Over 100 unusual stunts demonstrating cold fire, dust explosions, much more. Text explains scientific principles and stresses safety precautions. 128pp. 5⅜ x 8½. 67628-5

THE DEVELOPMENT OF MODERN CHEMISTRY, Aaron J. Ihde. Authoritative history of chemistry from ancient Greek theory to 20th-century innovation. Covers major chemists and their discoveries. 209 illustrations. 14 tables. Bibliographies. Indices. Appendices. 851pp. 5⅜ x 8½. 64235-6

CATALYSIS IN CHEMISTRY AND ENZYMOLOGY, William P. Jencks. Exceptionally clear coverage of mechanisms for catalysis, forces in aqueous solution, carbonyl- and acyl-group reactions, practical kinetics, more. 864pp. 5⅜ x 8½.
65460-5

THE HISTORICAL BACKGROUND OF CHEMISTRY, Henry M. Leicester. Evolution of ideas, not individual biography. Concentrates on formulation of a coherent set of chemical laws. 260pp. 5⅜ x 8½. 61053-5

A SHORT HISTORY OF CHEMISTRY, J. R. Partington. Classic exposition explores origins of chemistry, alchemy, early medical chemistry, nature of atmosphere, theory of valency, laws and structure of atomic theory, much more. 428pp. 5⅜ x 8½. (Available in U.S. only.) 65977-1

GENERAL CHEMISTRY, Linus Pauling. Revised 3rd edition of classic first-year text by Nobel laureate. Atomic and molecular structure, quantum mechanics, statistical mechanics, thermodynamics correlated with descriptive chemistry. Problems. 992pp. 5⅜ x 8½. 65622-5

Engineering

DE RE METALLICA, Georgius Agricola. The famous Hoover translation of greatest treatise on technological chemistry, engineering, geology, mining of early modern times (1556). All 289 original woodcuts. 638pp. 6¾ x 11. 60006-8

FUNDAMENTALS OF ASTRODYNAMICS, Roger Bate et al. Modern approach developed by U.S. Air Force Academy. Designed as a first course. Problems, exercises. Numerous illustrations. 455pp. 5⅜ x 8½. 60061-0

DYNAMICS OF FLUIDS IN POROUS MEDIA, Jacob Bear. For advanced students of ground water hydrology, soil mechanics and physics, drainage and irrigation engineering and more. 335 illustrations. Exercises, with answers. 784pp. 6⅛ x 9¼. 65675-6

ANALYTICAL MECHANICS OF GEARS, Earle Buckingham. Indispensable reference for modern gear manufacture covers conjugate gear-tooth action, gear-tooth profiles of various gears, many other topics. 263 figures. 102 tables. 546pp. 5⅜ x 8½. 65712-4

MECHANICS, J. P. Den Hartog. A classic introductory text or refresher. Hundreds of applications and design problems illuminate fundamentals of trusses, loaded beams and cables, etc. 334 answered problems. 462pp. 5⅜ x 8½. 60754-2

MECHANICAL VIBRATIONS, J. P. Den Hartog. Classic textbook offers lucid explanations and illustrative models, applying theories of vibrations to a variety of practical industrial engineering problems. Numerous figures. 233 problems, solutions. Appendix. Index. Preface. 436pp. 5⅜ x 8½. 64785-4

STRENGTH OF MATERIALS, J. P. Den Hartog. Full, clear treatment of basic material (tension, torsion, bending, etc.) plus advanced material on engineering methods, applications. 350 answered problems. 323pp. 5⅜ x 8½. 60755-0

A HISTORY OF MECHANICS, René Dugas. Monumental study of mechanical principles from antiquity to quantum mechanics. Contributions of ancient Greeks, Galileo, Leonardo, Kepler, Lagrange, many others. 671pp. 5⅜ x 8½. 65632-2

METAL FATIGUE, N. E. Frost, K. J. Marsh, and L. P. Pook. Definitive, clearly written, and well-illustrated volume addresses all aspects of the subject, from the historical development of understanding metal fatigue to vital concepts of the cyclic stress that causes a crack to grow. Includes 7 appendixes. 544pp. 5⅜ x 8½.　40927-9

STATISTICAL MECHANICS: Principles and Applications, Terrell L. Hill. Standard text covers fundamentals of statistical mechanics, applications to fluctuation theory, imperfect gases, distribution functions, more. 448pp. 5⅜ x 8½.　65390-0

THE VARIATIONAL PRINCIPLES OF MECHANICS, Cornelius Lanczos. Graduate level coverage of calculus of variations, equations of motion, relativistic mechanics, more. First inexpensive paperbound edition of classic treatise. Index. Bibliography. 418pp. 5⅜ x 8½.　65067-7

THE VARIOUS AND INGENIOUS MACHINES OF AGOSTINO RAMELLI: A Classic Sixteenth-Century Illustrated Treatise on Technology, Agostino Ramelli. One of the most widely known and copied works on machinery in the 16th century. 194 detailed plates of water pumps, grain mills, cranes, more. 608pp. 9 x 12.　28180-9

ORDINARY DIFFERENTIAL EQUATIONS AND STABILITY THEORY: An Introduction, David A. Sánchez. Brief, modern treatment. Linear equation, stability theory for autonomous and nonautonomous systems, etc. 164pp. 5⅜ x 8¼.　63828-6

ROTARY WING AERODYNAMICS, W. Z. Stepniewski. Clear, concise text covers aerodynamic phenomena of the rotor and offers guidelines for helicopter performance evaluation. Originally prepared for NASA. 537 figures. 640pp. 6⅛ x 9¼.　64647-5

INTRODUCTION TO SPACE DYNAMICS, William Tyrrell Thomson. Comprehensive, classic introduction to space-flight engineering for advanced undergraduate and graduate students. Includes vector algebra, kinematics, transformation of coordinates. Bibliography. Index. 352pp. 5⅜ x 8½.　65113-4

HISTORY OF STRENGTH OF MATERIALS, Stephen P. Timoshenko. Excellent historical survey of the strength of materials with many references to the theories of elasticity and structure. 245 figures. 452pp. 5⅜ x 8½.　61187-6

ANALYTICAL FRACTURE MECHANICS, David J. Unger. Self-contained text supplements standard fracture mechanics texts by focusing on analytical methods for determining crack-tip stress and strain fields. 336pp. 6⅛ x 9¼.　41737-9

Mathematics

HANDBOOK OF MATHEMATICAL FUNCTIONS WITH FORMULAS, GRAPHS, AND MATHEMATICAL TABLES, edited by Milton Abramowitz and Irene A. Stegun. Vast compendium: 29 sets of tables, some to as high as 20 places. 1,046pp. 8 x 10½.　61272-4

FUNCTIONAL ANALYSIS (Second Corrected Edition), George Bachman and Lawrence Narici. Excellent treatment of subject geared toward students with background in linear algebra, advanced calculus, physics and engineering. Text covers introduction to inner-product spaces, normed, metric spaces, and topological spaces; complete orthonormal sets, the Hahn-Banach Theorem and its consequences, and many other related subjects. 1966 ed. 544pp. 6⅛ x 9¼. 40251-7

ASYMPTOTIC EXPANSIONS OF INTEGRALS, Norman Bleistein & Richard A. Handelsman. Best introduction to important field with applications in a variety of scientific disciplines. New preface. Problems. Diagrams. Tables. Bibliography. Index. 448pp. 5⅜ x 8½. 65082-0

FAMOUS PROBLEMS OF GEOMETRY AND HOW TO SOLVE THEM, Benjamin Bold. Squaring the circle, trisecting the angle, duplicating the cube: learn their history, why they are impossible to solve, then solve them yourself. 128pp. 5⅜ x 8½. 24297-8

VECTOR AND TENSOR ANALYSIS WITH APPLICATIONS, A. I. Borisenko and I. E. Tarapov. Concise introduction. Worked-out problems, solutions, exercises. 257pp. 5⅝ x 8¼. 63833-2

THE ABSOLUTE DIFFERENTIAL CALCULUS (CALCULUS OF TENSORS), Tullio Levi-Civita. Great 20th-century mathematician's classic work on material necessary for mathematical grasp of theory of relativity. 452pp. 5⅝ x 8¼. 63401-9

AN INTRODUCTION TO ORDINARY DIFFERENTIAL EQUATIONS, Earl A. Coddington. A thorough and systematic first course in elementary differential equations for undergraduates in mathematics and science, with many exercises and problems (with answers). Index. 304pp. 5⅜ x 8½. 65942-9

FOURIER SERIES AND ORTHOGONAL FUNCTIONS, Harry F. Davis. An incisive text combining theory and practical example to introduce Fourier series, orthogonal functions and applications of the Fourier method to boundary-value problems. 570 exercises. Answers and notes. 416pp. 5⅜ x 8½. 65973-9

COMPUTABILITY AND UNSOLVABILITY, Martin Davis. Classic graduate-level introduction to theory of computability, usually referred to as theory of recurrent functions. New preface and appendix. 288pp. 5⅜ x 8½. 61471-9

ASYMPTOTIC METHODS IN ANALYSIS, N. G. de Bruijn. An inexpensive, comprehensive guide to asymptotic methods–the pioneering work that teaches by explaining worked examples in detail. Index. 224pp. 5⅜ x 8½ 64221-6

ESSAYS ON THE THEORY OF NUMBERS, Richard Dedekind. Two classic essays by great German mathematician: on the theory of irrational numbers; and on transfinite numbers and properties of natural numbers. 115pp. 5⅜ x 8½. 21010-3

APPLIED COMPLEX VARIABLES, John W. Dettman. Step-by-step coverage of fundamentals of analytic function theory—plus lucid exposition of five important applications: Potential Theory; Ordinary Differential Equations; Fourier Transforms; Laplace Transforms; Asymptotic Expansions. 66 figures. Exercises at chapter ends. 512pp. 5⅜ x 8½.   64670-X

INTRODUCTION TO LINEAR ALGEBRA AND DIFFERENTIAL EQUATIONS, John W. Dettman. Excellent text covers complex numbers, determinants, orthonormal bases, Laplace transforms, much more. Exercises with solutions. Undergraduate level. 416pp. 5⅜ x 8½.   65191-6

MATHEMATICAL METHODS IN PHYSICS AND ENGINEERING, John W. Dettman. Algebraically based approach to vectors, mapping, diffraction, other topics in applied math. Also generalized functions, analytic function theory, more. Exercises. 448pp. 5⅜ x 8¼.   65649-7

CALCULUS OF VARIATIONS WITH APPLICATIONS, George M. Ewing. Applications-oriented introduction to variational theory develops insight and promotes understanding of specialized books, research papers. Suitable for advanced undergraduate/graduate students as primary, supplementary text. 352pp. 5⅜ x 8½.   64856-7

COMPLEX VARIABLES, Francis J. Flanigan. Unusual approach, delaying complex algebra till harmonic functions have been analyzed from real variable viewpoint. Includes problems with answers. 364pp. 5⅜ x 8½.   61388-7

AN INTRODUCTION TO THE CALCULUS OF VARIATIONS, Charles Fox. Graduate-level text covers variations of an integral, isoperimetrical problems, least action, special relativity, approximations, more. References. 279pp. 5⅜ x 8½.   65499-0

CATASTROPHE THEORY FOR SCIENTISTS AND ENGINEERS, Robert Gilmore. Advanced-level treatment describes mathematics of theory grounded in the work of Poincaré, R. Thom, other mathematicians. Also important applications to problems in mathematics, physics, chemistry and engineering. 1981 edition. References. 28 tables. 397 black-and-white illustrations. xvii + 666pp. 6⅛ x 9¼.   67539-4

INTRODUCTION TO DIFFERENCE EQUATIONS, Samuel Goldberg. Exceptionally clear exposition of important discipline with applications to sociology, psychology, economics. Many illustrative examples; over 250 problems. 260pp. 5⅜ x 8½.   65084-7

NUMERICAL METHODS FOR SCIENTISTS AND ENGINEERS, Richard Hamming. Classic text stresses frequency approach in coverage of algorithms, polynomial approximation, Fourier approximation, exponential approximation, other topics. Revised and enlarged 2nd edition. 721pp. 5⅜ x 8½.   65241-6

INTRODUCTION TO NUMERICAL ANALYSIS (2nd Edition), F. B. Hildebrand. Classic, fundamental treatment covers computation, approximation, interpolation, numerical differentiation and integration, other topics. 150 new problems. 669pp. 5⅜ x 8½.   65363-3

THE FUNCTIONS OF MATHEMATICAL PHYSICS, Harry Hochstadt. Comprehensive treatment of orthogonal polynomials, hypergeometric functions, Hill's equation, much more. Bibliography. Index. 322pp. 5⅜ x 8½. 65214-9

THREE PEARLS OF NUMBER THEORY, A. Y. Khinchin. Three compelling puzzles require proof of a basic law governing the world of numbers. Challenges concern van der Waerden's theorem, the Landau-Schnirelmann hypothesis and Mann's theorem, and a solution to Waring's problem. Solutions included. 64pp. 5⅜ x 8½. 40026-3

CALCULUS REFRESHER FOR TECHNICAL PEOPLE, A. Albert Klaf. Covers important aspects of integral and differential calculus via 756 questions. 566 problems, most answered. 431pp. 5⅜ x 8½. 20370-0

THE PHILOSOPHY OF MATHEMATICS: An Introductory Essay, Stephan Körner. Surveys the views of Plato, Aristotle, Leibniz & Kant concerning propositions and theories of applied and pure mathematics. Introduction. Two appendices. Index. 198pp. 5⅜ x 8½. 25048-2

INTRODUCTORY REAL ANALYSIS, A.N. Kolmogorov, S. V. Fomin. Translated by Richard A. Silverman. Self-contained, evenly paced introduction to real and functional analysis. Some 350 problems. 403pp. 5⅜ x 8½. 61226-0

APPLIED ANALYSIS, Cornelius Lanczos. Classic work on analysis and design of finite processes for approximating solution of analytical problems. Algebraic equations, matrices, harmonic analysis, quadrature methods, much more. 559pp. 5⅜ x 8½. 65656-X

AN INTRODUCTION TO ALGEBRAIC STRUCTURES, Joseph Landin. Superb self-contained text covers "abstract algebra": sets and numbers, theory of groups, theory of rings, much more. Numerous well-chosen examples, exercises. 247pp. 5⅜ x 8½. 65940-2

SPECIAL FUNCTIONS, N. N. Lebedev. Translated by Richard Silverman. Famous Russian work treating more important special functions, with applications to specific problems of physics and engineering. 38 figures. 308pp. 5⅜ x 8½. 60624-4

QUALITATIVE THEORY OF DIFFERENTIAL EQUATIONS, V. V. Nemytskii and V.V. Stepanov. Classic graduate-level text by two prominent Soviet mathematicians covers classical differential equations as well as topological dynamics and ergodic theory. Bibliographies. 523pp. 5⅜ x 8½. 65954-2

NUMBER THEORY AND ITS HISTORY, Oystein Ore. Unusually clear, accessible introduction covers counting, properties of numbers, prime numbers, much more. Bibliography. 380pp. 5⅜ x 8½. 65620-9

THEORY OF MATRICES, Sam Perlis. Outstanding text covering rank, nonsingularity and inverses in connection with the development of canonical matrices under the relation of equivalence, and without the intervention of determinants. Includes exercises. 237pp. 5⅜ x 8½. 66810-X

INTRODUCTION TO ANALYSIS, Maxwell Rosenlicht. Unusually clear, accessible coverage of set theory, real number system, metric spaces, continuous functions, Riemann integration, multiple integrals, more. Wide range of problems. Undergraduate level. Bibliography. 254pp. 5⅜ x 8½. 65038-3

MODERN NONLINEAR EQUATIONS, Thomas L. Saaty. Emphasizes practical solution of problems; covers seven types of equations. ". . . a welcome contribution to the existing literature...."–*Math Reviews.* 490pp. 5⅜ x 8½. 64232-1

MATRICES AND LINEAR ALGEBRA, Hans Schneider and George Phillip Barker. Basic textbook covers theory of matrices and its applications to systems of linear equations and related topics such as determinants, eigenvalues and differential equations. Numerous exercises. 432pp. 5⅜ x 8½. 66014-1

MATHEMATICS APPLIED TO CONTINUUM MECHANICS, Lee A. Segel. Analyzes models of fluid flow and solid deformation. For upper-level math, science and engineering students. 608pp. 5⅜ x 8½. 65369-2

ELEMENTS OF REAL ANALYSIS, David A. Sprecher. Classic text covers fundamental concepts, real number system, point sets, functions of a real variable, Fourier series, much more. Over 500 exercises. 352pp. 5⅜ x 8½. 65385-4

AN INTRODUCTION TO MATRICES, SETS AND GROUPS FOR SCIENCE STUDENTS, G. Stephenson. Concise, readable text introduces sets, groups, and most importantly, matrices to undergraduate students of physics, chemistry, and engineering. Problems. 164pp. 5⅜ x 8½. 65077-4

SET THEORY AND LOGIC, Robert R. Stoll. Lucid introduction to unified theory of mathematical concepts. Set theory and logic seen as tools for conceptual understanding of real number system. 496pp. 5⅜ x 8¼. 63829-4

TENSOR CALCULUS, J.L. Synge and A. Schild. Widely used introductory text covers spaces and tensors, basic operations in Riemannian space, non-Riemannian spaces, etc. 324pp. 5⅜ x 8¼. 63612-7

ORDINARY DIFFERENTIAL EQUATIONS, Morris Tenenbaum and Harry Pollard. Exhaustive survey of ordinary differential equations for undergraduates in mathematics, engineering, science. Thorough analysis of theorems. Diagrams. Bibliography. Index. 818pp. 5⅜ x 8½. 64940-7

INTEGRAL EQUATIONS, F. G. Tricomi. Authoritative, well-written treatment of extremely useful mathematical tool with wide applications. Volterra Equations, Fredholm Equations, much more. Advanced undergraduate to graduate level. Exercises. Bibliography. 238pp. 5⅜ x 8½. 64828-1

FOURIER SERIES, Georgi P. Tolstov. Translated by Richard A. Silverman. A valuable addition to the literature on the subject, moving clearly from subject to subject and theorem to theorem. 107 problems, answers. 336pp. 5⅜ x 8½. 63317-9

POPULAR LECTURES ON MATHEMATICAL LOGIC, Hao Wang. Noted logician's lucid treatment of historical developments, set theory, model theory, recursion theory and constructivism, proof theory, more. 3 appendixes. Bibliography. 1981 edition. ix + 283pp. 5⅜ x 8½. 67632-3

CALCULUS OF VARIATIONS, Robert Weinstock. Basic introduction covering isoperimetric problems, theory of elasticity, quantum mechanics, electrostatics, etc. Exercises throughout. 326pp. 5⅜ x 8½. 63069-2

THE CONTINUUM: A Critical Examination of the Foundation of Analysis, Hermann Weyl. Classic of 20th-century foundational research deals with the conceptual problem posed by the continuum. 156pp. 5⅜ x 8½. 67982-9

CHALLENGING MATHEMATICAL PROBLEMS WITH ELEMENTARY SOLUTIONS, A. M. Yaglom and I. M. Yaglom. Over 170 challenging problems on probability theory, combinatorial analysis, points and lines, topology, convex polygons, many other topics. Solutions. Total of 445pp. 5⅜ x 8½. Two-vol. set.
Vol. I: 65536-9 Vol. II: 65537-7

A SURVEY OF NUMERICAL MATHEMATICS, David M. Young and Robert Todd Gregory. Broad self-contained coverage of computer-oriented numerical algorithms for solving various types of mathematical problems in linear algebra, ordinary and partial, differential equations, much more. Exercises. Total of 1,248pp. 5⅜ x 8½. Two volumes.
Vol. I: 65691-8 Vol. II: 65692-6

INTRODUCTION TO PARTIAL DIFFERENTIAL EQUATIONS WITH APPLICATIONS, E. C. Zachmanoglou and Dale W. Thoe. Essentials of partial differential equations applied to common problems in engineering and the physical sciences. Problems and answers. 416pp. 5⅜ x 8½. 65251-3

THE THEORY OF GROUPS, Hans J. Zassenhaus. Well-written graduate-level text acquaints reader with group-theoretic methods and demonstrates their usefulness in mathematics. Axioms, the calculus of complexes, homomorphic mapping, p-group theory, more. Many proofs shorter and more transparent than older ones. 276pp. 5⅜ x 8½. 40922-8

DISTRIBUTION THEORY AND TRANSFORM ANALYSIS: An Introduction to Generalized Functions, with Applications, A. H. Zemanian. Provides basics of distribution theory, describes generalized Fourier and Laplace transformations. Numerous problems. 384pp. 5⅜ x 8½. 65479-6

Math–Decision Theory, Statistics, Probability

ELEMENTARY DECISION THEORY, Herman Chernoff and Lincoln E. Moses. Clear introduction to statistics and statistical theory covers data processing, probability and random variables, testing hypotheses, much more. Exercises. 364pp. 5⅜ x 8½. 65218-1

STATISTICS MANUAL, Edwin L. Crow et al. Comprehensive, practical collection of classical and modern methods prepared by U.S. Naval Ordnance Test Station. Stress on use. Basics of statistics assumed. 288pp. 5⅜ x 8½. 60599-X

SOME THEORY OF SAMPLING, William Edwards Deming. Analysis of the problems, theory and design of sampling techniques for social scientists, industrial managers and others who find statistics important at work. 61 tables. 90 figures. xvii +602pp. 5⅜ x 8½. 64684-X

STATISTICAL ADJUSTMENT OF DATA, W. Edwards Deming. Introduction to basic concepts of statistics, curve fitting, least squares solution, conditions without parameter, conditions containing parameters. 26 exercises worked out. 271pp. 5⅜ x 8½. 64685-8

LINEAR PROGRAMMING AND ECONOMIC ANALYSIS, Robert Dorfman, Paul A. Samuelson and Robert M. Solow. First comprehensive treatment of linear programming in standard economic analysis. Game theory, modern welfare economics, Leontief input-output, more. 525pp. 5⅜ x 8½. 65491-5

DICTIONARY/OUTLINE OF BASIC STATISTICS, John E. Freund and Frank J. Williams. A clear concise dictionary of over 1,000 statistical terms and an outline of statistical formulas covering probability, nonparametric tests, much more. 208pp. 5⅜ x 8½. 66796-0

PROBABILITY: An Introduction, Samuel Goldberg. Excellent basic text covers set theory, probability theory for finite sample spaces, binomial theorem, much more. 360 problems. Bibliographies. 322pp. 5⅜ x 8½. 65252-1

GAMES AND DECISIONS: Introduction and Critical Survey, R. Duncan Luce and Howard Raiffa. Superb nontechnical introduction to game theory, primarily applied to social sciences. Utility theory, zero-sum games, n-person games, decision-making, much more. Bibliography. 509pp. 5⅜ x 8½. 65943-7

FIFTY CHALLENGING PROBLEMS IN PROBABILITY WITH SOLUTIONS, Frederick Mosteller. Remarkable puzzlers, graded in difficulty, illustrate elementary and advanced aspects of probability. Detailed solutions. 88pp. 5⅜ x 8½. 65355-2

PROBABILITY THEORY: A Concise Course, Y. A. Rozanov. Highly readable, self-contained introduction covers combination of events, dependent events, Bernoulli trials, etc. 148pp. 5⅜ x 8¼. 63544-9

STATISTICAL METHOD FROM THE VIEWPOINT OF QUALITY CONTROL, Walter A. Shewhart. Important text explains regulation of variables, uses of statistical control to achieve quality control in industry, agriculture, other areas. 192pp. 5⅜ x 8½. 65232-7

THE COMPLEAT STRATEGYST: Being a Primer on the Theory of Games of Strategy, J. D. Williams. Highly entertaining classic describes, with many illustrated examples, how to select best strategies in conflict situations. Prefaces. Appendices. 268pp. 5⅜ x 8½. 25101-2

Math–Geometry and Topology

ELEMENTARY CONCEPTS OF TOPOLOGY, Paul Alexandroff. Elegant, intuitive approach to topology from set-theoretic topology to Betti groups; how concepts of topology are useful in math and physics. 25 figures. 57pp. 5⅜ x 8½. 60747-X

COMBINATORIAL TOPOLOGY, P. S. Alexandrov. Clearly written, well-organized, three-part text begins by dealing with certain classic problems without using the formal techniques of homology theory and advances to the central concept, the Betti groups. Numerous detailed examples. 654pp. 5⅜ x 8½. 40179-0

EXPERIMENTS IN TOPOLOGY, Stephen Barr. Classic, lively explanation of one of the byways of mathematics. Klein bottles, Moebius strips, projective planes, map coloring, problem of the Koenigsberg bridges, much more, described with clarity and wit. 43 figures. 210pp. 5⅜ x 8½. 25933-1

CONFORMAL MAPPING ON RIEMANN SURFACES, Harvey Cohn. Lucid, insightful book presents ideal coverage of subject. 334 exercises make book perfect for self-study. 55 figures. 352pp. 5⅝ x 8¼. 64025-6

THE GEOMETRY OF RENÉ DESCARTES, René Descartes. The great work founded analytical geometry. Original French text, Descartes's own diagrams, together with definitive Smith-Latham translation. 244pp. 5⅜ x 8½. 60068-8

THE THIRTEEN BOOKS OF EUCLID'S ELEMENTS, translated with introduction and commentary by Sir Thomas L. Heath. Definitive edition. Textual and linguistic notes, mathematical analysis. 2,500 years of critical commentary. Unabridged. 1,414pp. 5⅜ x 8½. Three-vol. set.
Vol. I: 60088-2 Vol. II: 60089-0 Vol. III: 60090-4

GEOMETRY OF COMPLEX NUMBERS, Hans Schwerdtfeger. Illuminating, widely praised book on analytic geometry of circles, the Moebius transformation, and two-dimensional non-Euclidean geometries. 200pp. 5⅝ x 8¼. 63830-8

DIFFERENTIAL GEOMETRY, Heinrich W. Guggenheimer. Local differential geometry as an application of advanced calculus and linear algebra. Curvature, transformation groups, surfaces, more. Exercises. 62 figures. 378pp. 5⅜ x 8½. 63433-7

CURVATURE AND HOMOLOGY: Enlarged Edition, Samuel I. Goldberg. Revised edition examines topology of differentiable manifolds; curvature, homology of Riemannian manifolds; compact Lie groups; complex manifolds; curvature, homology of Kaehler manifolds. New Preface. Four new appendixes. 416pp. 5⅜ x 8½. 40207-X

TOPOLOGY, John G. Hocking and Gail S. Young. Superb one-year course in classical topology. Topological spaces and functions, point-set topology, much more. Examples and problems. Bibliography. Index. 384pp. 5⅝ x 8¼. 65676-4

LECTURES ON CLASSICAL DIFFERENTIAL GEOMETRY, Second Edition, Dirk J. Struik. Excellent brief introduction covers curves, theory of surfaces, fundamental equations, geometry on a surface, conformal mapping, other topics. Problems. 240pp. 5⅜ x 8½.   65609-8

Math–History of

A SHORT ACCOUNT OF THE HISTORY OF MATHEMATICS, W. W. Rouse Ball. One of clearest, most authoritative surveys from the Egyptians and Phoenicians through 19th-century figures such as Grassman, Galois, Riemann. Fourth edition. 522pp. 5⅜ x 8½.   20630-0

THE HISTORY OF THE CALCULUS AND ITS CONCEPTUAL DEVELOPMENT, Carl B. Boyer. Origins in antiquity, medieval contributions, work of Newton, Leibniz, rigorous formulation. Treatment is verbal. 346pp. 5⅜ x 8½.   60509-4

THE HISTORICAL ROOTS OF ELEMENTARY MATHEMATICS, Lucas N. H. Bunt, Phillip S. Jones, and Jack D. Bedient. Fundamental underpinnings of modern arithmetic, algebra, geometry and number systems derived from ancient civilizations. 320pp. 5⅜ x 8½.   25563-8

A HISTORY OF MATHEMATICAL NOTATIONS, Florian Cajori. This classic study notes the first appearance of a mathematical symbol and its origin, the competition it encountered, its spread among writers in different countries, its rise to popularity, its eventual decline or ultimate survival. Original 1929 two-volume edition presented here in one volume. xxviii+820pp. 5⅜ x 8½.   67766-4

GAMES, GODS & GAMBLING: A History of Probability and Statistical Ideas, F. N. David. Episodes from the lives of Galileo, Fermat, Pascal, and others illustrate this fascinating account of the roots of mathematics. Features thought-provoking references to classics, archaeology, biography, poetry. 1962 edition. 304pp. 5⅜ x 8½. (Available in U.S. only.)   40023-9

OF MEN AND NUMBERS: The Story of the Great Mathematicians, Jane Muir. Fascinating accounts of the lives and accomplishments of history's greatest mathematical minds–Pythagoras, Descartes, Euler, Pascal, Cantor, many more. Anecdotal, illuminating. 30 diagrams. Bibliography. 256pp. 5⅜ x 8½.   28973-7

HISTORY OF MATHEMATICS, David E. Smith. Nontechnical survey from ancient Greece and Orient to late 19th century; evolution of arithmetic, geometry, trigonometry, calculating devices, algebra, the calculus. 362 illustrations. 1,355pp. 5⅜ x 8½. Two-vol. set.   Vol. I: 20429-4   Vol. II: 20430-8

A CONCISE HISTORY OF MATHEMATICS, Dirk J. Struik. The best brief history of mathematics. Stresses origins and covers every major figure from ancient Near East to 19th century. 41 illustrations. 195pp. 5⅜ x 8½.   60255-9

Physics

OPTICAL RESONANCE AND TWO-LEVEL ATOMS, L. Allen and J. H. Eberly. Clear, comprehensive introduction to basic principles behind all quantum optical resonance phenomena. 53 illustrations. Preface. Index. 256pp. 5⅜ x 8½. 65533-4

ULTRASONIC ABSORPTION: An Introduction to the Theory of Sound Absorption and Dispersion in Gases, Liquids and Solids, A. B. Bhatia. Standard reference in the field provides a clear, systematically organized introductory review of fundamental concepts for advanced graduate students, research workers. Numerous diagrams. Bibliography. 440pp. 5⅜ x 8½. 64917-2

QUANTUM THEORY, David Bohm. This advanced undergraduate-level text presents the quantum theory in terms of qualitative and imaginative concepts, followed by specific applications worked out in mathematical detail. Preface. Index. 655pp. 5⅜ x 8½. 65969-0

ATOMIC PHYSICS (8th edition), Max Born. Nobel laureate's lucid treatment of kinetic theory of gases, elementary particles, nuclear atom, wave-corpuscles, atomic structure and spectral lines, much more. Over 40 appendices, bibliography. 495pp. 5⅜ x 8½. 65984-4

AN INTRODUCTION TO HAMILTONIAN OPTICS, H. A. Buchdahl. Detailed account of the Hamiltonian treatment of aberration theory in geometrical optics. Many classes of optical systems defined in terms of the symmetries they possess. Problems with detailed solutions. 1970 edition. xv + 360pp. 5⅜ x 8½. 67597-1

THIRTY YEARS THAT SHOOK PHYSICS: The Story of Quantum Theory, George Gamow. Lucid, accessible introduction to influential theory of energy and matter. Careful explanations of Dirac's anti-particles, Bohr's model of the atom, much more. 12 plates. Numerous drawings. 240pp. 5⅜ x 8½. 24895-X

ELECTRONIC STRUCTURE AND THE PROPERTIES OF SOLIDS: The Physics of the Chemical Bond, Walter A. Harrison. Innovative text offers basic understanding of the electronic structure of covalent and ionic solids, simple metals, transition metals and their compounds. Problems. 1980 edition. 582pp. 6⅛ x 9¼. 66021-4

HYDRODYNAMIC AND HYDROMAGNETIC STABILITY, S. Chandrasekhar. Lucid examination of the Rayleigh-Benard problem; clear coverage of the theory of instabilities causing convection. 704pp. 5⅜ x 8¼. 64071-X

INVESTIGATIONS ON THE THEORY OF THE BROWNIAN MOVEMENT, Albert Einstein. Five papers (1905–8) investigating dynamics of Brownian motion and evolving elementary theory. Notes by R. Fürth. 122pp. 5⅜ x 8½. 60304-0

THE PHYSICS OF WAVES, William C. Elmore and Mark A. Heald. Unique overview of classical wave theory. Acoustics, optics, electromagnetic radiation, more. Ideal as classroom text or for self-study. Problems. 477pp. 5⅜ x 8½. 64926-1

PHYSICAL PRINCIPLES OF THE QUANTUM THEORY, Werner Heisenberg. Nobel Laureate discusses quantum theory, uncertainty, wave mechanics, work of Dirac, Schroedinger, Compton, Wilson, Einstein, etc. 184pp. 5⅜ x 8½.   60113-7

ATOMIC SPECTRA AND ATOMIC STRUCTURE, Gerhard Herzberg. One of best introductions; especially for specialist in other fields. Treatment is physical rather than mathematical. 80 illustrations. 257pp. 5⅜ x 8½.   60115-3

AN INTRODUCTION TO STATISTICAL THERMODYNAMICS, Terrell L. Hill. Excellent basic text offers wide-ranging coverage of quantum statistical mechanics, systems of interacting molecules, quantum statistics, more. 523pp. 5⅜ x 8½.
                                                                                          65242-4

THEORETICAL PHYSICS, Georg Joos, with Ira M. Freeman. Classic overview covers essential math, mechanics, electromagnetic theory, thermodynamics, quantum mechanics, nuclear physics, other topics. First paperback edition. xxiii + 885pp. 5⅜ x 8½.   65227-0

PROBLEMS AND SOLUTIONS IN QUANTUM CHEMISTRY AND PHYSICS, Charles S. Johnson, Jr. and Lee G. Pedersen. Unusually varied problems, detailed solutions in coverage of quantum mechanics, wave mechanics, angular momentum, molecular spectroscopy, more. 280 problems plus 139 supplementary exercises. 430pp. 6½ x 9¼.   65236-X

THEORETICAL SOLID STATE PHYSICS, Vol. 1: Perfect Lattices in Equilibrium; Vol. II: Non-Equilibrium and Disorder, William Jones and Norman H. March. Monumental reference work covers fundamental theory of equilibrium properties of perfect crystalline solids, non-equilibrium properties, defects and disordered systems. Appendices. Problems. Preface. Diagrams. Index. Bibliography. Total of 1,301pp. 5⅜ x 8½. Two volumes.   Vol. I: 65015-4   Vol. II: 65016-2

A TREATISE ON ELECTRICITY AND MAGNETISM, James Clerk Maxwell. Important foundation work of modern physics. Brings to final form Maxwell's theory of electromagnetism and rigorously derives his general equations of field theory. 1,084pp. 5⅜ x 8½. Two-vol. set.   Vol. I: 60636-8   Vol. II: 60637-6

OPTICKS, Sir Isaac Newton. Newton's own experiments with spectroscopy, colors, lenses, reflection, refraction, etc., in language the layman can follow. Foreword by Albert Einstein. 532pp. 5⅜ x 8½.   60205-2

THEORY OF ELECTROMAGNETIC WAVE PROPAGATION, Charles Herach Papas. Graduate-level study discusses the Maxwell field equations, radiation from wire antennas, the Doppler effect and more. xiii + 244pp. 5⅜ x 8½.   65678-5

INTRODUCTION TO QUANTUM MECHANICS With Applications to Chemistry, Linus Pauling & E. Bright Wilson, Jr. Classic undergraduate text by Nobel Prize winner applies quantum mechanics to chemical and physical problems. Numerous tables and figures enhance the text. Chapter bibliographies. Appendices. Index. 468pp. 5⅜ x 8½.   64871-0

METHODS OF THERMODYNAMICS, Howard Reiss. Outstanding text focuses on physical technique of thermodynamics, typical problem areas of understanding, and significance and use of thermodynamic potential. 1965 edition. 238pp. 5⅜ x 8½. 69445-3

TENSOR ANALYSIS FOR PHYSICISTS, J. A. Schouten. Concise exposition of the mathematical basis of tensor analysis, integrated with well-chosen physical examples of the theory. Exercises. Index. Bibliography. 289pp. 5⅜ x 8½. 65582-2

RELATIVITY IN ILLUSTRATIONS, Jacob T. Schwartz. Clear nontechnical treatment makes relativity more accessible than ever before. Over 60 drawings illustrate concepts more clearly than text alone. Only high school geometry needed. Bibliography. 128pp. 6⅛ x 9¼. 25965-X

THE ELECTROMAGNETIC FIELD, Albert Shadowitz. Comprehensive undergraduate text covers basics of electric and magnetic fields, builds up to electromagnetic theory. Also related topics, including relativity. Over 900 problems. 768pp. 5⅝ x 8¼. 65660-8

GREAT EXPERIMENTS IN PHYSICS: Firsthand Accounts from Galileo to Einstein, edited by Morris H. Shamos. 25 crucial discoveries: Newton's laws of motion, Chadwick's study of the neutron, Hertz on electromagnetic waves, more. Original accounts clearly annotated. 370pp. 5⅜ x 8½. 25346-5

RELATIVITY, THERMODYNAMICS AND COSMOLOGY, Richard C. Tolman. Landmark study extends thermodynamics to special, general relativity; also applications of relativistic mechanics, thermodynamics to cosmological models. 501pp. 5⅜ x 8½. 65383-8

LIGHT SCATTERING BY SMALL PARTICLES, H. C. van de Hulst. Comprehensive treatment including full range of useful approximation methods for researchers in chemistry, meteorology and astronomy. 44 illustrations. 470pp. 5⅜ x 8½. 64228-3

STATISTICAL PHYSICS, Gregory H. Wannier. Classic text combines thermodynamics, statistical mechanics and kinetic theory in one unified presentation of thermal physics. Problems with solutions. Bibliography. 532pp. 5⅜ x 8½. 65401-X

Paperbound unless otherwise indicated. Available at your book dealer, online at **www.doverpublications.com**, or by writing to Dept. GI, Dover Publications, Inc., 31 East 2nd Street, Mineola, NY 11501. For current price information or for free catalogues (please indicate field of interest), write to Dover Publications or log on to **www.doverpublications.com** and see every Dover book in print. Dover publishes more than 500 books each year on science, elementary and advanced mathematics, biology, music, art, literary history, social sciences, and other areas.